Methods in Enzymology

Volume 175

CUMULATIVE SUBJECT INDEX

Volumes 135–139, 141–167

METHODS IN ENZYMOLOGY

EDITORS-IN-CHIEF

John N. Abelson   Melvin I. Simon

DIVISION OF BIOLOGY
CALIFORNIA INSTITUTE OF TECHNOLOGY
PASADENA, CALIFORNIA

FOUNDING EDITORS

Sidney P. Colowick and Nathan O. Kaplan

Methods in Enzymology

Volume 175

Cumulative Subject Index

Volumes 135–139, 141–167

ACADEMIC PRESS, INC.
Harcourt Brace Jovanovich, Publishers
San Diego New York Boston
London Sydney Tokyo Toronto

This book is printed on acid-free paper. ∞

ACADEMIC PRESS, INC.
San Diego, California 92101

United Kingdom Edition published by
ACADEMIC PRESS LIMITED
24–28 Oval Road, London NW1 7DX

LIBRARY OF CONGRESS CATALOG CARD NUMBER: 54-9110

ISBN 0-12-182076-9 (alk. paper)

PRINTED IN THE UNITED STATES OF AMERICA
90 91 92 93 9 8 7 6 5 4 3 2 1

Table of Contents

Preface

The idea for a cumulative index was recognized by the founding editors who prepared one for Volumes I through VI of *Methods of Enzymology* by weeding and interpolating from the entries that had been indexed in the individual volumes. As the series developed in both number and complexity, different individuals with different backgrounds served as volume indexers. Subsequently, the series was fortunate in having Dr. Martha G. Dennis and Dr. Edward A. Dennis accept the challenge of computerizing the data available from these individual indexes, and this effort resulted in Volumes 33, 75, and 95, which cover Volumes 1 through 80.

Although each of the three books produced with the aid of computerization provided an appropriate cumulative index, major problems were encountered. One was time, both expensive computer time and lag time before such efforts resulted in publication. The most important difficulty was that the compilers were hampered by the lack of uniformity in the indexing of the individual volumes, resulting in the need for much hand editing to achieve a reasonable collation. The products were very decent, if uneven, indexes that also contributed to the methodology of computerized indexing, albeit with much delay and great expense.

This cumulative index has been produced by the staff of Academic Press. The indexers have gone back to the text rather than to the individual volume indexes and, using a uniform set of guidelines, have culled the major topics, leading to five to ten entries for each article. Thus, if searching by name for one of the dozen substrates of an enzyme, the isolation of which is being presented, it probably will not be found in the index, although the assay substrate may be there. Nor will the specific inhibitors of the enzyme be itemized, although the topic of the enzyme's inhibition will form an entry. The index, then, is not complete, but should lead to the broad subject headings. Since there is a tendency to identify specific topics and methods with particular individuals, a contributor index is included. Finally, the complete table of contents of each of the volumes indexes is included.

Contents of Volumes 135–139, 141–167

VOLUME 135
Immobilized Enzymes and Cells (Part B)

Section I. Immobilization Techniques for Biocatalysts

1. Characterization of Immobilized Enzymes — K. Buchholz and J. Klein 3
2. A Survey of Enzyme Coupling Techniques — William H. Scouten 30
3. Tresyl Chloride-Activated Supports for Enzyme Immobilization — Kurt Nilsson and Klaus Mosbach 65
4. Colored Sulfonyl Chloride as an Activating Agent for Hydroxylic Matrices — William H. Scouten, Will van den Tweel, Hans Kranenburg, and Matthijs Dekker 79
5. Immobilization of Proteins and Ligands Using Chlorocarbonates — Talia Miron and Meir Wilchek 84
6. Polymers Bearing Isonitrile Functional Groups as Supports for Enzyme Immobilization — Leon Goldstein 90
7. 1,1′-Carbonyldiimidazole-Mediated Immobilization of Enzymes and Affinity Ligands — Milton T. W. Hearn 102
8. Immobilization of Enzymes on Transition Metal-Activated Supports — John F. Kennedy and J. M. S. Cabral 117
9. Covalent Immobilization of Proteins by Techniques Which Permit Subsequent Release — H. Robert Horton and Harold E. Swaisgood 130
10. Immobilization of Glycoenzymes through Carbohydrate Chains — G. P. Royer 141
11. Immobilization of Enzymes and Microorganisms by Radiation Polymerization — Koji Kawashima 146
12. Immobilization of Nucleoside Diphosphatase at the Allosteric Site Using Pyridoxal 5′-Phosphate Derivatives — Mitsuo Yamazaki and Norio Kawakita 154
13. Use of Monoclonal Antibodies for the Preparation of Highly Active Immobilized Enzymes — Beka Solomon, Zeev Hollander, Rela Koppel, and Ephraim Katchalski-Katzir 160

Section II. Immobilization Techniques for Cells/Organelles

14. Overview — P. Brodelius and K. Mosbach 173
15. Cell Immobilization in Calcium Alginate — C. Bucke 175
16. Immobilization of Cells in Carrageenan — Ichiro Chibata, Tetsuya Tosa, Tadashi Sato, and Isao Takata 189
17. Immobilization of Living Microbial Cells in Polyacrylamide Gel — G. K. Skryabin and K. A. Koshcheenko 198
18. Gel Entrapment of Whole Cells in Cross-Linked Prepolymerized Polyacrylamide–Hydrazide Gels — Amihay Freeman 216

19. Entrapment of Microbial and Plant Cells in Beaded Polymers
K. NILSSON, P. BRODELIUS, AND K. MOSBACH 222
20. Entrapment of Biocatalysts with Photo-Cross-Linkable Resin Prepolymers and Urethane Resin Prepolymers
SABURO FUKUI, KENJI SONOMOTO, AND ATSUO TANAKA 230
21. Immobilization of Microbial Cells in an Epoxy Carrier System
J. KLEIN AND B. KRESSDORF 252
22. Entrapment of Microbial Cells in Chitosan K. D. VORLOP AND J. KLEIN 259
23. Entrapment of Microbial Cells in Cellulose Gel
YU-YEN LINKO AND P. LINKO 268
24. Microbial Cells Entrapped in Cellulose Acetate Beads
SILVÍA GIOVENCO, WALTER MARCONI, AND PAOLO PANSOLLI 282
25. Immobilization of Enzymes and Microbial Cells in Gelatin
VINCENZO SCARDI 293
26. Methods for Selecting and Growing Mixed Cultures in Biofilm Fluidized Sand Beds IRVING J. DUNN 300
27. Immobilization of Microbial Cells by Adsorption to Solid Supports
PIERRE MONSAN, GILBERT DURAND, AND JEAN MARIE NAVARRO 307
28. Immobilization of Mycelial Cells on Celite
KODZO GBEWONYO, JURG MEIER, AND DANIEL I. C. WANG 318
29. Immobilization of Yeast Cells by Adsorption on Wood Chips
J. LAMPTEY AND M. MOO-YOUNG 333
30. Cell Immobilization by Covalent Linkage
VLADIMÍR JIRKŮ AND JAROSLAVA TURKOVÁ 341
31. Immobilization of Microbial Cells on Transition Metal-Activated Supports
J. M. S. CABRAL AND JOHN F. KENNEDY 357
32. Immobilization of Cells by Entrapment in Membrane Reactors
M. L. SHULER 372
33. Mammalian Cell Culture KJELL NILSSON 387
34. Flat Bed Hollow Fiber Reactors for Large-Scale Cultivation of Animal Cells
JOSEPH FEDER 393
35. Entrapment of Animal Cells
K. NILSSON, W. SCHEIRER, H. W. D. KATINGER, AND K. MOSBACH 399
36. Techniques for the Immobilization of Plant Cells
KEITH LINDSEY AND MICHAEL M. YEOMAN 410
37. Immobilization of Plant Protoplasts Using Microcarriers
CHRIS H. BORNMAN AND ANDERS ZACHRISSON 421
38. Immobilization of Plant Protoplasts by Entrapment
L. LINSEFORS AND P. BRODELIUS 434
39. Entrapment Techniques for Chloroplasts, Cyanobacteria, and Hydrogenases
P. E. GISBY, K. K. RAO, AND D. O. HALL 440
40. Immobilized Organelles in Cross-Linked Proteins
JEAN-NOËL BARBOTIN, MARIE-FRANÇOISE COCQUEMPOT, VÉRONIQUE LARRETA-GARDE, BRIGITTE THOMASSET, GÉRARD GELLF, JENNY D. CLEMENT-METRAL, AND DANIEL THOMAS 454

Section III. Application of Immobilized Enzymes/Cells to Fundamental Studies in Biochemistry

41. Overview L. ANDERSSON AND K. MOSBACH 475
42. Properties of Immobilized Subunits of 20β-Hydroxysteroid Dehydrogenase
GIACOMO CARREA AND PIERO PASTA 475

43. Use of Immobilized Subunits for the Purification of Oligomeric and Self-Associating Proteins — EMILIA CHIANCONE AND MAURIZIO GATTONI 484
44. Immobilized Subunits of Alkaline Phosphatase — SUSAN MCCRACKEN AND EDWARD MEIGHEN 492
45. Electron Paramagnetic Resonance Studies of Immobilized Chymotrypsin — JAMES E. BAILEY AND DOUGLAS S. CLARK 502
46. Nuclear Magnetic Resonance Studies of Immobilized Cells — H. J. VOGEL, P. BRODELIUS, H. LILJA, AND E. M. LOHMEIER-VOGEL 512
47. Enzyme Studies at Subzero Temperatures Using Immobilized Substrates — CLAUDE BALNY AND PIERRE DOUZOU 528
48. Kinetic Studies of Immobilized α-Chymotrypsin in Aprotic Solvents — MYRON L. BENDER 537
49. Spectroscopic Methods for Characterization of Immobilized Alcohol Dehydrogenase — HELGA SCHNEIDER-BERNLÖHR, HELMUT DIETRICH, AND MICHAEL ZEPPEZAUER 546
50. Oscillatory Phenomena in Immobilized Enzyme Systems — J. F. HERVAGAULT AND D. THOMAS 554
51. Chemical Stabilization of Conformational States of Aspartate Transcarbamoylase — CAROLINE A. ENNS AND WILLIAM W.-C. CHAN 569
52. Chemical Stabilization of Conformational States of dCMP Deaminase — MOSÈ ROSSI, C. A. RAIA, AND C. VACCARO 577
53. Reactivation of Immobilized Enzymes — V. V. MOZHAEV, I. V. BEREZIN, AND KAREL MARTINEK 586
54. Use of Immobilized Proteinases and Peptidases to Study Structural Changes in Proteins — HAROLD E. SWAISGOOD AND GEORGE L. CATIGNANI 596

VOLUME 136
IMMOBILIZED ENZYMES AND CELLS (PART C)

Section I. Multistep Enzyme Systems and Coenzymes

1. Immobilized Active Coenzymes — MATS-OLLE MÅNSSON AND KLAUS MOSBACH 3
2. Continuous Enzymatic Transformation in an Enzyme–Membrane Reactor with Simultaneous NADH Regeneration — MARIA-REGINA KULA AND CHRISTIAN WANDREY 9
3. Coimmobilized System of NAD with Dehydrogenases — YOSHIMITSU YAMAZAKI AND HIDEKATSU MAEDA 21
4. Polymerizable NAD Derivative and Model Enzyme Reactor with Recycling of Polyethylene Glycol-Bound NAD — HIROSUKE OKADA AND ITARU URABE 34
5. Adenosine 5′-Triphosphate Recycling in an Enzyme Reactor Based on Aqueous Two-Phase Systems — HIDEO SUZUKI AND YOSHIMITSU YAMAZAKI 45
6. Alcohol Dehydrogenase Coimmobilized with Its Coenzyme — KEITH J. LAIDLER AND M. ABDUL MAZID 56
7. Recycling of NAD(P) by Multienzyme Systems Immobilized by Microencapsulation in Artificial Cells — T. M. S. CHANG 67
8. Bioluminescent Assays Using Coimmobilized Enzymes — G. WIENHAUSEN, L. J. KRICKA, AND M. DELUCA 82

9. Nonseparation Enzyme Channeling Immunometric Assays
IAN GIBBONS, RICHARD ARMENTA, ROBERT K. DINELLO, AND EDWIN F. ULLMAN 93
10. Immobilized and Soluble Site-to-Site Directed Enzyme Complexes Composed of Alcohol Dehydrogenase and Lactate Dehydrogenase
NILS SIEGBAHN, MATS-OLLE MÅNSSON, AND KLAUS MOSBACH 103

Section II. Immobilized Enzymes/Cells in Organic Synthesis

11. Overview L. ANDERSSON, M.-O. MÅNSSON, AND K. MOSBACH 117
12. Enzymatic Production of Optically Active Compounds in Biphasic Aqueous–Organic Systems ALEXANDER M. KLIBANOV AND BERNARD CAMBOU 117
13. Two-Liquid Phase Biocatalytic Reactors
M. D. LILLY, S. HARBRON, AND T. J. NARENDRANATHAN 138
14. Enzyme-Catalyzed Steroid Transformations in Water–Organic Solvent Two-Phase Systems GIACOMO CARREA AND PIERO CREMONESI 150
15. Deblocking in Peptide Synthesis with Immobilized Carboxypeptidase Y
G. P. ROYER 157
16. Use of Immobilized *Achromobacter* Protease I for Semisynthesis of Human Insulin
KAZUYUKI MORIHARA, RYONOSUKE MUNEYUKI, AND TATSUSHI OKA 162
17. Application of Immobilized Aminopeptidases to the Sequential Hydrolysis of Proline-Containing Peptides G. FLEMINGER AND A. YARON 170
18. Peptide Synthesis Using Immobilized Proteases
HANS-DIETER JAKUBKE AND ANDREAS KÖNNECKE 178
19. Activity and Conformation of Enzymes in Reverse Micellar Solutions
P. L. LUISI AND B. STEINMANN-HOFMANN 188
20. Design of Reversed Micellar Media for the Enzymatic Synthesis of Apolar Compounds COLJA LAANE, RIET HILHORST, AND CEES VEEGER 216
21. Disaccharide Synthesis with Immobilized β-Galactosidase
PER-OLOF LARSSON, LARS HEDBYS, SIGFRID SVENSSON, AND KLAUS MOSBACH 230
22. Synthesis of Urea with Urease in Water–Organic Solvent Mixtures
LARRY G. BUTLER 234
23. Dextran Synthesis Using Immobilized *Leuconostoc mesenteroides* Dextransucrase P. MONSAN, F. PAUL, D. AURIOL, AND A. LOPEZ 239
24. Preparation of Immobilized Milk Xanthine Oxidase and Application in Organic Syntheses JOHANNES TRAMPER 254
25. Enzymatic Regeneration of Adenosine 5′-Triphosphate: Acetyl Phosphate, Phosphoenolpyruvate, Methoxycarbonyl Phosphate, Dihydroxyacetone Phosphate, 5-Phospho-α-D-ribosyl Pyrophosphate, Uridine-5′-diphosphoglucose
DEBBIE C. CRANS, ROMAS J. KAZLAUSKAS, BERNARD L. HIRSCHBEIN, CHI-HUEY WONG, OBSIDIANA ABRIL, AND GEORGE M. WHITESIDES 263
26. Equilibrium and Kinetically Controlled Synthesis with Enzymes: Semisynthesis of Penicillins and Peptides
VOLKER KASCHE, URSULA HAUFLER, AND LUTZ RIECHMANN 280
27. Optical Resolution of *dl*-Menthol by Entrapped Biocatalysts
SABURO FUKUI AND ATSUO TANAKA 293

28. Electroenzymatic and Electromicrobial Reduction: Preparation of Chiral Compounds
IORDANIS THANOS, JOHANN BADER, HELMUT GÜNTHER, STEFAN NEUMANN, FRIEDRICH KRAUSS, AND HELMUT SIMON 302
29. Long-Term Alkaloid Production by Immobilized Cells of *Claviceps purpurea*
BETTINA KOPP 317
30. Cell Immobilization in the Production of Patulin and Penicillin by *Penicillium urticae* and *Penicillium chrysogenum* G. M. GAUCHER AND L. A. BEHIE 329
31. Isolation and Immobilization of Strictosidine Synthase
URSULA PFITZNER AND MEINHART H. ZENK 342

Section III. Enzyme Engineering (Enzyme Technology)

32. Overview S. GESTRELIUS AND K. MOSBACH 353
33. Industrial-Scale Production and Application of Immobilized Glucose Isomerase
VILLY J. JENSEN AND SUSANNE RUGH 356
34. Industrial Operation of Immobilized Enzymes M. J. DANIELS 371
35. Large-Scale Production of Photo-Cross-Linkable Resin-Immobilized Yeast and Its Application to Industrial Ethanol Production
SHOGO NOJIMA AND TOMIAKI YAMADA 380
36. Large-Scale Preparation of Calcium Alginate-Immobilized Yeast Cells and Its Application to Industrial Ethanol Production
MINORU NAGASHIMA, MASAKI AZUMA, SADAO NOGUCHI, KEIICHI INUZUKA, AND HIROTOSHI SAMEJIMA 394
37. Regiospecific Interesterification of Triglyceride with Celite-Adsorbed Lipase
SHIGERU YAMANAKA AND TAKASHI TANAKA 405
38. Large-Scale Production and Application of Immobilized Lactase
J. L. BARET 411
39. Continuous Treatment of Ultrahigh-Temperature Sterilized Milk Using Immobilized Sulfhydryl Oxidase
HAROLD E. SWAISGOOD, VIOLETA G. JANOLINO, AND PAUL J. SKUDDER 423
40. Production of Isomaltulose Using Immobilized Microbial Cells
PETER S. J. CHEETHAM 432
41. Production of L-Malic Acid by Immobilized Microbial Cells
ICHIRO CHIBATA, TETSUYA TOSA, KOZO YAMAMOTO, AND ISAO TAKATA 455
42. Industrial Production of L-Aspartic Acid Using Polyurethane-Immobilized Cells Containing Aspartase MURRAY C. FUSEE 463
43. Continuous L-Alanine Production Using Two Different Immobilized Microbial Cell Preparations on an Industrial Scale
ICHIRO CHIBATA, TETSUYA TOSA, AND SATORU TAKAMATSU 472
44. Immobilized Aminotransferases for Amino Acid Production
J. DAVID ROZZELL 479
45. Phenylalanine Production via Polyazetidine-Immobilized *Escherichia coli*: Optimization of Cell Loading
G. J. CALTON, L. L. WOOD, AND M. L. CAMPBELL 497
46. Production of Aspartame by Immobilized Thermoase
KIYOTAKA OYAMA, SHIGEAKO IRINO, AND NORIO HAGI 503
47. Production of 5′-Ribonucleotides Using Immobilized 5′-Phosphodiesterase
REINHOLD KELLER, MERTEN SCHLINGMANN, AND ROLF WOERNLE 517

48. Acrylamide Production Method Using Immobilized Nitrilase-Containing Microbial Cells — ICHIRO WATANABE 523
49. Application of Immobilized *Thiobacillus ferrooxidans* for Large-Scale Treatment of Acid Mine Drainage — T. MURAYAMA, Y. KONNO, T. SAKATA, AND T. IMAIZUMI 530

VOLUME 137
IMMOBILIZED ENZYMES AND CELLS (PART D)

Section I. Analytical Applications with Emphasis on Biosensors

1. Introduction — B. DANIELSSON AND K. MOSBACH 3

A. Bioselective Electrodes

2. Enzyme Electrode Probes — GEORGE G. GUILBAULT 14
3. Coupled Enzyme Reactions in Enzyme Electrodes Using Sequence, Amplification, Competition, and Antiinterference Principles — FRIEDER W. SCHELLER, REINHARD RENNEBERG, AND FLORIAN SCHUBERT 29
4. Industrial Applications of Enzyme Electrodes Including Instrumentation — J. L. ROMETTE AND D. THOMAS 44
5. Multienzyme Membrane Electrodes — KENTARO YODA 61
6. Long-Term Implantation of Voltammetric Oxidase/Peroxide Glucose Sensors in the Rat Peritoneum — LELAND C. CLARK, JR., LINDA K. NOYES, ROBERT B. SPOKANE, RANJAN SUDAN, AND MARIAN L. MILLER 68
7. Amperometric Biosensors Based on Mediator-Modified Electrodes — ANTHONY P. F. TURNER 90
8. Immobilized Flavin Coenzyme Electrodes for Analytical Applications — LEMUEL B. WINGARD, JR. AND KRISHNA NARASIMHAN 103
9. Bioaffinity Sensors — YOSHIHITO IKARIYAMA AND MASUO AIZAWA 111

B. Hybrid, Cell, and Tissue Electrodes

10. Microbial Sensors for Estimation of Biochemical Oxygen Demand and Determination of Glutamate — MOTOHIKO HIKUMA AND TAKEO YASUDA 124
11. Hybrid Biosensors for Clinical Analysis and Fermentation Control — ISAO KARUBE 131
12. Bioselective Membrane Electrodes Using Tissue Materials as Biocatalysts — GARRY A. RECHNITZ 138
13. Organelle Electrodes — FLORIAN SCHUBERT AND FRIEDER W. SCHELLER 152

C. Bioluminescence-Immobilized Systems

14. Continuous-Flow Assays with Nylon Tube-Immobilized Bioluminescent Enzymes — ALDO RODA, STEFANO GIROTTI, SEVERINO GHINI, AND GIACOMO CARREA 161

15. Flow-Injection Analysis with Immobilized Chemiluminescent and Bioluminescent Columns
KALEVI KURKIJÄRVI, PEKKA TURUNEN, TIINA HEINONEN, OUTI KOLHINEN, RAIMO RAUNIO, ARNE LUNDIN, AND TIMO LÖVGREN 171

D. Thermistor Probes

16. Enzyme Thermistors BENGT DANIELSSON AND KLAUS MOSBACH 181
17. Flow-Injection Analysis for Automated Determination of β-Lactams Using Immobilized Enzyme Reactors with Thermistor or Ultraviolet Spectrophotometric Detection GEORG DECRISTOFORO 197
18. Biomedical Applications of the Enzyme Thermistor in Lipid Determinations
IKUO SATOH 217
19. Flow Enthalpimetric Determination of Glucose
NOBUTOSHI KIBA AND MOTOHISA FURUSAWA 225

E. Enzyme Transistors

20. Use of Hydrogen- and Ammonia-Sensitive Semiconductor Structures in Analytical Biochemistry: Enzyme Transistors
FREDRIK WINQUIST, BENGT DANIELSSON, INGEMAR LUNDSTRÖM, AND KLAUS MOSBACH 232
21. Enzymatically Sensitive Field Effect Transistors
STEVE CARAS AND JIŘÍ JANATA 247
22. Microenzyme Sensors ISAO KARUBE AND TOYOSAKA MORIIZUMI 255

F. Enzyme Reactors

23. Use of Immobilized Enzyme Column Reactors in Clinical Analysis
TAKASHI MURACHI AND MASAYOSHI TABATA 260
24. Use of Immobilized Enzyme Reactors in Flow-Injection Analysis
MAT H. HO 271
25. Routine Analysis with Immobilized Enzyme Nylon Tube Reactors
P. V. SUNDARAM 288

G. Continuous Monitoring

26. Enzyme Sensors for Fermentation Monitoring: Sample Handling
SVEN-OLOF ENFORS AND NEIL CLELAND 298
27. Enzyme Thermistors for Process Monitoring and Control
CARL FREDRIK MANDENIUS AND BENGT DANIELSSON 307
28. Portable Continuous Blood Glucose Analyzer HAKAN HAKANSON 319
29. Needle-Type Glucose Sensor
MOTOAKI SHICHIRI, RYUZO KAWAMORI, AND YOSHIMITSU YAMASAKI 326
30. Automated TELISA Procedure for Process Monitoring
STAFFAN BIRNBAUM, LEIF BÜLOW, BENGT DANIELSSON, AND KLAUS MOSBACH 334

H. Optical Methods

31. Solid-Phase Optoelectronic Biosensors
CHRISTOPHER R. LOWE AND MICHAEL J. GOLDFINCH 338

32. Optical Fiber Affinity Sensors JEROME S. SCHULTZ AND SOHRAB MANSOURI 349
33. Surface-Oriented Optical Methods for Biomedical Analysis
HANS ARWIN AND INGEMAR LUNDSTRÖM 366
34. Surface Immobilization Techniques in Combination with Ellipsometry
ULF JÖNSSON, MAGNUS MALMQVIST, GÖRAN OLOFSSON, AND INGER RÖNNBERG 381
35. Coupling of Biomolecules to Silicon Surfaces for Use in Ellipsometry and Other Related Techniques
CARL FREDRIK MANDENIUS, STEFAN WELIN, INGEMAR LUNDSTRÖM, AND KLAUS MOSBACH 388

I. Miscellaneous New Techniques

36. Construction of Dry Reagent Chemistries: Use of Reagent Immobilization and Compartmentalization Techniques BERT WALTER 394
37. Mass Spectrometry Combined with Immobilized Cells and Enzymes
JAMES C. WEAVER 421
38. Bioelectrocatalysis: Conductive and Semiconductive Matrices for Immobilized Enzymes S. D. VARFOLOMEEV 430

Section II. Medical Applications

39. Overview KLAUS MOSBACH 443
40. Medical Applications of Immobilized Proteins, Enzymes, and Cells
T. M. S. CHANG 444
41. Extracorporeal Systems for Adsorption of Antibodies in Hemophilia A and B
CHRISTIAN FREIBURGHAUS, STEN OHLSON, AND INGA MARIE NILSSON 458
42. Low Density Lipoprotein-Pheresis: Selective Immunoadsorption of Plasma Lipoproteins from Patients with Premature Atherosclerosis
T. S. PARKER AND J. F. STUDEBAKER 466
43. Enzyme Reactors for the Removal of Amino Acids from Plasma
GOTTFRIED SCHMER AND WAYNE L. CHANDLER 479
44. Dialysis Membranes Containing Asparaginase Entrapped in Microparticles
PETER EDMAN, ULF NYLÉN, AND INGVAR SJÖHOLM 491
45. Preparation of Protein A Immobilized on Collodion-Coated Charcoal and Plasma Perfusion System for Treatment of Cancer DAVID S. TERMAN 496
46. Immobilized Heparin Lyase System for Blood Deheparinization
HOWARD BERNSTEIN, VICTOR C. YANG, CHARLES L. COONEY, AND ROBERT LANGER 515
47. Clinical Applications of Urokinase-Treated Material
TAKESHI OHSHIRO, MOU CHUNG LIU, JUNICHI KAMBAYASHI, AND TAKESADA MORI 529
48. Application of Immobilized Enzymes for Biomaterials Used in Surgery
SATOSHI WATANABE, YASUHIKO SHIMIZU, TAKASHI TERAMATSU, TAKASHI MURACHI, AND TSUNETOSHI HINO 545
49. Immobilized Enzymes for Thrombolytic Therapy
V. P. TORCHILIN, A. V. MAKSIMENKO, AND A. V. MAZAEV 552
50. Soluble Enzyme–Albumin Conjugates: New Possibilities for Enzyme Replacement Therapy MARK J. POZNANSKY 566
51. Microencapsulation of Pancreatic Islet Cells: A Bioartificial Endocrine Pancreas
ANTHONY M. SUN 575

Section III. Novel Techniques for and Aspects of Immobilized Enzymes and Cells

52. Overview K. MOSBACH 583
53. Enzyme Stabilization by Immobilization PIERRE MONSAN AND DIDIER COMBES 584
54. Stabilization of Pyranose 2-Oxidase and Catalase by Chemical Modification ZE'EV SHAKED AND SIDNEY WOLFE 599
55. Stabilization of Enzymes by Intramolecular Cross-Linking Using Bifunctional Reagents KAREL MARTINEK AND V. P. TORCHILIN 615
56. Long-Term Stability of Nongrowing Immobilized Cells of *Clostridium acetobutylicum* Controlled by the Intermittent Nutrient Dosing Technique LENA HÄGGSTRÖM AND CECILIA FÖRBERG 626
57. Production of Proinsulin by Entrapped Bacteria with Control of Cell Division by Inhibitors of DNA Synthesis L. BÜLOW, S. BIRNBAUM, AND K. MOSBACH 632
58. Permeabilized and Immobilized Cells HANSRUEDI FELIX 637
59. Cells Coimmobilized with Enzymes BÄRBEL HAHN-HÄGERDAL 641
60. Affinity Immobilization BO MATTIASSON 647
61. Bioconversions in Aqueous Two-Phase Systems: An Alternative to Conventional Immobilization BO MATTIASSON 657
62. Biochemical Energy Conversion by Immobilized Photosynthetic Bacteria ISAO KARUBE AND SHUICHI SUZUKI 668
63. Immobilization of a Cytochrome *P*-450 Enzyme from *Saccharomyces cerevisiae* DAVID J. KING, MAHMOOD R. AZARI, AND ALAN WISEMAN 675
64. Enzyme Production by Immobilized Cells YU-YEN LINKO, G.-X. LI, LI-CHAN ZHONG, SUSAN LINKO, AND P. LINKO 686
65. Application of Immobilized Captured Microorganisms in Water Purification: An Overview J. D. BRYERS AND G. HAMER 697

VOLUME 138

COMPLEX CARBOHYDRATES (PART E)

Section I. Analytical Methods

1. Isolation and Purification of Glycosphingolipids by High-Performance Liquid Chromatography REIJI KANNAGI, KIYOHIRO WATANABE, AND SEN-ITIROH HAKOMORI 3
2. Microscale Methylation Analysis of Glycolipids Using Capillary Gas Chromatography–Chemical Ionization Mass Fragmentography with Selected Ion Monitoring STEVEN B. LEVERY AND SEN-ITIROH HAKOMORI 13
3. Reductive Cleavage of Permethylated Polysaccharides GARY R. GRAY 26
4. Oligosaccharide Structure by Two-Dimensional Proton Nuclear Magnetic Resonance Spectroscopy THEODORE A. W. KOERNER, JAMES H. PRESTEGARD, AND ROBERT K. YU 38
5. Direct Chemical Ionization Mass Spectrometry of Carbohydrates VERNON N. REINHOLD 59
6. BioGel P-4 Column Chromatography of Oligosaccharides: Effective Size of Oligosaccharides Expressed in Glucose Units AKIRA KOBATA, KATSUKO YAMASHITA, AND SEIICHI TAKASAKI 84

7. Separation of Benzoylated Oligosaccharides by Reversed-Phase High-Pressure Liquid Chromatography: Application to High-Mannose Type Oligosaccharides — PETER F. DANIEL 94
8. High-Pressure Liquid Chromatography Analysis of Neutral Glycosphingolipids: Perbenzoylated Mono-, Di-, Tri-, and Tetraglycosylceramides — M. D. ULLMAN AND R. H. MCCLUER 117
9. Short-Bed, Continuous Development, Thin-Layer Chromatography of Glycosphingolipids — WILLIAM W. YOUNG, JR. AND CHERYL A. BORGMAN 125
10. Analysis of Sialic Acids — ROLAND SCHAUER 132
11. Resialylated Erythrocytes for Assessment of the Specificity of Sialyl Oligosaccharide Binding Proteins — JAMES C. PAULSON AND GARY N. ROGERS 162
12. Detection of Polysialosyl-Containing Glycoproteins in Brain Using Prokaryotic-Derived Probes — FREDERIC A. TROY, PATRICK C. HALLENBECK, RONALD D. MCCOY, AND ERIC R. VIMR 169
13. Inositol-Containing Sphingolipids — ROGER A. LAINE AND THOMAS C.-Y. HSIEH 186
14. Antibodies against Cell Surface Carbohydrates: Determination of Antigen Structure — JOHN L. MAGNANI, STEVEN L. SPITALNIK, AND VICTOR GINSBURG 195
15. Immunostaining Free Oligosaccharides Directly on Thin-Layer Chromatograms — JOHN L. MAGNANI 208
16. Preparation of Total Nonacid Glycolipids for Overlay Analysis of Receptors for Bacteria and Viruses and for Other Studies — KARL-ANDERS KARLSSON 212
17. Overlay and Solid-Phase Analysis of Glycolipid Receptors for Bacteria and Viruses — KARL-ANDERS KARLSSON AND NICKLAS STRÖMBERG 220
18. Lectin Affinity Chromatography of Glycopeptides — ROBERTA K. MERKLE AND RICHARD D. CUMMINGS 232
19. Detection of Glycosaminoglycans with ^{125}I-Labeled Cytochrome *c* — PHYLLIS M. SAMPSON, RALPH HEIMER, AND ALFRED P. FISHMAN 260
20. Electrophoretic Separation of Lipopolysaccharide Monomers Differing in Polysaccharide Length — R. C. GOLDMAN AND L. LEIVE 267

Section II. Preparations

21. Proteoglycans: Isolation and Purification from Tissue Cultures — MASAKI YANAGISHITA, RONALD J. MIDURA, AND VINCENT C. HASCALL 279
22. Isolation of Sialyl Oligosaccharides and Sialyl Oligosaccharide Phosphates from Bovine Colostrum and Human Urine — JAAKO PARKKINEN AND JUKKA FINNE 289
23. Isolation and Purification of Gangliosides from Plasma — STEPHAN LADISCH AND BAIBA GILLARD 300
24. Affinity Purification of Oligosaccharides Using Monoclonal Antibodies — D. ZOPF, K. SCHROER, G. HANNSON, J. DAKOUR, AND A. LUNDBLAD 307
25. Fluorescent Gangliosides — SARAH SPIEGEL 313
26. Lysogangliosides: Synthesis and Use in Preparing Labeled Gangliosides — GÜNTER SCHWARZMANN AND KONRAD SANDHOFF 319
27. Chemical Deglycosylation of Glycoproteins — HAKIMUDDIN T. SOJAR AND OM P. BAHL 341
28. Enzymatic Deglycosylation of Glycoproteins — NAGESWARA T. THOTAKURA AND OM P. BAHL 350

29. Chemical Synthesis of Oligosaccharides HAROLD M. FLOWERS 359
30. Nondegradable Carbohydrate Ligands
Y. C. LEE, C. A. HOPPE, K. KAWAGACHI, AND P. LIN 404
31. Attachment of Oligosaccharide–Asparagine Derivatives to Proteins: Activation of Asparagine with Ninhydrin and Coupling to Protein by Reductive Amination
ARLENE J. MENCKE, DAVID T. CHEUNG, AND FINN WOLD 409
32. Covalent Attachment of Oligosaccharide–Asparagine Derivatives: Incorporation into Glutamine Residues with the Enzyme Transglutaminase
SAU-CHI BETTY YAN 413
33. Noncovalent Attachment of Oligosaccharide–Asparagine Derivatives to Proteins
VICTOR J. CHEN 418
34. Cluster Glycosides REIKO T. LEE AND Y. C. LEE 424
35. Labeling Glycoconjugates with Hydrazide Reagents
MEIR WILCHEK AND EDWARD A. BAYER 429
36. Biochemical Characterization of Animal Cell Glycosylation Mutants
PAMELA STANLEY 443
37. Isolation of Mutant Chinese Hamster Ovary Cells Defective in Endocytosis
APRIL R. ROBBINS AND CALVIN F. ROFF 458

Section III. Carbohydrate-Binding Proteins

38. Sulfatide-Binding Proteins DAVID D. ROBERTS 473
39. Mouse and Rat Monoclonal Antibodies Directed against Carbohydrates
JOHN L. MAGNANI 484
40. Human Monoclonal Antibodies Directed against Carbohydrates
STEVEN L. SPITALNIK 492
41. Low Molecular Weight Phosphomannosyl Receptor from Bovine Testes
JACK J. DISTLER AND GEORGE W. JOURDIAN 504
42. Soluble Lactose-Finding Lectins from Chicken and Rat
SAMUEL H. BARONDES AND HAKON LEFFLER 510
43. *Xenopus laevis* Lectins SAMUEL H. BARONDES AND MARIE M. ROBERSON 516
44. *O*-Acetylsialic Acid-Specific Lectin from the Crab *Cancer anntenarius*
M. H. RAVINDRANATH AND JAMES C. PAULSON 520
45. Properties of a Sialic Acid-Specific Lectin from the Slug *Limax flavus*
RONALD L. MILLER 527
46. *Vicia villosa* Lectins SHERIDA TOLLEFSEN AND ROSALIND KORNFELD 536
47. *Erythrina* Lectins HALINA LIS AND NATHAN SHARON 544
48. *Evonymus europaea* Lectin JERZY PETRYNIAK AND IRWIN J. GOLDSTEIN 552
49. Lectin from Hog Peanut, *Amphicarpaea bracteata*
M. MALIARIK, D. ROBERTS, AND I. J. GOLDSTEIN 561

Section IV. Biosynthesis

50. Glycosyltransferase Assay BO E. SAMUELSSON 567
51. Glycolipids
MANJU BASU, TRIPTI DE, KAMAL K. DAS, JOHN W. KYLE, HUNG-CHE CHON, ROBERT J. SCHAEPER, AND SUBHASH BASU 575
52. Acetyl-CoA:α-glucosaminide *N*-Acetyltransferase from Rat Liver
KAREN J. BAME AND LEONARD H. ROME 607

53. Sialic Acids: Metabolism of *O*-Acetyl Groups ROLAND SCHAUER 611
54. CMP-NeuNAc:Poly-α-2,8-Sialosyl Sialyltransferase in Neural Cell Membranes RONALD D. MCCOY AND FREDERIC A. TROY 627
55. Fungal 1,3-β-Glucan Synthase ENRICO CABIB AND MOHINDER S. KANG 637
56. Chitin Synthase from *Saccharomyces cerevisiae* ENRICO CABIB, MOHINDER S. KANG, AND JANICE AU-YOUNG 643
57. Dextransucrase: A Glucosyltransferase from *Streptococcus sanguis* ROBERT M. MAYER 649
58. Glycosylation Inhibitors for N-Linked Glycoproteins ALAN D. ELBEIN 661
59. Transport of Sugar Nucleotides into the Lumen of Vesicles Derived from Rat Liver Rough Endoplasmic Reticulum and Golgi Apparatus MARY PEREZ AND CARLOS B. HIRSCHBERG 709
60. Plant Cell Wall Polysaccharides That Activate Natural Plant Defenses PAUL D. BISHOP AND CLARENCE A. RYAN 715

Section V. Degradation

61. Enzymatic Diagnosis of Sphingolipidoses KUNIHIKO SUZUKI 727
62. Endo-β-*N*-acetylglucosaminidase H from *Streptomyces plicatus* ROBERT B. TRIMBLE, ROBERT J. TRUMBLY, AND FRANK MALEY 763
63. Peptide-N^4-(*N*-acetyl-β-glucosaminyl) Asparagine Amidase and Endo-β-*N*-acetylglucosaminidase from *Flavobacterium meningosepticum* ANTHONY L. TARENTINO AND THOMAS H. PLUMMER, JR. 770
64. α-Mannosidases I and II from *Aspergillus saitoi* AKIRA KOBATA AND JUNKO AMANO 779
65. Polysialic Acid Depolymerase BARTŁOMIEJ KWIATKOWSKI AND STEPHAN STIRM 786
66. Activator Proteins for Lysosomal Glycolipid Hydrolysis ERNST CONZELMANN AND KONRAD SANDHOFF 792
67. Preparation of Keyhole Limpet Oligosaccharyl Sulfatase PREMANAND V. WAGH, KALYAN R. ANUMULA, AND OM P. BAHL 816

VOLUME 139
CELLULAR REGULATORS (PART A: CALCIUM- AND CALMODULIN-BINDING PROTEINS)

Section I. Isolation and Characterization of Calcium-Binding Proteins

1. The Use of Melittin–Sepharose Chromatography for Gram-Preparative Purification of Calmodulin RANDALL L. KINCAID 3
2. Calcimedins: Purification and Production of Antibodies J. K. MATHEW, R. R. SCULLY, V. L. SMITH, E. BERNICKER, AND J. R. DEDMAN 19
3. Purification of Calelectrins THOMAS C. SÜDHOF AND DENNIS K. STONE 30
4. Calregulin: Purification, Cellular Localization, and Tissue Distribution NAVIN C. KHANNA, MASAAKI TOKUDA, AND DAVID MORTON WAISMAN 36

5. Purification and Characterization of *Acanthamoeba* Calcium-Binding Proteins
DAVID C. WYLIE AND THOMAS C. VANAMAN 50
6. Purification of Novel Calcium-Binding Proteins from Bovine Brain
MASAAKI TOKUDA, NAVIN C. KHANNA, AND DAVID MORTON WAISMAN 68
7. Calmodulin-Dependent Protein Phosphatase: Isolation of Subunits and Reconstitution to Holoenzyme DENNIS L. MERAT AND WAI YIU CHEUNG 79
8. Isolation and Characterization of a Novel 21-kDa Ca^{2+}-Binding Protein from Bovine Brain
JOHN R. MCDONALD, MICHAEL P. WALSH, WILLIAM D. MCCUBBIN, AND CYRIL M. KAY 88
9. Isolation and Characterization of Calmodulin-Dependent Myosin Heavy Chain Kinase from Intestinal Brush Border
JAMES P. RIEKER, HELENA SWANLJUNG-COLLINS, JUDITH MONTIBELLER, AND JIMMY H. COLLINS 105
10. Preparation and Properties of the Calmodulin-Binding Domain of Skeletal Muscle Myosin Light Chain Kinase
DONALD K. BLUMENTHAL AND EDWIN G. KREBS 115
11. Purification and Assay of Calspermin: A Heat-Stable Calmodulin-Binding Protein in Male Reproductive System and Central Nervous System
TOMIO ONO, YOSHINOBU KOIDE, CHIZUKO ISHINAMI, YUJI ARAI, AND KAMEJIRO YAMASHITA 127
12. Purification and Characterization of the 110-kDa Actin- and Calmodulin-Binding Protein from Intestinal Brush Border: A Myosin-Like ATPase
HELENA SWANLJUNG-COLLINS, JUDITH MONTIBELLER, AND JIMMY H. COLLINS 137
13. Immunoassay and Immunopurification of Macrophage Calmodulin-Binding Protein SETH J. ORLOW AND ORA M. ROSEN 148
14. Isolation, Localization, and Properties of the Oncodevelopmental Calcium-Binding Protein Oncomodulin JOHN P. MACMANUS AND LINDA M. BREWER 156
15. Isolation and Spectroscopic Analyses of S-100 Proteins and Their Interactions with Metal Ions RAJAM S. MANI AND CYRIL M. KAY 168
16. Affinity Labeling of the Nucleotide-Binding Site of Myosin Light Chain Kinases
JOEL C. COLBURN, PETER J. KENNELLY, EDWIN G. KREBS, AND JAMES T. STULL 188
17. Study of Calmodulin–Peptide Interactions by NMR Spectroscopy
RACHEL E. KLEVIT 197
18. Novel Ca^{2+}-Binding Proteins in Tumor Cells CLAUS W. HEIZMANN 207

Section II. Molecular Cloning of Calcium-Binding Proteins, cDNAs, and Genes

19. Isolation and Structural Analysis of the Chromosomal Gene for Chicken Calmodulin
PAUL EPSTEIN, ROSALIA C. M. SIMMEN, TOSHIO TANAKA, AND ANTHONY R. MEANS 217
20. The Calmodulin Gene of *Drosophila melanogaster*
KATHY BECKINGHAM, KEVIN E. DOYLE, AND JOHN F. MAUNE 230
21. Isolation of the Yeast Calmodulin Gene Using Synthetic Oligonucleotide Probes
TRISHA N. DAVIS AND JEREMY THORNER 248

22. Calmodulin from *Trypanosoma brucei:* Immunological Analysis and Genomic Organization LARRY RUBEN AND CURTIS L. PATTON 262
23. Dissection of Tissues for Characterizing Nucleic Acids from *Aplysia:* Isolation of the Structural Gene Encoding Calmodulin J. H. SCHWARTZ AND M. E. SWANSON 277
24. The Use of Synthetic Oligodeoxyribonucleotides in the Examination of Calmodulin Gene and Protein Structure and Function DANIEL M. ROBERTS, WARREN E. ZIMMER, AND D. MARTIN WATTERSON 290
25. Bacterial Expression Vectors for Calmodulin JOHN A. PUTKEY, PATRICIA V. DONNELLY, AND ANTHONY R. MEANS 303
26. Cloning of the Rat Parvalbumin Gene MARTIN W. BERCHTOLD 317
27. The Cloning and the Codon and Amino Acid Sequence of the Quail Slow/Cardiac Troponin C cDNA PETER C. MAISONPIERRE, KENNETH E. M. HASTINGS, AND CHARLES P. EMERSON, JR. 326
28. 9-kDa Vitamin D-Induced Intestinal Calcium-Binding Protein: Cloning and Characterization of a Specific cDNA MONIQUE THOMASSET, CLAUDE DESPLAN, AND CHRISTINE PERRET 337
29. Spec Proteins: Calcium-Binding Proteins in the Embryonic Ectoderm Cells of Sea Urchins WILLIAM H. KLEIN, SUSAN H. HARDIN, AND BRUCE P. BRANDHORST 354
30. Cloning and Evolution of Calcium-Dependent Protease, cDNA Cloning of a Novel Family of Calcium-Binding Proteins SHIGEO OHNO, YASUFUMI EMORI, HINDEMITSU SUGIHARA, SHINOBU IMAJOH, AND KOICH SUZUKI 363
31. Cloning of the Gene for Protein S: A Development-Specific Ca^{2+}-Binding Protein from *Myxococcus xanthus* SUMIKO INOUYE AND MASAYORI INOUYE 380

Section III. Reagents and Methods for the Study of Calcium-Binding Proteins

32. Site-Directed Antibodies to Vertebrate and Plant Calmodulins LINDA J. VAN ELDIK AND THOMAS J. LUKAS 393
33. Synthesis and Properties of $CAPP_1$–Calmodulin D. L. NEWTON, T. R. BURKE, JR., K. C. RICE, AND C. B. KLEE 405
34. Specific Chemical Modification as a Probe of Calmodulin Function DENNIS MANN AND THOMAS C. VANAMAN 417
35. Use of the ^{125}I-Labeled Protein Gel Overlay Technique to Study Calmodulin-Binding Proteins GAYLE R. SLAUGHTER AND ANTHONY R. MEANS 433
36. Interaction of Calmodulin and Other Calcium-Modulated Proteins with Gap Junctions ELLIOT L. HERTZBERG AND LINDA J. VAN ELDIK 445
37. Recognition and Characterization of Calmodulin-Binding Sequences in Peptides and Proteins SUSAN ERICKSON-VIITANEN AND WILLIAM F. DEGRADO 455
38. Immunolocalization of Calmodulin in Plants CHIN-TARNG LIN 478
39. Immunoperoxidase Staining of S-100 and Calmodulin in Tissue Sections and Cultured Cells MAMORU SANO, AKIKO SETO-OHSHIMA, SATOKO KITAJIMA, AND KANEFUSA KATO 495
40. Production and Characterization of a Monoclonal Antibody Cross-Reactive with Calmodulin, Calmodulin-Dependent Phosphodiesterase, and Protein Phosphatase MARTIN A. WINKLER, JOHN R. ZYSK, AND WAI YIU CHEUNG 505

41. Monoclonal Antibodies as Specific Probes for the Study of CaM-Regulatory System
JERRY H. WANG, TAKASHI SUZUKI, MARILYN MOOIBROEK, ANNE-MARIE ADACHI, RAJENDRA K. SHARMA, AND Y.-H. PETER LAM 518
42. Rat Calbindin D_{28K}: Purification, Quantitation, Immunocytochemical Localization, and Comparative Aspects
SYLVIA CHRISTAKOS, WILLIAM B. RHOTEN, AND SUSAN C. FELDMAN 534
43. Immunolocalization of Parvalbumin
CLAUS W. HEIZMANN AND MARCO R. CELIO 552
44. Transmembrane Ca^{2+} Signaling and a New Class of Inhibitors
HIROYOSHI HIDAKA AND TOSHIO TANAKA 570

Section IV. Regulation of Calcium and Calcium-Binding Proteins

45. Determination of the Three-Dimensional Structure of Vitamin D-Dependent Calcium-Binding Protein from Bovine Intestine
D. M. E. SZEBENYI AND K. MOFFAT 585
46. Molecular Structure of Troponin C and Its Implications for the Ca^{2+} Triggering of Muscle Contraction
OSNAT HERZBERG, JOHN MOULT, AND MICHAEL N. G. JAMES 610
47. X-Ray Diffraction Studies of Calmodulin
Y. SUDHAKAR BABU, CHARLES E. BUGG, AND WILLIAM J. COOK 632
48. Methods for Analyzing Bovine Papilloma Virus-Based Calmodulin Expression Vectors
COLIN D. RASMUSSEN, ROSALIA C. M. SIMMEN, ELIZABETH A. MACDOUGALL, AND ANTHONY R. MEANS 642
49. Ca^{2+}/Calmodulin Regulation of Prolactin Gene Expression
BRUCE A. WHITE AND CARTER BANCROFT 655
50. Calmodulin as an Activator and a Substrate of Methyltransferase Enzymes
FRANK L. SIEGEL, LYNDA S. WRIGHT, AND PAUL M. ROWE 667
51. Fluorometric and Flow Cytometric Analysis of the Cyclosporine–Calmodulin Interaction in T Lymphocytes PAUL M. COLOMBANI AND ALLAN D. HESS 677
52. Role of Autophosphorylation in Regulating Calmodulin-Dependent Protein Kinases PAUL T. KELLY AND SHIRISH SHENOLIKAR 690
53. DNA Repair in Human Cells: Methods for the Determination of Calmodulin Involvement PAUL A. CHARP 715
54. Calmodulin-Stimulated Estradiol Receptor–Tyrosine Kinase
FERDINANDO AURICCHIO, ANTIMO MIGLIACCIO, GABRIELLA CASTORIA, ANDREA ROTONDI, AND MARINA DI DOMENICO 731
55. Interaction of Calmodulin with Phosphofructokinase: Binding Studies and Evaluation of Enzymatic and Physicochemical Changes GEORG W. MAYR 745
56. Regulation of Parvalbumin Concentration in Mammalian Muscle
EKKEHARD LEBERER, GARY A. KLUG, UDO SEEDORF, AND DIRK PETTE 763
57. Reconstitution of Calmodulin-Sensitive Adenylate Cyclase from Bovine Brain with Phospholipids, Calmodulin, and β-Adrenergic Receptors
GARY B. ROSENBERG, ARDA-E-VIRAF M. MINOCHERHOMJEE, AND DANIEL R. STORM 776
58. Purification, Reconstitution, and Molecular Characterization of the Ca^{2+} Pump of Plasma Membranes V. NIGGLI, M. ZURINI, AND E. CARAFOLI 791

59. Ca^{2+} Regulation of Sperm Axonemal Motility
JOSEPH S. TASH AND ANTHONY R. MEANS 808
60. The Use of Microinjection and Video Microscopy for the Study of Calmodulin and Calcium in Living Cells MICHAEL SHELANSKI AND RAJIV RATAN 824
61. Calmodulin and Cold-Labile Microtubules YOUNG C. LEE AND J. WOLFF 834
62. Calmodulin and Actin Polymerization
ROBERT W. WALLACE AND GARY A. PIAZZA 846

VOLUME 141
CELLULAR REGULATORS (PART B: CALCIUM AND LIPIDS)

Section I. Measurement and Perturbation of Intracellular Calcium

1. Measurement of Calcium Release from Sarcoplasmic Reticulum of Skeletal Muscle: Effect of Calcium and Inositol 1,4,5-Trisphosphate
POMPEO VOLPE, FRANCESCO ZORZATO, TULLIO POZZAN, GIOVANNI SALVIATI, AND FRANCESCO DI VIRGILIO 3
2. Glycogen Phosphorylase: A Noninvasive Indicator of Cytoplasmic Calcium
M. W. WILDE, J. L. SLONCZEWSKI, M. CARSON, AND S. H. ZIGMOND 18
3. Preparation and Utilization of an Ion-Specific Calcium Minielectrode
ERWIN SIGEL AND HUBERT AFFOLTER 25
4. Measurement of Changes in Cellular Calcium Metabolism in Response to Thyrotropin-Releasing Hormone
ELIZABETH GERAS-RAAKA AND MARVIN C. GERSHENGORN 36
5. Flow Cytometric Analysis of Intracellular Calcium Mobilization
JOHN T. RANSOM, DAVID L. DIGIUSTO, AND JOHN CAMBIER 53
6. Activation of Calcium Channels by Maitotoxin
IVAN S. LOGIN, ALLAN M. JUDD, AND ROBERT M. MACLEOD 63

Section II. Inositol Phospholipids and Proximal Metabolites

A. Quantitation and Metabolism in Cells, Permeable Cells, and Cell-Free Systems

7. Measurement of Phosphoinositide Pools in WRK-1 Cells MARIE E. MONACO 83
8. Measurement of Subcellular Sites of Polyphosphoinositide Metabolism in Isolated Rat Hepatocytes
WILLIAM W. WELLS, MARK A. SEYFRED, CHARLES D. SMITH, AND MASAHIRO SAKAI 92
9. Measurement of Lipid Turnover in Response to Thyrotropin-Releasing Hormone
ATSUSHI IMAI AND MARVIN C. GERSHENGORN 100
10. Hormone-Regulated Phosphoinositide Turnover in Permeabilized Cells and Membranes THOMAS F. J. MARTIN 111
11. Measurement of Biosynthesis of *myo*-Inositol from Glucose 6-Phosphate
FRANK EISENBERG, JR. AND RANGANATHAN PARTHASARATHY 127
12. Measurement of Capillary Gas Chromatography of Mass Changes in *myo*-Inositol Trisphosphate SUSAN ERIKA RITTENHOUSE 143

13. Use of Lithium Ion in Measurement of Stimulated Pituitary Inositol Phospholipid Turnover WILLIAM R. HUCKLE AND P. MICHAEL CONN 149
14. Measurement of Inositol Phospholipid Metabolites by One-Dimensional Thin-Layer Chromatography WILLIAM V. ANDREWS AND P. MICHAEL CONN 156
15. Measurement of Inositol Phospholipid Metabolism in PC12 Pheochromocytoma Cells JOEL HORWITZ AND ROBERT L. PERLMAN 169
16. Measurement of Inositol Phospholipid Turnover in Platelets EDUARDO G. LAPETINA AND WOLFGANG SIESS 176
17. Neurotransmitter-Stimulated Inositol Phosphate Accumulation in Hippocampal Slices RODRIGO LABARCA, AARON JANOWSKY, AND STEVEN M. PAUL 192
18. Autoradiographic Localization of Newly Synthesized Inositol Phospholipids in the Superior Cervical Ganglion CONWELL H. ANDERSON, JOEL HORWITZ, AND ROBERT L. PERLMAN 201
19. Phosphatidylinositol Kinases from Rabbit Reticulocytes POLYGENA T. TUAZON AND JOLINDA A. TRAUGH 210

B. Regulation of Lipid Synthesis

20. Methods for Altering the Synthesis of Phosphatidylinositol and Phosphatidylglycerol by Type II Pneumonocytes JOHN E. BLEASDALE 222
21. Nutritional Regulation of Cellular Phosphatidylinositol BRUCE J. HOLUB AND CLARK M. SKEAFF 234
22. Modulation of Phosphatidylinositol Turnover by Liposomes Containing Phosphatidylinositol NABILA M. WASSEF AND CARL R. ALVING 244

C. Guanine Nucleotide-Binding Component

23. Hormonal Activation of Phosphoinositide Breakdown in the Presence of Guanine Nucleotides Using a Cell-Free System from Blowfly Salivary Glands JOHN N. FAIN AND IRENE LITOSCH 255
24. Guanine Nucleotide Regulatory Proteins in Receptor-Mediated Polyphosphoinositide Hydrolysis in Human Leukocytes CHARLES D. SMITH AND RALPH SNYDERMAN 261

Section III. Phorbol Esters and Diacylglycerols: Preparation, Quantitation, and Binding

25. Measurement of Phorbol Ester Receptors in Intact Cells and Subcellular Fractions SUSAN JAKEN 275
26. Mixed Micellar Assay for Phorbol Ester Binding YUSUF A. HANNUN AND ROBERT M. BELL 287
27. Quantitative Measurement of *sn*-1,2-Diacylglycerols JENNIFER E. PREISS, CARSON R. LOOMIS, ROBERT M. BELL, AND JAMES E. NIEDEL 294
28. Assays and Substrate Preparation for the Enzymatic Hydrolysis of Diacylglycerols and Phorbol Diesters MYLES C. CABOT, CLEMENT J. WELSH, ZU-CHUAN ZHANG, AND HOLLY CHABBOTT 301
29. Synthesis of Cell-Permeant Diacylglycerol Analogs for Structure–Function Analysis of Protein Kinase C and Other Enzymes BARRY R. GANONG AND ROBERT M. BELL 313

Section IV. Prostaglandins, Icosanoids, Leukotrienes, and Other Lipids

30. Measurement of Prostaglandins by Radioimmunoassay
WILLIAM B. CAMPBELL AND SERGIO R. OJEDA 323
31. Measurement of Prostaglandin Production by Isolated Tissues and Cells in Culture A. RICHARD WHORTON 341
32. Measurement of Icosanoid Precursor Uptake and Release by Intact Cells
MICHAEL LAPOSATA AND PHILIP W. MAJERUS 350
33. Leukotriene B_4 Catabolism: Quantitation of Leukotriene B_4 and Its ω-Oxidation Products by Reversed-Phase High-Performance Liquid Chromatography
STEVEN SHAK 355
34. Assay for the Leukotriene B_4 Receptor R. R. GORMAN AND A. H. LIN 372
35. Measurement of Key Enzyme Activities Involved in the Metabolism of Platelet Activating Factor
FRED SNYDER, MERLE BLANK, TEN-CHING LEE, MITCHELL ROBINSON, AND DONNA WOODARD 379

Section V. Protein Kinase C

36. Hormone- and Tumor Promoter-Induced Activation or Membrane Association of Protein Kinase C in Intact Cells
THOMAS P. THOMAS, RAYUDU GOPALAKRISHNA, AND WAYNE B. ANDERSON 399
37. Protein Kinase C-Mediated Phosphorylation in Intact Cells
LEE A. WITTERS AND PERRY J. BLACKSHEAR 412
38. Protein Kinase C Purification by High-Performance Liquid Chromatography: An Improved Method
MASAYOSHI GO, JUNKO KOUMOTO, USHIO KIKKAWA, AND YASUTOMI NISHIZUKA 424
39. Measurement of Synergistic Effects of Protein Kinase C Activators and Calcium Ionophore in Pituitary Cell Cultures
CYNTHIA HARRIS WOOGE AND P. MICHAEL CONN 429

Section VI. New Probes

40. Fluorescence Photobleaching Assay of Tumor Promoter 12-*O*-Tetradecanoylphorbol 13-Acetate Inhibition of Cell–Cell Communication
MELVIN SCHINDLER, JAMES E. TROSKO, AND MARGARET H. WADE 439
41. Fluorescence Redistribution after Photobleaching as a Tool for Dissecting the Control Elements of Nucleocytoplasmic Transport
MELVIN SCHINDLER AND LIAN-WEI JIANG 447
42. Tumoricidal Effects of Liposomes Containing Phosphatidylinositol or Phosphatidylcholine MARTI JETT AND CARL R. ALVING 459

Section VII. Cross-Index to Prior Volumes

43. Previously Published Articles from Methods in Enzymology Related to Cellular Regulators 469

VOLUME 142
Metabolism of Aromatic Amino Acids and Amines

Section I. Catabolism of the Aromatic Amino Acids

1. Phenylalanine 4-Monooxygenase from Rat Liver — Seymour Kaufman 3
2. Purification and Assay of Rat Liver Phenylalanine 4-Monooxygenase — Ross Shiman 17
3. Phenylalanine 4-Monooxygenase from Human Liver — Jean-Pierre Abita, Félicienne Blandin-Savoja, and Françoise Rey 27
4. Phenylalanine 4-Monooxygenase from Bovine Liver — Anne P. Døskeland, Stein Ove Døskeland, and Torgeir Flatmark 35
5. Phenylalanine 4-Monooxygenase from *Chromobacterium violaceum* — Hitoshi Fujisawa and Hiroyasu Nakata 44
6. *Chromobacterium violaceum* Phenylalanine 4-Monooxygenase — Stephen O. Pember, Joseph J. Villafranca, and Stephen J. Benkovic 50
7. Tyrosine 3-Monooxygenase from Bovine Adrenal Medulla — Toshiharu Nagatsu and Kazuhiro Oka 56
8. Tyrosine 3-Monooxygenase from Rat Adrenals — Hitoshi Fujisawa and Sachiko Okuno 63
9. Tyrosine 3-Monooxygenase from Rat Pheochromocytoma — A. William Tank and Norman Weiner 71
10. Tryptophan 5-Monooxygenase from Rat Brain Stem — Hitoshi Fujisawa and Hiroyasu Nakata 83
11. Tryptophan 5-Monooxygenase from Mouse Mastocytoma: High-Performance Liquid Chromatography Assay — Hiroyuki Hasegawa and Arata Ichiyama 88
12. Tryptophan 5-Monooxygenase from Mouse Mastocytoma Clone P815 — Hitoshi Fujisawa and Hiroyasu Nakata 93
13. Dihydropteridine Reductase from Sheep Liver — Seymour Kaufman 97
14. Dihydropteridine Reductase from Bovine Liver — Hiroyuki Hasegawa and Nobuo Nakanishi 103
15. NADPH-Specific Dihydropteridine Reductase from Bovine Liver — Hiroyuki Hasegawa and Nobuo Nakanishi 111
16. Use of a Naphthoquinone Adsorbent for the Isolation of Human Dihydropteridine Reductase — Frank A. Firgaira, R. G. H. Cotton, I. Jennings, and D. M. Danks 116
17. Dihydropteridine Reductase from Sheep Brain — K. G. Scrimgeour and S. Cheema-Dhadli 127
18. 4-Hydroxyphenylpyruvate Dioxygenase from Pig Liver — David J. Buckthal, Patrick A. Roche, Thomas J. Moorehead, Brian J. R. Forbes, and Gordon A. Hamilton 132
19. 4-Hydroxyphenylpyruvate Dioxygenase from Human Liver — Sven Lindstedt and Birgit Odelhög 139
20. 4-Hydroxyphenylpyruvate Dioxygenase from *Pseudomonas* — Sven Lindstedt and Birgit Odelhög 143
21. 4-Hydroxyphenylpyruvate Dioxygenase from Avian Liver — J. H. Fellman 148
22. Mammalian Monophenol Monooxygenase (Tyrosinase): Purification, Properties, and Reactions Catalyzed — Vincent J. Hearing, Jr. 154

23. Monophenol Monooxygenase from *Neurospora crassa* KONRAD LERCH 165
24. Aromatic-L-Amino Acid Decarboxylase THEODORE L. SOURKES 170
25. Aromatic-L-Amino Acid Decarboxylase from Pig Kidney
CARLA BORRI VOLTATTORNI, ANNA GIARTOSIO, AND CARLO TURANO 179
26. Indoleamine 2,3-Dioxygenase RYOTARO YOSHIDA AND OSAMU HAYAISHI 188
27. Purification and Properties of Tryptophan Side Chain Oxidase: Types I and II from *Pseudomonas* KATSUJI TAKAI AND OSAMU HAYAISHI 195
28. Kynurenine–Oxoglutarate Aminotransferase from Rat Kidney
MICHAEL C. TOBES 217
29. Arylformamidase from *Streptomyces parvulus*
EDWARD KATZ, DALE BROWN, AND M. J. M. HITCHCOCK 225
30. Tryptophan *C*-Methyltransferase of *Streptomyces flocculus*
MARILYN K. SPEEDIE 235
31. Phenylalanine Ammonia-lyase from the Yeast *Rhodotorula glutinis*
CREED W. ABELL AND RONG-SEN SHEN 242
32. L-Phenylalanine Ammonia-lyase from Soybean Cell Suspension Cultures
EVELYN A. HAVIR 248
33. Transamination of Aromatic Amino Acids in *Escherichia coli*
CHARALAMPOS MAVRIDES 253
34. Aromatic-Amino-Acid Aminotransferase from Small Intestine
TOMOO NOGUCHI 267
35. Aromatic-Amino-Acid-Glyoxylate Aminotransferase from Rat Liver
YOSHIKAZU TAKADA AND TOMOO NOGUCHI 273
36. Cathepsin T HENRY C. PITOT AND EIICHI GOHDA 279

Section II. Biosynthesis of the Aromatic Amino Acids

37. Synthesis of Tryptophan from Chorismate: Comparative Aspects
IRVING P. CRAWFORD 293
38. Anthranilate Synthase from Fluorescent Pseudomonads
IRVING P. CRAWFORD 300
39. Dehydroquinate Synthase from *Escherichia coli*, and Its Substrate 3-Deoxy-D-*arabino*-heptulosonic Acid 7-Phosphate
SHUJAATH MEHDI, JOHN W. FROST, AND JEREMY R. KNOWLES 306
40. Shikimate Dehydrogenase from *Escherichia coli*
SUBHENDU CHAUDHURI, IAN A. ANTON, AND JOHN R. COGGINS 315
41. 3-Dehydroquinate Dehydratase from *Escherichia coli*
SUBHENDU CHAUDHURI, KENNETH DUNCAN, AND JOHN R. COGGINS 320
42. The *arom* Multifunctional Enzyme from *Neurospora crassa*
JOHN R. COGGINS, MARTIN R. BOOCOCK, SUBHENDU CHAUDHURI, JOHN M. LAMBERT, JOHN LUMSDEN, GILLIAN A. NIMMO, AND D. DRUMMOND S. SMITH 325
43. 3-Phosphoshikimate 1-Carboxyvinyltransferase from *Escherichia coli*
ANN LEWENDON AND JOHN R. COGGINS 342
44. 3-Phosphoshikimate 1-Carboxyvinyltransferase from *Pisum sativum*
DAVID M. MOUSDALE AND JOHN R. COGGINS 348
45. Shikimate Kinases from *Escherichia coli* K12 ROBERT DE FEYTER 355
46. Chorismate Synthase: A Bifunctional Enzyme in *Neurospora crassa*
FRANK H. GAERTNER 362

47. Anthranilate Synthase–Anthranilate Phosphoribosyltransferase Complex and Subunits of *Salmonella typhimurium*
RONALD BAUERLE, JOHN HESS, AND SARAH FRENCH 366
48. Phosphoribosylanthranilate Isomerase–Indoleglycerol-Phosphate Synthase from *Escherichia coli*
KASPER KIRSCHNER, HALINA SZADKOWSKI, THEODORE S. JARDETZKY, AND VRENI HAGER 386
49. Tryptophan Synthase from *Escherichia coli* and *Salmonella typhimurium*
EDITH WILSON MILES, RONALD BAUERLE, AND SYED ASHRAFUDDIN AHMED 398
50. Tryptophanase from *Escherichia coli* B/1t7-A
IRAJ BEHBAHANI-NEJAD, J. L. DYE, AND CLARENCE H. SUELTER 414
51. Purification of Chorismate, Prephenate, and Arogenate by HPLC
JAMES A. CONNELLY AND DANIEL L. SIEHL 422
52. Chorismate Mutase–Prephenate Dehydratase from *Escherichia coli*
BARRIE E. DAVIDSON 432
53. Chorismate Mutase–Prephenate Dehydrogenase from *Escherichia coli*
BARRIE E. DAVIDSON AND GRAHAM S. HUDSON 440
54. Chorismate Mutase from Mung Bean and Sorghum
DAVID G. GILCHRIST AND JAMES A. CONNELLY 450
55. Chorismate Mutase from *Steptomyces aureofaciens* HELMUT GÖRISCH 463
56. The Postprephenate Biochemical Pathways to Phenylalanine and Tyrosine: An Overview R. JENSEN AND R. FISCHER 472
57. Prephenate Aminotransferase C. BONNER AND R. JENSEN 479
58. Arogenate Dehydrogenase C. BONNER AND R. JENSEN 488
59. Arogenate Dehydratase R. FISCHER AND R. JENSEN 495
60. Prephenate Dehydrogenase (Monofunctional) R. FISCHER AND R. JENSEN 503
61. Prephenate Dehydratase (Monofunctional) R. FISCHER AND R. JENSEN 507
62. Arogenate Dehydrogenase from *Phenylobacterium immobile*
FRANZ LINGENS, EBERHARD KELLER, AND BRIGITTE KELLER 513

Section III. Methods of Determination and Metabolism of the Aromatic Amines

63. Catecholamine Measurements in Plasma by High-Performance Liquid Chromatography with Electrochemical Detection PAUL HJEMDAHL 521
64. Catecholamine Metabolite Measurements in Mouse Brain Using High-Performance Liquid Chromatography and Electrochemical Detection: Comparison of a One-Column Technique with a Two-Column Switching Technique
ELISABETH MORIER-TEISSIER AND RICHARD RIPS 535
65. Enzymatic Radioimmunoassay of Epinephrine, Norepinephrine, Metanephrine, and Normetanephrine WILLIAM J. RAUM 550
66. Quantitation of Catecholamine Metabolites by Gas Chromatography–Mass Spectrometry with Selected Ion Monitoring
JERRY J. WARSH, DAMODAR D. GODSE, AND PETER P. LI 571
67. Arylalkylamine *N*-Acetyltransferase from Mammalian Pineal Gland
M. A. A. NAMBOODIRI, R. DUBBELS, AND D. C. KLEIN 583
68. Hydroxyindole *O*-Methyltransferase
DAVID SUGDEN, VALENTIN CEÑA, AND DAVID C. KLEIN 590
69. Dopamine β-Monooxygenase from Bovine Adrenal Medulla
TORBJØRN LJONES 596

70. Dopamine β-Monooxygenase from Human Plasma RONALD P. FRIGON 603
71. Avian Dopamine β-Monooxygenase ROGER M. WEPPELMAN 608
72. Assay and Purification of Liver Monoamine Oxidase MOUSSA B. H. YOUDIM AND MEIR TENNE 617
73. Preparation of the Flavin-Containing Aromatic Amine Oxidases of Human Placenta and Beef Liver JAMES I. SALACH AND WALTER WEYLER 627
74. Monoamine Oxidase A and B from Human Liver and Brain CREED W. ABELL 638
75. Aromatic Amine Dehydrogenase from *Alcaligenes faecalis* MITSUHIRO NOZAKI 650
76. Norepinephrine *N*-Methyltransferase from Rabbit Adrenal Glands RAY W. FULLER 655
77. Amine *N*-Methyltransferases SHERRY S.ANSHER AND WILLIAM B. JAKOBY 660
78. Indoleamine *N*-Methyltransferase from Rabbit Lung RAFFAELE PORTA 668

VOLUME 143
SULFUR AND SULFUR AMINO ACIDS

Section I. Separation and Analysis

A. Inorganic Sulfur and Selenium Compounds

1. Sulfate: Turbidimetric and Nephelometric Methods BO SÖRBO 3
2. Sulfate Determination: Ion Chromatography DARRYL M. SULLIVAN 7
3. Sulfite Determination: Sulfite Oxidase HANS-OTTO BEUTLER 11
4. Sulfite Determination: Fuchsin Method FRANZ-JOSEF LEINWEBER AND KENNETH J. MONTY 15
5. Sulfite and Thiosulfate: Using *S*-(2-Amino-2-carboxyethylsulfonyl)-L-cysteine TOSHIHIKO UBUKA 17
6. Thiocyanate and Thiosulfate JOHN WESTLEY 22
7. Sulfane Sulfur JOHN L. WOOD 25
8. Sulfide Determination: Ion-Specific Electrode RANDOLPH T. WEDDING 29
9. Hydrogen Selenide and Methylselenol HOWARD E. GANTHER AND R. J. KRAUS 32
10. Polarography of Sulfur Compounds HOWARD ADLER AND JOHN WESTLEY 38

B. Thiols and Disulfides

11. Spectrophotometric Assay of Thiols PETER C. JOCELYN 44
12. Fluorometric Assay of Thiols with Fluorobenzoxadiazoles KAZUHIRO IMAI AND TOSHIMASA TOYO'OKA 67
13. Thiol Labeling with Bromobimanes NECHAMA S. KOSOWER AND EDWARD M. KOSOWER 76
14. Determination of Low-Molecular-Weight Thiols Using Monobromobimane Fluorescent Labeling and High-Performance Liquid Chromatography ROBERT C. FAHEY AND GERALD L. NEWTON 85
15. Purification of Thiols from Biological Samples GERALD L. NEWTON AND ROBERT C. FAHEY 96
16. High-Performance Liquid Chromatography of Thiols and Disulfides: Dinitrophenol Derivatives MARC W. FARISS AND DONALD J. REED 101

17. High-Performance Liquid Chromatography of Hepatic Thiols with Electrochemical Detection — EUGENE G. DEMASTER AND BETH REDFERN 110
18. Analysis for Disulfide Bonds in Peptides and Proteins — THEODORE W. THANNHAUSER, YASUO KONISHI, AND HAROLD A. SCHERAGA 115
19. Assay of Enzyme-Catalyzed Oxygen-Dependent Disulfide Bond Formation — MARK X. SLIWKOWSKI AND HAROLD E. SWAISGOOD 119
20. Cellular Protein-Mixed Disulfides — MARJORIE F. LOU, LAWRENCE L. POULSEN, AND DANIEL M. ZIEGLER 124
21. Measurement of Thiol–Disulfide Interchange Reactions and Thiol pK_a Values — JANETTE HOUK, RAJEEVA SINGH, AND GEORGE M. WHITESIDES 129

C. Other Organic Sulfur and Selenium Compounds

22. Cysteine and Cystine: High-Performance Liquid Chromatography of *o*-Phthalaldehyde Derivatives — JOHN D. H. COOPER AND D. C. TURNELL 141
23. Cystine: Binding Protein Assay — MARGARET SMITH, CLEMENT E. FURLONG, ALICE A. GREENE, AND JERRY A. SCHNEIDER 144
24. Selenocysteine — NOBUYOSHI ESAKI AND KENJI SODA 148
25. Cysteamine and Cystamine — MICHAEL W. DUFFEL, DANIEL J. LOGAN, AND DANIEL M. ZIEGLER 149
26. Cysteinesulfinic Acid, Hypotaurine, and Taurine: Reversed-Phase High-Performance Liquid Chromatography — MARTHA H. STIPANUK, LAWRENCE L. HIRSCHBERGER, AND JAMES DE LA ROSA 155
27. Cysteinesulfinic Acid: Fuchsin Method — FRANZ-JOSEF LEINWEBER AND KENNETH J. MONTY 160
28. Cysteinesulfinic Acid and Cysteic Acid: High-Performance Liquid Chromatography — K. KURIYAMA AND Y. TANAKA 164
29. Resolution of Cysteine and Methionine Enantiomers — OWEN W. GRIFFITH AND ERNEST B. CAMPBELL 166
30. Isethionic Acid — JACK H. FELLMAN 172
31. 3-Mercaptopyruvate, 3-Mercaptolactate, and Mercaptoacetate — BO SÖRBO 178
32. Hypotaurine Aminotransferase Assay: [2-^{3}H]Hypotaurine — JACK H. FELLMAN 183
33. Penicillamine — ELISABETH M. WOLF-HEUSS 186
34. *S*-Adenosylmethionine and Its Sulfur Metabolites — RICHARD K. GORDON, GEORGE A. MIURA, TERESA ALONSO, AND PETER K. CHIANG 191
35. Trimethylselenonium Ion — HOWARD E. GANTHER, R. J. KRAUS, AND S. J. FOSTER 195
36. Sulfotransferase Assays — SENGODA G. RAMASWAMY AND WILLIAM B. JAKOBY 201
37. Arylsulfatases: Colorimetric and Fluorometric Assays — ALEXANDER B. ROY 207

Section II. Preparative Methods

A. Preparation of Specific Metabolites

38. *β*-Sulfopyruvate — OWEN W. GRIFFITH AND CATHERINE L. WEINSTEIN 221
39. Alanine Sulfodisulfane — JACK H. FELLMAN 223
40. Thetin — H. KONDO AND MAKOTO ISHIMOTO 227

41. Cysteine S-Conjugates
PATRICK HAYDEN, VALENTINE H. SCHAEFFER, GERALD LARSEN, AND JAMES L. STEVENS 228
42. Chromogenic and Fluorigenic Substrates for Sulfurtransferases
MARY R. BURROUS, JOHN LANE, AIKO WESTLEY, AND JOHN WESTLEY 235
43. Selenocysteine HIDEHIKO TANAKA AND KENJI SODA 240
44. Selenodjenkolic Acid HIDEHIKO TANAKA AND KENJI SODA 243
45. *Te*-Phenyltellurohomocysteine and *Te*-Phenyltellurocysteine
NOBUYOSHI ESAKI AND KENJI SODA 245

B. General Preparative Techniques

46. Chemical Reduction of Disulfides PETER C. JOCELYN 246
47. Electrochemical Reduction of Disulfides HAROLD KADIN 257
48. Formation of Disulfides with Diamide
NECHAMA S. KOSOWER AND EDWARD M. KOSOWER 264
49. Sulfinic Acids from Disulfides: L-[^{35}S]Cysteinesulfinic Acid
OWEN W. GRIFFITH AND CATHERINE L. WEINSTEIN 270
50. Amino Acid Sulfones: *S*-Benzyl-DL-α-methylcysteine Sulfone
OWEN W. GRIFFITH 274
51. Sulfonic Acids: L-Homocysteinesulfonic Acid
MICHAEL A. HAYWARD, ERNEST B. CAMPBELL, AND OWEN W. GRIFFITH 279
52. Sulfoxides CARL R. JOHNSON AND MARK P. WESTRICK 281
53. Amino Acid Sulfoximines: α-Ethylmethionine Sulfoximine
OWEN W. GRIFFITH 286
54. Preparation of Sulfur and Selenium Amino Acids with Microbial Pyridoxal Phosphate Enzymes NOBUYOSHI ESAKI AND KENJI SODA 291

C. Nutritional Methods

55. Construction of Assay Diets for Sulfur-Containing Amino Acids
DAVID H. BAKER 297
56. Production of Selenium Deficiency in the Rat RAYMOND F. BURK 307
57. Intracellular Delivery of Cysteine
MARY E. ANDERSON AND ALTON MEISTER 313

Section III. Enzymes

A. Inorganic Sulfur and Sulfate

58. Overview: Inorganic Sulfur and Sulfate Activation
JEROME A. SCHIFF AND TEKCHAND SAIDHA 329
59. Sulfate-Activating Enzymes
IRWIN H. SEGEL, FRANCO RENOSTO, AND PETER A. SEUBERT 334
60. Thiosulfate Reductase
THOMAS R. CHAUNCEY, LAWRENCE C. UHTEG AND JOHN WESTLEY 350
61. Adenylylsulfate–Ammonia Adenylyltransferase from *Chlorella*
HEINZ FANKHAUSER, JEROME A. SCHIFF, LEONARD J. GARBER, AND TEKCHAND SAIDHA 354
62. Sulfatases from *Helix pomatia* ALEXANDER B. ROY 361

B. Sulfur Amino Acids: Mammalian Systems

63. Mammalian Sulfur Amino Acid Metabolism: An Overview
OWEN W. GRIFFITH 366
64. Adenosylhomocysteinase (Bovine) PETER K. CHIANG 377
65. Betaine–Homocysteine *S*-Methyltransferase (Human)
WILLIAM E. SKIBA, MARILYN S. WELLS, JOHN H. MANGUM, AND WILLIAM M. AWAD, JR. 384
66. Cystathionine β-Synthase (Human) JAN P. KRAUS 388
67. Cysteine Dioxygenase KENJI YAMAGUCHI AND YU HOSOKAWA 399
68. Cysteinesulfinate Decarboxylase
CATHERINE L. WEINSTEIN AND OWEN W. GRIFFITH 404
69. Cysteamine Dioxygenase RUSSELL B. RICHERSON AND DANIEL M. ZIEGLER 410
70. Selenocysteine β-Lyase (Porcine) NOBUYOSHI ESAKI AND KENJI SODA 415

C. Sulfur Amino Acids: Plants

71. Sulfur Amino Acids of Plants: An Overview JOHN GIOVANELLI 419
72. 1-Aminocyclopropane-1-carboxylate Synthase
DOUGLAS O. ADAMS AND SHANG FA YANG 426
73. Adenosylhomocysteinase from Yellow Lupine
ANDRZEJ GURANOWSKI AND HIERONIM JAKUBOWSKI 430
74. S-Substituted L-Cysteine Sulfoxide Lyase from Shiitake Mushroom
KYODEN YASUMOTO AND KIMIKAZU IWAMI 434
75. Cystine Lyase from Cabbage IVAN K. SMITH AND DAVID I. HALL 439
76. Cystathionine β-Lyase from Spinach JOHN GIOVANELLI 443
77. D-Cysteine Desulfhydrase from Spinach AHLERT SCHMIDT 449

D. Sulfur Amino Acids: Microbial Systems

78. Microbial Sulfur Amino Acids: An Overview KENJI SODA 453
79. L-Methionine γ-Lyase from *Pseudomonas putida* and *Aeromonas*
NOBUYOSHI ESAKI AND KENJI SODA 459
80. *O*-Acetylhomoserine Sulfhydrylase from *Schizosaccharomyces pombe*
SHUZO YAMAGATA 465
81. *O*-Acetylhomoserine Sulfhydrylase from *Brevibacterium flavum*
ISAMU SHIIO AND HACHIRO OZAKI 470
82. *O*-Acetylserine Sulfhydrylase from *Bacillus sphaericus*
TORU NAGASAWA AND HIDEAKI YAMADA 474
83. *O*-Acetyl-L-Serine-*O*-Acetyl-L-Homoserine Sulfhydrylase from *Saccharomyces cerevisiae*
SHUZO YAMAGATA 478
84. Cystathionine β-Lyase from *Escherichia coli* JACK R. UREN 483
85. Cystathionine γ-Lyase from *Streptomyces phaeochromogenes*
TORU NAGASAWA, HIROSHI KANZAKI, AND HIDEAKI YAMADA 486
86. Selenocysteine β-Lyase from *Citrobacter freundii*
NOBUYOSHI ESAKI AND KENJI SODA 493
87. Taurine Dehydrogenase HIROYUKI KONDO AND MAKOTO ISHIMOTO 496
88. ω-Amino Acid–Pyruvate Aminotransferase
KAZUO YONAHA, SEIZEN TOYAMA, AND KENJI SODA 500

E. Enzymes Active with Disulfides

89. Sulfhydryl Oxidase from Milk
HAROLD E. SWAISGOOD AND H. ROBERT HORTON 504
90. Sulfhydryl Oxidase from Rat Skin LOWELL A. GOLDSMITH 510
91. Asparagusate Reductase HIROSHI YANAGAWA 516

Appendix: Index of Related Articles in Other Volumes of Methods in Enzymology 523

VOLUME 144

STRUCTURAL AND CONTRACTILE PROTEINS (PART D: EXTRACELLULAR MATRIX)

Biochemistry of the Major Components of the Extracellular Matrix

A. Collagens

1. The Collagens: An Overview and Update
EDWARD J. MILLER AND STEFFEN GAY 3
2. Isolation and Characterization of Collagens and Procollagens
DONALD K. FURUTO AND EDWARD J. MILLER 41
3. Structure and Expression of Collagen Genes
BENOIT DE CROMBRUGGHE AND AZRIEL SCHMIDT 61
4. Relative Transcription Rates and mRNA Levels for the Two Chains of Type I Procollagen ANNE S. OLSEN AND DARWIN J. PROCKOP 74
5. *In Vitro* Translation of Procollagen Messenger RNAs SHERRILL L. ADAMS 84
6. Recent Developments in Posttranslational Modification: Intracellular Processing
KARI I. KIVIRIKKO AND RAILI MYLLYLÄ 96
7. Collagen Cross-Linking Amino Acids DAVID EYRE 115
8. Catabolism and Turnover of Collagens: Collagenases
HENNING BIRKEDAL-HANSEN 140

B. Elastins

9. Elastin: An Overview JOEL ROSENBLOOM 172
10. Isolation and Characterization of Insoluble and Soluble Elastins
NORMAL T. SOSKEL, TERRIL B. WOLT, AND LAWRENCE B. SANDBERG 196
11. Regulation of Elastin Synthesis in Organ and Cell Culture
JEFFREY M. DAVIDSON AND GREGORY C. SEPHEL 214
12. Modulation of Elastin Synthesis: *In Vitro* Models ROBERT P. MECHAM 232
13. Immunology of Elastin DAVID S. WRENN AND ROBERT P. MECHAM 246
14. Molecular Cloning and Gene Structure of Elastins JOEL ROSENBLOOM 259
15. Elastin Degradation
MICHAEL J. BANDA, ZENA WERB, AND JAMES H. MCKERROW 288

C. Proteoglycans

16. Proteoglycans: An Overview DICK HEINEGÅRD AND YNGVE SOMMARIN 305
17. Isolation and Characterization of Proteoglycans
DICK HEINEGÅRD AND YNGVE SOMMARIN 319

18. Biosynthesis of Cartilage Proteoglycan and Link Protein
JAMES H. KIMURA, TAMAYUKI SHINOMURA, AND EUGENE J.-M. A. THONAR 372
19. Analysis of Membrane-Associated Proteoglycans
MAGNUS HÖÖK, LENA KJELLÉN, AND ANNE WOODS 394
20. Isolation and Characterization of the Link Proteins
JOHN R. BAKER AND PETER J. NEAME 401
21. Catabolism and Turnover of Proteoglycans
CHRISTOPHER J. HANDLEY AND MARGARET A. CAMPBELL 412

D. Glycoproteins

22. Preparation of Placental (Fetal Tissue) Fibronectin and Its Carbohydrates
ROGER A. LAINE, SUSAN J. FISHER, AND BETTY C. R. ZHU 420
23. Purification and Characterization of Vitronectin
ERKKI RUOSLAHTI, SHINTARO SUZUKI, EDWARD G. HAYMAN, CHARLES R. ILL, AND MICHAEL D. PIERSCHBACHER 430
24. Isolation and Characterization of Thrombospondin
SAMUEL A. SANTORO AND WILLIAM A. FRAZIER 438
25. Isolation and Analysis of cDNA and Genomic Clones of Fibronectin and Its Receptor
RICHARD O. HYNES, JEAN E. SCHWARZBAUER, AND JOHN W. TAMKUN 447
26. Molecular Cloning of Laminin
D. P. BARLOW, J. H. MCVEY, AND B. L. M. HOGAN 464
27. Arginine-Glycine-Aspartic Acid Adhesion Receptors
ROBERT PYTELA, MICHAEL D. PIERSCHBACHER, SCOTT ARGRAVES, SHINTARO SUZUKI AND ERKKI RUOSLAHTI 475
28. Isolation and Characterization of Laminin Receptors
KLAUS VON DER MARK AND GUNDULA RISSE 490

VOLUME 145

STRUCTURAL AND CONTRACTILE PROTEINS (PART E: EXTRACELLULAR MATRIX)

Section I. Physical and Immunohistochemical Methods

1. Electron Microscopy and Other Physical Methods for the Characterization of Extracellular Matrix Components: Laminin, Fibronectin, Collagen IV, Collagen VI, and Proteoglycans JÜRGEN ENGEL AND HEINZ FURTHMAYR 3
2. Electron Immunohistochemistry of the Extracellular Matrix: An Overview
ANTONIO MARTINEZ-HERNANDEZ 78
3. Methods for Electron Immunohistochemistry
ANTONIO MARTINEZ-HERNANDEZ 103
4. Specific Methods for Electron Immunohistochemistry
PETER S. AMENTA AND ANTONIO MARTINEZ-HERNANDEZ 133
5. Characterization and Isolation of Poly- and Monoclonal Antibodies against Collagen for Use in Immunohistochemistry STEFFEN GAY AND JO-DAVID FINE 148

Section II. Genetic Anomalies of the Extracellular Matrix

6. Electrophoretic Isolation and Peptide Mapping of Collagen Types from Microsamples of Tissue
JOHN H. LILLIE, JOYCE A. M. WOOTTON, DONALD K. MACCALLUM, STEVEN W. MCKELVEY, AND RONALD R. MINOR 171
7. Two-Dimensional Assays of Peptide Fragments
WILLIAM G. COLE, DANNY CHAN, AND JOHN F. BATEMAN 183
8. Restriction Fragment Length Polymorphisms PETROS TSIPOURAS 205
9. Characterization of a Pro-α2(I) Collagen Gene Mutation by Nuclease S_1 Mapping
TAINA PIHLAJANIEMI AND JEANNE C. MYERS 213
10. Analysis of Cytoplasmic and Nuclear Messenger RNA in Fibroblasts from Patients with Type I Osteogenesis Imperfecta
CONNIE GENOVESSE AND DAVID ROWE 223
11. R-Loop Analysis of Procollagen Messenger RNA for the Assessment of Human Collagen Mutations WOUTER J. DE WET 235

Section III. Biochemistry of the Extracellular Matrix of Specific Tissues

A. Mineralized Tissues (Bone and Dentin)

12. Mineralized Tissues: An Overview WILLIAM T. BUTLER 255
13. Proteoglycans of Mineralized Matrices CHARLES W. PRINCE 261
14. Bone Glycoproteins
LARRY W. FISHER, PAMELA GEHRON ROBEY, MARIAN F. YOUNG, AND JOHN D. TERMINE 269
15. Dentin-Specific Proteins WILLIAM T. BUTLER 290
16. Bone Cell Cultures J. SODEK AND F. A. BERKMAN 303
17. Hormonal Influences on Bone Cells
T. J. MARTIN, K. W. NG, N. C. PARTRIDGE, AND S. A. LIVESEY 324

B. Other Tissues

18. Cartilage DICK HEINEGÅRD AND MATS PAULSSON 336
19. Basement Membranes
RUPERT TIMPL, MATS PAULSSON, MARIE DZIADEK, AND SAKUHEI FUJIWARA 363
20. Analysis of Extracellular Matrix Proteins in Biological Fluids
LEILA RISTELI AND JUHA RISTELI 391

VOLUME 146
PEPTIDE GROWTH FACTORS (PART A)

Section I. Epidermal Growth Factor

1. Purification of Human Epidermal Growth Factor by Monoclonal Antibody Affinity Chromatography
ROBERT A. HARPER, JAMES PIERCE, AND C. RICHARD SAVAGE, JR. 3

2. Derivation of Monoclonal Antibody to Human Epidermal Growth Factor
KATSUZO NISHIKAWA, YOSHINO YOSHITAKE, AND SHIGERU IKUTA 11
3. Design, Chemical Synthesis, and Molecular Cloning of a Gene for Human Epidermal Growth Factor MICKEY S. URDEA 22
4. Purification of Polypeptide Growth Factors from Milk
YUEN SHING, SUSAN DAVIDSON, AND MICHAEL KLAGSBRUN 42
5. Biosynthesis of the Epidermal Growth Factor Receptor in Cultured Cells
ANN MANGELSDORF SODERQUIST AND GRAHAM CARPENTER 49
6. Derivation and Assay of Biological Effects of Monoclonal Antibodies to Epidermal Growth Factor Receptors
J. DENRY SATO, ANH D. LE, AND TOMOYUKI KAWAMOTO 63
7. Purification of Functionally Active Epidermal Growth Factor Receptor Protein Using a Competitive Antagonist Monoclonal Antibody and Competitive Elution with Epidermal Growth Factor
GORDON N. GILL AND WOLFGANG WEBER 82
8. *In Vitro* of A431 Human Epidermoid Carcinoma DAVID BARNES 89

Section II. Transforming Growth Factors

9. Radioreceptor Assays for Transforming Growth Factors
CHARLES A. FROLIK AND JOSEPH E. DE LARCO 95
10. Purification of Type-α Transforming Growth Factor from Transformed Cells
JOAN MASSAGUÉ 103
11. Separation of Melanoma Growth Stimulatory Activity and Human Type-α Transforming Growth Factor
ANN RICHMOND, H. GREG THOMAS, AND ROBERT G. B. ROY 112
12. Solid-Phase Synthesis of Type-α Transforming Growth Factor
JAMES P. TAM 127
13. Identification of Receptor Proteins for Type-α Transforming Growth Factor
JOAN MASSAGUÉ 143
14. Purification of Type-β Transforming Growth Factor from Human Platelets
RICHARD K. ASSOIAN 153
15. Isolation of the BSC-1 Monkey Kidney Cell Growth Inhibitor
ROBERT W. HOLLEY, JULIA H. BALDWIN, AND SYBIL GREENFIELD 163
16. An Assay for Type-β Transforming Growth Factor Receptor
LALAGE M. WAKEFIELD 167
17. Identification of Receptors for Type-β Transforming Growth Factor
JOAN MASSAGUÉ 174

Section III. Somatomedin/Insulin-Like Growth Factors

18. Solid-Phase Synthesis of Insulin-Like Growth Factor I
DONALD YAMASHIRO AND CHOH HAO LI 199
19. Derivation of Monoclonal Antibodies to Human Somatomedin C/Insulin-Like Growth Factor I
G. YANCEY GILLESPIE, JUDSON J. VAN WYK, LOUIS E. UNDERWOOD, AND MARJORIE E. SVOBODA 207
20. Radioimmunoassay of Somatomedin C/Insulin-Like Growth Factor I
RICHARD W. FURLANETTO AND JEAN M. MARINO 216

21. Estimation of Tissue Concentrations of Somatomedin C/Insulin-Like Growth Factor I — A. JOSEPH D'ERCOLE AND LOUIS E. UNDERWOOD 227
22. Somatomedin C/Insulin-Like Growth Factor I Receptors on Human Mononuclear Cells and the IM-9 Lymphoid Cell Line — RAYMOND L. HINTZ 234
23. Human Insulin-Like Growth Factor I and II Messenger RNA: Isolation of Complementary DNA and Analysis of Expression — LESLIE B. RALL, JAMES SCOTT, AND GRAEME I. BELL 239
24. Radioligand Assays for Insulin-Like Growth Factor II — WILLIAM H. DAUGHADAY 248
25. Purification of Rat Insulin-Like Growth Factor II — LAWRENCE A. GREENSTEIN, LYNNE A. GAYNES, JOYCE A. ROMANUS, LILLY LEE, MATTHEW M. RECHLER, AND S. PETER NISSLEY 259
26. Purification of Multiplication-Stimulating Activity Carrier Protein — GARY L. SMITH, RUSSETTE M. LYONS, RICHARD N. HARKINS, AND DANIEL J. KNAUER 270

Section IV. Bone and Cartilage Growth Factors

27. Purification of Bovine Skeletal Growth Factor — JOHN C. JENNINGS, SUBBURAMAN MOHAN, AND DAVID J. BAYLINK 281
28. Preparation and Bioassay of Bone Morphogenetic Protein and Polypeptide Fragments — MARSHALL R. URIST, J. J. CHANG, A. LIETZE, Y. K. HUO, A. G. BROWNELL, AND R. J. DELANGE 294
29. Preparation of Cartilage-Derived Factor — FUJIO SUZUKI, YUJI HIRAKI, AND YUKIO KATO 313
30. Purification of Cartilage-Derived Growth Factors — JOACHIM SASSE, ROBERT SULLIVAN, AND MICHAEL KLAGSBRUN 320

Section V. Techniques for the Study of Growth Factor Activity: Assays, Phosphorylation, and Surface Membrane Effects

31. Assay of Mitogen-Induced Effects on Cellular Incorporation of Precursors for Scavenger, *de Novo*, and Net DNA Synthesis — ROBERT B. DICKSON, SUSAN AITKEN, AND MARC E. LIPPMAN 329
32. Soft Agar Growth Assays for Transforming Growth Factors and Mitogenic Peptides — ANGIE RIZZINO 341
33. Assay of Growth Factor-Stimulated Tyrosine Kinases Using Synthetic Peptide Substrates — LINDA J. PIKE 353
34. Separation of Multiple Phosphorylated Forms of 40 S Ribosomal Protein S6 by Two-Dimensional Polyacrylamide Gel Electrophoresis — MICHEL SIEGMANN AND GEORGE THOMAS 362
35. Phosphopeptide Analysis of 40 S Ribosomal Protein S6 — JORGE MARTIN-PÉREZ AND GEORGE THOMAS 369
36. Growth Factor Effects on Membrane Transport: Uptake Studies Using Cell Cultures and Isolated Membrane Vesicles — JULIA E. LEVER 376
37. Measurement of Ion Flux and Concentration in Fibroblastic Cells — STANLEY A. MENDOZA AND ENRIQUE ROZENGURT 384

38. Growth Factor Stimulation of Sugar Uptake
WENDELYN H. INMAN AND SIDNEY P. COLOWICK 399
39. Assay of Growth Factor Stimulation of Fluid-Phase Endocytosis
H. STEVEN WILEY AND DANA N. MCKINLEY 402

VOLUME 147
PEPTIDE GROWTH FACTORS (PART B)

Section I. Platelet-Derived Growth Factor

1. Purification of Human Platelet-Derived Growth Factor
CARL-HENRIK HELDIN, ANN JOHNSSON, BO EK, SUZANNE WENNERGREN, LARS RÖNNSTRAND, ANNET HAMMACHER, BRIAN FAULDERS, ÅKE WASTESON, AND BENGT WESTERMARK 3
2. A Radioreceptor Assay for Platelet-Derived Growth Factor JAI PAL SINGH 13
3. Structural and Functional Identification of Platelet-Derived Growth Factor-Like Proteins Produced by Mammalian Cells
HARRY N. ANTONIADES AND PANAYOTIS PANTAZIS 22
4. Purification of Fibroblast-Derived Growth Factor
P. STROOBANT, M. D. WATERFIELD, AND E. ROZENGURT 40
5. Identification and Assay of Platelet-Derived Growth Factor-Binding Proteins
ELAINE W. RAINES AND RUSSELL ROSS 48
6. Differential Colony Hybridization: Molecular Cloning from a Zero Data Base
BRENT H. COCHRAN, PETER ZUMSTEIN, JOHN ZULLO, BARRETT ROLLINS, MARK MERCOLA, AND CHARLES D. STILES 64
7. Identification of Platelet-Derived Growth Factor-Modulated Proteins
CHARLES D. SCHER AND W. JACKSON PLEDGER 85

Section II. Angiogenesis, Endothelial and Fibroblast Growth Factors

8. Purification of Endothelial Cell Growth Factors by Heparin Affinity Chromatography
MICHAEL KLAGSBRUN, ROBERT SULLIVAN, SANDRA SMITH, RUSSELL RYBKA, AND YUEN SHING 95
9. Isolation and Characterization of Acidic and Basic Fibroblast Growth Factor
D. GOSPODAROWICZ 106
10. Purification and Characterization of Acidic Fibroblast Growth Factor
KENNETH A. THOMAS 120
11. Assay of Capillary Endothelial Cell Migration BRUCE R. ZETTER 135
12. Spectrophotometric Assay for the Quantitation of Cell Migration in the Boyden Chamber Chemotaxis Assay GARY R. GROTENDORST 144
13. Isolation of Two Proteinase Inhibitors from Human Hepatoma Cells That Stimulate Human Endothelial Cell Growth
W. L. MCKEEHAN AND K. A. MCKEEHAN 152
14. Identification of Cell Surface Proteins Sensitive to Proteolysis by Thrombin
JAMES A. THOMPSON AND DENNIS D. CUNNINGHAM 157

Section III. Nerve and Glial Growth Factors

15. Two-Site Enzyme Immunoassay for Nerve Growth Factor
SIGRUN KORSCHING AND HANS THOENEN 167
16. Derivation of Monoclonal Antibody to Nerve Growth Factor
KENNETH E. NEET, MICHAEL W. FANGER, AND THOMAS J. BARIBAULT 186
17. Isolation of Complementary DNA Encoding Mouse Nerve Growth Factor and Epidermal Growth Factor
JAMES SCOTT, MARK J. SELBY, AND GRAEME I. BELL 194
18. PC12 Pheochromocytoma Cells: Culture, Nerve Growth Factor Treatment, and Experimental Exploitation
LLOYD A. GREENE, JOHN M. ALETTA, ADRIANA RUKENSTEIN, AND STEVEN H. GREEN 207
19. Assay and Isolation of Glial Growth Factor from the Bovine Pituitary
JEREMY P. BROCKES 217
20. Purification of Glia Maturation Factor RAMON LIM AND JOYCE F. MILLER 255

Section IV. Transferrin, Erythropoietin, and Related Factors

21. Isolation of Transferrin Receptor from Human Placenta
PAUL A. SELIGMAN AND ROBERT H. ALLEN 239
22. Receptor Assay with Radiolabeled Transferrin
JOHN H. WARD AND JERRY KAPLAN 247
23. Radioimmunoassay of Transferrin Receptor
SUSAN E. TONIK AND HOWARD H. SUSSMAN 253
24. Monoclonal Antibodies to Transferrin Receptor and Assay of Their Biological Effects
IAN S. TROWBRIDGE, JAYNE F. LESLEY, DERRICK DOMINGO, ROBERTA SCHULTE, CAROL SAUVAGE, AND HANS-GEORG RAMMENSEE 265
25. Molecular Cloning of Receptor Genes by Transfection
ALAN MCCLELLAND, MICHAEL E. KAMARCK, AND FRANK H. RUDDLE 280
26. Transferrin: Assay of Myotrophic Effects and Method of Immunocytochemical Localization GEORGE J. MARKELONIS AND TAE H. OH 291
27. Cell Culture Assay of Biological Activity of Lactoferrin and Transferrin
SHUICHI HASHIZUME, KAZUHIKO KURODA, AND HIROKI MURAKAMI 302
28. Biological Activity of Human Plasma Copper-Binding Growth Factor Glycyl-L-histidyl-L-lysine LOREN PICKART AND STEVE LOVEJOY 314
29. Isolation of Human Erythropoietin with Monoclonal Antibodies
RYUZO SASAKI, SHIN-ICHI YANAGAWA, AND HIDEO CHIBA 328
30. Large-Scale Procurement of Erythropoietin-Responsive Erythroid Cells: Assay for Biological Activity of Erythropoietin
STEPHEN T. SAWYER, MARK J. KOURY, AND MAURICE C. BONDURANT 340

Section V. Techniques for the Study of Growth Factor Activity: Genetic Approaches and Biological Effects

31. Isolation of Mitogen-Specific Nonproliferative Variant Cell Lines
HARVEY R. HERSCHMAN 355
32. Use of Variant Cell Lines and Cell Hybrids for the Study of Hormone and Growth Factor Action DANIEL S. STRAUS 369

33. Conjugation of Peptide Growth Factors with Toxin to Isolate Cell Variants Involved in Endocytosis and Mitogenic Response NOBUYOSHI SHIMIZU 382
34. Transmembrane Delivery of Polypeptide Growth Factors Using Their Conjugates with Cell Surface Binding Proteins
FUMIAKI ITO AND NOBUYOSHI SHIMIZU 387
35. Kinetic Analysis of Cell Growth Factor–Nutrient Interactions
W. L. MCKEEHAN AND K. A. MCKEEHAN 392
36. Characterization of Growth Factor-Modulated Events Regulating Cellular Proliferation MAUREEN A. HARRINGTON AND W. JACKSON PLEDGER 400
37. Assays of Growth Factor Stimulation of Extracellular Matrix Biosynthesis
WILLIAM R. KIDWELL 407
38. Measurement of Growth Factor-Induced Changes in Intracellular pH
DAN CASSEL AND PAUL ROTHENBERG 415
39. Detection of Proteins Induced by Growth Regulators
MARIT NILSEN-HAMILTON AND RICHARD T. HAMILTON 427

VOLUME 148
PLANT CELL MEMBRANES

Section I. Cells, Protoplasts, and Liposomes

1. Techniques of Cell Suspension Culture
RICHARD BLIGNY AND JEAN-JACQUES LEGUAY 3
2. Culture and Characteristics of Green Plant Cells
COLIN DALTON AND IAN MACKENZIE 16
3. Preparation of Protoplasts from Plant Tissues for Organelle Isolation
MIKIO NISHIMURA, IKUKO HARA-NISHIMURA, AND TAKASHI AKAZAWA 27
4. Interaction of Plant Protoplast and Liposome TOSHIYUKI NAGATA 34
5. Liposomes as Carriers for the Transfer and Expression of Nucleic Acids into Higher Plant Protoplasts MICHEL CABOCHE AND PAUL F. LURQUIN 39
6. Interspecific Transfer of Partial Nuclear Genomic Information by Protoplast Fusion OTTO SCHIEDER 45

Section II. Vacuoles and Tonoplasts

7. Isolation of Mature Vacuoles of Higher Plants: General Principles, Criteria for Purity and Integrity GEORGE J. WAGNER 55
8. Isolation of Vacuoles and Tonoplast from Protoplasts
ALAIN M. BOUDET AND GILBERT ALIBERT 74
9. Preparation of Tonoplast Vesicles from Isolated Vacuoles
ROBERT T. LEONARD 82
10. Biochemical and Enzymatic Components of a Vacuolar Membrane: Tonoplast of Lutoids from *Hevea* Latex J. D'AUZAC, H. CHRESTIN, AND B. MARIN 87
11. Characterization of Tonoplast Enzyme Activities and Transport
BENJAMIN JACOBY 105
12. Preparation of Tonoplast Vesicles: Applications to H^+-Coupled Secondary Transport in Plant Vacuoles
EDUARDO BLUMWALD, PHILIP A. REA, AND RONALD J. POOLE 115

13. Purification and Characterization of the Tonoplast H^+-Translocating ATPase
STEPHEN K. RANDALL AND HEVEN SZE 123
14. High-Performance Liquid Chromatography for Simultaneous Kinetic Measurements of Adenine Nucleotides in Isolated Vacuoles
M. HILL, A. DUPAIX, P. VOLFIN, A. KURKDJIAN, AND B. ARRIO 132

Section III. Plastids

15. Isolation of Intact Chloroplasts: General Principles and Criteria of Integrity
DAVID A. WALKER, ZORAN G. CEROVIC, AND SIMON P. ROBINSON 145
16. Isolation of Plastids in Density Gradients of Percoll and Other Silica Sols
CARL A. PRICE, JOHN C. CUSHMAN, LETICIA R. MENDIOLA-MORGENTHALER, AND ELLEN M. REARDON 157
17. Purification of Chloroplasts Using Silica Sols K. W. JOY AND W. R. MILLS 179
18. Separation of Chloroplasts and Cytosol from Protoplasts
SIMON P. ROBINSON 188
19. Use of Thermolysin to Probe the Cytosolic Surface of the Outer Envelope Membrane from Plastids
JACQUES JOYARD, ALBERT-JEAN DORNE, AND ROLAND DOUCE 195
20. Characterization of Plastid Polypeptides from the Outer and Inner Envelope Membranes
JACQUES JOYARD, MARYSE A. BLOCK, JACQUES COVÈS, CLAUDE ALBAN, AND ROLAND DOUCE 206
21. Isolation of Amyloplasts from Suspension Cultures of Soybean
TOM AP REES AND FRASER D. MACDONALD 218
22. Isolation of Amyloplasts from Developing Endosperm of Maize (*Zea mays L.*)
JACK C. SHANNON, EDGARDO ECHEVERRIA, AND CHARLES BOYER 226
23. Isolation of Plastids from Buds of Cauliflower (*Brassica oleracea L.*)
ETIENNE-PASCAL JOURNET 234
24. Isolation of Membranous Chromoplasts from Daffodil Flowers
BODO LIEDVOGEL 241
25. Structure and Function of the Inner Membrane Systems in Etioplasts
SATORU MURAKAMI 246
26. Isolation Procedures for Inside-Out Thylakoid Vesicles
HANS-ERIK ÅKERLUND AND BERTIL ANDERSSON 252
27. Characterization of Chloroplast Cytochromes
DEREK S. BENDALL AND STEPHEN A. ROLFE 259
28. Cell-Free Reconstitution of Protein Transport into Chloroplasts
MICHAEL L. MISHKIND, KAREN L. GREER, AND GREGORY W. SCHMIDT 274
29. Polar Lipids of Chloroplast Membranes
DAVID J. CHAPMAN AND JAMES BARBER 294
30. Isolation and Reconstitution of Thylakoid Lipids
SALLIE G. SPRAGUE AND L. ANDREW STAEHELIN 319
31. Long-Chain Fatty Acid Synthesis and Utilization by Isolated Chloroplasts
GRATTAN ROUGHAN 327
32. Phosphatidylglycerol Synthesis in Chloroplast Membranes
J. BRIAN MUDD, JAEN E. ANDREWS, AND S. A. SPARACE 338
33. Galactolipid Biosynthesis in Chloroplast Membranes
NORA W. LEM AND JOHN P. WILLIAMS 346

34. Chlorophylls and Carotenoids: Pigments of Photosynthetic Biomembranes HARTMUT K. LICHTENTHALER 350
35. α-Tocopherol and Plastoquinone Synthesis in Chloroplast Membranes JÜRGEN SOLL 383
36. Solubilization and Reconstitution of Carotenogenic Enzymes from Daffodil Chromoplast Membranes Using 3-[(3-Cholamidopropyl)dimethylammonio]-1-propane Sulfonate P. BEYER 392

Section IV. Mitochondria

37. Isolation of Plant Mitochondria: General Principles and Criteria of Integrity ROLAND DOUCE, JACQUES BOURGUIGNON, RENAUD BROUQUISSE, AND MICHEL NEUBURGER 403
38. Purification of Plant Mitochondria on Silica Sol Gradients ANTHONY L. MOORE AND MICHAEL O. PROUDLOVE 415
39. Isolation of Mitochondria from Leaves of C_3, C_4, and Crassulacean Acid Metabolism Plants GERALD E. EDWARDS AND PER GARDESTRÖM 421
40. Separation of Spinach Leaf Mitochondria according to Surface Properties: Partition in Aqueous Polymer Two-Phase Systems PER GARDESTRÖM AND INGEMAR ERICSON 434
41. Isolation of Submitochondrial Particles with Different Polarities IAN M. MØLLER, ANNIKA C. LIDÉN, INGEMAR ERICSON, AND PER GARDESTRÖM 442
42. Isolation of the Outer Membrane of Plant Mitochondria CARMEN A. MANNELLA 453
43. Characterization of Channels Isolated from Plant Mitochondria MARCO COLOMBINI 465
44. Phosphoglycerides of Mitochondrial Membranes JOHN L. HARWOOD 475
45. Ubiquinone Biosynthesis in Plant Mitochondria FRIEDHELM LÜTKE-BRINKHAUS AND HANS KLEINIG 486
46. Purification of Complexes II and IV from Plant Mitochondria MASAYOSHI MAESHIMA, TSUKAHO HATTORI, AND TADASHI ASAHI 491

Section V. Peroxisomes and Glyoxysomes

47. Isolation of Glyoxysomes and Purification of Glyoxysomal Membranes EUGENE L. VIGIL, TUNG K. FANG, AND ROBERT P. DONALDSON 505
48. Peroxisomes and Fatty Acid Degradation BERNT GERHARDT 516
49. Proteins and Phospholipids of Glyoxysomal Membranes from Castor Bean HARRY BEEVERS AND ELMA GONZÁLEZ 526

Section VI. Nuclei, Endoplasmic Reticulum, and Plasma Membrane

50. Isolation of Nuclei from Soybean Suspension Cultures PAUL KEIM 535
51. Isolation of the Plasma Membrane: Membrane Markers and General Principles DONALD P. BRISKIN, ROBERT T. LEONARD, AND THOMAS K. HODGES 542

52. Preparation of High-Purity Plasma Membranes
CHRISTER LARSSON, SUSANNE WIDELL, AND PER KJELLBOM 558
53. Possible Approaches to Surface Labeling of the Plasma Membrane
J. L. HALL 568
54. Isolation of Endoplasmic Reticulum: General Principles, Enzymatic Markers, and Endoplasmic Reticulum-Bound Polysomes J. MICHAEL LORD 576
55. Phosphoglyceride Synthesis in Endoplasmic Reticulum
THOMAS S. MOORE, JR. 585

Section VII. General Physical and Biochemical Methods

56. Electron Microscopy of Plant Cell Membranes JEAN-PIERRE CARDE 599
57. Two-Dimensional Electrophoresis in the Analysis and Preparation of Cell Organelle Polypeptides R. REMY AND F. AMBARD-BRETTEVILLE 623
58. Plant Membrane Sterols: Isolation, Identification, and Biosynthesis
MARIE-ANDRÉE HARTMANN AND PIERRE BENVENISTE 632
59. Separation of Molecular Species of Plant Glycolipids and Phospholipids by High-Performance Liquid Chromatography J. KESSELMEIER AND E. HEINZ 650
60. Movement of Phospholipids between Membranes: Purification of a Phospholipid Transfer Protein from Spinach Leaf
JEAN-CLAUDE KADER, CHANTAL VERGNOLLE, AND PAUL MAZLIAK 661
61. Pesticides and Lipid Synthesis in Plant Membranes C. ANDING 667
62. Rapid Filtration Technique for Metabolite Fluxes across Cell Organelles
YVES DUPONT AND MARIE-JOSÉ MOUTIN 675
63. Application of Nuclear Magnetic Resonance Methods to Plant Tissues
R. G. RATCLIFFE 683

VOLUME 149
DRUG AND ENZYME TARGETING (PART B)

Section I. Cell Targeting Techniques

1. Biochemical Methods for the Study of Receptor-Mediated Endocytosis
JOE HARFORD AND RICHARD D. KLAUSNER 3
2. High-Efficiency Gene Transfer into Cells
DOUGLAS J. JOLLY, JIING-KUAN YEE, AND THEODORE FRIEDMANN 10
3. Lectin-Specific Targeting of Lysosomal Enzymes to Reticuloendothelial Cells
GARY J. MURRAY 25
4. Osmotic Lysis of Pinosomes MARTIN RECHSTEINER 42

Section II. Liposome Carriers

5. Multiple Emulsions as Targetable Delivery Systems
STANLEY S. DAVIS AND IAN M. WALKER 51
6. High-Pressure Continuous-Flow System for Drug Entrapment in Liposomes
E. MAYHEW, S. CONROY, J. KING, R. LAZO, G. NIKOLOPOULUS, A. SICILIANO, AND W. J. VAIL 64

7. Temperature- and pH-Sensitive Liposomes for Drug Targeting
M. B. YATVIN, I.-M. TEGMO-LARSSON, AND W. H. DENNIS 77
8. pH-Sensitive Immunoliposomes
LEAF HUANG, JEROME CONNOR, AND CHEN-YEN WANG 88
9. Photolabile Liposomes as Carriers CHARLES PIDGEON AND C. A. HUNT 99
10. Covalent Attachment of Proteins to Liposomes TIMOTHY D. HEATH 111
11. Use of Avidin–Biotin Technology for Liposome Targeting
B. RIVNAY, E. A. BAYER, AND M. WILCHEK 119
12. Complement-Dependent Phagocytosis of Liposomes by Macrophages
NABILA M. WASSEF AND CARL R. ALVING 124
13. Interaction of Liposomes with Cells TIMOTHY D. HEATH 135
14. Uptake of Liposome-Encapsulated Agents
FRANCIS C. SZOKA, JR. AND CHRISTOPHER S. JONES 143
15. Liposome-Mediated Macrophage Activities
RICHARD KIRSH AND GEORGE POSTE 147
16. Liposomes as Carriers for *in Vivo* Gene Transfer and Expression
CLAUDE NICOLAU, ALAIN LEGRAND, AND ELISABETH GROSSE 157
17. Liposome Entrapment for Delivery of Genetic Material to Plant Protoplasts
TOSHIYUKI NAGATA 176
18. Liposome-Encapsulated Hemoglobin as an Artificial Oxygen-Carrying System
MARTHA C. FARMER AND BRUCE P. GABER 184
19. Use of Liposomes as Agglutination-Enhancement Agents in Diagnostic Tests
FRANCIS J. MARTIN AND VIOLA T. KUNG 200

Section III. Cellular Carriers

20. Erythrosomes: Erythrocyte Cytoskeletons Coated with Exogenous Phospholipid as an Encapsulating System CHAN Y. JUNG 217
21. Hypotonic Hemolysis Methods for Entrapment of Agents in Resealed Erythrocytes GARRET M. IHLER AND HANNAH CHI-WAN TSANG 221
22. High-Efficiency Entrapment of Enzymes in Resealed Red Cell Ghosts by Dialysis
GEORGE L. DALE 229
23. Dialysis Method for Entrapment of Proteins into Resealed Red Blood Cells
JOHN R. DELOACH 235
24. Large-Scale Entrapment of Drugs into Resealed Red Blood Cells Using a Continuous-Flow Dialysis System
C. ROPARS, G. AVENARD, AND M. CHASSAIGNE 242
25. Electric Modification of Membrane Permeability for Drug Loading into Living Cells TIAN YOW TSONG 248
26. Drug-Induced Endocytosis and Entrapment in Red Cells and Ghosts
STANLEY L. SCHRIER 260
27. Preparation of White Resealable Erythrocyte Ghosts PHILLIP G. WOOD 271
28. Phospholipid Asymmetry of Loaded Red Cells
ROBERT A. SCHLEGEL, JON A. REED, LESLIE MCEVOY, LOURDES ALGARIN, AND PATRICK WILLIAMSON 281
29. Red Cell-Mediated Microinjection
ROBERT A. SCHLEGEL AND LESLIE MCEVOY 293
30. Magnetically Responsive Erythrocyte "Ghosts"
U. SPRANDEL, D.-J. LANZ, AND W. VON HÖRSTEN 301
31. Use of Platelets as Drug Carriers for the Treatment of Hematologic Diseases
YEON S. AHN, WILLIAM J. HARRINGTON, AND RAVINDRA MYLVAGANAM 312

VOLUME 150

IMMUNOCHEMICAL TECHNIQUES (PART K: *IN VITRO* MODELS OF B AND T CELL FUNCTIONS AND LYMPHOID CELL RECEPTORS)

Section I. Methods for the Stimulation of Lymphocytes

1. T Cell Mitogens and Polyclonal B Cell Activators
GIOVANNI DI SABATO, JANICE M. HALL, AND LUANN THOMPSON 3
2. Cross-Linked Derivatives of Concanavalin A
TOSHIAKI OSAWA AND MASATOSHI BEPPU 17
3. Biochemical Changes Taking Place in Mitogen-Treated Lymphocytes
CHARLES W. PARKER 29
4. Allogeneic and Autologous Mixed Lymphocyte Reactions
JACK R. BATTISTO AND MEHER M. DUSTOOR 83
5. Stimulation of Lymphocytes with Antibodies to Surface Membrane-Bound Antigen Receptors NICHOLAS CHIORAZZI 92
6. Stimulation of Lymphocytes with Periodate of Neuraminidase plus Galactose Oxidase WILLIAM E. BOWERS 105
7. Stimulation of Lymphocytes with Proteolytic Enzymes THOMAS L. VISCHER 109
8. Stimulation of Lymphocytes with Zinc Ions
H. KIRCHNER AND MARGARITA SALAS 112
9. Enzyme Immunoassay Using a Monoclonal Antibody against 5-Bromo-2-deoxyuridine for the Assessment of Lymphoid Cell Proliferation
THÉRÈSE TERNYNCK, THOMAS PORTSMANN, AND STRATIS AVRAMEAS 117
10. T Cell Responses Studied in a Basal Serum-Free Medium
MERRIL J. GERSTEN AND MELVIN COHN 129
11. Chemically Defined Medium for the Growth of Lymphocytes
KATHLEEN SHIVE MATTHEWS, FLORA PETTIT, JOCELYNE BOGHOSSIAN, AND WILLIAM SHIVE 134
12. Proliferating Cell Nuclear Antigen: Cyclin
KENJI OGATA, JULIO E. CELIS, AND ENG M. TAN 147

Section II. *In Vitro* Models and Assays of B and T Lymphocyte Differentiation and Function

13. Mouse Inbred and Congenic Strains JAN KLEIN AND DAGMAR KLEIN 163
14. Splenic Focus Assay BARBARA G. FROSCHER AND NORMAN R. KLINMAN 196
15. Direct and Indirect Plaque Assays IVAN LEFKOVITS 209
16. Protein A Plaque Assay FRITZ MELCHERS 217
17. Poly(L-lysine) Plaque Assay for the Measurement of Antigen-Activated Human B Lymphocytes HANS-MICHAEL DOSCH AND ERWIN W. GELFAND 223
18. Growth of Antibody-Producing Cell Clones in Microcultures
IVAN LEFKOVITS 240
19. Adherent Layer-Dependent Development of B Cell Progenitor in Semisolid Agar
CHRISTOPHER J. PAIGE AND ROLAND H. GISLER 251
20. Growth of B Cell Colonies in Double-Layer Agar Cultures
CHRISTOPHER J. PAIGE, HELMUT SAUTER, AND GILLIAN E. WU 257
21. Culture in Liquid Medium of Single, Hapten-Specific, Antibody-Producing B Lymphocytes BEVERLEY L. PIKE 265

22. Long-Term Culture of Murine Bone Marrow Precursors of B Lymphocytes
C. A. WHITLOCK AND O. N. WITTE 275
23. Cloning of Mitogen- and Antigen-Reactive B Lymphocytes on Filter Paper Disks: Phenotypic and Genotypic Analysis of B Cell Colonies
GARNETT KELSOE 287
24. Production of Antibodies *in Vitro* in Cultures of Murine Lymphocytes
TOSHIKI ISHIHARA AND C. GARRISON FATHMAN 304
25. *In Vitro* Production of Antibody in Cultures of Human Peripheral Blood Lymphocytes DALE T. UMETSU AND RAIF S. GEHA 309
26. Purification and Characterization of Antigen-Binding Virgin and Memory B Cells
C. D. MYERS, V. M. SANDERS, E. YEFENOF, K. G. OLIVER, J. W. UHR, AND E. S. VITETTA 316
27. Murine T Cell Clones
ALEXANDRA LIVINGSTONE AND C. GARRISON FATHMAN 325
28. Cloning of Human Alloreactive T Cells
JORG GORONZY, CORNELIA WEYAND, AND C. GARRISON FATHMAN 333
29. Cloning with Antigens and Interleukin 2 of Murine T Lymphocytes Having Distinct Functions
LOUIS A. MATIS, MICHAEL BOOKMAN, AND STEVEN A. ROSENBERG 342
30. Karyotypic Analysis of T Cell Clones and Hybrids JUDITH P. JOHNSON 352
31. Generation of Terminal Deoxynucleotidyltransferase-Positive Lymphoid Precursor Cells *in Vitro* EUGENE S. MEDLOCK AND IRVING GOLDSCHNEIDER 363

Section III. Receptors on Lymphoid Cells

32. Receptors on Lymphoid Cells: An Overview
JOHN J. MARCHALONIS AND ROBERT M. GALBRAITH 377
33. Strategies for the Isolation of Cell Surface Receptors of Lymphoid Cells
ROBERT E. CONE 388
34. Nonpermeant Covalent Labels in Analytical Studies of Lymphocyte Membrane Proteins GREGORY W. WARR AND JOHN J. MARCHALONIS 399
35. Electron Spin Resonance Spectroscopy in the Study of Lymphoid Cell Receptors
CYRIL C. CURTAIN, FRANK D. LOONEY, AND LARRY M. GORDON 418
36. Air-Driven Ultracentrifuge for Sedimentation Equilibrium and Binding Studies
GEOFFREY J. HOWLETT 447
37. Lectins in the Isolation of Receptors on Lymphocytes
JOHN J. MARCHALONIS 463
38. Fluorescence Flow Cytometry in the Study of Lymphoid Cell Receptors
DAVID M. SEGAL, JULIE A. TITUS, AND DAVID A. STEPHANY 478
39. Photoaffinity Labeling in the Study of Lymphoid Cell β-Adrenergic Receptors
ARNOLD E. RUOHO, RICHARD B. CLARK, ROSS D. FELDMAN, AND ABBAS RASHIDBAIGI 492
40. Membrane-Impermeant Cross-Linking Reagents: Application to the Study of the Cell Surface Receptor for IgE
JAMES V. STAROS, WILLIAM T. LEE, AND DANIEL H. CONRAD 503
41. Fc_ε Receptors CHARLES W. PARKER 512
42. Immunoglobulin G Fc Receptors of Human Leukocytes
CLARK L. ANDERSON AND R. JOHN LOONEY 524
43. E-Rosette Receptor in Human Lymphocytes
MARCUS BRAUN, CLAUDIO ANASETTI, PAUL J. MARTIN, AND JOHN A. HANSEN 536

44. Virus Receptors on Lymphoid Cells
GLEN R. NEMEROW AND NEIL R. COOPER 548
45. Clq Receptor BERHANE GHEBREHIWET 558
46. Human Receptor for C3b/C4b: Complement Receptor Type I
WINNIE W. WONG AND DOUGLAS T. FEARON 579
47. The Complement Receptor Type 2 and Factor H Receptors
BARRY L. MYONES AND GORDON D. ROSS 586
48. Human Complement Receptor Type 3
M. AMIN ARNAOUT, MARK W. PIERCE, NAVA DANA, AND LINDA K. CLAYTON 602
49. C5a Receptor R. HUEY AND T. E. HUGLI 615
50. Antigen-Specific T Cell Receptors
JOHN J. MARCHALONIS AND SAMUEL F. SCHLUTER 628
51. Non-MHC-Restricted T Cell Antigen-Binding Proteins ROBERT E. CONE 666
52. Human Interleukin 2 Receptor
WARNER C. GREENE AND WARREN J. LEONARD 682
53. Receptors of Insulin and Growth Hormone on Lymphoid Cells
MAXINE A. LESNIAK, JOSE A. HEDO, GEORGE GRUNBERGER, BERNICE MARCUS-SAMUELS, JESSE ROTH, AND PHILLIP GORDEN 701
54. Lymphoid Receptors for Transferrin ROLAND A. NEWMAN 723
55. Isolation of Genes Encoding Proteins of Immunological Importance
DOUGLAS R. CAVENER 746

VOLUME 151
MOLECULAR GENETICS OF MAMMALIAN CELLS

Section I. Cell Lines Useful for Genetic Analysis

1. Chinese Hamster Ovary Cells MICHAEL M. GOTTESMAN 3
2. Cultured S49 Mouse T Lymphoma Cells
THEODOOR VAN DAALEN WETTERS AND PHILIP COFFINO 9
3. Liver Cell Lines G. J. DARLINGTON 19
4. HeLa Cell Lines SHIN-ICHI AKIYAMA 38
5. Somatic Cell Genetic Analysis of Myelomas and Hybridomas
DEBORAH FRENCH, THERESA KELLY, SUSAN BUHL, AND MATTHEW D. SCHARFF 50
6. Mouse Teratocarcinoma Cells
HEDWIG JAKOB AND JEAN-FRANÇOIS NICOLAS 66

Section II. Special Techniques for Mutant Selection

7. Amplification of Genes in Somatic Mammalian Cells
ROBERT T. SCHIMKE, DAVID S. ROOS, AND PETER C. BROWN 85
8. Replica Plating and Detection of Antigen Variants SUSANNAH GAL 104
9. Drug-Resistant Mutants: Selection and Dominance Analysis
MICHAEL M. GOTTESMAN 113
10. Suicide Selection of Mammalian Cell Mutants
DAVID PATTERSON AND CHARLES A. WALDREN 121

11. Use of Tritium-Labeled Precursors to Select Mutants
JACQUES POUYSSÉGUR AND ARLETTE FRANCHI 131
12. Construction of Immunotoxins Using *Pseudomonas* Exotoxin A
DAVID J. P. FITZGERALD 139
13. Isolation of Temperature-Sensitive Mutants
JOSEPH HIRSCHBERG AND MENASHE MARCUS 145
14. Fluorescence-Activated Cell Sorting of Hybrid and Transfected Cells
MICHAEL E. KAMARCK 150

Section III. Genetic Mapping and Analysis

15. Isozyme Identification of Chromosomes in Interspecific Somatic Cell Hybrids
MICHAEL J. SICILIANO AND BILLIE F. WHITE 169
16. Principles of Electrofusion and Electropermeabilization
ULRICH ZIMMERMANN AND HOWARD B. URNOVITZ 194
17. Cell Enucleation, Cybrids, Reconstituted Cells, and Nuclear Hybrids
JERRY W. SHAY 221
18. Toxin–Antitoxin Selection for Isolating Heterokaryons and Cell Hybrids
WOODRING E. WRIGHT 237
19. Chromosome Sorting by Flow Cytometry
MARTY BARTHOLDI, JULIE MEYNE, KEVIN ALBRIGHT, MARY LUEDEMANN, EVELYN CAMPBELL, DOUGLAS CHRITTON, LARRY L. DEAVEN, AND L. SCOTT CRAM 252
20. Methods for Chromosome Banding of Human and Experimental Tumors *in Vitro*
JEFFREY M. TRENT AND FLOYD H. THOMPSON 267
21. *In Situ* Hybridization of Metaphase and Prometaphase Chromosomes
SUSAN L. NAYLOR, JOHN R. MCGILL, AND BERNHARD U. ZABEL 279
22. Gene Mapping with Sorted Chromosomes
ROGER V. LEBO AND BARRY D. BRUCE 292
23. Transfer and Selective Retention of Single Specific Human Chromosomes via Microcell-Mediated Chromosome Transfer
PAUL J. SAXON AND ERIC J. STANBRIDGE 313

Section IV. Isolation and Detection of Mutant Genes

24. Strategies for Isolation of Mutant Genes MICHAEL M. GOTTESMAN 329
25. Use of in-Gel DNA Renaturation for Detection and Cloning of Amplified Genes
IGOR B. RONINSON 332
26. Detection and Characterization of Specific mRNA by Microinjection and Complementation of Mutant Cells
PIN-FANG LIN, DAVID B. BROWN, PAT MURPHY, MASARU YAMAIZUMI, AND FRANK H. RUDDLE 371
27. Use of the CAT Reporter Gene for Optimization of Gene Transfer into Eukaryotic Cells C. MICHAEL FORDIS AND BRUCE H. HOWARD 382
28. Cosmid Vector Systems for Genomic DNA Cloning
MARY MCCORMICK, MAX E. GOTTESMAN, GEORGE A. GAITANARIS, AND BRUCE H. HOWARD 397
29. Preparation of a Genomic Cosmid Library
ROBERT FLEISCHMANN, MARY MCCORMICK, AND BRUCE H. HOWARD 405

30. Colony Screening of Genomic Cosmid Libraries BRUCE R. TROEN 416
31. Rescue of Genes Transferred from Cosmid Libraries YUN-FAI LAU 426
32. Phage-Mediated Transduction of cDNA Expression Libraries into Mammalian Cells HIROTO OKAYAMA 434
33. Sib Selection MARY McCORMICK 445
34. Repetitive Cloning of Mutant Genes Using Locus-Specific Sticky Ends LAWRENCE A. CHASIN, ADELAIDE M. CAROTHERS, AND GAIL URLAUB 449
35. Strategies for Mapping and Cloning Macroregions of Mammalian Genomes CASSANDRA L. SMITH, SIMON K. LAWRANCE, GERALD A. GILLESPIE, CHARLES R. CANTOR, SHERMAN M. WEISSMAN, AND FRANCIS S. COLLINS 461
36. Insertional Mutagenesis to Isolate Genes STEPHEN P. GOFF 489
37. Microcloning of Mammalian Metaphase Chromosomes JAN-ERIK EDSTRÖM, ROLF KAISER, AND DAN RÖHME 503

Section V. Gene Regulation in Tissue Culture

38. Antisense RNA Inhibition of Endogenous Genes J. TIMOTHY STOUT AND C. THOMAS CASKEY 519
39. *In Situ* Detection of Transcription in Transfected Cells Using Biotin-Labeled Molecular Probes GILBERT H. SMITH 530
40. RNA Detection and Localization in Cells and Tissue Sections by *in Situ* Hybridization of ^{35}S-Labeled RNA Probes MARY E. HARPER AND LISA M. MARSELLE 539
41. Electrophoretic Assay for DNA-Binding Proteins ALEXANDER VARSHAVSKY 551

VOLUME 152

GUIDE TO MOLECULAR CLONING TECHNIQUES

Section I. Requirements for a Molecular Biology Laboratory

1. Equipping the Laboratory DAPHNE D. BLUMBERG 3
2. Creating a Ribonuclease-Free Environment DAPHNE D. BLUMBERG 20
3. Safety with ^{32}P- and ^{35}S-Labeled Compounds ROBERT A. ZOON 25

Section II. General Methods for Isolating and Characterizing Nucleic Acids

4. Large- and Small-Scale Phenol Extractions DONALD M. WALLACE 33
5. Precipitation of Nucleic Acids DONALD M. WALLACE 41
6. Quantifying ^{32}P-Labeled and Unlabeled Nucleic Acids SHELBY L. BERGER 49
7. Autoradiograms: ^{35}S and ^{32}P WILLIAM M. BONNER 55
8. Electrophoresis in Agarose and Acrylamide Gels RICHARD C. OGDEN AND DEBORAH A. ADAMS 61

Section III. Enzymatic Techniques and Recombinant DNA Technology

9. Nick Translation JUDY L. MEINKOTH AND GEOFFREY M. WAHL 91
10. Enzymes for Modifying and Labeling DNA and RNA FABIO COBIANCHI AND SAMUEL H. WILSON 94

Section IV. Restriction Enzymes

11. Properties and Uses of Restriction Endonucleases JOAN E. BROOKS 113
12. Restriction and Modification *in Vivo* by *Escherichia coli* K12 ELISABETH A. RALEIGH 130

Section V. Growth and Maintenance of Bacteria

13. Practical Aspects of Preparing Phage and Plasmid DNA: Growth, Maintenance, and Storage of Bacteria and Bacteriophage HARVEY MILLER 145

Section VI. Genomic Cloning

14. Host Strains That Alleviate Underrepresentation of Specific Sequences: Overview ARLENE R. WYMAN AND KENNETH F. WERTMAN 173
15. Isolation of Genomic DNA BERNHARD G. HERRMANN AND ANNA-MARIA FRISCHAUF 180
16. Digestion of DNA: Size Fractionation ANNA-MARIA FRISCHAUF 183
17. Construction and Characterization of a Genomic Library in λ ANNA-MARIA FRISCHAUF 190
18. Cloning Large Segments of Genomic DNA Using Cosmid Vectors ANTHONY G. DILELLA AND SAVIO L. C. WOO 199

Section VII. Preparation and Characterization of RNA

19. Preparation and Characterization of RNA: Overview SHELBY L. BERGER 215
20. Isolation of RNA Using Guanidinium Salts RAYMOND J. MACDONALD, GALVIN H. SWIFT, ALAN E. PRZYBYLA, AND JOHN M. CHIRGWIN 219
21. Isolation of Cytoplasmic RNA: Ribonucleoside–Vanadyl Complexes SHELBY L. BERGER 227
22. Isolation and Analysis of Nuclear RNA JOSEPH R. NEVINS 234
23. Isolation of Messenger RNA from Membrane-Bound Polysomes BERNARD M. MECHLER 241
24. Use of Antibodies to Obtain Specific Polysomes DENNIS C. LYNCH 248
25. Purification and Fractionation of Poly(A)$^+$ RNA ALLAN JACOBSON 254
26. Determination of the Molar Concentration of Messenger RNA MARC S. KRUG AND SHELBY L. BERGER 262
27. Translation in Cell-Free Systems ROSEMARY JAGUS 267

28. Microinjection into *Xenopus* Oocytes: Equipment
MICHAEL J. M. HITCHCOCK, EDWARD I. GINNS, AND CAROL J. MARCUS-SEKURA 276
29. Preparation of Oocytes for Microinjection of RNA and DNA
CAROL J. MARCUS-SEKURA AND MICHAEL J. M. HITCHCOCK 284
30. Translation of Messenger RNA in Injected Frog Oocytes
DOUGLAS A. MELTON 288
31. Characterization of *in Vitro* Translation Products ROSEMARY JAGUS 296

Section VIII. Preparation of cDNA and the Generation of cDNA Libraries

32. Preparation of cDNA and the Generation of cDNA Libraries: Overview
ALAN R. KIMMEL AND SHELBY L. BERGER 307
33. First-Strand cDNA Synthesis Primed with Oligo(dT)
MARC S. KRUG AND SHELBY L. BERGER 316
34. Second-Strand cDNA Synthesis: Classical Method UELI GUBLER 325
35. Second-Strand cDNA Synthesis: mRNA Fragments as Primers
UELI GUBLER 330
36. Purification of Large Double-Stranded Fragments
WILLIAM H. ESCHENFELDT AND SHELBY L. BERGER 335
37. Homopolymeric Tailing
WILLIAM H. ESCHENFELDT, ROBERT S. PUSKAS, AND SHELBY L. BERGER 337
38. Adaptors, Linkers, and Methylation
RAY WU, TIYUN WU, AND ANURADHA RAY 343
39. Directional cDNA Cloning in Plasmid Vectors by Sequential Addition of Oligonucleotide Linkers
DAVID M. HELFMAN, JOHN C. FIDDES, AND DOUGLAS HANAHAN 349
40. Cloning cDNA into λgt10 and λgt11
JERRY JENDRISAK, RICHARD A. YOUNG, AND JAMES DOUGLAS ENGEL 359
41. Full-Length cDNA Clones: Vector-Primed cDNA Synthesis
PRESCOTT L. DEININGER 371

Section IX. Selection of Clones from Libraries

42. Selection of Clones from Libraries: Overview ALAN R. KIMMEL 393
43. Molecular Hybridization of Immobilized Nucleic Acids: Theoretical Concepts and Practical Considerations
GEOFFREY M. WAHL, SHELBY L. BERGER, AND ALAN R. KIMMEL 399
44. Amplification, Storage, and Replication of Libraries
G. VOGELI AND P. S. KAYTES 407
45. Screening Colonies or Plaques with Radioactive Nucleic Acid Probes
GEOFFREY M. WAHL AND SHELBY L. BERGER 415
46. Isolation of Differentially Expressed Genes THOMAS D. SARGENT 423
47. Oligonucleotide Probes for the Screening of Recombinant DNA Libraries
R. BRUCE WALLACE AND C. GARRETT MIYADA 432
48. Gene Cloning Based on Long Oligonucleotide Probes WILLIAM I. WOOD 443

49. Hybridization of Genomic DNA to Oligonucleotide Probes in the Presence of Tetramethylammonium Chloride
ANTHONY G. DILELLA AND SAVIO L. C. WOO 447
50. Use of Antibodies to Screen cDNA Expression Libraries Prepared in Plasmid Vectors DAVID M. HELFMAN AND STEPHEN H. HUGHES 451
51. Gene Isolation by Screening λgt11 Libraries with Antibodies
ROBERT C. MIERENDORF, CHRIS PERCY, AND RICHARD A. YOUNG 458
52. Gene Isolation by Retroviral Tagging STEPHEN P. GOFF 469
53. Isolation of Genes by Complementation in Yeast MARK D. ROSE 481

Section X. Identification and Characterization of Specific Clones

54. Identification and Characterization of Specific Clones: Strategy for Confirming the Validity of Presumptive Clones ALAN R. KIMMEL 507
55. Subcloning JONATHAN R. GREENE AND LEONARD GUARENTE 512
56. DNA Sequencing: Chemical Methods
BARBARA J. B. AMBROSE AND REYNALDO C. PLESS 522
57. Sequencing DNA with Dideoxyribonucleotides as Chain Terminators: Hints and Strategies for Big Projects WAYNE M. BARNES 538
58. Direct Sequencing of Denatured Plasmid DNA
ROBERT C. MIERENDORF AND DIANA PFEFFER 556
59. Sequencing of RNA Transcripts Synthesized *in Vitro* from Plasmids Containing Bacteriophage Promoters ROBERT C. MIERENDORF AND DIANA PFEFFER 563
60. Hybrid Selection of mRNA and Hybrid Arrest of Translation
ROSEMARY JAGUS 567
61. Northern and Southern Blots
GEOFFREY M. WAHL, JUDY L. MEINKOTH, AND ALAN R. KIMMEL 572
62. Fast Blots: Immobilization of DNA and RNA from Cells
CARL COSTANZI AND DAVID GILLESPIE 582
63. Identification of Genomic Sequences Corresponding to cDNA Clones
NIKOLAUS A. SPOEREL AND FOTIS C. KAFATOS 588
64. Isolation of Full-Length Genes: Walking the Chromosome
NIKOLAUS A. SPOEREL AND FOTIS C. KAFATOS 598
65. Cosmid Vectors for Genomic Walking and Rapid Restriction Mapping
GLEN A. EVANS AND GEOFFREY M. WAHL 604
66. Mapping of Gene Transcripts by Nuclease Protection Assays and cDNA Primer Extension FRANK J. CALZONE, ROY J. BRITTEN, AND ERIC H. DAVIDSON 611
67. A Molecular Titration Assay to Measure Transcript Prevalence Levels
JAMES J. LEE AND NANCY A. COSTLOW 633
68. Demonstration of Tissue-Specific Gene Expression by *in Situ* Hybridization
LYNNE M. ANGERER, KATHLEEN H. COX, AND ROBERT C. ANGERER 649
69. Expression, Identification, and Characterization of Recombinant Gene Products in *Escherichia coli* ALLAN R. SHATZMAN AND MARTIN ROSENBERG 661
70. Heterologous Gene Expression in Yeast GRANT A. BITTER 673
71. Use of Eukaryotic Expression Technology in the Functional Analysis of Cloned Genes BRYAN R. CULLEN 684
72. Identification of Regulatory Elements of Cloned Genes with Functional Assays
NADIA ROSENTHAL 704

73. Interaction of Protein with DNA *in Vitro*
LOTHAR HENNIGHAUSEN AND HENRYK LUBON 721
74. *In Vivo* Footprinting of Specific Protein–DNA Interactions
P. DAVID JACKSON AND GARY FELSENFELD 735

VOLUME 153
RECOMBINANT DNA (PART D)

Section I. Vectors for Cloning DNA

1. Production of Single-Stranded Plasmid DNA
JEFFREY VIEIRA AND JOACHIM MESSING 3
2. pKUN, Vectors for the Separate Production of Both DNA Strands of Recombinant Plasmids
RUUD N. H. KONINGS, ELS J. M. VERHOEVEN, AND BEN P. H. PEETERS 12
3. Restriction Site Bank Vectors for Cloning in Gram-Negative Bacteria and Yeast
JOHN DAVISON, MICHEL HEUSTERSPREUTE, AND FRANÇOISE BRUNEL 34
4. Plasmids for the Selection and Analysis of Prokaryotic Promoters
JÜRGEN BROSIUS AND JAMES R. LUPSKI 54
5. A λ DNA Protocol Based on Purification of Phage on DEAE–Cellulose
CYNTHIA HELMS, JAMES E. DUTCHIK, AND MAYNARD V. OLSON 69
6. Double *cos* Site Vectors: Simplified Cosmid Cloning PAUL BATES 82
7. Improved *in Vitro* Packaging of λ DNA SUSAN M. ROSENBERG 95
8. λ Phage Vectors—EMBL Series
A. M. FRISCHAUF, N. MURRAY, AND H. LEHRACH 103
9. Plasmid and Phage Vectors for Gene Cloning and Analysis in *Streptomyces*
DAVID A. HOPWOOD, MERVYN J. BIBB, KEITH F. CHATER, AND TOBIAS KIESER 116
10. Cosmid Shuttle Vectors for Cloning and Analysis of *Streptomyces* DNA
R. NAGARAJA RAO, M. A. RICHARDSON, AND S. KUHSTOSS 166
11. Host–Vector Systems for Gene Cloning in Cyanobacteria
C. J. KUHLEMEIER AND G. A. VAN ARKEL 199
12. Genetic Engineering of the Cyanobacterial Chromosome
SUSAN S. GOLDEN, JUDY BRUSSLAN, AND ROBERT HASELKORN 215
13. Conjugal Transfer of Plasmids to Cyanobacteria
TERESA THIEL AND C. PETER WOLK 232
14. Site-Directed Chromosomal Rearrangements in Yeast
RICHARD T. SUROSKY AND BIK-KWOON TYE 243
15. Improved Vectors for Plant Transformation: Expression Cassette Vectors and New Selectable Markers
S. G. ROGERS, H. J. KLEE, R. B. HORSCH, AND R. T. FRALEY 253
16. Vectors for Cloning in Plant Cells
R. DEBLAERE, A. REYNAERTS, H. HÖFTE, J.-P. HERNALSTEENS, J. LEEMANS, AND M. VAN MONTAGU 277
17. Binary Ti Vectors for Plant Transformation and Promoter Analysis
GYNHEUNG AN 292

18. Detection of Monocot Transformation via *Agrobacterium tumefaciens*
PAUL J. J. HOOYKAAS AND ROB A. SCHILPEROORT 305
19. Direct Gene Transfer to Protoplasts of Dicotyledonous and Monocotyledonous Plants by a Number of Methods, Including Electroporation
RAYMOND D. SHILLITO AND INGO POTRYKUS 313
20. Uptake of DNA and RNA into Cells Mediated by Electroporation
W. H. R. LANGRIDGE, B. J. LI, AND A. A. SZALAY 336
21. Electroporation of DNA and RNA into Plant Protoplasts
MICHAEL FROMM, JUDY CALLIS, LOVERINE P. TAYLOR, AND VIRGINIA WALBOT 351
22. Cloning Vectors of Mitochondrial Origin for Eukaryotes
CHRISTINE LANG-HINRICHS AND ULF STAHL 366

Section II. Vectors for Expression of Cloned Genes

23. Short Homopeptide Leader Sequences Enhanced Production of Human Proinsulin in *Escherichia coli*
WING L. SUNG, FEI-L. YAO, AND SARAN A. NARANG 385
24. Expression of Bovine Growth Hormone Derivatives in *Escherichia coli* and the Use of the Derivatives to Produce Natural Sequence Growth Hormone by Cathepsin C Cleavage
HANSEN M. HSIUNG AND WARREN C. MACKELLAR 390
25. Expression of Eukaryotic Genes in *Escherichia coli* with a Synthetic Two-Cistron System
BRIGITTE E. SCHONER, RAMA M. BELAGAJE, AND RONALD G. SCHONER 401
26. Expression of Heterologous Unfused Protein in *Escherichia coli*
ERIK REMAUT, ANNE MARMENOUT, GUUS SIMONS, AND WALTER FIERS 416
27. Directing Ribosomes to a Single mRNA Species: A Method to Study Ribosomal RNA Mutations and Their Effects on Translation of a Single Messenger in *Escherichia coli*
ANNA HUI, PARKASH JHURANI, AND HERMAN A. DE BOER 432
28. New Expression Vectors for Identifying and Testing Signal Structures for Initiation and Termination of Transcription
KLAUS SCHNEIDER AND CHRISTOPH F. BECK 452
29. Synthesis and Sequence-Specific Proteolysis of Hybrid Proteins Produced in *Escherichia coli* KIYOSHI NAGAI AND HANS CHRISTIAN THØGERSEN 461
30. Expression Plasmid Containing the λ P_L Promoter and cI857 Repressor
GREGORY MILMAN 482
31. Expression and Secretion of Foreign Proteins in *Escherichia coli*
GUY D. DUFFAUD, PAUL E. MARCH, AND MASAYORI INOUYE 492
32. Engineering for Protein Secretion in Gram-Positive Bacteria SHING CHANG 507
33. Expression and Secretion Vectors for Yeast
GRANT A. BITTER, KEVIN M. EGAN, RAYMOND A. KOSKI, MATTHEW O. JONES, STEVEN G. ELLIOTT, AND JAMES C. GIFFIN 516
34. Vaccinia Virus as an Expression Vector
ANTONIA PICCINI, MARION E. PERKUS, AND ENZO PAOLETTI 545

VOLUME 154
RECOMBINANT DNA (PART E)

Section I. Methods for Cloning cDNA

1. High-Efficiency Cloning of Full-Length cDNA; Construction and Screening of cDNA Expression Libraries for Mammalian Cells
H. OKAYAMA, M. KAWAICHI, M. BROWNSTEIN, F. LEE, T. YOKOTA, AND K. ARAI 3
2. A Method for Cloning Full-Length cDNA in Plasmid Vectors
GISELA HEIDECKER AND JOACHIM MESSING 28
3. An Efficient Vector–Primer cDNA Cloning System DANNY C. ALEXANDER 41
4. Use of Primer–Restriction End Adapters in cDNA Cloning
CHRISTOPHER COLECLOUGH 64

Section II. Identification of Cloned Genes and Mapping of Genes

5. Mapping of RNA Using S_1 Nuclease and Synthetic Oligonucleotides
A. A. REYES AND R. BRUCE WALLACE 87
6. Oligonucleotide Hybridization Techniques
C. GARRETT MIYADA AND R. BRUCE WALLACE 94
7. λgt 11: Gene Isolation with Antibody Probes and Other Applications
MICHAEL SNYDER, STEPHEN ELLEDGE, DOUGLAS SWEETSER, RICHARD A. YOUNG, AND RONALD W. DAVIS 107
8. Searching for Clones with Open Reading Frames
MARK R. GRAY, GAIL P. MAZZARA, PRANHITHA REDDY, AND MICHAEL ROSBASH 129
9. Use of Open Reading Frame Expression Vectors GEORGE M. WEINSTOCK 156
10. 5-Fluoroorotic Acid as a Selective Agent in Yeast Molecular Genetics
JEF D. BOEKE, JOSHUA TRUEHEART, GEORGES NATSOULIS, AND GERALD R. FINK 164
11. Transposon Tn5 Mutagenesis to Map Genes FRANS J. DE BRUIJN 175
12. Site-Directed Tn5 and Transplacement Mutagenesis: Methods to Identify Symbiotic Nitrogen Fixation Genes in Slow-Growing *Rhizobium*
JOHN D. NOTI, MITTUR N. JAGADISH, AND ALADAR A. SZALAY 197

Section III. Chemical Synthesis and Analysis of Oligodeoxynucleotides

13. Simultaneous Synthesis and Biological Applications of DNA Fragments: An Efficient and Complete Methodology
RONALD FRANK, ANDREAS MEYERHANS, KONRAD SCHWELLNUS, AND HELMUT BLÖCKER 221
14. The Segmented Paper Method: DNA Synthesis and Mutagenesis by Rapid Microscale "Shotgun Gene Synthesis"
HANS W. DJURHUUS MATTHES, ADRIEN STAUB, AND PIERRE CHAMBON 250

15. Chemical Synthesis of Deoxyoligonucleotides by the Phosphoramidite Method
M. H. CARUTHERS, A. D. BARONE, S. L. BEAUCAGE, D. R. DODDS, E. F. FISHER, L. J. MCBRIDE, M. MATTEUCCI, Z. STABINSKY, AND J.-Y. TANG 287
16. An Automated DNA Synthesizer Employing Deoxynucleoside 3′-Phosphoramidites
SUZANNA J. HORVATH, JOSEPH R. FIRCA, TIM HUNKAPILLER, MICHAEL W. HUNKAPILLER, AND LEROY HOOD 314

Section IV. Site-Specific Mutagenesis and Protein Engineering

17. Oligonucleotide-Directed Mutagenesis: A Simple Method Using Two Oligonucleotide Primers and a Single-Stranded DNA Template
MARK J. ZOLLER AND MICHAEL SMITH 329
18. Oligonucleotide-Directed Construction of Mutations via Gapped Duplex DNA
WILFRIED KRAMER AND HANS-JOACHIM FRITZ 350
19. Rapid and Efficient Site-Specific Mutagenesis without Phenotypic Selection
THOMAS A. KUNKEL, JOHN D. ROBERTS, AND RICHARD A. ZAKOUR 367
20. Improved Oligonucleotide-Directed Mutagenesis Using M13 Vectors
PAUL CARTER 382
21. Site-Specific Mutagenesis to Modify the Human Tumor Necrosis Factor Gene
DAVID F. MARK, ALICE WANG, AND COREY LEVENSON 403
22. An Improved Method to Obtain a Large Number of Mutants in a Defined Region of DNA
RICHARD PINE AND P. C. HUANG 415
23. Molecular Mechanics and Dynamics in Protein Design
PETER KOLLMAN AND W. F. VAN GUNSTEREN 430
24. The Use of Structural Profiles and Parametric Sequence Comparison in the Rational Design of Polypeptides
SÁNDOR PONGOR 450
25. Structure–Function Analysis of Proteins through the Design, Synthesis, and Study of Peptide Models
JOHN W. TAYLOR AND E. T. KAISER 473
26. Effect of Point Mutations on the Folding of Globular Proteins
C. ROBERT MATTHEWS 498
27. Structure and Thermal Stability of Phage T4 Lysozyme
TOM ALBER AND BRIAN W. MATTHEWS 511

VOLUME 155
RECOMBINANT DNA (PART F)

Section I. Restriction Enzymes

1. Restriction Enzyme *Hgi*CI Characterization of the 6-Nucleotide Staggered Cut Sequence and Its Application in Mismatch Cloning
MANFRED KRÖGER AND GERD HOBOM 3
2. *Rsr*II: A Restriction Endonuclease with a Heptanucleotide Recognition Sequence
C. DAVID O'CONNOR, JEREMY N. B. WALKER, AND JON R. SAUNDERS 11

3. *Not*I and *Sfi*I: Restriction Endonucleases with Octanucleotide Recognition Sequences — BO-QIN QIANG AND IRA SCHILDKRAUT 15
4. Site-Specific Cleavage of DNA at 8-, 9-, and 10-bp Sequences — MICHAEL MCCLELLAND 22
5. Purification and Assay of Type II DNA Methylases — MICHAEL NELSON AND MICHAEL MCCLELLAND 32
6. The Use of DNA Methylases to Alter the Apparent Recognition Specificities of Restriction Endonucleases — MICHAEL NELSON AND IRA SCHILDKRAUT 41

Section II. Rapid Methods for DNA Sequence Analysis

7. Random Cloning and Sequencing by the M13/Dideoxynucleotide Chain Termination Method — A. T. BANKIER, K. M. WESTON, AND B. G. BARRELL 51
8. The Use of DNase I, Buffer Gradient Gel, and ^{35}S Label for DNA Sequencing — GUO-FAN HONG 93
9. pEMBL: A New Family of Single-Stranded Plasmids for Sequencing DNA — LUCIANA DENTE AND RICCARDO CORTESE 111
10. DNA Sequencing: A New Strategy to Create Ordered Deletions, Modified M13 Vector, and Improved Reaction Conditions for Sequencing by Dideoxy Chain Termination Method — TAPAN K. MISRA 119
11. Construction and Use of pBR322 Plasmids That Yield Single-Stranded DNA for Sequencing — ROBERT ZAGURSKY AND KIRK BAUMEISTER 139
12. Unidirectional Digestion with Exonuclease III in DNA Sequence Analysis — STEVEN HENIKOFF 156
13. DNA Sequencing Using α-Thiodeoxynucleotides — S. LABEIT, H. LEHRACH, AND R. S. GOODY 166
14. Use of Transposon-Promoted Deletions in DNA Sequence Analysis — ASAD AHMED 177
15. A Rapid Single-Stranded Cloning, Sequencing, Insertion, and Deletion Strategy — RODERIC M. K. DALE AND AMY ARROW 204
16. A New and Simple Rapid Method for Sequencing DNA — ZE-GUO PENG AND RAY WU 214
17. A Systematic Approach to Chemical DNA Sequencing by Subcloning in pGV451 and Derived Vectors — GUIDO VOLCKAERT 231
18. Direct Transfer Electrophoresis Used for DNA Sequencing — FRITZ M. POHL AND STEPHEN BECK 250
19. The Synthesis and Use of Fluorescent Oligonucleotides in DNA Sequence Analysis — LLOYD M. SMITH, ROBERT J. KAISER, JANE Z. SANDERS, AND LEROY E. HOOD 260
20. Solid-Phase Methods for Sequencing of Oligodeoxynucleotides and DNA — ANDRÉ ROSENTHAL, RUDOLF JUNG, AND HANS-DIETER HUNGER 301

Section III. Miscellaneous Methods

21. Specific Synthesis of DNA *in Vitro* via a Polymerase-Catalyzed Chain Reaction — KARY B. MULLIS AND FRED A. FALOONA 335
22. Visual Assay for Chromosome Ploidy — DOUGLAS KOSHLAND AND PHILIP HIETER 351

23. Isolation of Amino Acid-Specific tRNA by High-Performance Liquid Chromatography
SUSUMU NISHIMURA, NOBUKO SHINDO-OKADA, AND PAMELA F. CRAIN 373
24. Analysis of Modified Nucleosides and Nucleotide Sequence of tRNA
YOSHIYUKI KUCHINO, NAOHIRO HANYU, AND SUSUMU NISHIMURA 379
25. *In Vitro* RNA Synthesis with SP6 RNA Polymerase
P. A. KRIEG AND D. A. MELTON 397
26. A T5 Promoter-Based Transcription–Translation System for the Analysis of Proteins *in Vitro* and *in Vivo*
HERMANN BUJARD, REINER GENTZ, MICHAEL LANZER, DIETRICH STUEBER, MICHAEL MUELLER, IBRAHIM IBRAHIMI, MARIE-THERESE HAEUPTLE, AND BERNHARD DOBBERSTEIN 416
27. Use of Boronate-Containing Gels for Electrophoretic Analysis of Both Ends of RNA Molecules GABOR L. IGLOI AND HANS KÖSSEL 433
28. Purification, Specific Fragmentation, and Separation of Large DNA Molecules
CASSANDRA L. SMITH AND CHARLES R. CANTOR 449
29. Orthogonal-Field-Alternation Gel Electrophoresis
GEORGES F. CARLE AND MAYNARD V. OLSON 468
30. Computational Simulation of DNA Melting and Its Application to Denaturing Gradient Gel Electrophoresis
LEONARD S. LERMAN AND KAREN SILVERSTEIN 482
31. Detection and Localization of Single Base Changes by Denaturing Gradient Gel Electrophoresis
RICHARD M. MYERS, TOM MANIATIS, AND LEONARD S. LERMAN 501
32. An Agarose Gel Electrophoresis Assay for the Detection of DNA-Binding Activities in Yeast Cell Extracts
JUDITH BERMAN, SHLOMO EISENBERG, AND BIK-KWOON TYE 528
33. Hydroxyl Radical Footprinting: A High-Resolution Method for Mapping Protein–DNA Contacts
THOMAS D. TULLIUS, BETH A. DOMBROSKI, MAIR E. A. CHURCHILL, AND LAURANCE KAM 537
34. Mutagenesis with Degenerate Oligonucleotides: An Efficient Method for Saturating a Defined DNA Region with Base Pair Substitutions
DAVID E. HILL, ARNOLD R. OLIPHANT, AND KEVIN STRUHL 558
35. The Use of Random-Sequence Oligonucleotides for Determining Consensus Sequences ARNOLD R. OLIPHANT AND KEVIN STRUHL 568
36. Computer Programs for Analyzing DNA and Protein Sequences
F. I. LEWITTER AND W. P. RINDONE 582

VOLUME 156
BIOMEMBRANES (PART P: ATP-DRIVEN PUMPS AND RELATED TRANSPORT: THE NA,K-PUMP)

1. Overview: The Na,K-Pump JENA CHRISTIAN SKOU 1

Section I. Preparation of Na^+,K^+-ATPase and Subunits

2. Purification of Na^+,K^+-ATPase: Enzyme Sources, Preparative Problems, and Preparation from Mammalian Kidney PETER LETH JØRGENSEN 29

3. Preparation of Membrane Na^+,K^+-ATPase from Rectal Glands of *Squalus Acanthias* JENS CHRISTIAN SKOU AND MIKAEL ESMANN 43
4. Purification of Na^+,K^+-ATPase from the Supraorbital Salt Gland of the Duck THOMAS WOODWARD SMITH 46
5. Preparation of Na^+,K^+-ATPase from Brine Shrimp GARY L. PETERSON AND LOWELL E. HOKIN 48
6. Preparation of the $\alpha(+)$ Isozyme of the Na^+,K^+-ATPase from Mammalian Axolemma KATHLEEN J. SWEADNER 65
7. Solubilization of Na^+,K^+-ATPase MIKAEL ESMANN 72
8. Crystallization of Membrane-Bound NA^+,K^+-ATPase in Two Dimensions ELISABETH SKRIVER, ARVID B. MAUNSBACH, HANS HERBERT, AND PETER L. JØRGENSEN 80
9. Preparation of Antibodies to Na^+,K^+-ATPase and Its Subunits WILLIAM J. BALL, JR., TERENCE L. KIRLEY, AND LOIS K. LANE 87

Section II. Assay of Na^+,K^+-ATPase Activity

10. ATPase and Phosphatase Activity of Na^+,K^+-ATPase: Molar and Specific Activity, Protein Determination MIKAEL ESMANN 105
11. Coupled Assay of Na^+,K^+-ATPase Activity JENS G. NØRBY 116
12. Identification and Quantitation of Na^+,K^+-ATPase by Back-Door Phosphorylation MARILYN D. RESH 119

Section III. Reconstitution of Na,K-Pump Activity

13. Reconstitution of Na,K-Pump Activity by Cholate Dialysis: Sidedness and Stoichiometry STANLEY M. GOLDIN, MICHAEL FORGAC, AND GILBERT CHIN 127
14. Reconstitution of the Na,K-Pump by Freeze–Thaw Sonication: Estimation of Coupling Ratio and Electrogenicity LOWELL E. HOKIN AND JOHN F. DIXON 141
15. Incorporation of $C_{12}E_8$-Solubilized Na^+,K^+-ATPase into Liposomes: Determination of Sidedness and Orientation FLEMMING CORNELIUS 156

Section IV. Analysis of the Pump Cycle

16. Measurement of Na^+ and K^+ Transport and Na^+,K^+-ATPase Activity in Inside-Out Vesicles from Mammalian Erythrocytes RHODA BLOSTEIN 171
17. Measurement of Active and Passive Na^+ and K^+ Fluxes in Reconstituted Vesicles S. J. D. KARLISH 179

Section V. Measurement of Ligand Binding and Distance between Ligands

18. Measurement of Binding of ATP and ADP to Na^+,K^+-ATPase JENS G. NØRBY AND JØRGEN JENSEN 191
19. Interaction of Cardiac Glycosides with Na^+,K^+-ATPase EARL T. WALLICK AND ARNOLD SCHWARTZ 201

20. Estimation of Na,K-Pump Numbers and Turnover in Intact Cells with ^{3}[H]Ouabain — SETH R. HOOTMAN AND STEPHEN A. ERNST 213
21. Measurement of Binding of Na^+, K^+, and Rb^+ to Na^+,K^+-ATPase by Centrifugation Methods — HIDEO MATSUI AND HARUO HOMAREDA 229
22. Estimating Affinities for Physiological Ligands and Inhibitors by Kinetic Studies on Na^+,K^+-ATPase and Its Partial Activities — JOSEPH D. ROBINSON 236
23. Inhibition of Translocation Reactions by Vanadate — LUIS BEAUGÉ 251

Section VI. Measurements of Conformational States of Na^+,K^+-ATPase

24. Use of Formycin Nucleotides, Intrinsic Protein Fluorescence, and Fluorescein Isothiocyanate-Labeled Enzymes for Measurement of Conformational States of Na^+,K^+-ATPase — S. J. D. KARLISH 271
25. Eosin as a Fluorescence Probe for Measurement of Conformational States of Na^+,K^+-ATPase — J. C. SKOU AND MIKAEL ESMANN 278
26. Rapid Ion-Exchange Technique for Measuring Rates of Release of Occluded Ions — I. M. GLYNN, D. E. RICHARDS, AND L. A. BEAUGÉ 281

Section VII. Modification of Na^+,K^+-ATPase

27. Proteolytic Cleavage as a Tool for Studying Structure and Conformation of Pure Membrane-Bound Na^+,K^+-ATPase — PETER L. JØRGENSEN AND ROBERT A. FARLEY 291
28. Irreversible and Reversible Modification of SH Groups and Effect on Catalytic Activity — WILHELM SCHONER, MARION HASSELBERG, AND RALF KISON 302
29. Photoaffinity Labeling with ATP Analogs — WILHELM SCHONER AND GEORGIOS SCHEINER-BOBIS 312
30. Affinity Labeling of the Digitalis-Binding Site — BERNARD ROSSI AND MICHEL LAZDUNSKI 323
31. Determination of Quaternary Structure of an Active Enzyme Using Chemical Cross-Linking with Glutaraldehyde — WILLIAM S. CRAIG 333
32. Use of Cross-Linking Reagents for Detection of Subunit Interactions of Membrane-Bound Na^+,K^+-ATPase — WU-HSIUNG HUANG, SHAM S. KAKAR, SANKARIDRUG M. PERIYASAMY, AND AMIR ASKARI 345

Section VIII. Magnetic Resonance Studies of Na^+,K^+-ATPase

33. Nuclear Magnetic Resonance Investigations of Na^+,K^+-ATPase — CHARLES M. GRISHAM 353
34. Electron Spin Resonance Investigations of Na^+,K^+-ATPase — MIKAEL ESMANN 371

Section IX. Biogenesis and Membrane Assembly

35. Molecular Cloning of Na^+,K^+-ATPase α Subunit Gene Using Antibody Probes — JAY W. SCHNEIDER, ROBERT W. MERCER, EDWARD J. BENZ, JR., AND ROBERT LEVENSON 379

36. Preparation and Use of Monoclonal Antibodies to Na^+,K^+-ATPase
MICHAEL KASHGARIAN AND DANIEL BIEMESDERFER 392

Section X. Microscopy of Na^+,K^+-ATPase

37. Histochemical Localization of Na^+,K^+-ATPase
HIROSHI MAYAHARA AND KAZUO OGAWA 417
38. Analysis of Na^+,K^+-ATPase by Electron Microscopy
ARVID B. MAUNSBACH, ELISABETH SKRIVER, AND PETER L. JØRGENSEN 430

VOLUME 157

BIOMEMBRANES (PART Q: ATP-DRIVEN PUMPS AND RELATED TRANSPORT: CALCIUM, PROTON, AND POTASSIUM PUMPS)

Section I. CA^{2+} Fluxes and Regulation

1. Membrane Transport of Calcium: An Overview ERNESTO CARAFOLI 3

A. Complex Systems and Subcellular Fractions Involved in Ca^{2+} Regulation

2. Measurement of Ca^{2+} Release in Skinned Fibers from Skeletal Muscle
M. ENDO AND M. IINO 12
3. Isolation and Characterization of Sarcolemmal Vesicles from Rabbit Fast Skeletal Muscle STEVEN SEILER AND SIDNEY FLEISCHER 26
4. Isolation of Sarcoplasmic Reticulum Fractions Referable to Longitudinal Tubules and Junctional Terminal Cisternae from Rabbit Skeletal Muscle
ALICE CHU, MARK C. DIXON, AKITSUGU SAITO, STEVEN SEILER, AND SIDNEY FLEISCHER 36
5. Isolation of the Junctional Face Membrane of Sarcoplasmic Reticulum
BRIAN COSTELLO, CHRISTOPHER CHADWICK, AND SIDNEY FLEISCHER 46
6. Isolation of Triads from Skeletal Muscle
R. D. MITCHELL, PHILIP PALADE, AKITSUGU SAITO, AND SIDNEY FLEISCHER 51
7. Isolation of Transverse Tubule Membranes from Skeletal Muscle: Ion Transport Activity, Reformation of Triad Junctions, and Isolation of Junctional Spanning Protein of Triads
ANTHONY H. CASWELL, NEIL R. BRANDT, J-P. BRUNSCHWIG, AND RICHARD M. KAWAMOTO 68
8. Rapid Preparation of Canine Cardiac Sarcolemmal Vesicles by Sucrose Flotation
LARRY R. JONES 85
9. Isolation of Canine Cardiac Sarcoplasmic Reticulum
BRIAN K. CHAMBERLAIN AND SIDNEY FLEISCHER 91
10. Junctional and Longitudinal Sarcoplasmic Reticulum of Heart Muscle
MAKOTO INUI, SHERRY WANG, AKITSUGU SAITO, AND SIDNEY FLEISCHER 100
11. Regulation of Ca^{2+}-Pump from Cardiac Sarcoplasmic Reticulum
MICHIHIKO TADA, MASAAKI KADOMA, MAKOTO INUI, AND JUN-ICHI FUJII 107

B. Characterization of Ca^{2+}-Pumps and Modulators from Various Sources

12. Kinetic and Equilibrium Characterization of an Energy-Transducing Enzyme and Its Partial Reactions
GIUSEPPE INESI, MARK KURZMACK, AND DAVID LEWIS 154
13. Approaches to Studying the Mechanisms of ATP Synthesis in Sarcoplasmic Reticulum LEOPOLDO DE MEIS 190
14. Fluorimetric Detection and Significance of Conformational Changes in Ca^{2+}-ATPase Y. DUPONT, F. GUILLAIN, AND J. J. LACAPERE 206
15. Synthesis of ATP from Ca^{2+} Gradient by Sarcoplasmic Reticulum Ca^{2+} Transport ATPase E. FASSOLD AND WILHELM HASSELBACH 220
16. Occluded Ca^{2+} HARUHIKO TAKISAWA AND MADOKA MAKINOSE 228
17. Modified Membrane Filtration Methods for Ligand Binding on ATP-Driven Pumps during ATP Hydrolysis
MOTONORI YAMAGUCHI AND TAKAHIDE WATANABE 233
18. Active Cation Pumping of Na^+,K^+-ATPase and Sarcoplasmic Reticulum Ca^{2+}-ATPase Induced by an Electric Field TIAN YOW TSONG 240
19. Chemical Derivatization of Ca^{2+}-Pump Protein from Skeletal Muscle with N-Substituted Maleimides and 5-(2-Iodoacetamidoethyl)aminonaphthalene 1-Sulfonate
MASAO KAWAKITA, KIMIKO YASUOKA-YABE, KIMIKO SAITO-NAKATSUKA, AKIKO BABA, AND TETSURO YAMASHITA 251
20. Use of Detergents to Solubilize the Ca^{2+}-Pump Protein as Monomers and Defined Oligomers
JESPER V. MØLLER, MARC LE MAIER, AND JENS P. ANDERSEN 261
21. Analysis of Two-Dimensional Crystals of Ca^{2+}-ATPase in Sarcoplasmic Reticulum
K. A. TAYLOR, L. DUX, S. VARGA, H. P. TING-BEALL, AND A. MARTONOSI 271
22. cDNA Cloning of Sarcoplasmic Reticulum Proteins
CHRISTOPHER J. BRANDL, LARRY FLIEGEL, AND DAVID H. MACLENNAN 289
23. Reconstitution of Skeletal Muscle Sarcoplasmic Reticulum Membranes: Strategies for Varying the Lipid/Protein Ratio
LIN HYMEL AND SIDNEY FLEISCHER 302
24. Reconstitution of Calcium Pumping of Cardiac Sarcoplasmic Reticulum
MAKOTO INUI AND SIDNEY FLEISCHER 314
25. Purification and Crystallization of Calcium-Binding Protein from Skeletal Muscle Sarcoplasmic Reticulum
ANDREAS MAURER, MASASHI TANAKA, TAKAYUKI OZAWA, AND SIDNEY FLEISCHER 321
26. Purification of Calmodulin and Preparation of Immobilized Calmodulin
THOMAS J. LUKAS AND D. MARTIN WATTERSON 328
27. Purification, Reconstitution, and Regulation of Plasma Membrane Ca^{2+}-Pumps
JOHN T. PENNISTON, ADELAIDA G. FILOTEO, CAROL S. MCDONOUGH, AND ERNESTO CARAFOLI 340
28. Isolation and Reconstitution of Ca^{2+}-Pump from Human and Porcine Platelets
WILLIAM L. DEAN 351
29. Purification of Phospholamban from Canine Cardiac Sarcoplasmic Reticulum Vesicles by Use of Sulfhydryl Group Affinity Chromatography
LARRY R. JONES, ADAM D. WEGENER, AND HEATHER K. B. SIMMERMAN 360

30. Modification of Phospholipid Environment in Sarcoplasmic Reticulum Using Nonspecific Phospholipid Transfer Protein
JOËL LUNARDI, PAUL DEFOOR, AND SIDNEY FLEISCHER 369

C. Ca^{2+} and Other Channels

31. Computer Programs for Calculating Total from Specified Free or Free from Specified Total Ionic Concentrations in Aqueous Solutions Containing Multiple Metals and Ligands ALEXANDRE FABIATO 378
32. Ionic Permeability of Isolated Muscle Sarcoplasmic Reticulum and Liver Endoplasmic Reticulum Vesicles GERHARD MEISSNER 417
33. Permeability of Sarcoplasmic Reticulum
MICHIKI KASAI AND KAZUKI NUNOGAKI 437
34. Measurement of Calcium Release in Isolated Membrane Systems: Coupling between the Transverse Tubule and Sarcoplasmic Reticulum
NORIAKI IKEMOTO, DO HAN KIM, AND BOZENA ANTONIU 469
35. Techniques for Observing Calcium Channels from Skeletal Muscle Sarcoplasmic Reticulum in Planar Lipid Bilayers
JEFFREY S. SMITH, ROBERTO CORONADO, AND GERHARD MEISSNER 480
36. Purification of Ca^{2+} Release Channel (Ryanodine Receptor) from Heart and Skeletal Muscle Sarcoplasmic Reticulum
MAKOTO INUI AND SIDNEY FLEISCHER 490
37. Measurement of Sodium–Calcium Exchange Activity in Plasma Membrane Vesicles JOHN P. REEVES 505

Section II. ATP-Driven Proton Pumps

38. Plasma Membrane ATPase from the Yeast *Schizosaccharomyces pombe*
JEAN-PIERRE DUFOUR, ANTOINE AMORY, AND ANDRÉ GOFFEAU 513
39. Plasma Membrane ATPase from the Yeast *Saccharomyces cerevisiae*
ANDRÉ GOFFEAU AND JEAN-PIERRE DUFOUR 528
40. H^+-ATPase from Plasma Membranes of *Saccharomyces cerevisiae* and *Avena sativa* Roots: Purification and Reconstitution RAMÓN SERRANO 533
41. Purification of Yeast Vacuolar Membrane H^+-ATPase and Enzymological Discrimination of Three ATP-Driven Proton Pumps in *Saccharomyces cerevisiae*
ETSUKO UCHIDA, YOSHINORI OHSUMI, AND YASUHIRO ANRAKU 544
42. Purification of Vacuolar Membranes, Mitochondria, and Plasma Membranes from *Neurospora crassa* and Modes of Discriminating among the Different H^+-ATPases EMMA JEAN BOWMAN AND BARRY J. BOWMAN 562
43. Large-Scale Purification of Plasma Membrane H^+-ATPase from a Cell Wall-Less Mutant of *Neurospora crassa* GENE A. SCARBOROUGH 574
44. H^+-ATPase from Vacuolar Membranes of Higher Plants
ALAN B. BENNETT, ROGER A. LEIGH, AND ROGER M. SPANSWICK 579
45. Preparation and Acidification Activity of Lysosomes and Lysosomal Membranes
DONALD L. SCHNEIDER AND JEAN CHIN 591
46. Analysis of Endosome and Lysosome Acidification *in Vitro*
CYNTHIA J. GALLOWAY, GARY E. DEAN, RENATA FUCHS, AND IRA MELLMAN 601
47. Proton ATPases in Golgi Vesicles and Endoplasmic Reticulum: Characterization and Reconstitution of Proton Pumping
ANSON LOWE AND QAIS AL-AWQATI 611

48. Chromaffin Granule Proton Pump
NATHAN NELSON, SHULAMIT CIDON, AND YOSHINORI MORIYAMA 619
49. Proton Pump of Clathrin-Coated Vesicles
XIAO-SONG XIE, DENNIS K. STONE, AND EFRAIM RACKER 634

Section III. ATP-Driven K^+ Pumps

50. Preparation of Gastric H^+,K^+-ATPase
EDD C. RABON, WHA BIN IM, AND GEORGE SACHS 649
51. Genetics of Kdp, the K^+-Transport ATPase of *Escherichia coli*
JAMES W. POLAREK, MARK O. WALDERHAUG, AND WOLFGANG EPSTEIN 655
52. K^+-ATPase from *Escherichia coli*: Isolation and Characterization
ANNETTE SIEBERS, LESZEK WIECZOREK, AND KARLHEINZ ALTENDORF 668
53. Purification of the ATPase of *Streptococcus faecalis*
MARC SOLIOZ AND PETER FÜRST 680

Addendum

Addendum to Article 30
JOËL LUNARDI, PAUL DEFOOR, AND SIDNEY FLEISCHER 693

VOLUME 158
METALLOBIOCHEMISTRY (PART A)

Section I. Sample Preparation

1. Preparation of Metal-Free Water
JAMES F. RIORDAN AND BERT L. VALLEE 3
2. Elimination of Adventitious Metals BARTON HOLMQUIST 6
3. Metal-Free Dialysis Tubing DAVID S. AULD 13
4. Metal-Free Chromatographic Media MARK T. MARTIN 15
5. Preparation of Metal-Free Enzymes FRED W. WAGNER 21
6. Metal-Buffered Systems JOHN O. BAKER 33
7. Standards for Metal Analysis CLAUDE VEILLON 56
8. Methods for Metal Substitution DAVID S. AULD 71
9. Preparation of Metal-Hybrid Enzymes
WOLFGANG MARET AND MICHAEL ZEPPEZAUER 79
10. Introduction of Exchange-Inert Metal Ions into Enzymes
HAROLD E. VAN WART 95
11. Use of Chelating Agents to Inhibit Enzymes DAVID S. AULD 110

Section II. Analytical Techniques

12. Atomic Absorption Spectrometry WALTER SLAVIN 117
13. Multielement Atomic Absorption Methods of Analysis
JAMES M. HARNLEY AND DONITA L. GARLAND 145

14. Ion Microscopy in Biology and Medicine
SUBHASH CHANDRA AND GEORGE H. MORRISON 157
15. Flame Atomic Emission Spectrometry TERENCE H. RISBY 180
16. Inductively Coupled Plasma-Emission Spectrometry KAREN A. WOLNIK 190
17. Inductively Coupled Plasma-Mass Spectrometry JOSÉ A. OLIVARES 205
18. Atomic Fluorescence Spectrometry ROBERT G. MICHEL 222
19. Electrochemical Methods of Analysis JANET OSTERYOUNG 243
20. Neutron Activation Analysis JACQUES VERSIECK 267

Section III. Analysis of Metals

21. Aluminum
JOHN SAVORY, SUE BROWN, ROGER L. BERTHOLF, NANCY MENDOZA, AND MICHAEL R. WILLS 289
22. Measurement of Total Calcium in Biological Fluids: Flame Atomic Absorption Spectrometry GEORGE N. BOWERS, JR. AND THEODORE C. RAINS 302
23. Measurement of Ionized Calcium in Biological Fluids: Ion-Selective Electrode Method SALVADOR F. SENA AND GEORGE N. BOWERS, JR. 320
24. Chromium CLAUDE VEILLON 334
25. Determination of Cobalt by Atomic Absorption Spectrometry
ROBERT SHAPIRO AND MARK T. MARTIN 344
26. Measurement of Copper in Biological Samples by Flame or Electrothermal Atomic Absorption Spectrometry MERLE A. EVENSON 351
27. Rapid Colorimetric Micromethod for the Quantitation of Complexed Iron in Biological Samples WAYNE W. FISH 357
28. Atomic Absorption Spectrometry of Magnesium
MARK T. MARTIN AND ROBERT SHAPIRO 365
29. Molybdenum JEAN L. JOHNSON 371
30. Nickel Analysis by Electrothermal Atomic Absorption Spectrometry
F. WILLIAM SUNDERMAN, JR., SIDNEY M. HOPFER, AND M. CRISTINA CRISOSTOMO 382
31. Determination of Selenium in Biological Matrices S. A. LEWIS 391
32. Vanadium DONNA M. MARTIN AND N. DENNIS CHASTEEN 402
33. Determination of Zinc in Biological Samples by Atomic Absorption Spectrometry K. H. FALCHUK, K. L. HILT, AND B. L. VALLEE 422

VOLUME 159

INITIATION AND TERMINATION OF CYCLIC NUCLEOTIDE ACTION

Section I. Cyclic Nucleotide Cascades

1. Role of cAMP in Cyclic Cascade Regulation
EMILY SHACTER, EARL R. STADTMAN, STEWART R. JURGENSEN, AND P. BOON CHOCK 3
2. Theoretical Considerations of the Regulatory Steps in the cAMP Cascade System
STÉPHANE SWILLENS, JEAN-MARIE BOEYNAEMS, AND JACQUES E. DUMONT 19
3. Yeast Mutants Altered in the cAMP Cascade System
TATSUO ISHIKAWA, KUNIHIRO MATSUMOTO, AND ISAO UNO 27

Section II. Assays of Cyclic Nucleotide Levels, Turnover, and Transport

4. Improvements in the Automated Radioimmunoassay for cAMP or cGMP
GARY BROOKER 45
5. cAMP Turnover in Intact Cells R. BARBER AND R. W. BUTCHER 50
6. Monitoring Cyclic Nucleotide Metabolism in Intact Cells by ^{18}O Labeling
TIMOTHY F. WALSETH, RICHARD M. GRAEFF, AND NELSON D. GOLDBERG 60
7. Purification and Assay of cAMP, cGMP, and Cyclic Nucleotide Analogs in Cells Treated with Cyclic Nucleotide Analogs
JACKIE D. CORBIN, THOMAS W. GETTYS, PETER F. BLACKMORE, STEPHEN J. BEEBE, SHARRON H. FRANCIS, DAVID B. GLASS, J. BRUCE REDMON, VIRENDER S. SHEORAIN, AND LESLIE R. LANDISS 74
8. cAMP Export and Its Regulation by Prostaglandin A_1
LAURENCE L. BRUNTON AND LYNN E. HEASLEY 83

Section III. Cyclic Nucleotide Action

A. Cyclic Nucleotide Action in Mammals

9. Use of Antibody–Sepharose Columns to Study Hormonal Activation of cAMP-Dependent Protein Kinase Isozymes
RONALD EKANGER AND STEIN OVE DØSKELAND 97
10. Selective Activation of the cAMP-Dependent Protein Kinase Isoenzymes
S. A. LIVESEY AND T. J. MARTIN 105
11. Use of Synergistic Pairs of Site-Selective cAMP Analogs in Intact Cells
STEPHEN J. BEEBE, PETER F. BLACKMORE, TED D. CHRISMAN, AND JACKIE D. CORBIN 118
12. cAMP-Dependent Protein Kinase Regulatory Subunit Phosphorylation in Intact Cells MARC C. MUMBY AND CLAY W. SCOTT 139
13. Ammonium Sulfate Precipitation Assay for the Study of Cyclic Nucleotide Binding to Proteins STEIN OVE DØSKELAND AND DAGFINN ØGREID 147
14. cGMP-Dependent Protein Kinase Activation in Intact Tissues
RONALD R. FISCUS AND FERID MURAD 150
15. cAMP Analog Antagonists of cAMP Action
LYNNE H. PARKER BOTELHO, JOHN D. ROTHERMEL, ROBERT V. COOMBS, AND BERND JASTORFF 159
16. Peptide Inhibitors of cAMP-Dependent Protein Kinase
BRUCE E. KEMP, HEUNG-CHIN CHENG, AND DONAL A. WALSH 173
17. Type II cAMP-Dependent Protein Kinase Regulatory Subunit-Binding Protein
SUZANNE M. LOHMANN, PIETRO DE CAMILLI, AND ULRICH WALTER 183
18. Use of Digitonin-Permeabilized Adipocytes for cAMP Studies
ROBERT A. MOONEY 193
19. Purification of cAMP-Free and cAMP-Bound Forms of Bovine Heart cAMP-Dependent Protein Kinase Holoenzyme
CHARLES E. COBB AND JACKIE D. CORBIN 202
20. Preparation of Regulatory Subunits from Bovine Heart cAMP-Dependent Protein Kinase by a Nondenaturing Method
MARK SEVILLE AND J. JOHN HOLBROOK 208
21. Purification of *Drosophila* cAMP-Dependent Protein Kinase
JOHN L. FOSTER AND LINDA M. HALL 215

22. Using Immunocolloidal Gold Electron Microscopy to Investigate cAMP-Dependent Protein Kinase Cellular Compartmentalization
RICHARD A. JUNGMANN, MICHAEL R. KUETTEL, STEPHEN P. SQUINTO, AND JOANNA KWAST-WELFELD 225

23. Direct Cytochemical Localization of the Free Catalytic Subunit of cAMP-Dependent Protein Kinase CRAIG V. BYUS AND WILLIAM H. FLETCHER 236

24. Direct Cytochemical Localization of Regulatory Subunit of cAMP-Dependent Protein Kinase Using Fluoresceinated Catalytic Subunit
WILLIAM H. FLETCHER, TAMAKO A. ISHIDA, SCOTT M. VAN PATTEN, AND DONAL A. WALSH 255

B. Nonmammalian cAMP Binding Proteins

25. Identification and Ligand-Induced Modification of the cAMP Receptor in *Dictyostelium* PETER KLEIN, ANNE THEIBERT, AND PETER DEVREOTES 267

26. Crystallizing Catabolite Gene Activator Protein with cAMP for Structural Analysis IRENE T. WEBER 278

27. Isolation of a *Trypanosoma* cAMP-Binding Protein Which Is Not a Regulatory Subunit of cAMP-Dependent Protein Kinase
RAFAEL RANGEL-ALDAO, EDMUNDO CAYAMA, OSCAR ALLENDE, AND FRANCISCO TRIANA 286

C. Molecular Genetic Approaches

28. Application of Molecular Genetic Techniques to the cAMP-Dependent Protein Kinase System
G. STANLEY MCKNIGHT, MICHAEL D. UHLER, CHRISTOPHER H. CLEGG, LESLAY A. CORRELL, AND GARY G. CADD 299

29. Cloning of cDNA for the Catalytic Subunit of cAMP-Dependent Protein Kinase
MARK O. SHOWERS AND RICHARD A. MAURER 311

30. Molecular Cloning of cDNA for a Hormone-Regulated Isoform of the Regulatory Subunit of Type II cAMP-Dependent Protein Kinase from Rat Ovaries
TORE JAHNSEN, LARS HEDIN, VINCE J. KIDD, TANYA SCHULZ, AND JOANNE S. RICHARDS 318

31. Expression and Mutagenesis of the Regulatory Subunit of cAMP-Dependent Protein Kinase in *Escherichia coli*
LAKSHMI D. SARASWAT, MARCIN FILUTOWICS, AND SUSAN TAYLOR 325

Section IV. Protein Phosphatases

32. Methods to Distinguish Various Types of Protein Phosphatase Activity
DAVID L. BRAUTIGAN AND CAROL L. SHRINER 339

33. Preparation of Protein Phosphatase-Resistant Substrates Using Adenosine 5′-*O*-(γ-Thio)-triphosphate
HENG-CHUN LI, PAUL F. SIMONELLI, AND LING-JUN HUAN 346

34. Preparation of Low-Molecular-Weight Forms of Rabbit Muscle Protein Phosphatase ANTHONY DEGUZMAN AND ERNEST Y. C. LEE 356

35. Purification of Glycogen-Bound High-Molecular-Weight Phosphoprotein Phosphatase from Rabbit Skeletal Muscle BALWANT S. KHATRA 368

36. Purification and Characterization of Phosphorylase Phosphatase from Rabbit Skeletal Muscle
STEVEN J. MCNALL, LISA M. BALLOU, EMMA VILLA-MORUZZI, AND EDMOND H. FISCHER 377
37. Protein Phosphatase-1 and Protein Phosphatase-2A from Rabbit Skeletal Muscle
PHILIP COHEN, SUSANA ALEMANY, BRIAN A. HEMMINGS, THERESE J. RESINK, PETER STRÅOLFORS, AND H. Y. LIM TUNG 390
38. Protein Phosphatase-2B from Rabbit Skeletal Muscle: A Ca^{2+}-Dependent, Calmodulin-Stimulated Enzyme
ALEXANDER A. STEWART AND PHILIP COHEN 409
39. Protein Phosphatase-2C from Rabbit Skeletal Muscle and Liver: An Mg^{2+}-Dependent Enzyme CLARE H. MCGOWAN AND PHILIP COHEN 416
40. Protein Phosphatase Inhibitor-1 and Inhibitor-2 from Rabbit Skeletal Muscle
PHILIP COHEN, J. GORDON FOULKES, CHARLES F. B. HOLMES, GILLIAN A. NIMMO, AND NICHOLAS K. TONKS 427
41. Purification of an Mg^{2+}-Dependent Protein Phosphatase
SHIGERU TSUIKI, AKIRA HIRAGA, KUNIMI KIKUCHI, AND SHINRI TAMURA 437
42. Purification of Smooth Muscle Myosin Phosphatase from Turkey Gizzard
MARY D. PATO AND EWA KERC 446

Section V. General Methods for Studies of Phosphodiesterases

43. Assay of Cyclic Nucleotide Phosphodiesterase Using Radiolabeled and Fluorescent Substrates RANDALL L. KINCAID AND VINCENT C. MANGANIELLO 457
44. Analysis of Phosphodiesterase Reaction Mixtures by High-Performance Liquid Chromatography D. MARTIN WATTERSON AND THOMAS J. LUKAS 471
45. Cyclic Nucleotide Phosphodiesterase Activity: Histochemical and Cytochemical Methods SATOKI UENO AND MANFRED UECK 477
46. Methylxanthine Inhibitors of Phosphodiesterases
JACK N. WELLS AND JERRY R. MILLER 489
47. Acylpeptide Inhibitors of Phosphodiesterase Produced by *Bacillus subtilis*
KUNIAKI HOSONO 497
48. Selective Inhibitors of Specific Phosphodiesterases in Intact Adipocytes
VINCENT MANGANIELLO, EVA DEGERMAN, AND MARTHA ELKS 504
49. Cyclic Nucleotide Analogs Used to Study Phosphodiesterase Catalytic and Allosteric Sites CHRISTOPHE ERNEUX AND FRANÇOISE MIOT 520
50. cAMP Analogs Used to Study Low-K_m, Hormone-Sensitive Phosphodiesterase
STEPHEN J. BEEBE, ALFREDA BEASLEY-LEACH, AND JACKIE D. CORBIN 531

Section VI. Methods for Isolation and Studies of Various Phosphodiesterase Isoenzymes

A. Calmodulin-Stimulated Phosphodiesterase

51. Purification of Calmodulin-Stimulated Cyclic Nucleotide Phosphodiesterase by Monoclonal Antibody Affinity Chromatography
R. SCOTT HANSEN, HARRY CHARBONNEAU, AND JOSEPH A. BEAVO 543
52. Purification and Properties of Calmodulin-Activated Cyclic Nucleotide Phosphodiesterase from Mammalian Brain
RANDALL L. KINCAID AND MARTHA VAUGHAN 557

53. Purification of Calmodulin-Stimulated Phosphodiesterase by Affinity Chromatography on Calmodulin Fragment 1–77 Linked to Sepharose
GIULIO DRAETTA AND CLAUDE B. KLEE 573
54. Isolation of Bovine Brain Calmodulin-Dependent Cyclic Nucleotide Phosphodiesterase Isozymes RAJENDRA K.SHARMA AND JERRY H. WANG 582
55. Estimating the Association of Phosphodiesterase with Calmodulin in Intact Cells
JERRY R. MILLER AND JACK N. WELLS 594
56. Preparation of Fluorescent, Cross-Linking, and Biotinylated Calmodulin Derivatives and Their Use in Studies of Calmodulin-Activated Phosphodiesterase and Protein Phosphatase
RANDALL L. KINCAID, MELVIN L. BILLINGSLEY, AND MARTHA VAUGHAN 605
57. Preparation, Characterization, and Properties of Affinity-Purified Antibodies to Calmodulin-Dependent Cyclic Nucleotide Phosphodiesterase and the Protein Phosphatase Calcineurin RANDALL L. KINCAID 627
58. Selective Inhibitors of Calmodulin-Dependent Phosphodiesterase and Other Enzymes HIROYOSHI HIDAKA, MASAKI INAGAKI, MASAKATSU NISHIKAWA, AND TOSHIO TANAKA 652
59. Heat-Sensitive Inhibitor of Calmodulin-Regulated Cyclic Nucleotide Phosphodiesterase
EUGENIA JEDLICKI, CATHERINE ALLENDE, AND JORGE E. ALLENDE 660
60. Preparation and Use of Iodinated Calmodulin for Studies of Calmodulin-Binding Proteins LINDA J. VAN ELDIK 667
61. High-Affinity, Calmodulin-Dependent Isoforms of Cyclic Nucleotide Phosphodiesterase in Rat Testis KENNETH PURVIS AND HALLGEIR RUI 675

B. cGMP-Binding Phosphodiesterases

62. Isolation and Comparison of Bovine Heart cGMP-Inhibited and cGMP-Stimulated Phosphodiesterases
S. A. HARRISON, N. BEIER, T. J. MARTINS, AND J. A. BEAVO 685
63. Purification of Rod Outer Segment GTP-Binding Protein Subunits and cGMP Phosphodiesterase by Single-Step Column Chromatography
AKIO YAMAZAKI, MASAHIRO TATSUMI, AND MARK W. BITENSKY 702
64. Platelet cGMP-Binding Phosphodiesterase
PAVEL HAMET AND JOHANNE TREMBLAY 710
65. Purification of cGMP-Binding Protein Phosphodiesterase from Rat Lung
SHARRON H. FRANCIS AND JACKIE D. CORBIN 722
66. Photoaffinity Labeling of High-Affinity cGMP-Specific Noncatalytic Binding Sites on cGMP Phosphodiesterase of Rod Outer Segments
AKIO YAMAZAKI, MARK W. BITENSKY, AND JOHN E. CASNELLIE 730

C. High-Affinity cAMP Phosphodiesterases

67. Liver Low-K_m, Hormone-Sensitive Phosphodiesterase E. G. LOTEN 736
68. Insulin-Sensitive cAMP Phosphodiesterase in Rat Adipose Tissue
TETSURO KONO 745
69. Isolation and Characterization of Insulin-Stimulated, High-Affinity cAMP Phosphodiesterases from Rat Liver
MILES D. HOUSLAY, NIGEL J. PYNE, AND MICHAEL E. COOPER 751

70. Preparation of Dog Kidney High-Affinity cAMP Phosphodiesterase
W. J. THOMPSON, C.-C. SHEN, AND S. J. STRADA 760
71. Brain Low-K_m cAMP Phosphodiesterase
PHILIPPE DE MAZANCOURT AND YVES GIUDICELLI 766

D. Nonmammalian Cyclic Nucleotide Phosphodiesterases

72. Purification of cAMP Phosphodiesterase from Platelets
PAUL G. GRANT AND ROBERT W. COLMAN 772
73. Zinc-Containing Cyclic Nucleotide Phosphodiesterases from Baker's Yeast
JOHN LONDESBOROUGH AND KARI SUORANTA 777
74. Mutational Analysis of Phosphodiesterase in *Drosophila* RONALD L. DAVIS 786

VOLUME 160

BIOMASS (PART A: CELLULOSE AND HEMICELLULOSE)

Section I. Cellulose

A. Preparation of Cellulosic Substrates

1. Increasing the Availability of Cellulose in Biomass Materials
T. M. WOOD AND J. N. SADDLER 3
2. Cadoxen Solvolysis of Cellulose CHRISTINE M. LADISCH 11
3. Preparation of Crystalline, Amorphous, and Dyed Cellulose Substrates
THOMAS M. WOOD 19
4. Preparation of Cellodextrins
A. N. PEREIRA, M. MOBEDSHAHI, AND M. R. LADISCH 26
5. Preparation of Cellodextrins: Another Perspective GEORG SCHMID 38
6. Fluorogenic and Chromogenic Glycosides as Substrates and Ligands of Carbohydrases
HERMAN VAN TILBEURGH, FRANK G. LOONTIENS, CLEMENT K. DE BRUYNE, AND MARC CLAEYSSENS 45
7. Use of Complex Formation between Congo Red and Polysaccharides in Detection and Assay of Polysaccharide Hydrolases
PETER J. WOOD, JAMES D. ERFLE, AND RONALD M. TEATHER 59
8. Soluble, Dye-Labeled Polysaccharides for the Assay of Endohydrolases
BARRY V. MCCLEARY 74

B. Assays for Cellulolytic Enzymes

9. Methods for Measuring Cellulase Activities
THOMAS M. WOOD AND K. MAHALINGESHWARA BHAT 87
10. Cellulase Assay Based on Cellobiose Dehydrogenase
GIORGIO CANEVASCINI 112
11. Nephelometric and Turbidometric Assay for Cellulase
T.-M. ENARI AND M.-L. NIKU-PAAVOLA 117
12. Selective Assay for Exo-1,4-β-glucanases
M. V. DESHPANDE, L. G. PETTERSSON, AND K.-E. ERIKSSON 126

13. Viscosimetric Determination of Carboxymethyl-cellulase Activity
MICHAEL A. HULME 130
14. Staining Techniques for the Detection of the Individual Components of Cellulolytic Enzyme Systems MICHAEL P. COUGHLAN 135

C. Chromatographic Methods for Carbohydrates

15. Liquid Chromatography of Carbohydrate Monomers and Oligomers
J. K. LIN, B. J. JACOBSON, A. N. PEREIRA, AND M. R. LADISCH 145
16. Quantitative Thin-Layer Chromatography of Sugars, Sugar Acids, and Polyalcohols R. KLAUS AND W. FISCHER 159
17. High-Performance Thin-Layer Chromatography of Starch, Cellulose, Xylan, and Chitin Hydrolyzates LANDIS W. DONER 176

D. Miscellaneous Methods for Cellulolytic Enzymes

18. Screening of Prokaryotes for Cellulose- and Hemicellulose-Degrading Enzymes
DIETER KLUEPFEL 180
19. Chromatographic Separation of Cellulolytic Enzymes
PETER TOMME, SHIELA MCCRAE, THOMAS M. WOOD, AND MARC CLAEYSSENS 187
20. Morphological Aspects of Wood Degradation ROBERT A. BLANCHETTE 193

E. Purification of Cellulose-Degrading Enzymes
Cellulases

21. Cellulases of *Pseudomonas fluorescens* var. *cellulosa*
KUNIO YAMANE AND HIROSHI SUZUKI 200
22. Cellulases of *Cellulomonas uda*
KATSUMI NAKAMURA AND KUMPEI KITAMURA 211
23. Cellulase of *Ruminococcus albus* THOMAS M. WOOD 216
24. Cellulase of *Trichoderma koningii* THOMAS M. WOOD 221
25. Cellulases of *Trichoderma reesei* MARTIN SCHÜLEIN 234
26. Cellulases of a Mutant Strain of *Trichoderma viride* QM 9414
A. G. J. VORAGEN, G. BELDMAN, AND F. M. ROMBOUTS 243
27. Cellulases from *Eupenicillium javanicum*
MITSUO TANAKA, MASAYUKI TANIGUCHI, RYUICHI MATSUNO, AND TADASHI KAMIKUBO 251
28. Cellulase of *Aspergillus niger* GENTARO OKADA 259
29. Cellulase and Hemicellulase from *Aspergillus fumigatus* Fresenius
JOHN C. STEWART AND JOHN HEPTINSTALL 264
30. Cellulases of *Aspergillus aculeatus*
SAWAO MURAO, REIICHIRO SAKAMOTO, AND MOTOO ARAI 274
31. Cellulases from *Thermoascus aurantiacus*
MAXWELL G. SHEPHERD, ANTHONY L. COLE, AND CHOW CHIN TONG 300
32. 1,4-β-D-Glucan Cellobiohydrolase from *Sclerotium rolfsii*
JAI C. SADANA AND RAJKUMAR V. PATIL 307
33. Cellulases of *Thermomonospora fusca* DAVID B. WILSON 314
34. Cellulases of *Humicola insolens* and *Humicola grisea*
SHINSAKU HAYASHIDA, KAZUYOSHI OHTA, AND KAIGUO MO 323

35. Cellulase–Hemicellulase Complex of *Phoma hibernica*
HENRYK URBANEK AND JADWIGA ZALEWSKA-SOBCZAK 332
36. Cellulases from *Sporocytophaga myxococcoides* JOSTEIN GOKSØYR 338
37. Cellulases in *Phaseolus vulgaris* MARY L. DURBIN AND LOWELL N. LEWIS 342

1,4-β-D-Glucan Glucanohydrolases (Endocellulases)

38. Endoglucanase from *Clostridium thermocellum*
THOMAS K. NG AND J. G. ZEIKUS 351
39. Crystalline Endoglucanase D of *Clostridium thermocellum* Overproduced in *Escherichia coli*
PIERRE BÉGUIN, GWENNAËL JOLIFF, MICHEL JUY, ADOLFO G. AMIT, JACQUELINE MILLET, ROBERTO J. POLJAK, AND JEAN-PAUL AUBERT 355
40. Isolation of 1,4-β-D-Glucan 4-Glucanohydrolases of *Talaromyces emersonii*
MICHAEL P. COUGHLAN AND AIDAN P. MOLONEY 363
41. Endo-1,4-β-glucanases of *Sporotrichum pulverulentum*
K.-E. ERIKSSON AND B. PETTERSSON 368
42. Carboxymethylcellulase from *Sclerotium rolfsii* WILLIAM A. LINDNER 376
43. β-Glucanases from *Pisum sativum* GORDON MACLACHLAN 382

1-4-β-D-Glucan Cellobiohydrolases (Exocellulases)

44. Cellobiosidase from *Ruminococcus albus*
KUNIO OHMIYA AND SHOICHI SHIMIZU 391
45. Cellobiohydrolases of *Penicillium pinophilum* THOMAS M. WOOD 398
46. Exocellulase of *Irpex lacteus* (*Polyporus tulipiferae*)
TAKAHISA KANDA AND KAZUTOSI NISIZAWA 403

1,4-β-D-Glucosidases

47. β-Glucosidase from *Ruminococcus albus*
KUNIO OHMIYA AND SHOICHI SHIMIZU 408
48. 1,4-β-Glucosidases of *Sporotrichum pulverulentum*
V. DESHPANDE AND K.-E. ERIKSSON 415
49. β-D-Glucosidases from *Sclerotium rolfsii*
JAI C. SADANA, RAJKUMAR V. PATIL, AND JAIPRAKASH G. SHEWALE 424
50. Purification and Assay of β-Glucosidase from *Schizophyllum commune*
AMY C. LO, GORDON WILLICK, ROGER BERNIER, JR., AND MICHEL DESROCHERS 432
51. Purification of β-D-Glucoside Glucohydrolases of *Talaromyces emersonii*
MICHAEL P. COUGHLAN AND ANTHONY MCHALE 437

Cellobiose Dehydrogenases

52. Cellobiose Dehydrogenase from *Sporotrichum thermophile*
GIORGIO CANEVASCINI 443
53. Cellobiose Dehydrogenase from *Sclerotium rolfsii*
JAI C. SADANA AND RAJKUMAR V. PATIL 448
54. Cellobiose Dehydrogenase Produced by *Monilia* sp. R. F. H. DEKKER 454
55. Cellobiose Dehydrogenase (Quinone)
U. WESTERMARK AND K.-E. ERIKSSON 463

Assays of Cellulolytic Enzymes and Miscellaneous Enzymes Involved in Cellulolysis

56. Cellobiose Phosphorylase from *Cellvibrio gilvus* TAKASHI SASAKI 468
57. Cellulosomes from *Clostridium thermocellum* RAPHAEL LAMED AND EDWARD A. BAYER 472
58. Macrocellulase Complexes and Yellow Affinity Substance from *Clostridium thermocellum* LARS G. LJUNGDAHL, MICHAEL P. COUGHLAN, FRANK MAYER, YUTAKA MORI, HIROMI HON-NAMI, AND KOYU HON-NAMI 483
59. Acid Proteases from *Sporotrichum pulverulentum* K.-E. ERIKSSON AND B. PETTERSSON 500

Section II. Hemicellulose

A. Preparation of Substrates for Hemicellulases

60. Purification of (1→3),(1→4)-β-D-Glucan from Barley Flour BARRY V. MCCLEARY 511
61. Synthesis of β-D-Mannopyranosides for the Assay of β-D-Mannosidase and Exo-β-D-mannanase BARRY V. MCCLEARY 515
62. Enzymatic Preparation of β-1,4-Mannooligosaccharides and β-1,4-Glucomannooligosaccharides ISAO KUSAKABE AND RIHEI TAKAHASHI 518
63. Carob and Guar Galactomannans BARRY V. MCCLEARY 523
64. Xylobiose and Xylooligomers JÜRGEN PULS, ANNEGRET BORCHMANN, DIETER GOTTSCHALK, AND JÜRGEN WIEGEL 528
65. Ramazol Brilliant Blue–Xylan: A Soluble Chromogenic Substrate for Xylanases PETER BIELY, DANA MISLOVIČOVÁ, AND RUDOLF TOMAN 536
66. Preparation of L-Arabinan and 1,5-L-Arabinan KIYOSHI TAGAWA AND AKIRA KAJI 542

B. Analysis of β-Glucan and Enzyme Assays

67. Measurement of (1→3),(1→4)-β-D-Glucan BARRY V. MCCLEARY, I. SHAMEER, AND M. GLENNIE-HOLMES 545
68. Measurement of Acetylxylan Esterase in *Streptomyces* K. G. JOHNSON, J. D. FONTANA, AND C. R. MACKENZIE 551
69. α-4-*O*-Methyl-D-glucuronidase Component of Xylanolytic Complexes J. D. FONTANA, M. GEBARA, M. BLUMEL, H. SCHNEIDER, C. R. MACKENZIE, AND K. G. JOHNSON 560

C. Purification of Hemicellulose-Degrading Enzymes

70. Lichenase from *Bacillus subtilis* BARRY V. MCCLEARY 572
71. Purification of β-D-Glucosidase from *Aspergillus niger* BARRY V. MCCLEARY AND JOAN HARRINGTON 575
72. Exo-1,4-β-mannanase from *Aeromonas hydrophila* TOSHIYOSHI ARAKI AND MANABU KITAMIKADO 583
73. Exo-β-D-mannanase from *Cyamopsis tetragonolobus* Guar Seed BARRY V. MCCLEARY 589

74. β-D-Mannanase BARRY V. MCCLEARY 596
75. β-Mannanase of *Streptomyces* ISAO KUSAKABE AND RIHEI TAKAHASHI 611
76. β-D-Mannosidase from *Helix pomatia* BARRY V. MCCLEARY 614
77. α-Mannanase from *Rhodococcus erythropolis*
I. YA. ZACHAROVA, V. Y. TAMM, AND I. N. PAVLOVA 620
78. α-D-Galactosidase from Lucerne and Guar Seed BARRY V. MCCLEARY 627
79. Xylanase of *Bacillus pumilus* HIROSUKE OKADA AND ATSUHIKO SHINMYO 632
80. Xylanase of *Cryptococcus albidus* PETER BIELY AND MÁRIA VRŠANSKÁ 638
81. Xylanases of *Streptomyces*
TUNEO YASUI, MASAKI MARUI, ISAO KUSAKABE, AND KOTOYOSHI NAKANISHI 648
82. Xylanases of Alkalophilic Thermophilic *Bacillus*
TERUHIKO AKIBA AND KOKI HORIKOSHI 655
83. Xylanase A of *Schizophyllum commune* L. JURASEK AND M. G. PAICE 659
84. Xylanases and β-Xylosidase of *Trichoderma lignorum*
MICHAEL JOHN AND JÜRGEN SCHMIDT 662
85. Xylanase of *Malbranchea pulchella* var. *sulfurea*
MASARU MATSUO AND TUNEO YASUI 671
86. Xylanase of *Talaromyces byssochlamydoides*
SHINSAKU HAYASHIDA, KAZUYOSHI OHTA, AND KAIGUO MO 675
87. 1,4-β-D-Xylan Xylohydrolase of *Sclerotium rolfsii* ANIL H. LACHKE 679
88. β-Xylosidases of Several Fungi MASARU MATSUO AND TUNEO YASUI 684
89. β-Xylosidase/β-Glucosidase of *Chaetomium trilaterale*
TUNEO YASUI AND MASARU MATSUO 696
90. Acetylxylan Esterase of *Schizophyllum commune*
P. BIELY, C. R. MACKENZIE, AND H. SCHNEIDER 700
91. α-L-Arabinofuranosidase from *Aspergillus niger*
KIYOSHI TAGAWA AND AKIRA KAJI 707
92. α-L-Arabinofuranosidase from *Scopolia japonica*
TSUNEKO UCHIDA AND MICHIO TANAKA 712
93. Arabinogalactanase of *Bacillus subtilis* var. *amylosacchariticus*
TAKEHIKO YAMAMOTO AND SHIGENORI EMI 719

VOLUME 161
BIOMASS (PART B: LIGNIN, PECTIN, AND CHITIN)

Section I. Lignin

A. Preparation of Substrates for Ligninases

1. Isolation of Lignin JOHN R. OBST AND T. KENT KIRK 3
2. Lignin–Carbohydrate Complexes from Various Sources
JUN-ICHI AZUMA AND KOSHIJIMA TETSUO 12
3. [^{14}C]-Lignin-Labeled Lignocelluloses and ^{14}C-Labeled Milled Wood Lignins: Preparation, Characterization, and Uses
RONALD L. CRAWFORD AND DON L. CRAWFORD 18
4. Preparation of Dioxane Lignin Fractions by Acidolysis BERNARD MONTIES 31
5. Acid-Precipitable Polymeric Lignin: Production and Analysis
DON L. CRAWFORD AND ANTHONY L. POMETTO III 35

6. Chemical Synthesis of Lignin Alcohols and Model Lignins Enriched with Carbon Isotopes KONRAD HAIDER, HARTMUT KERN, AND LUDGER ERNST 47
7. Synthesis of Lignols and Related Compounds FUMIAKI NAKATSUBO 57
8. Synthetic ^{14}C-Labeled Lignins T. KENT KIRK AND GÖSTA BRUNOW 65

B. Assays for Ligninases

9. Use of Polymeric Dyes in Lignin Biodegradation Assays MICHAEL H. GOLD, JEFFREY K. GLENN, AND MARGARET ALIC 74
10. Ligninolytic Activity of *Phanerochaete chrysosporium* Measured as Ethylene Production from α-Keto-γ-methylthiolbutyric Acid ROBERT L. KELLEY 79
11. Assays for Extracellular Aromatic Methoxyl-Cleaving Enzymes for the White Rot Fungus *Phanerochaete chrysosporium* VAN-BA HUYNH, RONALD L. CRAWFORD, AND ANDRZEJ PASZCZYŃSKI 83

C. Chemical Methods for Characterization of Lignin

12. Lignin Determination T. KENT KIRK AND JOHN R. OBST 87
13. Chemical Degradation Methods for Characterization of Lignins MITSUHIKO TANAHASHI AND TAKAYOSHI HIGUCHI 101
14. Characterization of Lignin by Oxidative Degradation: Use of Gas Chromatography-Mass Spectrometry Technique CHEN-LOUNG CHEN 110
15. Characterization of Lignin by ^{1}H and ^{13}C NMR Spectroscopy CHEN-LOUNG CHEN AND DANIELLE ROBERT 137

D. Chromatographic Methods for Lignin and Related Compounds

16. Gas–Liquid Chromatography of Aromatic Fragments from Lignin Degradation ANTHONY L. POMETTO III AND DON L. CRAWFORD 175
17. High-Performance Liquid Chromatography of Aromatic Fragments from Lignin Degradation ANTHONY L. POMETTO III AND DON L. CRAWFORD 183
18. Conventional and High-Performance Size-Exclusion Chromatography of Graminaceous Lignin–Carbohydrate Complexes WOLFGANG ZIMMERMANN, ALISTAIR PATERSON, AND PAUL BRODA 191
19. Analysis of Lignin Degradation Intermediates by Thin-Layer Chromatography and Gas Chromatography-Mass Spectrometry TOSHIAKI UMEZAWA AND TAKAYOSHI HIGUCHI 200

E. Nucleic Acid Preparations Related to Lignin Degradation

20. Preparation and Characterization of DNA from Lignin-Degrading Fungi UTE RAEDER AND PAUL BRODA 211
21. Preparation and Characterization of mRNA from Ligninolytic Fungi RICHARD HAYLOCK AND PAUL BRODA 221
22. Use of Synthetic Oligonucleotide Probes for Identifying Ligninase cDNA Clones YI-ZHENG ZHANG AND C. ADINARAYANA REDDY 228

F. Purification of Lignin-Degrading Enzymes

23. Lignin Peroxidase of *Phanerochaete chrysosporium* MING TIEN AND T. KENT KIRK 238
24. Lignin-Depolymerizing Activity of *Streptomyces* DON L. CRAWFORD AND ANTHONY L. POMETTO III 249

25. Manganese Peroxidase from *Phanerochaete chrysosporium*
MICHAEL H. GOLD AND JEFFREY K. GLENN 258
26. Manganese Peroxidase of *Phanerochaete chrysosporium*: Purification
ANDRZEJ PASZCZYŃSKI, RONALD L. CRAWFORD, AND VAN-BA HUYNH 264
27. NAD(P)H Dehydrogenase (Quinone) from *Sporotrichum pulverulentum*
JOHN A. BUSWELL AND KARL-ERIK ERIKSSON 271
28. Vanillate Hydroxylase from *Sporotrichum pulverulentum*
JOHN A. BUSWELL AND KARL-ERIK ERIKSSON 274
29. 4-Methoxybenzoate Monooxygenase from *Pseudomonas putida*: Isolation, Biochemical Properties, Substrate Specificity, and Reaction Mechanisms of the Enzyme Components
FRITHJOF-HANS BERNHARDT, ECKHARD BILL, ALFRED XAVER TRAUTWEIN, AND HANS TWILFER 281
30. Vanillate *O*-Demethylase from *Pseudomonas* Species
JOHN A. BUSWELL AND DOUGLAS W. RIBBONS 294
31. Purification of Coniferyl Alcohol Dehydrogenase from *Rhodococcus erythropolis*
E. JAEGER 301
32. Glucose Oxidase of *Phanerochaete chrysosporium*
ROBERT L. KELLEY AND C. ADINARAYANA REDDY 307
33. Pyranose 2-Oxidase from *Phanerochaete chrysosporium*
J. VOLC AND KARL-ERIK ERIKSSON 316
34. Methanol Oxidase of *Phanerochaete chrysosporium*
KARL-ERIK ERIKSSON AND ATSUMI NISHIDA 322

Section II. Pectin

A. Assays for Pectin-Degrading Enzymes

35. Assay Methods for Pectic Enzymes
ALAN COLLMER, JEFFREY L. RIED, AND MARK S. MOUNT 329

B. Purification of Pectin-Degrading Enzymes

36. Protopectinase from Yeasts and a Yeastlike Fungus TAKUO SAKAI 335
37. Pectin Lyase from *Phoma medicaginis* var. *pinodella* D. PITT 350
38. Pectinesterases from *Phytophthora infestans* HELGA FÖRSTER 355
39. Polygalacturonase from *Corticium rolfsii*
KIYOSHI TAGAWA AND AKIRA KAJI 361
40. Isozymes of Pectinesterase and Polygalacturonase from *Botrytis cinerea* Pers.
ABEL SCHEJTER AND LIONEL MARCUS 366
41. Galacturan 1,4-α-Galacturonidase from Carrot *Daucus carota* and Liverwort *Marchantia polymorpha* HARUYOSHI KONNO 373
42. Endopectate Lyase from *Erwinia aroideae* HARUYOSHI KONNO 381
43. High-Performance Liquid Chromatography of Pectic Enzymes
OTAKAR MIKEŠ AND LUBOMÍRA REXOVÁ-BENKOVÁ 385

Section III. Chitin

A. Preparation of Substrates for Chitin-Degrading Enzymes

44. Chitin Solutions and Purification of Chitin PAUL R. AUSTIN 403
45. Water-Soluble Glycol Chitin and Carboxymethylchitin SHIGEHIRO HIRANO 408

46. Isolation of Oligomeric Fragments of Chitin by Preparative High-Performance Liquid Chromatography KEVIN B. HICKS 410
47. Preparation of Crustacean Chitin KENZO SHIMAHARA AND YASUYUKI TAKIGUCHI 417

B. Assay for Chitin-Degrading Enzymes

48. Assay for Chitinase Using Tritiated Chitin ENRICO CABIB 424
49. Viscosimetric Assay for Chitinase AKIRA OHTAKARA 426
50. Colorimetric Assay for Chitinase THOMAS BOLLER AND FELIX MAUCH 430

C. Analytical Methods for Chitin

51. Physical Methods for the Determination of Chitin Structure and Conformation JOHN BLACKWELL 435
52. Determination of the Degree of Acetylation of Chitin and Chitosan DONALD H. DAVIES AND ERNEST R. HAYES 442
53. Determination of Molecular-Weight Distribution of Chitosan by High-Performance Liquid Chromatography ARNOLD C. M. WU 447
54. Analysis of Chitooligosaccharides and Reduced Chitooligosaccharides by High-Performance Liquid Chromatography AKIRA OHTAKARA AND MASARU MITSUTOMI 453
55. Enzymatic Determination of Chitin ENRICO CABIB AND ADRIANA SBURLATI 457

D. Purification of Chitin-Degrading Enzymes

56. Chitinase from *Serratia marcescens* ENRICO CABIB 460
57. Chitinase and β-*N*-Acetylhexosaminidase from *Pycnoporus cinnabarinus* AKIRA OHTAKARA 462
58. Chitinase from *Neurospora crassa* ANGEL ARROYO-BEGOVICH 471
59. Chitinase from *Verticillium albo-atrum* G. F. PEGG 474
60. Chitinase from *Phaseolus vulgaris* Leaves THOMAS BOLLER, ANNETTE GEHRI, FELIX MAUCH, AND URS VÖGELI 479
61. Chitinase from Tomato *Lycopersicon esculentum* G. F. PEGG 484
62. Chitinase–Chitobiase from Soybean Seeds and Puffballs JOHN P. ZIKAKIS AND JOHN E. CASTLE 490
63. Endochitinase from Wheat Germ ENRICO CABIB 498
64. Chitosanase from *Bacillus* Species YASUSHI UCHIDA AND AKIRA OHTAKARA 501
65. Chitosanase from *Streptomyces griseus* AKIRA OHTAKARA 505
66. Chitin Deacetylase YOSHIO ARAKI AND EIJI ITO 510
67. Poly(*N*-acetylgalactosamine) Deacetylase YOSHIO ARAKI AND EIJI ITO 514
68. Chitin Deacetylase from *Colletotrichum lindemuthianum* HEINRICH KAUSS AND BÄRBEL BAUCH 518
69. *N,N'*-Diacetylchitobiase of *Vibrio harveyi* RAFAEL W. SOTO-GIL, LISA C. CHILDERS, WILLIAM H. HUISMAN, A. STEPHEN DAHMS, MEHRDAD JANNATIPOUR, FARAH HEDJRAN, AND JUDITH W. ZYSKIND 524

VOLUME 162
IMMUNOCHEMICAL TECHNIQUES (PART L: CHEMOTAXIS AND INFLAMMATION)

Section I. Chemotaxis

A. Methods for the Study of Chemotaxis

1. Chemotaxis: An Overview — PETER C. WILKINSON AND WENDY S. HASTON 3
2. Visual Methods for Measuring Leukocyte Locomotion — WENDY S. HASTON AND PETER C. WILKINSON 17
3. Micropore Filter Methods for Leukocyte Chemotaxis — PETER C. WILKINSON 38
4. Agarose Method for Human Neutrophil Chemotaxis — ROBERT D. NELSON AND MICHAEL J. HERRON 50
5. Chromium-51 Radioimmunoassay for Chemotaxis — JOHN I. GALLIN 59
6. Orientation Chamber in Chemotaxis — SALLY H. ZIGMOND 65
7. Neutrophil Aggregation and Chemokinesis Assays — ANTHONY W. FORD-HUTCHINSON AND JILLIAN F. EVANS 72
8. Semiautomated Measurement of Neutrophil Migration with an Image Analyzer — CARLOS C. DAUGHADAY, ALAN N. BOHRER, AND ISAIAS SPILBERG 79
9. Concentration Gradients of Chemotactic Factors in Chemotaxis Assays — DOUGLAS A. LAUFFENBURGER, ROBERT T. TRANQUILLO, AND SALLY H. ZIGMOND 85
10. Measurement of Leukocyte Chemotaxis *in Vivo* — MICHAEL D. P. BOYLE, MICHAEL J. P. LAWMAN, ADRIAN P. GEE, AND MICHAEL YOUNG 101
11. Necrotaxis and Galvanotaxis — HANS GRULER 115

B. Methods for the Study of Chemoattractants and Biochemistry of Chemotaxis

12. Chemotactic Factors: An Overview — PETER C. WILKINSON 127
13. Synthesis of Chemotactic Peptides — NATESA MUTHUKUMARASWAMY AND RICHARD J. FREER 132
14. Natural Mediators of Leukocyte Chemotaxis — HIDEO HAYASHI, MITSUO HONDA, YASUHIRO MIBU, SHUNSUKE YAMAMOTO, AND MITSUOMI HIRASHIMA 140
15. Eosinophilic Chemotactic Factors of Anaphylaxis — JOHN A. SMITH 170
16. High-Molecular-Weight Eosinophil Chemotactic Factor from *Schistosoma japonicum* Eggs — MAKOTO OWHASHI AND YUKIFUMI NAWA 174
17. Denatured Proteins as Chemotactic Agents: Mitogens as Lymphocyte Locomotion Activators — PETER C. WILKINSON 180
18. Crystal-Induced Chemotactic Factor — ISAIAS SPILBERG AND ASHWIN K. BHATT 193
19. Chemotactic Factors of Bacterial Origin — WAYNE A. MARASCO AND PETER A. WARD 198
20. Chemotactic Fragments of Fibronectin — NORMAN E. WIKNER AND RICHARD A. F. CLARK 214

21. Inhibitors of Chemotaxis
EUFRONIO G. MADERAZO, CHARLES L. WORONICK, AND PETER A. WARD 223
22. Leukocyte Chemoattractant Receptors
MARILYN C. PIKE AND RALPH SNYDERMAN 236
23. Contractile Proteins in Leukocytes
MARK J. DINUBILE AND FREDERICK S. SOUTHWICK 246
24. Chemoattractant-Induced Membrane Phenomena of Phagocytes
CHARLES D. SMITH AND RALPH SNYDERMAN 271
25. Biochemical Changes in Leukocytes in Response to Chemoattractants
MARGRITH W. VERGHESE AND RALPH SNYDERMAN 279
26. Ionic Events Relevant to Neutrophil Activation
PAUL H. NACCACHE AND RAMADAN I. SHA'AFI 283

Section II. Inflammation

A. Methods for the Study of the Cellular Phenomena of Inflammation and Experimental Models of Inflammation

27. Cellular and Vascular Phenomena in Inflammation
ANDREW C. ISSEKUTZ AND THOMAS B. ISSEKUTZ 301
28. Carrageenan Foot Edema Test IVAN G. OTTERNESS AND PETER F. MOORE 320
29. Sponge Implants as Models PHILIP J. BAILEY 327
30. Carrageenan Pleurisy MARCIA L. BLIVEN AND IVAN G. OTTERNESS 334
31. Adjuvant Arthritis
JOEL D. TAUROG, DENNIS C. ARGENTIERI, AND RICHARD A. MCREYNOLDS 339
32. Type II Collagen-Induced Arthritis in Rats
SUSAN C. RIDGE, ARNOLD L. ORONSKY, AND S. S. KERWAR 355
33. Collagen-Induced Arthritis in the Mouse PAUL H. WOOLEY 361
34. Streptococcal Cell Wall-Induced Arthritis in Rats
SUSAN C. RIDGE, JOHN B. ZABRISKIE, ARNOLD L. ORONSKY, AND S. S. KERWAR 373
35. Chronic Arthritis in Rabbits HUGO E. JASIN 379
36. Murine Lupus SYAMAL K. DATTA 385
37. Experimental Allergic Encephalomyelitis ROBERT H. SWANBORG 413
38. Experimental Glomerulonephritis
DAVID J. SALANT AND ANDREY V. CYBULSKY 421
39. Autoimmune Orchitis and Thyroiditis PIERLUIGI E. BIGAZZI 461
40. Arthus Phenomenon PHILIP J. BAILEY AND DANIEL S. FLETCHER 478
41. Experimental Serum Sickness BERNICE NOBLE AND JAN R. BRENTJENS 484
42. Mast Cells and Mast Cell Products FRANK M. GRAZIANO 501
43. Basophils in Allergic and Inflammatory Reactions FRANK M. GRAZIANO 522
44. Human Neutrophil Degranulation DANIEL G. WRIGHT 538

B. Methods for the Study of Complement in Chemotaxis and Inflammation

45. Membrane Attack Complex Proteins C5b-6, C7, C8, and C9 of Human Complement ALFRED F. ESSER AND JAMES M. SODETZ 551
46. C3, C5 Components and C3a, C4a, and C5a Fragments of the Complement System JARMILA JANATOVA 579

47. C1, C4, and C2 Components of the Classical Pathway of Complement and Regulatory Proteins — IRMA GIGLI AND FRANCISCO A. TAUSK 626
48. Alternative Pathway of Complement — MICHAEL K. PANGBURN 639
49. Human C5a Anaphylatoxin: Gene Synthesis, Expression, and Recovery of Biologically Active Material from *Escherichia coli* — ARTHUR E. FRANKE, GLENN C. ANDREWS, NORMA P. STIMLER-GERARD, CRAIG J. GERARD, AND HENRY J. SHOWELL 653

VOLUME 163
IMMUNOCHEMICAL TECHNIQUES (PART M: CHEMOTAXIS AND INFLAMMATION)

Section I. Methods for the Study of the Biochemistry of Inflammation

1. Mediators of Inflammation: An Overview — ALLEN P. KAPLAN AND MICHAEL SILVERBERG 3
2. Phospholipase A_2 and Lipocortins, Antiphospholipase Proteins — MASSIMO DI ROSA, LIDIA SAUTEBIN, AND ROSA CARNUCCIO 23
3. Human Eosinophil Lysophospholipase — PETER F. WELLER 31
4. Platelet-Activating Factor — ROBERT L. WYKLE, JOSEPH T. O'FLAHERTY, AND MICHAEL J. THOMAS 44
5. Bovine Hageman Factor and Its Fragments — KAZUO FUJIKAWA 54
6. Human Hageman Factor and Its Fragments — MICHAEL SILVERBERG AND ALLEN P. KAPLAN 68
7. Measurement of Contact-Activation Patterns of Human Plasma Samples with Synthetic Substrates — GERARD DOOIJEWAARD, CORNELIS KLUFT, AND JOLANDA J. L. VAN IERSEL 80
8. Prekallikrein — MICHAEL SILVERBERG AND ALLEN P. KAPLAN 85
9. Latent Kallikrein from Human Urine — JOCELYN SPRAGG 96
10. Human Tissue Kallikrein — REINHARD GEIGER AND WERNER MISKA 102
11. Rat Pancreas Kallikrein — G. S. BAILEY 115
12. Rat Urinary Kallikrein — JULIE CHAO AND LEE CHAO 128
13. Immunochemical Viewing of Kallikrein in Tissues — TORILL BERG 143
14. Use of ^{125}I-Labeled Aprotinin in the Assay of Glandular Kallikreins — JAMES W. RYAN 160
15. Synthetic Substrates for the Assay of Prekallikrein and Factor XII — CORNELIS KLUFT 170
16. C_1 Inhibitor: The Predominant Inhibitor of Plasma Kallikrein — MARC SCHAPIRA, ARIANE DE AGOSTINI, AND ROBERT W. COLMAN 179
17. Arginine Carboxypeptidase and Its Inhibitors — JAMES W. RYAN 186
18. Angiotensin-Converting Enzyme, Dipeptidyl Carboxypeptidase I, and Its Inhibitors — JAMES W. RYAN 194
19. Kinin Receptors — D. REGOLI AND J. BARABÉ 210
20. Cloning of cDNAs and Genomic DNAs for High-Molecular-Weight and Low-Molecular-Weight Kininogens — NAOMI HITAMURA, HIROYUKI NAWA, YOSHIO TAKAGAKI, SUMIKO FURUTO-KATO, AND SHIGETADA NAKANISHI 230

21. Human Kininogens
WERNER MÜLLER-ESTERL, DAVID A. JOHNSON, GUY SALVESEN, AND ALAN J. BARRETT 240
22. Separation of Kinins by High-Performance Liquid Chromatography
FRANZ FIEDLER AND REINHARD GIEGER 257
23. Synthesis of Bradykinin Analogs G. DRAPEAU AND D. REGOLI 263
24. T-Kinin and T-Kininogen
LOWELL M. GREENBAUM AND HIROSHI OKAMOTO 272
25. Kinin Antagonists J. BARABÉ AND D. REGOLI 282
26. Plasminogen Activators and Their Inhibitors
DAVID J. LOSKUTOFF AND RAYMOND R. SCHLEEF 293
27. Tissue-Type Plasminogen Activator and Fast-Acting Plasminogen Activator Inhibitor in Plasma J. H. VERHEIJEN 302
28. Human Neutrophil and Mast Cell Proteases Implicated in Inflammation
RICHARD F. REST 309
29. *In Vitro* Models for the Study of Oxidant-Induced Injury of Cells in Inflammation
INGRID U. SCHRAUFSTÄTTER, W. A. HALSEY, JR., P. A. HYSLOP, AND C. G. COCHRANE 328
30. Leukotriene B_4 Receptors on Rat and Human Neutrophil Membranes
ANTHONY W. FORD-HUTCHINSON, STELLA CHARLESON, AND JILLIAN F. EVANS 340
31. 5- and 15(ω-6)-Lipoxygenases from Human Polymorphonuclear Leukocytes
ROY J. SOBERMAN 344
32. Leukotriene B_4 20-Hydroxylase of Human Polymorphonuclear Leukocytes
ROY J. SOBERMAN AND RICHARD T. OKITA 349
33. Leukotriene C_4 Synthase from Rat Basophilic Leukemia Cell Microsomes
ROY J. SOBERMAN AND TANIHIRO YOSHIMOTO 353
34. *In Vivo* and *in Vitro* Delayed Hypersensitivity
MARIANNE BROOME POWELL AND MARK I. GREENE 357

Section II. Methods for the Study of Acute-Phase Reactants

35. The Acute Phase Response: An Overview IRVING KUSHNER 373
36. C-Reactive Protein STEPHEN S. MACINTYRE 383
37. α_1-Antitrypsin PHILIPPE ARNAUD AND COLETTE CHAPUIS-CELLIER 400
38. α_1-Acid Glycoprotein
PHILIPPE ARNAUD, LAURENT MIRIBEL, AND ANNE FRANCOISE ROUX 418
39. α_2-HS Glycoprotein PHILIPPE ARNAUD, L. MIRIBEL, AND D. L. EMERSON 431
40. Ceruloplasmin
PHILIPPE ARNAUD, ELISABETTA GIANAZZA, AND LAURENT MIRIBEL 441
41. Haptoglobin
BARBARA H. BOWMAN, DON R. BARNETT, JEAN B. LUM, AND FUNMEI YANG 452
42. Fibrinogen
GERALD M. FULLER, ROBERT J. BUNZEL, AND JAMES E. NESBITT 474
43. α_2-Macroglobulin
HIROMI ISHIBASHI, KAZUHIRO HAYASHIDA, AND HIDEO OKUBO 485
44. Endogenous Pyrogens CHARLES A. DINARELLO 495

45. Serum Amyloid A Protein
EARL P. BENDITT, NILS ERIKSEN, AND RICK L. MEEK 510
46. Amyloid P Component MARTHA SKINNER AND ALAN S. COHEN 523
47. Hemopexin URSULA MULLER-EBERHARD 536
48. Electrophoretic Analysis of Acute-Phase Plasma Proteins HEINZ BAUMANN 566
49. Hepatocyte-Stimulating Factor
BARRY M. R. N. J. WOLOSKI AND GERALD M. FULLER 594

Section III. Methods for the Study of Repair Processes in Inflammation

50. Hepatic Granuloma as a Model of Inflammation and Repair: An Overview
SHARON M. WAHL 605
51. Fibroblasts and Myofibroblasts ALBERT K. HARRIS 623
52. Collagen Production by Fibroblasts MARIA AGELLI AND SHARON M. WAHL 642
53. Elastin M. GABRIELLA GIRO AND JEFFREY M. DAVIDSON 656
54. Collagen Metabolism
ANNE BUCKLEY, KENNETH E. HILL, AND JEFFREY M. DAVIDSON 674
55. Fibroblast Chemoattractants
ARNOLD E. POSTLETHWAITE AND ANDREW H. KANG 694
56. Growth Factors Involved in Repair Processes: An Overview DAVID BARNES 707
57. Lymphocyte- and Macrophage-Derived Growth Factors SHARON M. WAHL 715
58. Connective Tissue Activating Peptides
C. WILLIAM CASTOR AND ANTONIO R. CABRAL 731

VOLUME 164
RIBOSOMES

Section I. Electron Microscopy

1. Studying Ribosome Structure by Electron Microscopy and Computer-Image Processing
JOACHIM FRANK, MICHAEL RADERMACHER, TERENCE WAGENKNECHT, AND ADRIANA VERSCHOOR 3
2. Statistical Image Analysis of Electron Micrographs of Ribosomal Subunits
GEORGE HARAUZ, EGBERT BOEKEMA, AND MARIN VAN HEEL 35
3. Structural Analysis of Ribosomes by Scanning Transmission Electron Microscopy
MILOSLAV BOUBLIK, GERRIT T. OOSTERGETEL, VALSAN MANDIYAN, JAMES F. HAINFELD, AND JOSEPH S. WALL 49
4. Identification of Protein–Protein Cross-Links within the *Escherichia coli* Ribosome by Immunoblotting Techniques
GEORG STÖFFLER, BERNHARD REDL, JAN WALLECZEK, AND MARINA STÖFFLER-MEILICKE 64
5. Electron Microscopy Studies of Ribosomal RNA
IVAN N. SHATSKY AND VICTOR D. VASILIEV 76

Section II. Other Biophysical Methods

6. Crystallographic and Image Reconstruction Studies on Ribosomal Particles from Bacterial Sources — A. YONATH AND H. G. WITTMANN 95
7. Neutron-Scattering Topography of Proteins of the Small Ribosomal Subunit — MALCOLM S. CAPEL AND V. RAMAKRISHNAN 117
8. Preparation and Activity Measurements of Deuterated 50S Subunits for Neutron-Scattering Analysis — PETRA NOWOTNY, VOLKER NOWOTNY, HELGA VOSS, AND KNUD H. NIERHAUS 131
9. Nuclear Magnetic Resonance Techniques for Studying Structure and Function of Ribosomes — V. N. BUSHUEV AND A. T. GUDKOV 148
10. Preparation of 5S RNA-Related Materials for Nuclear Magnetic Resonance and Crystallography Studies — PETER B. MOORE, STEVEN ABO, BETTY FREEBORN, DANIEL T. GEWIRTH, NEOCLES B. LEONTIS, AND GRACE SUN 158
11. Fluorescence Labeling and Isolation of Labeled RNA and Ribosomal Proteins — O. W. ODOM, H.-Y. DENG, AND BOYD HARDESTY 174
12. Isolation and Characterization of Colicin Fragments of Bacterial 16S Ribosomal RNA — PETER H. VAN KNIPPENBERG AND HANS A. HEUS 188

Section III. Protein–RNA Interactions

13. Physical Studies of Ribosomal Protein–RNA Interactions — DAVID E. DRAPER, INGRID C. DECKMAN, AND JAILAXMI V. VARTIKAR 203
14. Preparation of Specific Ribosomal RNA Fragments — DAVID E. DRAPER, SUSAN A. WHITE, AND JOANNE M. KEAN 221
15. Physical Studies of the Interaction of *Escherichia coli* Translational Initiation Factor 3 Protein with Ribosomal RNA — ERIC WICKSTROM AND LANCE G. LAING 238
16. Isolation of Fragments of Ribosomal Proteins That Recognize rRNA — LI-MING CHANGCHIEN, RICHARD CONRAD, AND GARY R. CRAVEN 258
17. Isolation of Kinetic Intermediates in *in Vitro* Assembly of the *Escherichia coli* Ribosome Using Cibacron Blue F3GA — DIPAK B. DATTA, LI-MING CHANGCHIEN, AND GARY R. CRAVEN 270
18. Total Reconstitution of 70S Ribosomes from *Escherichia coli* — ROLF LIETZKE AND KNUD H. NIERHAUS 278

Section IV. Cross-Linking and Affinity-Labeling Methods

19. Intra-RNA and RNA–Protein Cross-Linking Techniques in *Escherichia coli* Ribosomes — RICHARD BRIMACOMBE, WOLFGANG STIEGE, APOSTOLOS KYRIATSOULIS, AND PETER MALY 287
20. RNA–Protein Cross-Linking — A. EXPERT-BEZANÇON AND C. CHIARUTTINI 310
21. Isolation and Identification of RNA Cross-Links — PAUL L. WOLLENZIEN 319
22. Psoralen Cross-Linking of Ribosomal RNA — SAMUEL E. LIPSON AND JOHN E. HEARST 330
23. Affinity Labeling of Ribosomes — BARRY S. COOPERMAN 341

24. Photoaffinity Labeling of Peptidyltransferase
ERNST KUECHLER, GÜNTER STEINER, AND ANDREA BARTA 361
25. Affinity Labeling of tRNA-Binding Sites on Ribosomes
JAMES OFENGAND, ROBERT DENMAN, KELVIN NURSE, ARNOLD LIEBMAN, DAVID MALAREK, ANTONINO FOCELLA, AND GLADYS ZENCHOFF 372

Section V. Chemical and Enzymatic Probing Methods

26. Probing Ribosome Structure and Function Using Short Oligodeoxyribonucleotides
WALTER E. HILL, DAVID G. CAMP, WILLIAM E. TAPPRICH, AND ANCHALEE TASSANAKAJOHN 401
27. Extension Inhibition Analysis of Translation Initiation Complexes
DIETER HARTZ, DAVID S. MCPHEETERS, ROBERT TRAUT, AND LARRY GOLD 419
28. Hot Tritium Bombardment Technique for Ribosome Surface Topography
M. M. YUSUPOV AND A. S. SPIRIN 426
29. Surface Topography of Ribosomal RNA
ALEXEY A. BOGDANOV, NINA V. CHICHKOVA, ALEXEY M. KOPYLOV, ALEXANDER S. MANKIN, AND EVGENY A. SKRIPKIN 440
30. Enzymatic and Chemical Probing of Ribosomal RNA–Protein Interactions
JAN CHRISTIANSEN AND ROGER GARRETT 456
31. Use of α-Sarcin to Analyze Ribosomal RNA–Protein Interactions
PAUL W. HUBER AND IRA G. WOOL 468
32. Use of RNA–DNA Hybridization in Chemical Probing of Large RNA Molecules
BARBARA J. VAN STOLK AND HARRY F. NOLLER 475
33. Structural Analysis of RNA Using Chemical and Enzymatic Probing Monitored by Primer Extension
SETH STERN, DANESH MOAZED, AND HARRY F. NOLLER 481

Section VI. Immunological Methods

34. Antibody Probes of Ribosomal RNA
DOHN G. GLITZ, PAULINE A. CANN, LINDA S. LASATER, AND HELEN MCKUSKIE OLSON 493
35. Localization of Ribosomal Proteins on the Surface of Ribosomal Subunits from *Escherichia coli* Using Immunoelectron Microscopy
MARINA STÖFFLER-MEILICKE AND GEORG STÖFFLER 503

Section VII. Isolation of Ribosomal Proteins

36. Reversed-Phase High-Performance Liquid Chromatography of Ribosomal Proteins
BARRY S. COOPERMAN, CARL J. WEITZMANN, AND MELISSA A. BUCK 523
37. Ion-Exchange High-Performance Liquid Chromatographic Separation of Ribosomal Proteins
MALCOLM S. CAPEL, DIPAK B. DATTA, CONCEPCION R. NIERRAS, AND GARY R. CRAVEN 532

38. Ribosomal Proteins from Archaebacteria High-Performance Liquid Chromatographic Purification for Microsequence Analysis
ROZA MARIA KAMP AND BRIGITTE WITTMANN-LIEBOLD 542

Section VIII. Ribosome Function and Kinetics

39. Analysis of 40S Ribosomal Protein S6 Phosphorylation during the Mitogenic Response JOACHIM KRIEG, ANDRÉE R. OLIVIER, AND GEORGE THOMAS 575
40. Pre-Steady-State Kinetic Studies on Ribosomal Translocation
JAMES M. ROBERTSON, HARALD PAULSEN, AND WOLFGANG WINTERMEYER 581
41. Quantitative Indicator Assay of tRNA Binding to the Ribosomal P and A Sites
ROLAND LILL AND WOLFGANG WINTERMEYER 597
42. Measurement of Translational Kinetic Parameters
MÅNS EHRENBERG AND C. G. KURLAND 611
43. Ribosomal Synthesis of Polypeptides from Aminoacyl-tRNA without Polynucleotide Template
A. S. SPIRIN, N. V. BELITSINA, AND G. Z. YUSUPOVA (TNALINA) 631
44. Mg^{2+}/NH_4^+/Polyamine System for Polyuridine-Dependent Polyphenylalanine Synthesis with Near *in Vivo* Characteristics
ANDREAS BARTETZKO AND KNUD H. NIERHAUS 650
45. Parameters for the Preparation of *Escherichia coli* Ribosomes and Ribosomal Subunits Active in tRNA Binding
HANS-JÖRG RHEINBERGER, UTE GEIGENMÜLLER, MARKUS WEDDE, AND KNUD H. NIERHAUS 658

Section IX. Genetics

46. Antibiotic Resistance Mutations in Ribosomal RNA Genes of *Escherichia coli*
CURT D. SIGMUND, MOHAMED ETTAYEBI, ANGELA BORDEN, AND EDWARD A. MORGAN 673
47. Analysis of Plasmid-Coded Ribosomal RNA by Maxicell Techniques
DAVID K. JEMIOLO, ROLF STEEN, MICHAEL J. R. STARK, AND ALBERT E. DAHLBERG 691
48. Temperature-Sensitive Mutants with Alterations in Ribosomal Protein L24 and Isolation of Intra- and Extragenic Suppressor Mutants
JOACHIM SCHNIER AND KAYOKO NISHI 706
49. Primer-Directed Deletions in 5S Ribosomal RNA
BIRTE VESTER, JAN EGEBJERG, ROGER GARRETT, AND JAN CHRISTIANSEN 710
50. 5S RNA Structure and Function H. U. GÖRINGER AND R. WAGNER 721
51. Cloning and Identification of Ribosomal Protein Genes in Chloroplast DNA
ANASTASIA PROMBONA, YASUNARI OGIHARA, AND ALAP R. SUBRAMANIAN 748

Section X. Mathematical and Computer Analysis of Ribosomal Sequences

52. Computer Analysis of Nucleic Acid Sequences MICHAEL S. WATERMAN 765
53. Phylogenetic Analysis Using Ribosomal RNA GARY J. OLSEN 793

VOLUME 165
MICROBIAL TOXINS: TOOLS IN ENZYMOLOGY

Section I. Preparation of Toxins

A. Gram-Positive Cocci

1. Preparation and Purification of Staphylococcal Alpha Toxin
SIDNEY HARSHMAN, NANCY SUGG, AND PAUL CASSIDY 3
2. Preparation and Purification of γ-Hemolysin of Staphylococci
MICHEL PLOMMET 8
3. Purification and Assay of Staphylococcal Δ-Lysin
T. HARRY BIRKBECK AND JOHN H. FREER 16
4. Purification and Crystallization of Staphylococcal Leukocidin
MASATOSHI NODA AND IWAO KATO 22
5. Purification of Epidermolytic Toxin of *Staphylococcus aureus*
JOYCE C. S. DE AZAVEDO, CHRISTOPHER J. BAILEY, AND JOHN P. ARBUTHNOTT 32
6. Preparation of Toxic Shock Syndrome Toxin-1
DEBRA A. BLOMSTER-HAUTAMAA AND PATRICK M. SCHLIEVERT 37
7. Purification of Staphylococcal Enterotoxins
JOHN J. IANDOLO AND RODNEY K. TWETEN 43
8. Production, Purification, and Assay of Streptolysin O
JOSEPH E. ALOUF AND CHRISTIANE GEOFFROY 52
9. Production, Purification, and Assay of Streptolysin S
JOSEPH E. ALOUF AND CATHERINE LORIDAN 59
10. Production, Purification, and Assay of Streptococcal Erythrogenic Toxin
CHRISTIANE GEOFFROY AND JOSEPH E. ALOUF 64

B. Gram-Positive Bacilli

11. Diphtheria Toxin: Purification and Properties
STEPHEN F. CARROLL, JOSEPH T. BARBIERI, AND R. JOHN COLLIER 68
12. Isolation and Characterization of the *Botulinum* Neurotoxins
LANCE L. SIMPSON, JAMES J. SCHMIDT, AND JOHN L. MIDDLEBROOK 76
13. Purification of Tetanus Toxin and Its Major Peptides JOHN P. ROBINSON 85
14. Purification of Alpha Toxin from *Clostridium perfringens*: Phospholipase C
COLETTE JOLIVET-REYNAUD, HERVÉ MOREAU, AND JOSEPH E. ALOUF 91
15. Production, Purification, and Assay of *Clostridium perfringens* Enterotoxin
JAMES L. MCDONEL AND BRUCE A. MCCLANE 94
16. Production and Purification of Anthrax Toxin STEPHEN H. LEPPLA 103

C. Gram-Negative Cocci

17. Production and Isolation of Neissereal IgA Proteases ANDREW G. PLAUT 117
18. Purification of Pertussis Toxin V. Y. PERERA 120

D. Gram-Negative Bacilli

19. Purification of *Escherichia coli* Heat-Stable Enterotoxin
RICHARD N. GREENBERG AND ABDUL M. K. SAEED 126

20. Purification of *Escherichia coli* α-Hemolysin
G. A. BOHACH, S. J. CAVALIERI, AND I. S. SNYDER 137
21. Preparation of *Pseudomonas* Exotoxin A
KENNETH J. KOZAK AND CATHARINE B. SAELINGER 147
22. Shiga Toxin: Production and Purification
GERALD T. KEUSCH, ARTHUR DONOHUE-ROLFE, MARY JACEWICZ, AND ANNE V. KANE 152
23. Preparation of *Yersinia pestis* Plague Murine Toxin THOMAS C. MONTIE 162

E. *Vibrios*

24. Production and Purification of Cholera Toxin JOHN J. MEKALANOS 169
25. Cytolytic Toxins of *Vibrio vulnificus* and *Vibrio damsela*
ARNOLD S. KREGER, MAHENDRA H. KOTHARY, AND LARRY D. GRAY 176
26. Thermostable Direct Hemolysin of *Vibrio parahaemolyticus*
YOSHIFUMI TAKEDA 189
27. Aerolysin from *Aeromonas hydrophila*
J. THOMAS BUCKLEY AND S. PETER HOWARD 193
28. Preparation of Hemolysin from *Aeromonas*
ÅSA LJUNGH AND TORKEL WADSTRÖM 200

F. Immunotoxins

29. Preparation of Diphtheria Toxin Fragment A Coupled to Hormone
THOMAS N. OELTMANN AND RONALD G. WILEY 204

Section II. Assay of Toxins

30. Assay of Hemolytic Toxins ALAN W. BERNHEIMER 213
31. Diphtheria Toxin: Quantification and Assay
STEPHEN F. CARROLL AND R. JOHN COLLIER 218
32. Use of Exotoxin A to Inhibit Protein Synthesis CATHARINE B. SAELINGER 226
33. Shiga Toxin as Inhibitor of Protein Synthesis
ARTHUR DONOHUE-ROLFE, MARY JACEWICZ, AND GERALD T. KEUSCH 231
34. [^{32}P]ADP-Ribosylation of Proteins Catalyzed by Cholera Toxin and Related Heat-Labile Enterotoxins D. MICHAEL GILL AND MARILYN WOOLKALIS 235
35. Assay and Purification of Cytosolic Factor Required for Cholera Toxin Activity
MARILYN WOOLKALIS, D. MICHAEL GILL, AND JENIFER COBURN 246
36. Selective Destruction of Vagal Sensory Neurons Using *Shigella* Cytotoxin
RONALD G. WILEY, ARTHUR DONOHUE-ROLFE, GERALD T. KEUSCH, AND THOMAS N. OELTMANN 250
37. Measurement of Neurotoxic Actions on Mammalian Nerve Impulse Conduction
JONATHAN J. LIPMAN 254
38. Electroencephalogram as a Tool for Assaying Neurotoxicity
JONATHAN J. LIPMAN 270
39. Assay of Pore-Forming Toxins in Cultured Cells Using Radioisotopes
MONICA THELESTAM 278
40. Incorporation of Toxin Pores into Liposomes
ROSWITHA FÜSSLE, JØRGEN TRANUM-JENSEN, ANDREAS SZIEGOLEIT, AND SUCHARIT BHAKDI 285

41. Assay Methods for Alpha Toxin from *Clostridium perfringens*: Phospholipase C
COLETTE JOLIVET-REYNAUD, HERVÉ MOREAU, AND JOSEPH E. ALOUF 293
42. Phospholipase Activity of Bacterial Toxins
LARRY W. DANIEL, LYNN KING, AND MARTHA KENNEDY 298
43. Detection of Endotoxin by Enhancement with Toxic Shock Syndrome Toxin-1 (TSST-1) G. A. BOHACH AND P. M. SCHLIEVERT 302
44. Ileal Loop Fluid Accumulation Test for Diarrheal Toxins
MERLIN S. BERGDOLL 306
45. Monkey Feeding Test for Staphyloccocal Enterotoxin
MERLIN S. BERGDOLL 324
46. Assays for Epidermolytic Toxin of *Staphylococcus aureus*
JOYCE C. S. DE AZAVEDO AND JOHN P. ARBUTHNOTT 333
47. Immunochemical Assays for Toxic Shock Syndrome Toxin-1
PATRICK M. SCHLIEVERT 339
48. Chick Embryo Assay for Staphylococcal Toxic Shock Syndrome Toxin-1
JOYCE C. S. DE AZAVEDO, ROGER N. LUCKEN, AND JOHN P. ARBUTHNOTT 344
49. Photoreactive Lipids for the Study of Membrane-Penetrating Toxins
CESARE MONTECUCCO 347
50. Electron Microscopy: Assays Involving Negative Staining
JØRGEN TRANUM-JENSEN 357
51. Electron Microscopy: Assays Involving Freeze-Fracture and Freeze-Etching
JØRGEN TRANUM-JENSEN 374
52. Two-Dimensional Crystals of Tetanus Toxin
J. REIDLER AND J. P. ROBINSON 389

Addenda

Addendum to Article 20
G. A. BOHACH, S. J. CAVALIERI, AND I. S. SNYDER 399
Addendum to Article 22
GERALD T. KEUSCH, ARTHUR DONOHUE-ROLFE, MARY JACEWICZ, AND ANNE V. KANE 399

VOLUME 166
BRANCHED-CHAIN AMINO ACIDS

Section I. Analytical and Synthetic Methods

1. Determination of Branched-Chain Amino and Keto Acids with Leucine Dehydrogenase GEOFFREY LIVESEY AND PATRICIA LUND 3
2. Analysis of Phosphorylation Sites on Branched-Chain 2-Keto Acid Dehydrogenase Complex
STEPHEN J. YEAMAN, KENNETH G. COOK, ANDREW P. BRADFORD, AND SHEELAGH M. A. JONES 10
3. High-Performance Liquid Chromatographic Separation of (3*R*)- and (3*S*)-β-Leucine Using Marfey's Reagent
D. JOHN ABERHART AND JACQUES-ALAIN COTTING 14
4. High-Performance Liquid Chromatographic Separation of α- and β-Leucine
D. JOHN ABERHART 17

5. Quantitative High-Performance Liquid Chromatographic Analysis of Branched-Chain 2-Keto Acids in Biological Samples
ANGELA L. GASKING, WILLIAM T. E. EDWARDS, ANTHONY HOBSON-FROHOCK, MARINOS ELIA, AND GEOFFREY LIVESEY 20
6. Determination of Branched-Chain 2-Hydroxy and 2-Keto Acids by Mass Spectrometry ORVAL A. MAMER AND JANE A. MONTGOMERY 27
7. Measurement of Plasma and Tissue Levels of Branched-Chain α-Keto Acids by Gas–Liquid Chromatography
PAMELA L. CROWELL, ROBERT H. MILLER, AND ALFRED E. HARPER 39
8. Determination of Methylmalonic Acid in Biological Fluids by Mass Spectrometry
JANE A. MONTGOMERY AND ORVAL A. MAMER 47
9. Analysis of Acyl-Coenzyme A Esters in Biological Samples
BARBARA E. CORKEY 55
10. Determination of Short-Chain Coenzyme A Compounds by Reversed-Phase High-Performance Liquid Chromatography
M. TODD KING, PAUL D. REISS, AND NEAL W. CORNELL 70
11. Radiochemical High-Performance Liquid Chromatography Methods for the Study of Branched-Chain Amino Acid Metabolism
KIM BARTLETT AND ANTHONY G. CAUSEY 79
12. Synthesis of α-Isopropylmalate, β-Isopropylmalate, and Dimethylcitraconate
JOHN V. SCHLOSS, RON MAGOLDA, AND MARK EMPTAGE 92
13. Utilization of Sulfometuron Methyl, an Acetolactate Synthase Inhibitor, in Molecular Biological and Metabolic Studies of Plants and Microbes
ROBERT A. LAROSSA AND TINA K. VAN DYK 97
14. Antibodies against Branched-Chain α-Keto Acid Dehydrogenase Proteins for Use in Defining Human Mutations and Gene Isolation
DEAN J. DANNER, LOUIS J. ELSAS, AND STUART LITWER 108
15. Inhibition of Branched-Chain α-Keto Acid Dehydrogenase Kinase by α-Chloroisocaproate
ROBERT A. HARRIS, MARTHA J. KUNTZ, AND RONALD SIMPSON 114

Section II. Enzyme Assay

16. Measurement of Relative Carbon Flux in α- and β-Keto Pathways of Leucine Metabolism J. MICHAEL POSTON 127
17. Assay of Leucine 2,3-Aminomutase J. MICHAEL POSTON 130
18. Enzyme Assays with Mutant Cell Lines of Maple Syrup Urine Disease
DAVID T. CHUANG AND RODY P. COX 135
19. Assays for E1 and E2 Components of Branched-Chain Keto Acid Dehydrogenase Complex DAVID T. CHUANG 146
20. Mutant Isovaleryl-CoA Dehydrogenase in Isovaleric Acidemia Cells: Assay of Activity and Molecular Characterization
YASUYUKI IKEDA AND KAY TANAKA 155
21. Assay of Branched-Chain α-Keto Acid Dehydrogenase Kinase in Mitochondrial Extracts and Purified Branched-Chain α-Keto Acid Dehydrogenase Complexes JOSEPH ESPINAL, MARK BEGGS, AND PHILIP J. RANDLE 166
22. Assay of Total Complex and Activity State of Branched-Chain α-Keto Acid Dehydrogenase Complex and of Activator Protein in Mitochondria, Cells, and Tissues
PHILIP A. PATSTON, JOSEPH ESPINAL, MARK BEGGS, AND PHILIP J. RANDLE 175

23. Determination of Activity and Activity State of Branched-Chain α-Keto Acid Dehydrogenase in Rat Tissues
GARY W. GOODWIN, BEI ZHANG, RALPH PAXTON, AND ROBERT A. HARRIS 189
24. Estimation of Branched-Chain α-Keto Acid Dehydrogenase Activation in Mammalian Tissues
KEVIN P. BLOCK, R. PAUL AFTRING, MARIA GORDON BUSE, AND ALFRED E. HARPER 201
25. Assay of 3-Methylglutaconyl-CoA Hydratase KENNETH M. GIBSON 214
26. Assay of 3-Hydroxy-3-Methylglutaryl-CoA Lyase KENNETH M. GIBSON 219
27. Permeabilized Cell and Radiochemical Assays for β-Isopropylmalate Dehydrogenase LILLIE L. SEARLES AND JOSEPH M. CALVO 225
28. Rapid Assay of Acetolactate Synthase in Permeabilized Bacteria
JULIUS H. JACKSON 230
29. Assay of Products of Acetolactate Synthase
NATAN GOLLOP, ZE'EV BARAK, AND DAVID M. CHIPMAN 234
30. Detection of the Acetolactate Synthase Isozymes I and III of *Escherichia coli* K12 MAURILIO DE FELICE, GIOVANNA GRIFFO, CARMINE T. LAGO, DANILA LIMAURO, AND EZIO RICCA 241
31. Transport of Branched-Chain Amino Acids in *Escherichia coli*
TAMMY K. ANTONUCCI AND DALE L. OXENDER 245
32. Transport of Branched-Chain Amino Acids and Their Corresponding 2-Keto Acids by Mammalian Cells
MICHAEL S. KILBERG AND MARY E. HANDLOGTEN 252
33. Leucine-tRNA Ligase Complexes ARNOLD HAMPEL AND RICHARD TRITZ 260

Section III. Enzymes

34. Purification of Branched-Chain-Amino-Acid Aminotransferase from Pig Heart
T. K. KORPELA 269
35. Pancreatic Branched-Chain-Amino-Acid Aminotransferase RYO KIDO 275
36. Isolation and Characterization of Leucine Dehydrogenase from *Bacillus subtilis*
GEOFFREY LIVESEY AND PATRICIA LUND 282
37. Purification and Assay of α-Ketoisocaproate Dioxygenase from Rat Liver
P. J. SABOURIN AND L. L. BIEBER 288
38. Isolation of Branched-Chain α-Keto Acid Dehydrogenase as Active Complex from Bovine Liver DEAN J. DANNER AND SUE C. HEFFELFINGER 298
39. Purification, Resolution, and Reconstitution of Branched-Chain 2-Keto Acid Dehydrogenase Complex from Bovine Kidney
KENNETH G. COOK AND STEPHEN J. YEAMAN 303
40. Branched-Chain α-Keto Acid Dehydrogenase Complex from Bovine Kidney
FLORA H. PETTIT AND LESTER J. REED 309
41. Branched-Chain α-Keto Acid Dehydrogenase and Its Kinase from Rabbit Liver and Heart RALPH PAXTON 313
42. Branched-Chain α-Keto Acid Dehydrogenase Phosphatase and Its Inhibitor Protein from Bovine Kidney ZAHI DAMUNI AND LESTER J. REED 321
43. Isolation and Properties of the Branched-Chain 2-Keto Acid and Pyruvate Dehydrogenase Multienzyme Complex from *Bacillus subtilis*
RICHARD N. PERHAM AND PETER N. LOWE 330
44. Purification of Branched-Chain Keto Acid Dehydrogenase and Lipoamide Dehydrogenase-Valine from *Pseudomonas* JOHN R. SOKATCH 342

45. Cloning of Genes for Branched-Chain Keto Acid Dehydrogenase in *Pseudomonas putida* P. J. SYKES AND JOHN R. SOKATCH 350
46. 2-Methyl Branched-Chain Acyl-CoA Dehydrogenase from Rat Liver YASUYUKI IKEDA AND KAY TANAKA 360
47. Isovaleryl-CoA Dehydrogenase from Rat Liver YASUYUKI IKEDA AND KAY TANAKA 374
48. Purification of Methylmalonate-Semialdehyde Dehydrogenase from *Pseudomonas aeruginosa* PAO KENNETH HATTER AND JOHN R. SOKATCH 389
49. D-Methylmalonyl-CoA Hydrolase ROBIN J. KOVACHY, SALLY P. STABLER, AND ROBERT H. ALLEN 393
50. DL-Methylmalonyl-CoA Epimerase from Rat Liver SALLY P. STABLER AND ROBERT H. ALLEN 400
51. L-Methylmalonyl-CoA Mutase from Human Placenta J. FRED KOLHOUSE, SALLY P. STABLER, AND ROBERT H. ALLEN 407
52. α-Isopropylmalate Synthase from Yeast GUNTER B. KOHLHAW 414
53. Isopropylmalate Dehydratase from Yeast GUNTER B. KOHLHAW 423
54. β-Isopropylmalate Dehydrogenase from Yeast GUNTER B. KOHLHAW 429
55. Purification and Assays of Acetolactate Synthase I from *Escherichia coli* K12 LILLIAN EOYANG AND PHILIP M. SILVERMAN 432
56. Acetolactate Synthase Isozyme II from *Salmonella typhimurium* JOHN V. SCHLOSS AND DREW E. VAN DYK 445
57. Acetolactate Synthase Isozyme III from *Escherichia coli* ZE'EV BARAK, JOSEPH M. CALVO, AND JOHN V. SCHLOSS 455

Section IV. Use of Animals and Animal Organs in the Study of Branched-Chain Amino Acid Metabolism

58. Amino Acid Control of Intracellular Protein Degradation GLENN E. MORTIMORE AND A. REETA PÖSÖ 461
59. Use of Rat Hindquarter Preparations in Studies of Branched-Chain Amino Acid Metabolism E. JACK DAVIS, SUNG-HEE CHO LEE, ØYSTEIN SPYDEVOLD, AND JON BREMER 476
60. Measurement of Branched-Chain α-Keto Acid Dehydrogenase Flux Rates in Perfused Heart and Liver DAVID S. LAPOINTE, ELLEN HILDEBRANDT, DENIS B. BUXTON, TARUN B. PATEL, PARVAN P. WAYMACK, AND MERLE S. OLSON 484

VOLUME 167
CYANOBACTERIA

Section I. Isolation, Identification, and Culturing of Cyanobacteria

1. Isolation and Purification of Cyanobacteria ROSMARIE RIPPKA 3
2. Recognition and Identification of Cyanobacteria ROSMARIE RIPPKA 28
3. Culturing Methods for Cyanobacteria RICHARD W. CASTENHOLZ 68
4. Purification and Storage of Nitrogen-Fixing Filamentous Cyanobacteria C. PETER WOLK 93

5. Thermophilic Cyanobacteria: Special Problems RICHARD W. CASTENHOLZ 96
6. Isolation and Growth of Marine Planktonic Cyanobacteria
JOHN B. WATERBURY AND JOANNE M. WILLEY 100
7. Isolation and Culture of Marine Nitrogen-Fixing Unicellular Cyanobacteria *Synechococcus* AKIRA MITSUI AND SHULI CAO 105
8. Symbiotic Associations J. C. MEEKS 113
9. Cell Stability and Storage S. EDWARD STEVENS, JR. 122
10. New Taxonomic Methods: DNA/DNA Hybridization
W. T. STAM AND B. K. STULP 125
11. 16 S Ribosomal RNA Cataloging
ERKO STACKEBRANDT AND WOLFGANG LUDWIG 132
12. Reverse Transcriptase Sequencing of Ribosomal RNA for Phylogenetic Analysis
DAVID J. LANE, KATHARINE G. FIELD, GARY J. OLSEN, AND NORMAN R. PACE 138
13. Protein Sequences as Taxonomic Probes of Cyanobacteria
ALASTAIR AITKEN 145

Section II. Ultrastructure, Inclusions, and Differentiation

14. Fine Structure of Cyanobacteria G. STANIER (COHEN-BAZIRE) 157
15. Cell Walls and External Layers
JÜRGEN WECKESSER AND UWE-JOHANNES JÜRGENS 173
16. Cyanobacterial Fimbriae TIMO VAARA AND MARTTI VAARA 189
17. Inclusions: Granules of Polyglucose, Polyphosphate, and Poly-β-hydroxybutyrate J. M. SHIVELY 195
18. Inclusions: Carboxysomes J. M. SHIVELY 204
19. Inclusions: Cyanophycin MARY MENNES ALLEN 207
20. Gas Vesicles: Chemical and Physical Properties P. K. HAYES 213
21. Cellular Differentiation: Akinetes MICHAEL HERDMAN 222
22. Cellular Differentiation: Hormogonia and Baeocytes
MICHAEL HERDMAN AND ROSMARIE RIPPKA 232

Section III. Membranes, Pigments, Redox Reactions, and Nitrogen Fixation

23. Isolation of Cyanobacterial Plasma Membranes
NORIO MURATA AND TATSUO OMATA 245
24. Membrane Lipids NAOKI SATO AND NORIO MURATA 251
25. Rapid Permeabilization of *Anacystis nidulans* to Electrolytes
GEORGE C. PAPAGEORGIOU 259
26. Photosystem I and Photosystem II Preparations from Thermophilic *Synechococcus* SAKAE KATOH 263
27. Prenylquinones SAKAE KATOH AND KAZUHIKO SATOH 269
28. Oxygen-Evolving Photosystem II Particles from *Phormidium laminosum*
DEREK S. BENDALL, JANE M. BOWES, ALISON C. STEWART, AND MARY E. TAYLOR 272
29. *In Vivo* Effect of Calcium on Photosystem II
JERRY J. BRAND AND DAVID W. BECKER 280
30. Photosystem II–Phycobilisome Complex Preparations
E. GANTT, J. D. CLEMENT-METRAL, AND B. M. CHERESKIN 286

31. Phycobiliproteins ALEXANDER N. GLAZER 291
32. Phycobilisomes ALEXANDER N. GLAZER 304
33. Phycobilisome Stability TETZUYA KATOH 313
34. Complementary Chromatic Adaptation: Physiological Conditions and Action Spectra N. TANDEAU DE MARSAC AND J. HOUMARD 318
35. *In Vitro* Carotenoid Biosynthesis in *Aphanocapsa* GERHARD SANDMANN 329
36. Carotene Oxygenase in *Microcystis* FRIEDRICH JÜTTNER 336
37. Cytochrome b_6/f Complex in Cyanobacteria: Common Components of Respiratory and Photosynthetic Electron Transport Systems RICHARD MALKIN 341
38. Preparation of Coupling Factor F_1-ATPase from Cyanobacteria DAVID B. HICKS 350
39. Respiratory Proton Extrusion and Plasma Membrane Energization G. A. PESCHEK, W. H. NITSCHMANN, AND T. CZERNY 361
40. Anoxygenic Photosynthetic Electron Transport SHIMSHON BELKIN, YOSEPHA SHAHAK, AND ETANA PADAN 380
41. Soluble Cytochromes and Ferredoxins HIROSHI MATSUBARA AND KEISHIRO WADA 387
42. Thioredoxin System PETER ROWELL, ALLAN J. DARLING, AND WILLIAM D. P. STEWART 410
43. Ferredoxin/Thioredoxin System NANCY A. CRAWFORD, BOIHON C. YEE, MICHEL DROUX, DONALD E. CARLSON, AND BOB B. BUCHANAN 415
44. Electron Paramagnetic Resonance Characterization of Iron–Sulfur Proteins R. CAMMACK 427
45. Characterization of Cytochrome-*c* Oxidase in Isolated and Purified Plasma and Thylakoid Membranes from Cyanobacteria GÜNTER A. PESCHEK, VÉRONIQUE MOLITOR, MARIA TRNKA, MARNIK WASTYN, AND WALTER ERBER 437
46. Electron Paramagnetic Resonance-Detectable Cu^{2+} in *Synechococcus* 6301 and 6311: aa_3-Type Cytochrome-*c* Oxidase of Cytoplasmic Membrane IAN V. FRY AND GÜNTER A. PESCHEK 450
47. Nitrogen and Hydrogen Metabolism: Induction and Measurement HELMAR ALMON AND PETER BÖGER 459
48. Measurement of Hydrogen Exchange and Nitrogen Uptake by Mass Spectrometry BENT B. JENSEN AND RAYMOND P. COX 467
49. Nitrogen Fixation in Cyanobacterial Mats LUCAS J. STAL 474
50. Nitrogen Fixation by Synchronously Growing Unicellular Aerobic Nitrogen-Fixing Cyanobacteria AKIRA MITSUI AND SHUZO KUMAZAWA 484
51. Isolation of Metabolically Active Heterocysts from Cyanobacteria RUSSELL L. SMITH, CHASE VAN BAALEN, AND F. ROBERT TABITA 490
52. Electron Donation to Nitrogenase in Heterocysts HERMANN BOTHE AND GABRIELE NEUER 496
53. Hydrogenases: Isolation and Assay K. K. RAO AND D. O. HALL 501
54. Measurement of Bioenergetics Phenomena in Cyanobacteria with Magnetic Resonance Techniques ROLF J. MEHLHORN AND RICHARD SULLIVAN 509

Section IV. Physiology and Metabolism

55. Buoyancy and Suspension of Planktonic Cyanobacteria R. L. OLIVER AND A. E. WALSBY 521
56. Osmotic Adjustment: Organic Solutes ROBERT H. REED 528

57. Inorganic Carbon Uptake by Cyanobacteria
AARON KAPLAN, YEHOUDA MARCUS, AND LEONORA REINHOLD 534
58. Carbon Metabolism: Studies with Radioactive Tracers *in Vivo*
JAMES A. BASSHAM AND SCOTT E. TAYLOR 539
59. Transport of Carbohydrates into Cyanobacteria
ARNOLD J. SMITH AND JOHN E. MORE 551
60. Ammonium Transport JANE GIBSON 557
61. Cation Transport in Cyanobacteria
ETANA PADAN AND ALEXANDER VITTERBO 561
62. Sulfur Metabolism in Cyanobacteria AHLERT SCHMIDT 572
63. Cyclic Nucleotides MICHAEL HERDMAN AND KHALIL ELMORJANI 584
64. Cyanobacterial DNA-Dependent RNA Polymerase
GEORGE BORBELY AND GEORGE J. SCHNEIDER 592
65. Extracellular Proteins D. J. SCANLAN AND N. G. CARR 599
66. Geosmin Production HELGA NAES 605
67. Quantitative Trace Analysis of Volatile Organic Compounds
FRIEDRICH JÜTTNER 609
68. Cyanobacterial Flocculants YESHAYA BAR-OR AND MOSHE SHILO 616
69. Cyanobacterial Heat-Shock Proteins and Stress Responses
GEORGE BORBELY AND GYULA SURANYI 622
70. Immobilization Methods for Cyanobacteria in Solid Matrices
M. BROUERS, D. J. SHI, AND D. O. HALL 629

Section V. General Physical Methods

71. Microsensors BO BARKER JØRGENSEN AND NIELS PETER REVSBECH 639
72. Determination of Turgor Pressure and Other Cell–Solute Relations by Using Gas Vesicles as Pressure Probes A. E. WALSBY 660
73. Energetic Status of Cyanobacteria: ^{31}P-NMR Spectra
LESTER PACKER, SUSAN SPATH, RICHARD BLIGNY, JEAN-BAPTISTE MARTIN, AND CLAUDE ROBY 667
74. Electron Spin Resonance Oximetry
SHIMSHON BELKIN, ROLF J. MEHLHORN, AND LESTER PACKER 670
75. Determination of pH Gradients in Intact Cyanobacteria by Electron Spin Resonance Spectroscopy SHIMSHON BELKIN AND LESTER PACKER 677
76. Mass Spectrometric Measurement of Photosynthetic and Respiratory Oxygen Exchange BERNARD DIMON, PIERRE GANS, AND GILLES PELTIER 686
77. Laser Light Scattering Techniques for Determining Size and Surface Charges of Membrane Vesicles from Cyanobacteria
MARIE-EMMANUEL RIVIÈRE, GEORGES JOHANNIN, DIDIER GAMET, VÉRONIQUE MOLITOR, GÜNTER A. PESCHEK, AND BERNARD ARRIO 691

Section VI. Molecular Genetics

78. DNA Transformation RONALD D. PORTER 703
79. Transformation in *Synechocystis* PCC 6714 and 6803: Preparation of Chromosomal DNA FRANÇOISE JOSET 712
80. Mutagenesis of Cyanobacteria by Classical and Gene-Transfer-Based Methods
SUSAN S. GOLDEN 714

81. Selection for Mutants in Energetic Pathways: Enrichment Procedures
FRANÇOISE JOSET 728
82. Use of Reporter Genes in Cyanobacteria DEVORAH FRIEDBERG 736
83. Conjugal Transfer of DNA to Cyanobacteria
JEFF ELHAI AND C. PETER WOLK 747
84. Isolation of Genes Encoding Components of Photosynthetic Apparatus
DONALD A. BRYANT AND NICOLE TANDEAU DE MARSAC 755
85. Construction of Specific Mutations in Photosystem II Photosynthetic Reaction Center by Genetic Engineering Methods in *Synechocystis* 6803
JOHN G. K. WILLIAMS 766
86. Isolation and Characterization of Genes for Cytochrome b_6/f Complex
TOIVO KALLAS AND RICHARD MALKIN 779
87. Genes for ATP Synthase in *Synechococcus* 6301
ALISON L. COZENS AND JOHN E. WALKER 794
88. Tests on *nif* Probes and DNA Hybridizations
CLAUDINE FRANCHE AND THIERRY DAMERVAL 803
89. Cyanobacterial Genetic Tools: Current Status
JEAN HOUMARD AND NICOLE TANDEAU DE MARSAC 808

Addendum

Addendum to Article 22 MICHAEL MERDMAN AND ROSMARIE RIPPKA 851

METHODS IN ENZYMOLOGY

VOLUME I. Preparation and Assay of Enzymes
Edited by SIDNEY P. COLOWICK AND NATHAN O. KAPLAN

VOLUME II. Preparation and Assay of Enzymes
Edited by SIDNEY P. COLOWICK AND NATHAN O. KAPLAN

VOLUME III. Preparation and Assay of Substrates
Edited by SIDNEY P. COLOWICK AND NATHAN O. KAPLAN

VOLUME IV. Special Techniques for the Enzymologist
Edited by SIDNEY P. COLOWICK AND NATHAN O. KAPLAN

VOLUME V. Preparation and Assay of Enzymes
Edited by SIDNEY P. COLOWICK AND NATHAN O. KAPLAN

VOLUME VI. Preparation and Assay of Enzymes (*Continued*)
Preparation and Assay of Substrates
Special Techniques
Edited by SIDNEY P. COLOWICK AND NATHAN O. KAPLAN

VOLUME VII. Cumulative Subject Index
Edited by SIDNEY P. COLOWICK AND NATHAN O. KAPLAN

VOLUME VIII. Complex Carbohydrates
Edited by ELIZABETH F. NEUFELD AND VICTOR GINSBURG

VOLUME IX. Carbohydrate Metabolism
Edited by WILLIS A. WOOD

VOLUME X. Oxidation and Phosphorylation
Edited by RONALD W. ESTABROOK AND MAYNARD E. PULLMAN

VOLUME XI. Enzyme Structure
Edited by C. H. W. HIRS

VOLUME XII. Nucleic Acids (Parts A and B)
Edited by LAWRENCE GROSSMAN AND KIVIE MOLDAVE

VOLUME XIII. Citric Acid Cycle
Edited by J. M. LOWENSTEIN

VOLUME XIV. Lipids
Edited by J. M. LOWENSTEIN

VOLUME XV. Steroids and Terpenoids
Edited by RAYMOND B. CLAYTON

VOLUME XVI. Fast Reactions
Edited by KENNETH KUSTIN

VOLUME XVII. Metabolism of Amino Acids and Amines (Parts A and B)
Edited by HERBERT TABOR AND CELIA WHITE TABOR

VOLUME XVIII. Vitamins and Coenzymes (Parts A, B, and C)
Edited by DONALD B. MCCORMICK AND LEMUEL D. WRIGHT

VOLUME XIX. Proteolytic Enzymes
Edited by GERTRUDE E. PERLMANN AND LASZLO LORAND

VOLUME XX. Nucleic Acids and Protein Synthesis (Part C)
Edited by KIVIE MOLDAVE AND LAWRENCE GROSSMAN

VOLUME XXI. Nucleic Acids (Part D)
Edited by LAWRENCE GROSSMAN AND KIVIE MOLDAVE

VOLUME XXII. Enzyme Purification and Related Techniques
Edited by WILLIAM B. JAKOBY

VOLUME XXIII. Photosynthesis (Part A)
Edited by ANTHONY SAN PIETRO

VOLUME XXIV. Photosynthesis and Nitrogen Fixation (Part B)
Edited by ANTHONY SAN PIETRO

VOLUME XXV. Enzyme Structure (Part B)
Edited by C. H. W. HIRS AND SERGE N. TIMASHEFF

VOLUME XXVI. Enzyme Structure (Part C)
Edited by C. H. W. HIRS AND SERGE N. TIMASHEFF

VOLUME XXVII. Enzyme Structure (Part D)
Edited by C. H. W. HIRS AND SERGE N. TIMASHEFF

VOLUME XXVIII. Complex Carbohydrates (Part B)
Edited by VICTOR GINSBURG

VOLUME XXIX. Nucleic Acids and Protein Synthesis (Part E)
Edited by LAWRENCE GROSSMAN AND KIVIE MOLDAVE

VOLUME XXX. Nucleic Acids and Protein Synthesis (Part F)
Edited by KIVIE MOLDAVE AND LAWRENCE GROSSMAN

VOLUME XXXI. Biomembranes (Part A)
Edited by SIDNEY FLEISCHER AND LESTER PACKER

VOLUME XXXII. Biomembranes (Part B)
Edited by SIDNEY FLEISCHER AND LESTER PACKER

VOLUME XXXIII. Cumulative Subject Index Volumes I–XXX
Edited by MARTHA G. DENNIS AND EDWARD A. DENNIS

VOLUME XXXIV. Affinity Techniques (Enzyme Purification: Part B)
Edited by WILLIAM B. JAKOBY AND MEIR WILCHEK

VOLUME XXXV. Lipids (Part B)
Edited by JOHN M. LOWENSTEIN

VOLUME XXXVI. Hormone Action (Part A: Steroid Hormones)
Edited by BERT W. O'MALLEY AND JOEL G. HARDMAN

VOLUME XXXVII. Hormone Action (Part B: Peptide Hormones)
Edited by BERT W. O'MALLEY AND JOEL G. HARDMAN

VOLUME XXXVIII. Hormone Action (Part C: Cyclic Nucleotides)
Edited by JOEL G. HARDMAN AND BERT W. O'MALLEY

VOLUME XXXIX. Hormone Action (Part D: Isolated Cells, Tissues, and Organ Systems)
Edited by JOEL G. HARDMAN AND BERT W. O'MALLEY

VOLUME XL. Hormone Action (Part E: Nuclear Structure and Function)
Edited by BERT W. O'MALLEY AND JOEL G. HARDMAN

VOLUME XLI. Carbohydrate Metabolism (Part B)
Edited by W. A. WOOD

VOLUME XLII. Carbohydrate Metabolism (Part C)
Edited by W. A. WOOD

VOLUME XLIII. Antibiotics
Edited by JOHN H. HASH

VOLUME XLIV. Immobilized Enzymes
Edited by KLAUS MOSBACH

VOLUME XLV. Proteolytic Enzymes (Part B)
Edited by LASZLO LORAND

VOLUME XLVI. Affinity Labeling
Edited by WILLIAM B. JAKOBY AND MEIR WILCHEK

VOLUME XLVII. Enzyme Structure (Part E)
Edited by C. H. W. HIRS AND SERGE N. TIMASHEFF

VOLUME XLVIII. Enzyme Structure (Part F)
Edited by C. H. W. HIRS AND SERGE N. TIMASHEFF

VOLUME XLIX. Enzyme Structure (Part G)
Edited by C. H. W. HIRS AND SERGE N. TIMASHEFF

VOLUME L. Complex Carbohydrates (Part C)
Edited by VICTOR GINSBURG

VOLUME LI. Purine and Pyrimidine Nucleotide Metabolism
Edited by PATRICIA A. HOFFEE AND MARY ELLEN JONES

VOLUME LII. Biomembranes (Part C: Biological Oxidations)
Edited by SIDNEY FLEISCHER AND LESTER PACKER

VOLUME LIII. Biomembranes (Part D: Biological Oxidations)
Edited by SIDNEY FLEISCHER AND LESTER PACKER

VOLUME LIV. Biomembranes (Part E: Biological Oxidations)
Edited by SIDNEY FLEISCHER AND LESTER PACKER

Volume LV. Biomembranes (Part F: Bioenergetics)
Edited by Sidney Fleischer and Lester Packer

Volume LVI. Biomembranes (Part G: Bioenergetics)
Edited by Sidney Fleischer and Lester Packer

Volume LVII. Bioluminescence and Chemiluminescence
Edited by Marlene A. DeLuca

Volume LVIII. Cell Culture
Edited by William B. Jakoby and Ira Pastan

Volume LIX. Nucleic Acids and Protein Synthesis (Part G)
Edited by Kivie Moldave and Lawrence Grossman

Volume LX. Nucleic Acids and Protein Synthesis (Part H)
Edited by Kivie Moldave and Lawrence Grossman

Volume 61. Enzyme Structure (Part H)
Edited by C. H. W. Hirs and Serge N. Timasheff

Volume 62. Vitamins and Coenzymes (Part D)
Edited by Donald B. McCormick and Lemuel D. Wright

Volume 63. Enzyme Kinetics and Mechanism (Part A: Initial Rate and Inhibitor Methods)
Edited by Daniel L. Purich

Volume 64. Enzyme Kinetics and Mechanism (Part B: Isotopic Probes and Complex Enzyme Systems)
Edited by Daniel L. Purich

Volume 65. Nucleic Acids (Part I)
Edited by Lawrence Grossman and Kivie Moldave

Volume 66. Vitamins and Coenzymes (Part E)
Edited by Donald B. McCormick and Lemuel D. Wright

Volume 67. Vitamins and Coenzymes (Part F)
Edited by Donald B. McCormick and Lemuel D. Wright

VOLUME 68. Recombinant DNA
Edited by RAY WU

VOLUME 69. Photosynthesis and Nitrogen Fixation (Part C)
Edited by ANTHONY SAN PIETRO

VOLUME 70. Immunochemical Techniques (Part A)
Edited by HELEN VAN VUNAKIS AND JOHN J. LANGONE

VOLUME 71. Lipids (Part C)
Edited by JOHN M. LOWENSTEIN

VOLUME 72. Lipids (Part D)
Edited by JOHN M. LOWENSTEIN

VOLUME 73. Immunochemical Techniques (Part B)
Edited by JOHN J. LANGONE AND HELEN VAN VUNAKIS

VOLUME 74. Immunochemical Techniques (Part C)
Edited by JOHN J. LANGONE AND HELEN VAN VUNAKIS

VOLUME 75. Cumulative Subject Index Volumes XXXI, XXXII, XXXIV–LX
Edited by EDWARD A. DENNIS AND MARTHA G. DENNIS

VOLUME 76. Hemoglobins
Edited by ERALDO ANTONINI, LUIGI ROSSI-BERNARDI, AND EMILIA CHIANCONE

VOLUME 77. Detoxication and Drug Metabolism
Edited by WILLIAM B. JAKOBY

VOLUME 78. Interferons (Part A)
Edited by SIDNEY PESTKA

VOLUME 79. Interferons (Part B)
Edited by SIDNEY PESTKA

VOLUME 80. Proteolytic Enzymes (Part C)
Edited by LASZLO LORAND

VOLUME 81. Biomembranes (Part H: Visual Pigments and Purple Membranes, I)
Edited by LESTER PACKER

VOLUME 82. Structural and Contractile Proteins (Part A: Extracellular Matrix)
Edited by LEON W. CUNNINGHAM AND DIXIE W. FREDERIKSEN

VOLUME 83. Complex Carbohydrates (Part D)
Edited by VICTOR GINSBURG

VOLUME 84. Immunochemical Techniques (Part D: Selected Immunoassays)
Edited by JOHN J. LANGONE AND HELEN VAN VUNAKIS

VOLUME 85. Structural and Contractile Proteins (Part B: The Contractile Apparatus and the Cytoskeleton)
Edited by DIXIE W. FREDERIKSEN AND LEON W. CUNNINGHAM

VOLUME 86. Prostaglandins and Arachidonate Metabolites
Edited by WILLIAM E. M. LANDS AND WILLIAM L. SMITH

VOLUME 87. Enzyme Kinetics and Mechanism (Part C: Intermediates, Stereochemistry, and Rate Studies)
Edited by DANIEL L. PURICH

VOLUME 88. Biomembranes (Part I: Visual Pigments and Purple Membranes, II)
Edited by LESTER PACKER

VOLUME 89. Carbohydrate Metabolism (Part D)
Edited by WILLIS A. WOOD

VOLUME 90. Carbohydrate Metabolism (Part E)
Edited by WILLIS A. WOOD

VOLUME 91. Enzyme Structure (Part I)
Edited by C. H. W. HIRS AND SERGE N. TIMASHEFF

VOLUME 92. Immunochemical Techniques (Part E: Monoclonal Antibodies and General Immunoassay Methods)
Edited by JOHN J. LANGONE AND HELEN VAN VUNAKIS

VOLUME 93. Immunochemical Techniques (Part F: Conventional Antibodies, Fc Receptors, and Cytotoxicity)
Edited by JOHN J. LANGONE AND HELEN VAN VUNAKIS

VOLUME 94. Polyamines
Edited by HERBERT TABOR AND CELIA WHITE TABOR

VOLUME 95. Cumulative Subject Index Volumes 61–74, 76–80
Edited by EDWARD A. DENNIS AND MARTHA G. DENNIS

VOLUME 96. Biomembranes [Part J: Membrane Biogenesis: Assembly and Targeting (General Methods; Eukaryotes)]
Edited by SIDNEY FLEISCHER AND BECCA FLEISCHER

VOLUME 97. Biomembranes [Part K: Membrane Biogenesis: Assembly and Targeting (Prokaryotes, Mitochondria, and Chloroplasts)]
Edited by SIDNEY FLEISCHER AND BECCA FLEISCHER

VOLUME 98. Biomembranes (Part L: Membrane Biogenesis: Processing and Recycling)
Edited by SIDNEY FLEISCHER AND BECCA FLEISCHER

VOLUME 99. Hormone Action (Part F: Protein Kinases)
Edited by JACKIE D. CORBIN AND JOEL G. HARDMAN

VOLUME 100. Recombinant DNA (Part B)
Edited by RAY WU, LAWRENCE GROSSMAN, AND KIVIE MOLDAVE

VOLUME 101. Recombinant DNA (Part C)
Edited by RAY WU, LAWRENCE GROSSMAN, AND KIVIE MOLDAVE

VOLUME 102. Hormone Action (Part G: Calmodulin and Calcium-Binding Proteins)
Edited by ANTHONY R. MEANS AND BERT W. O'MALLEY

VOLUME 103. Hormone Action (Part H: Neuroendocrine Peptides)
Edited by P. MICHAEL CONN

VOLUME 104. Enzyme Purification and Related Techniques (Part C)
Edited by WILLIAM B. JAKOBY

VOLUME 105. Oxygen Radicals in Biological Systems
Edited by LESTER PACKER

VOLUME 106. Posttranslational Modifications (Part A)
Edited by FINN WOLD AND KIVIE MOLDAVE

VOLUME 107. Posttranslational Modifications (Part B)
Edited by FINN WOLD AND KIVIE MOLDAVE

VOLUME 108. Immunochemical Techniques (Part G: Separation and Characterization of Lymphoid Cells)
Edited by GIOVANNI DI SABATO, JOHN J. LANGONE, AND HELEN VAN VUNAKIS

VOLUME 109. Hormone Action (Part I: Peptide Hormones)
Edited by LUTZ BIRNBAUMER AND BERT W. O'MALLEY

VOLUME 110. Steroids and Isoprenoids (Part A)
Edited by JOHN H. LAW AND HANS C. RILLING

VOLUME 111. Steroids and Isoprenoids (Part B)
Edited by JOHN H. LAW AND HANS C. RILLING

VOLUME 112. Drug and Enzyme Targeting (Part A)
Edited by KENNETH J. WIDDER AND RALPH GREEN

VOLUME 113. Glutamate, Glutamine, Glutathione, and Related Compounds
Edited by ALTON MEISTER

VOLUME 114. Diffraction Methods for Biological Macromolecules (Part A)
Edited by HAROLD W. WYCKOFF, C. H. W. HIRS, AND SERGE N. TIMASHEFF

VOLUME 115. Diffraction Methods for Biological Macromolecules (Part B)
Edited by HAROLD W. WYCKOFF, C. H. W. HIRS, AND SERGE N. TIMASHEFF

VOLUME 116. Immunochemical Techniques (Part H: Effectors and Mediators of Lymphoid Cell Functions)
Edited by GIOVANNI DI SABATO, JOHN J. LANGONE, AND HELEN VAN VUNAKIS

VOLUME 117. Enzyme Structure (Part J)
Edited by C. H. W. HIRS AND SERGE N. TIMASHEFF

VOLUME 118. Plant Molecular Biology
Edited by ARTHUR WEISSBACH AND HERBERT WEISSBACH

VOLUME 119. Interferons (Part C)
Edited by SIDNEY PESTKA

VOLUME 120. Cumulative Subject Index Volumes 81–94, 96–101

VOLUME 121. Immunochemical Techniques (Part I: Hybridoma Technology and Monoclonal Antibodies)
Edited by JOHN J. LANGONE AND HELEN VAN VUNAKIS

VOLUME 122. Vitamins and Coenzymes (Part G)
Edited by FRANK CHYTIL AND DONALD B. MCCORMICK

VOLUME 123. Vitamins and Coenzymes (Part H)
Edited by FRANK CHYTIL AND DONALD B. MCCORMICK

VOLUME 124. Hormone Action (Part J: Neuroendocrine Peptides)
Edited by P. MICHAEL CONN

VOLUME 125. Biomembranes (Part M: Transport in Bacteria, Mitochondria, and Chloroplasts: General Approaches and Transport Systems)
Edited by SIDNEY FLEISCHER AND BECCA FLEISCHER

VOLUME 126. Biomembranes (Part N: Transport in Bacteria, Mitochondria, and Chloroplasts: Protonmotive Force)
Edited by SIDNEY FLEISCHER AND BECCA FLEISCHER

VOLUME 127. Biomembranes (Part O: Protons and Water: Structure and Translocation)
Edited by LESTER PACKER

VOLUME 128. Plasma Lipoproteins (Part A: Preparation, Structure, and Molecular Biology)
Edited by JERE P. SEGREST AND JOHN J. ALBERS

VOLUME 129. Plasma Lipoproteins (Part B: Characterization, Cell Biology, and Metabolism)
Edited by JOHN J. ALBERS AND JERE P. SEGREST

VOLUME 130. Enzyme Structure (Part K)
Edited by C. H. W. HIRS AND SERGE N. TIMASHEFF

VOLUME 131. Enzyme Structure (Part L)
Edited by C. H. W. HIRS AND SERGE N. TIMASHEFF

VOLUME 132. Immunochemical Techniques (Part J: Phagocytosis and Cell-Mediated Cytotoxicity)
Edited by GIOVANNI DI SABATO AND JOHANNES EVERSE

VOLUME 133. Bioluminescence and Chemiluminescence (Part B)
Edited by MARLENE DELUCA AND WILLIAM D. MCELROY

VOLUME 134. Structural and Contractile Proteins (Part C: The Contractile Apparatus and the Cytoskeleton)
Edited by RICHARD B. VALLEE

VOLUME 135. Immobilized Enzymes and Cells (Part B)
Edited by KLAUS MOSBACH

VOLUME 136. Immobilized Enzymes and Cells (Part C)
Edited by KLAUS MOSBACH

VOLUME 137. Immobilized Enzymes and Cells (Part D)
Edited by KLAUS MOSBACH

VOLUME 138. Complex Carbohydrates (Part E)
Edited by VICTOR GINSBURG

VOLUME 139. Cellular Regulators (Part A: Calcium- and Calmodulin-Binding Proteins)
Edited by ANTHONY R. MEANS AND P. MICHAEL CONN

VOLUME 140. Cumulative Subject Index Volumes 102–119, 121–134

VOLUME 141. Cellular Regulators (Part B: Calcium and Lipids)
Edited by P. MICHAEL CONN AND ANTHONY R. MEANS

VOLUME 142. Metabolism of Aromatic Amino Acids and Amines
Edited by SEYMOUR KAUFMAN

VOLUME 143. Sulfur and Sulfur Amino Acids
Edited by WILLIAM B. JAKOBY AND OWEN GRIFFITH

VOLUME 144. Structural and Contractile Proteins (Part D: Extracellular Matrix)
Edited by LEON W. CUNNINGHAM

VOLUME 145. Structural and Contractile Proteins (Part E: Extracellular Matrix)
Edited by LEON W. CUNNINGHAM

VOLUME 146. Peptide Growth Factors (Part A)
Edited by DAVID BARNES AND DAVID A. SIRBASKU

VOLUME 147. Peptide Growth Factors (Part B)
Edited by DAVID BARNES AND DAVID A. SIRBASKU

VOLUME 148. Plant Cell Membranes
Edited by LESTER PACKER AND ROLAND DOUCE

VOLUME 149. Drug and Enzyme Targeting (Part B)
Edited by RALPH GREEN AND KENNETH J. WIDDER

VOLUME 150. Immunochemical Techniques (Part K: *In Vitro* Models of B and T Cell Functions and Lymphoid Cell Receptors)
Edited by GIOVANNI DI SABATO

VOLUME 151. Molecular Genetics of Mammalian Cells
Edited by MICHAEL M. GOTTESMAN

VOLUME 152. Guide to Molecular Cloning Techniques
Edited by SHELBY L. BERGER AND ALAN R. KIMMEL

VOLUME 153. Recombinant DNA (Part D)
Edited by RAY WU AND LAWRENCE GROSSMAN

VOLUME 154. Recombinant DNA (Part E)
Edited by RAY WU AND LAWRENCE GROSSMAN

VOLUME 155. Recombinant DNA (Part F)
Edited by RAY WU

VOLUME 156. Biomembranes (Part P: ATP-Driven Pumps and Related Transport: The Na,K-Pump)
Edited by SIDNEY FLEISCHER AND BECCA FLEISCHER

VOLUME 157. Biomembranes (Part Q: ATP-Driven Pumps and Related Transport: Calcium, Proton, and Potassium Pumps)
Edited by SIDNEY FLEISCHER AND BECCA FLEISCHER

VOLUME 158. Metalloproteins (Part A)
Edited by JAMES F. RIORDAN AND BERT L. VALLEE

VOLUME 159. Initiation and Termination of Cyclic Nucleotide Action
Edited by JACKIE D. CORBIN AND ROGER A. JOHNSON

VOLUME 160. Biomass (Part A: Cellulose and Hemicellulose)
Edited by WILLIS A. WOOD AND SCOTT T. KELLOGG

VOLUME 161. Biomass (Part B: Lignin, Pectin, and Chitin)
Edited by WILLIS A. WOOD AND SCOTT T. KELLOGG

VOLUME 162. Immunochemical Techniques (Part L: Chemotaxis and Inflammation)
Edited by GIOVANNI DI SABATO

VOLUME 163. Immunochemical Techniques (Part M: Chemotaxis and Inflammation)
Edited by GIOVANNI DI SABATO

VOLUME 164. Ribosomes
Edited by HARRY F. NOLLER, JR., AND KIVIE MOLDAVE

VOLUME 165. Microbial Toxins: Tools for Enzymology
Edited by SIDNEY HARSHMAN

VOLUME 166. Branched-Chain Amino Acids
Edited by ROBERT A. HARRIS AND JOHN R. SOKATCH

VOLUME 167. Cyanobacteria
Edited by LESTER PACKER AND ALEXANDER N. GLAZER

VOLUME 168. Hormone Action (Part K: Neuroendocrine Peptides)
Edited by P. MICHAEL CONN

VOLUME 169. Platelets: Receptors, Adhesion, Secretion (Part A)
Edited by JACK HAWIGER

VOLUME 170. Nucleosomes
Edited by PAUL M. WASSARMAN AND ROGER D. KORNBERG

VOLUME 171. Biomembranes (Part R: Transport Theory: Cells and Model Membranes)
Edited by SIDNEY FLEISCHER AND BECCA FLEISCHER

VOLUME 172. Biomembranes (Part S: Transport: Membrane Isolation and Characterization)
Edited by SIDNEY FLEISCHER AND BECCA FLEISCHER

VOLUME 173. Biomembranes [Part T: Cellular and Subcellular Transport: Eukaryotic (Nonepithelial) Cells]
Edited by SIDNEY FLEISCHER AND BECCA FLEISCHER

VOLUME 174. Biomembranes [Part U: Cellular and Subcellular Transport: Eukaryotic (Nonepithelial) Cells]
Edited by SIDNEY FLEISCHER AND BECCA FLEISCHER

VOLUME 175. Cumulative Subject Index Volumes 135–139, 141–167

VOLUME 176. Nuclear Magnetic Resonance (Part A: Spectral Techniques and Dynamics)
Edited by NORMAN J. OPPENHEIMER AND THOMAS L. JAMES

VOLUME 177. Nuclear Magnetic Resonance (Part B: Structure and Mechanism)
Edited by NORMAN J. OPPENHEIMER AND THOMAS L. JAMES

VOLUME 178. Antibodies, Antigens, and Molecular Mimicry
Edited by JOHN J. LANGONE

VOLUME 179. Complex Carbohydrates (Part F)
Edited by VICTOR GINSBURG

VOLUME 180. RNA Processing (Part A: General Methods)
Edited by JAMES E. DAHLBERG AND JOHN N. ABELSON

VOLUME 181. RNA Processing (Part B: Specific Methods)
Edited by JAMES E. DAHLBERG AND JOHN N. ABELSON

VOLUME 182. Guide to Protein Purification
Edited by MURRAY P. DEUTSCHER

VOLUME 183. Molecular Evolution: Computer Analysis of Protein and Nucleic Acid Sequences
Edited by RUSSELL F. DOOLITTLE

VOLUME 184. Avidin–Biotin Technology
Edited by MEIR WILCHEK AND EDWARD A. BAYER

VOLUME 185. Gene Expression Technology
Edited by DAVID V. GOEDDEL

VOLUME 186. Oxygen Radicals in Biological Systems (Part B: Oxygen Radicals and Antioxidants)
Edited by LESTER PACKER AND ALEXANDER N. GLAZER

VOLUME 187. Arachidonate-Related Lipid Mediators
Edited by ROBERT C. MURPHY AND FRANK A. FITZPATRICK

VOLUME 188. Hydrocarbons and Methylotrophy
Edited by MARY E. LIDSTROM

VOLUME 189. Retinoids (Part A: Molecular and Metabolic Aspects)
Edited by LESTER PACKER

VOLUME 190. Retinoids (Part B: Cell Differentiation and Clinical Applications)
Edited by LESTER PACKER

VOLUME 191. Biomembranes (Part V: Cellular and Subcellular Transport: Epithelial Cells)
Edited by SIDNEY FLEISCHER AND BECCA FLEISCHER

VOLUME 192. Biomembranes (Part W: Cellular and Subcellular Transport: Epithelial Cells)
Edited by SIDNEY FLEISCHER AND BECCA FLEISCHER

VOLUME 193. Mass Spectrometry
Edited by JAMES A. MCCLOSKEY

VOLUME 194. Guide to Yeast Genetics and Molecular Biology
Edited by CHRISTINE GUTHRIE AND GERALD R. FINK

VOLUME 195. Adenylyl Cyclase, G Proteins, and Guanylyl Cyclase (in preparation)
Edited by ROGER A. JOHNSON AND JACKIE D. CORBIN

VOLUME 196. Molecular Motors and the Cytoskeleton (in preparation)
Edited by RICHARD B. VALLEE

VOLUME 197. Phospholipases (in preparation)
Edited by EDWARD A. DENNIS

Subject Index

Boldface numerals indicate volume number.

A

Absorption spectroscopy
  nucleic acids, **152**, 49
Acanthamoeba castellanii
  purification of calmodulin, **139**, 50
Acer pseudoplatanus, *see* Sycamore
Acetate kinase
  and hexokinase, in glucose 6-phosphate production, **136**, 52
  immobilized
    dihydroxyacetone phosphate synthesis with, **136**, 277
    glucose 6-phosphate synthesis with, **136**, 279
    sn-glycerol 3-phosphate synthesis with, **136**, 276
Acetic acid
  with EDTA, in isolation of plague murine toxin subunits, **165**, 168
Acetobacter
  coupling to hydrous transition metal oxides, **135**, 358
Acetolactate synthase
  assay in permeabilized bacteria, **166**, 230
  inhibitor sulfometuron methyl
    hypersensitive microbial mutants, isolation, **166**, 105
    resistant microbial and plant mutants, isolation, **166**, 103
  isozyme I
    assay, **166**, 101, 241, 436
    properties, **166**, 440
    purification from *E. coli*, **166**, 438
  isozyme II
    assay, **166**, 101, 446
    properties, **166**, 451
    purification from *Salmonella typhimurium*, **166**, 448
  isozyme III
    assay, **166**, 101, 241, 455
    properties, **166**, 456
    purification from *E. coli*, **166**, 456
  reaction products, assay, **166**, 234
  structural genes, isolation from microbes and plants, **166**, 101
Acetolysis
  mannan, HPLC analysis of products, **138**, 112
Acetone
  production from glucose in aqueous two-phase system, **137**, 663
Acetylation
  glycosphingolipids, **138**, 9
  xylans, **160**, 552
    evaluation, **160**, 554
Acetylcholine receptors
  transferrin effects, myotrophic assay, **147**, 298
Acetylcholinesterase
  assay in sarcolemmal vesicles, **157**, 32
  immobilized system, oscillatory phenomena, **135**, 559
  transferrin effects, myotrophic assay, **147**, 296
Acetyl-CoA:α-glucosaminide *N*-acetyltransferase
  assay, **138**, 608
  properties, **138**, 610
  purification from rat liver, **138**, 609
Acetylene
  reduction, in cyanobacterial nitrogenase assay, **167**, 475
N-Acetylgalactosamine-4-sulfatase
  diagnostic assay for multiple sulfatase deficiency, **138**, 745
N-Acetylgalactosaminyltransferase
  glycosphingolipid-synthesizing forms, assays, **138**, 594
N-Acetylglucosamine
  removal from glucocerebrosidase before cell targeting, **149**, 35
N-Acetyl-β-glucosaminidase
  assay, **161**, 494, 524
  detergent removal from *E. coli* cells, **161**, 526
  levels in sponge implants, measurement, **162**, 330
  properties, **161**, 527
  purification from
    puffballs, **161**, 494
    soybean seeds, **161**, 492
    Vibrio harveyi, **161**, 527

N-Acetylglucosaminyltransferase
glycosphingolipid-synthesizing forms, assays, **138**, 597
β-*N*-Acetylhexosaminidase
assay, **161**, 463
diagnostic assay for G_{M2}-gangliosidosis, **138**, 755
glucocerebrosidase digestion before cell targeting, **149**, 35
as marker for chromosome detection in somatic cell hybrids, **151**, 185
properties, **161**, 467
purification from *Pycnoporus cinnabarinus*, **161**, 464
O-Acetylhomoserine sulfhydrylase
assay, **143**, 465, 470
properties, **143**, 468, 473
purification from
Brevibacterium flavum, **143**, 471
Schizosaccharomyces pombe, **143**, 466
4-Acetylphenyl sulfate
arylsulfatase assay with, **143**, 215
Acetyl phosphate
enzymatic synthesis, **136**, 267
O-Acetylserine–*O*-acetylhomoserine sulfhydrylase
assay, **143**, 479
properties, **143**, 482
purification from yeast, **143**, 480
O-Acetylserine sulfhydrylase, *see* *O*-Acetylserine (thiol)-lyase
O-Acetylserine (thiol)-lyase
assay, **143**, 474
properties, **143**, 477
purification from *Bacillus sphaericus*, **143**, 475
Acetylserotonin methyltransferase
amino acid composition, **142**, 593
assay, **142**, 591
properties, **142**, 594
purification from pineal gland, **142**, 592
Acetylxylan esterase
assay, **160**, 701
in *Streptomyces* species, **160**, 551
detection in gels, **160**, 702
properties, **160**, 706
purification from *Schizophyllum commune*, **160**, 705
Achromobacter proteinase I
immobilized
assay, **136**, 163
preparation, **136**, 165
in preparation of
human insulin, **136**, 168
porcine 30^{B}-dealanine insulin, **136**, 167
Acidemia
isovaleric, mutant isovaleryl-CoA dehydrogenase assay and properties in cells, **166**, 155
α_1-Acid glycoprotein
assay, **163**, 420
characterization, **163**, 422
chromosomal location and molecular biology, **163**, 430
electrophoretic separation of variants, **163**, 424
genetic polymorphism, **163**, 427
purification from human plasma, **163**, 420
Acidification
endosomal, fluorescence assay, **157**, 607
lysosomal
fluorescence assay, **157**, 607
radiochemical assay with gel filtration, **157**, 596
Acidolysis
in assay for lignin depolymerization by *Streptomyces*, **161**, 256
lignin analysis by, **161**, 93
lignin characterization by, **161**, 103
in preparation of dioxane lignin fractions, **161**, 31
Acid phosphatase
gelatin-immobilized, properties, **135**, 297
–glucose oxidase electrode, description, **137**, 37
as marker for chromosome detection in somatic cell hybrids, **151**, 177
Acid protease
assay, **160**, 501
properties, **160**, 506
purification from *Sporotrichum pulverulentum*, **160**, 501
Acrylamide
production by immobilized nitrilase-containing bacteria, **136**, 523
Actin
binding proteins, purification from intestinal brush border and properties, **139**, 137

in leukocytes
biochemistry, **162**, 246
intracellular behavior, **162**, 265
microfilament network, regulation, **162**, 247
purification, **162**, 250
role in locomotion, **162**, 246
modulating proteins, purification from macrophages, **162**, 254
polymerization, calmodulin effects, **139**, 852
properties in presence of modulating proteins, **162**, 260
purification from human platelets, **139**, 847
α-Actinin
preparation from macrophages, **162**, 258
Acumentin
preparation from macrophages, **162**, 257
Acylation
fatty, antibodies, for immunoliposome preparation, **149**, 89
Acyl-CoA esters
gradient system for resolution, **166**, 83
HPLC analysis in tissue samples, **166**, 55
preparation for HPLC, **166**, 86
Acyl-CoA oxidase
peroxisomal, assay, **148**, 523
Acyl-CoA synthetase
peroxisomal, assay, **148**, 521
Acylpeptides
cAMP phosphodiesterase-inhibiting
assay, **159**, 497
properties, **159**, 501
purification from *Bacillus subtilis*, **159**, 498
structure, **159**, 499
Acylsphingosine deacylase
diagnostic assay for Farber disease, **138**, 734
Adapters
oligonucleotide, ligation to cDNA, **152**, 348
primer–restriction end, in cDNA cloning, **154**, 64
Adenine nucleotides
in plant vacuoles, HPLC, **148**, 132
Adenine phosphoribosyltransferase
as marker for chromosome detection in somatic cell hybrids, **151**, 178
Adenosine cyclic 3′,5′-phosphorothioate
Rp diastereomer
inhibition of cAMP-dependent protein kinase, **159**, 163
synthesis, **159**, 160
Adenosine deaminase
as marker for chromosome detection in somatic cell hybrids, **151**, 178
Adenosine kinase
as marker for chromosome detection in somatic cell hybrids, **151**, 179
Adenosine 3′-phosphate 5′-phosphosulfate
carrier-free, ^{35}S-labeled, preparation, **143**, 348
unlabeled, preparation, **143**, 349
Adenosine 5′-phosphosulfate
carrier-free, ^{35}S-labeled, preparation, **143**, 348
Adenosylhomocysteinase
assay, **143**, 378, 430
properties, **143**, 382, 432
purification from
bovine liver, **143**, 380
yellow lupine, **143**, 431
S-Adenosylmethionine
and sulfur metabolites, HPLC, **143**, 192
Adenylate cyclase
in anthrax toxin assay, **165**, 113
assay in
cyanobacteria, **167**, 588
permeabilized adipocytes, **159**, 200
in bone cells, effects of hormones, **145**, 330
calmodulin-sensitive
purification from bovine brain, **139**, 777
reconstitution with
β-adrenergic receptors, **139**, 788
calmodulin, **139**, 787
multilamellar liposomes, **139**, 780
in skeletal muscle triads, assay, **157**, 68
in yeast cAMP cascade system mutants, assay, **159**, 35
Adenylate kinase
immobilized, 5-phospho-α-D-ribosyl pyrophosphate synthesis with, **136**, 278
as marker for chromosome detection in somatic cell hybrids, **151**, 179
thiol–disulfide interchange in, thiol pK_a values, **143**, 134

Adenylylsulfate–ammonia adenylyltransferase
assay, **143**, 354
properties, **143**, 359
purification from *Chlorella pyrenoidosa*, **143**, 357
Adenylylsulfate kinase
assays, **143**, 342
catalyzed reaction, **143**, 334
purification from *Penicillium chrysogenum*, **143**, 344
Adipocytes
digitonin-permeabilized, in cAMP studies, **159**, 193
effects of specific phosphodiesterase inhibitors, **159**, 504
rat, isolation of low-K_m cAMP phosphodiesterase, **159**, 534
synergistic responses to site-selective cAMP analogs, **159**, 123
Adjuvants
Freund's-type, for induction of arthritis, preparation, **162**, 341
Adoptive transfer
experimental allergic encephalomyelitis in rats, **162**, 418
ADP (adenosine 5′-diphosphate)
–ATP transport in mitochondria, analytical technique, **148**, 680
binding to Na^+,K^+-ATPase, assay
centrifugation method, **156**, 200
rate of dialysis method, **156**, 193
exchange reactions in sarcoplasmic reticulum vesicles, **157**, 182
phosphoryls, isolation, **159**, 68
role in coupling chemical reactions and transport of Na^+, **156**, 18
ADP-ribosylation
cholera toxin-catalyzed, analysis, **165**, 235
ADP-ribosyltransferase
diphtheria toxin activity as, assay, **165**, 222
Pseudomonas exotoxin A activity as, assay, **165**, 226
Adrenal gland
bovine, purification of
chromaffin granules, **157**, 620
dopamine β-monooxygenase, **142**, 599
tyrosine 3-monooxygenase, **142**, 60
chicken, purification of dopamine β-monooxygenase, **142**, 610
rabbit, purification of norepinephrine *N*-methyltransferase, **142**, 657
rat, purification of tyrosine 3-monooxygenase, **142**, 66
β-Adrenergic receptors
adenylate cyclase reconstitution with, **139**, 788
photoaffinity labeling study in lymphoid cells, **150**, 492
Adriamycin, *see* Doxorubicin
Adsorption
antibodies in hemophilia, extracorporeal systems, **137**, 458
chitinase on chitin, **161**, 460
enzyme immobilization by, **135**, 58
yeast cells
on solid supports
applications, **135**, 317
kinetics, **135**, 310
methodology, **135**, 308
on wood chips
in ethanol fermentation, **135**, 340
flow velocity effects, **135**, 338
methodology, **135**, 334
Aerolysin
hemolytic assay, **165**, 193
production, **165**, 194
properties, **165**, 197
purification, **165**, 195
Aeromonas
A. hydrophila
diarrheal toxin, ileal loop test, **165**, 320
purification of
aerolysin, **165**, 195
exo-1,4-β-mannanase, **160**, 585
hemolysin, **165**, 201
purification of methionine γ-lyase, **143**, 462
Affinity chromatography
α_1-acid glycoprotein, **163**, 421
adenylate cyclase, **139**, 777
Amphicarpaea bracteata lectin, **138**, 562
α_1-antitrypsin, **163**, 408
bone glycoproteins, **145**, 276
Ca^{2+}-ATPase, **139**, 793
calelectrins, **139**, 32
calmodulin, **139**, 10, 51

calmodulin-binding proteins, **139**, 851
calmodulin-kinase II, **139**, 709
calmodulin-stimulated phosphodiesterase, **159**, 573
calspermin, **139**, 131
Cancer antennarius lectin, **138**, 521
cartilage-derived factor, **146**, 316
cartilage-derived growth factor, **146**, 323
cellobiohydrolases, **160**, 188
cellooligomers, **160**, 43
ceruloplasmin, **163**, 446
cGMP-regulated phosphodiesterase, **159**, 692
chitinase, **161**, 492
chitobiase, **161**, 492
with chlorocarbonate-activated Sepharose, **135**, 89
chondrosarcoma-derived growth factor, **147**, 101
C1q receptor, **150**, 571
dihydropteridine reductase, **142**, 121
diphtheria toxin fragment A–hormone hybrid, **165**, 207
DNA-directed RNA polymerase, **167**, 595
endocellulase C, **160**, 192
endothelial cell growth factors, **147**, 97
Erythrina lectins, **138**, 546
Evonymus europaea lectin, **138**, 555
factor XII, **163**, 58
fibroblast growth factor, **147**, 110
fibronectin, **144**, 426
α-galactosidase, **160**, 629, 630
β-glucosidases, **160**, 224
glycopeptides
  concanavalin A–Sepharose system, **138**, 244
  Datura stramonium agglutinin–agarose system, **138**, 254
  E_4-phytohemagglutinin–agarose system, **138**, 249
  lentil lectin–Sepharose system, **138**, 248
  L_4-phytohemagglutinin–agarose system, **138**, 251
  overview, **138**, 232
  pea lectin–Sepharose system, **138**, 246
  Ricinus communis agglutinin I–Sepharose system, **138**, 256
haptoglobin, **163**, 459
hemopexin, **163**, 540
hepatoma-derived growth factor, **147**, 104
α_2-HS glycoprotein, **163**, 434
hypothalamus-derived growth factor, **147**, 105
immuno-, *see* Immunoaffinity chromatography
in isolation of human basophils, **162**, 525
lactose-binding lectins, **138**, 510
Limax flavus agglutinin, **138**, 529
lipocortins, **163**, 28
leukocyte elastase at subzero temperatures, **135**, 535
lymphocyte surface receptors, **150**, 467
macrocellulase complexes, **160**, 487
β-mannanase, **160**, 600
microtubule-associated proteins, **139**, 839
MSA carrier protein, **146**, 273
myosin phosphatase, **159**, 450
Na^+,K^+-ATPase, **156**, 77
pancreatic elastase at subzero temperatures, **135**, 531
pectinesterase, **161**, 368
pertussis toxin, **165**, 123
phospholamban, **157**, 364
phosphomannosyl receptor 2, **138**, 507
polygalacturonase, **161**, 372
protein A, **137**, 459
proteoglycan link protein, **144**, 405
sialoproteins, **138**, 534
S-100 protein, **139**, 14
sulfhydryl oxidase, **143**, 508
T-cell receptors, **150**, 673
transferrin, **147**, 292
transferrin receptors, **147**, 244, 254; **150**, 732
with tresyl-silica-coupled alcohol dehydrogenase, **135**, 77
vanillate hydroxylase, **161**, 278
Vicia villosa lectins, **138**, 538
Xenopus laevis lectins, **138**, 516
Affinity immobilization
  with antigen–antibody interactions, **137**, 653
  with concanavalin A–Sepharose, **137**, 650
Affinity labeling
  digitalis-binding site in Na^+,K^+-ATPase, **156**, 323

epidermal growth factor receptors, **146**, 146
growth hormone receptors, **150**, 713
insulin receptors, **150**, 711, 714
Na^+,K^+-ATPase, **156**, 310
nucleotide-binding site of myosin light chain kinase, **139**, 188
photo-, *see* Photoaffinity labeling
ribosomes, **164**, 341
transforming growth factor receptors
type α, **146**, 146
type β, **146**, 184
tRNA binding sites on ribosomes, **164**, 372

Agar
double-layer, B-cell cultures in, **150**, 257
entrapment of chloroplasts and cyanobacteria, **135**, 446
salt bridge in electrochemical reduction cell, preparation, **143**, 260
semisolid, adherent layer-dependent development of B-cell progenitors in, **150**, 251
supported nylon netting, plant cell entrapment, **135**, 414

Agarose
activation with
cyanogen bromide, **135**, 50
dabsyl chloride, **135**, 80
dansyl chloride, **135**, 81
diabsyl chloride, **135**, 82
dipsyl chloride, **135**, 81
Lissamine Rhodamine B sulfonyl chloride, **135**, 81
pentafluorobenzenesulfonyl chloride, **135**, 82
bead-forming procedure, **135**, 223
carboxyalkyl, coupling of glycoenzyme–ethylenediamine adducts, **135**, 145
entrapped animal cells
concentration, **135**, 404
growth monitoring, **135**, 405
preparation, **135**, 403
immobilization of plant protoplasts, **135**, 436
tresyl chloride-activated
chymotrypsin coupling to, **135**, 69
preparation, **135**, 67
trypsin inhibitor coupling to, **135**, 72

Agarose gel assay
for chemotaxis, attractant concentration gradients, **162**, 93
neutrophil chemotaxis
addition of cells and chemoattractants to wells, **162**, 54
cutting of wells, **162**, 52
gel preparation, **162**, 51
incubation period and termination, **162**, 55
migration patterns, **162**, 55
quantification of migratory functions, **162**, 56
special applications, **162**, 58
troubleshooting, **162**, 58

Agarose gels
cell immobilization in, **135**, 227
DNA renaturation for detection and cloning of amplified genes, **151**, 332
β-D-glucanase and β-D-xylanase detection in, **160**, 69
for nucleic acid electrophoresis
alkaline, **152**, 79
formaldehyde-containing, **152**, 82
methylmercuric hydroxide-containing, **152**, 81
nondenaturing, **152**, 77
in preparation of intact genomic DNA, **155**, 449
in specific fragmentation of genomic DNA, **155**, 463

Agarose microdroplet assay
neutrophil chemokinesis, **162**, 75

Age
effect on testicular isoforms of cyclic nucleotide phosphodiesterase, **159**, 679

Agglutination assay
with ligand-bearing liposomes
antinuclear antibodies, **149**, 210
antistreptococcal antibodies, **149**, 211
rheumatoid factors, **149**, 209
thrombospondin, **144**, 445

Agglutinins
in glycopeptide affinity chromatography
Datura stramonium-derived forms, **138**, 254
Ricinus communis-derived forms, **138**, 256

Agmenellum quadruplicatum
ESR oximetry, **167**, 671
purification of C-phycocyanin, **167**, 301
Agrobacterium tumefaciens
Ti plasmid, derived plant transformation vectors, **153**, 254
transformation of monocots, detection, **153**, 305
Air sac technique
for chemotaxis assays *in vivo*
advantages and limitations, **162**, 114
cellular infiltration as function of time, **162**, 112
general features, **162**, 107
procedures, **162**, 108
reproducibility, **162**, 111
role of C5a in chemotactic response, **162**, 112
Akinetes
cyanobacterial
germination, **167**, 228
identification, **167**, 223
isolation and purification, **167**, 226
mutational loss, **167**, 231
production in culture, **167**, 223
storage, **167**, 227
L-Alanine
industrial production by immobilized bacteria, **136**, 472
Alanine aminotransferase
assay with dry reagent chemistries, **137**, 414
immobilization and assay, **136**, 488, 490
Alanine sulfodisulfane
properties, **143**, 224
synthesis, **143**, 223
Albumin
crosslinked, immobilization of organelles, **135**, 457, 471
–enzyme conjugates, for enzyme replacement therapy, **137**, 566
protective effect on detergent inactivation of Na^+,K^+-ATPase, **156**, 76
serum, *see* Serum albumin
thiols, fluorometric assay, **143**, 74
Alcaligenes
A. eutrophus
in preparation of (*R*)-propanediol, **136**, 316
purification of prephenate dehydrogenase, **142**, 504
A. faecalis, purification of aromatic amine dehydrogenase, **142**, 651
Alcohol dehydrogenase
coimmobilized with NAD
application to ethanol analyzer, **136**, 31
assay, **136**, 28
preparation, **136**, 26, 56
coupling to 1,1′-carbonyldiimidazole-activated support, **135**, 115
–lactate dehydrogenase site-to-site complex
assays, **136**, 108
characterization, **136**, 110
immobilized system, **136**, 104
soluble system, **136**, 106
metal-depleted, preparation, **158**, 32
metal-hybrid, preparation, **158**, 87
in multienzymatic NADH recycling, **136**, 70, 74, 77
reactor
ethanol assay with, **137**, 276
preparation, **137**, 276
Sepharose-bound
absorption spectrophotometry, **135**, 547
fluorometry, **135**, 552
tresyl-silica-coupled
affinity chromatography with, **135**, 77
characterization, **135**, 73
preparation, **135**, 69
Alcohols
primary racemic, enzymatic preparation, **136**, 121
secondary racemic, enzymatic preparation, **136**, 127
Alditol acetates
partially methylated, GC–MS, **138**, 19
Algae
red, purification of phycobilisomes, **167**, 305
Algin
available types, **135**, 177
entrapment of cyanobacteria, **167**, 630
gel formation, **135**, 179
and poly(L-lysine), islet microencapsulation with, **137**, 577

solutions
cell dispersion in, **135**, 181
preparation, **135**, 181
Alkaline phosphatase
assay in bone cell cultures, **145**, 324
immobilized subunits
catalytic activity, **135**, 492
preparation, **135**, 493
properties, **135**, 499
structural analysis, **135**, 494
metal-hybrid, preparation, **158**, 89
removal of 5′-phosphates from DNA, **152**, 99
treatment of linearized vector DNA, **152**, 203
Alkaloids
production by immobilized *Claviceps purpurea*, **136**, 317
vinca, platelets as carriers
clinical applications, **149**, 316
overview, **149**, 312
prospects for use, **149**, 324
Alkanal monooxygenase (FMN-linked)
coimmobilized with other enzymes, bioluminescent assays with, **136**, 82
1-Alkyl-2-acetylglycerol cholinephosphotransferase
assay, **141**, 387
properties, **141**, 388
1-Alkyl-2-acetylglycerophosphocholine esterase
assay, **141**, 393
properties, **141**, 393
1-Alkylglycerophosphate acyltransferase
assay, **141**, 385
properties, **141**, 386
1-Alkylglycerophosphocholine acetyltransferase
assay, **141**, 382
properties, **141**, 384
1-Alkyl-2-lyso-glycerophosphocholine arachidonoyltransacylase
assay, **141**, 391
properties, **141**, 392
Allergic reactions
mediators from mast cells and/or basophils, overview, **163**, 3
Allophycocyanin
absorption spectra, **167**, 319
assay in cyanobacterial cells, **167**, 322
purification from
Anabaena spp., **167**, 295
Fremyella diplosiphon, **167**, 298
Alpha toxin
clostridial
hemolytic assay, **165**, 294
properties, **165**, 94
purification, **165**, 91
titrimetric assay, **165**, 297
turbidimetric assay, **165**, 296
staphylococcal
purification, **165**, 3
purity assessment, **165**, 7
Alumina, *see* Aluminum oxide
Aluminum
analytical techniques
atomic absorption spectrometry, **158**, 128, 293
atomic emission spectrometry, **158**, 291
chemical and physicochemical methods, **158**, 290
neutron activation analysis, **158**, 291
X-ray fluorescence, **158**, 290
contamination of, control during analysis, **158**, 301
specimen collection, **158**, 299
localization in tissues, **158**, 298
spectrometric analysis in
feces, **158**, 296
serum, **158**, 295
tissues, **158**, 296
urine, **158**, 296
Aluminum oxide
acid washing, **142**, 552
catecholamine radioimmunoassay with, **142**, 559
Amberlite
thermoase immobilization with, **136**, 507
Amides
–lithium chloride system, as solvent for chitin, **161**, 404
Amidination
pyranose oxidase, **137**, 602
Amine dehydrogenase
aromatic, *see* Aromatic amine dehydrogenase
Amine *N*-methyltransferase
assay, **142**, 661
properties, **142**, 666

purification from rabbit liver
type A, **142**, 662
type B, **142**, 665
Amine oxidase (flavin-containing)
assay, **142**, 619, 628, 647
immunocytochemical localization in brain, **142**, 649
multiple forms, **142**, 618
properties, **142**, 618, 636, 638, 648
purification from mammalian liver, **142**, 624
solubilization, **142**, 623
type A, purification from
human liver, **142**, 648
human placenta, **142**, 631
type B
–antibody complex, characterization, **142**, 647
purification from
bovine liver, **142**, 633
human liver, **142**, 645
Amine sulfotransferase
ion-pair extraction, **143**, 203
radiochemical assay, **143**, 205
Amino acid analysis
with HPLC, penicillamine, **143**, 189
Amino acid analyzer
in assay of ω-amino acid–pyruvate aminotransferase, **143**, 500
Amino acid composition
aerolysin, **165**, 199
arogenate dehydrogenase, **142**, 518
bone morphogenetic protein and polypeptides, **146**, 301
calcium-binding protein, 21-kDa form, **139**, 91
calmodulins, **139**, 58
cartilage-derived factor, **146**, 319
cellobiosidase, **160**, 396
cellulases, **160**, 322
connective tissue-activating peptides I and Ib, **163**, 747
damselysin, **165**, 188
E. coli heat-stable enterotoxin, **165**, 134
elastin
porcine soluble and mature forms, **163**, 657
vertebrate aortic forms, **144**, 177
endo-1,4-β-glucanases, **160**, 375
factor XII, **163**, 65
β-glucosidase, **160**, 413
hemopexins, **163**, 550
hydroxyindole *O*-methyltransferase, **142**, 593
isovaleryl-CoA dehydrogenase, **166**, 385
kinin analogs, **163**, 271
2-methyl branched-chain acyl-CoA dehydrogenase, **166**, 370
MSA carrier protein, **146**, 276
phenylalanine 4-monooxygenase
bovine hepatic enzyme, **142**, 42
Chromobacterium violaceum enzyme, **142**, 49
human hepatic enzyme, **142**, 33
phospholamban, **157**, 366
proliferating cell nuclear antigen/cyclin, **150**, 153
proteoglycan link protein, **144**, 407
protopectinases, **161**, 345
tissue kallikreins, **163**, 137
toxic shock syndrome toxin-1, **165**, 42
vulnificolysin, **165**, 182
xylanases, **160**, 652
ω-Amino acid–pyruvate aminotransferase
assay, **143**, 500
properties, **143**, 502
purification from *Pseudomonas*, **143**, 501
Amino acids
branched-chain
enzymatic assay, **166**, 3
metabolism
radiochemical HPLC analysis, **166**, 79
studies with rat hindquarter preparations, **166**, 476
transport in
E. coli, assay, **166**, 245
mammalian cells, assay, **166**, 252
cellular transport, defective mutants, isolation, **151**, 136
collagen-crosslinking
assays, **144**, 120
general discussion, **144**, 115
regulation of intracellular protein degradation, **166**, 461
removal from plasma, enzyme reactors for, **137**, 479
selenium-containing, metabolism in microorganisms, **143**, 459

sulfur-containing
analytical resolution, **143**, 167
assay diets for studies, **143**, 301
metabolism
in microorganisms, **143**, 453
overview, **143**, 366
in plants, **143**, 419
and synthesis, in cyanobacteria, **167**, 579
related articles in other volumes, **143**, 523
synthesis, postprephenate pathways to phenylalanine and tyrosine, overview, **142**, 472
Amino acid sequence
amino-terminal
cellulases, **160**, 322
Na^+,K^+-ATPase α subunit, **156**, 64
protopectinases, **161**, 344
branched-chain 2-keto acid dehydrogenase phosphorylation sites, **166**, 14
connective tissue-activating peptide III, **163**, 746
C-reactive protein subunit, **163**, 391
cyanobacterial ferredoxins, **167**, 400
cyanobacterial gas vesicle proteins, **167**, 218
cytochromes c_6, comparisons between cyanobacteria and chloroplasts, **167**, 151
E. coli heat-stable enterotoxin, **165**, 135
eosinophil chemotactic factors, **162**, 173
fibroblast growth factor, **147**, 132
hemopexin and heme peptide, **163**, 550
interleukin-2 receptor, **150**, 699
Na^+,K^+-ATPase, **156**, 15
oncomodulin, **139**, 209
partial
Acanthamoeba calmodulin, **139**, 61
Aplysia calmodulin, **139**, 289
phospholamban, **157**, 150
plastocyanins, comparisons between cyanobacteria and chloroplasts, **167**, 148
prediction with DNA subclones, **152**, 512
prexylanase, **160**, 637
protein S, **139**, 388
slow/cardiac troponin C, **139**, 331, 335
vitronectin, **144**, 434
Aminoacylase
as marker for chromosome detection in somatic cell hybrids, **151**, 180
(*p*-Aminobenzyldiazonio)ouabain
synthesis, **156**, 329
S-(2-Amino-2-carboxyethylsulfonyl)-L-cysteine
sulfite and thiosulfate assays with, **143**, 17
1-Aminocyclopropane-1-carboxylate synthase
assay, **143**, 426
properties, **143**, 429
purification from tomato, **143**, 428
5′-Amino-5′-deoxythymidine
synthesis, **155**, 270
Aminohexane-Sepharose 4B
–glucomannan, preparation for β-mannanase purification, **160**, 597
α-Aminoisobutyric acid
radiolabeled
in assay of pore-forming toxins in cultured cells, **165**, 280
in selection of amino acid transport-defective mutants, **151**, 136
uptake by fibroblast membrane vesicles, assay, **146**, 380
6-Aminopenicillanic acid
production from benzylpenicillin in aqueous two-phase system, **137**, 665
Aminopeptidase
apoenzyme
assay, **158**, 28
preparation, **158**, 27
titration, **158**, 29
Aminopeptidase P
assay, **136**, 171, 173
immobilization and properties, **136**, 172
sequential peptide hydrolysis with, **136**, 174
4-(Aminosulfonyl)-7-fluoro-2,1,3-benzoxadiazole
in fluorometric assay of thiols, **143**, 71
Aminotransferases
advantages and disadvantages as biocatalysts, **136**, 481
assays, **136**, 485
immobilization, **136**, 485
overview, **136**, 479
sources, **136**, 483

usefulness in amino acid production, **136**, 483
Ammonia
assay with enzyme transistor, **137**, 241
Ammonium
–Mg^{2+}–polyamine system, for poly(Phe) synthesis with near *in vivo* features, **164**, 650
quaternary bases, solvolysis of cellulose, **160**, 12
quaternary compounds in cyanobacteria, NMR analysis, **167**, 532
uptake by cyanobacteria
direct assay, **167**, 558
indirect assay, **167**, 559
Ammonium 7-fluoro-2,1,3-benzoxadiazole 4-sulfonate
in fluorometric assay of thiols, **143**, 67
Ammonium fumarate
effect on bacterial L-aspartic acid production in reactor, **136**, 470
Ammonium reineckate, *see* Reinecke salt
Ammonium sulfate
precipitation of protein-bound cyclic nucleotides, assay, **159**, 147
in purification of hemopexin, **163**, 543
Amniotic fluid
extracellular matrix proteins, analysis, **145**, 396
AMP (adenosine 5′-monophosphate)
endogenous, α-phosphoryl isolation, **159**, 69
Amphomycin
as inhibitor of lipid-linked saccharide pathway, **138**, 672
Ampicillin
synthesis with immobilized penicillin amidase, **136**, 284
Amplification
DNA libraries, **152**, 407
genes, *see* Gene amplification
α-Amylase
and amylo-1,6-glucosidase, starch degradation in aqueous two-phase system, **137**, 661
assay, **137**, 691
with biosensor, **137**, 64
with dye-labeled CM-amylose, **160**, 81
oligosaccharides, TLC analysis, **160**, 175
production by *Aspergillus*, **137**, 689, 693
Amylo-1,6-glucosidase
and α-amylase, starch degradation in aqueous two-phase system, **137**, 661
Amyloid A protein
serum, *see* Serum amyloid A protein
Amyloid P component
chemical characterization, **163**, 525
isolation from human tissues, **163**, 528
quantitation, **163**, 532
serum, *see* Serum amyloid P component
Amyloplasts
isolation from
maize endosperm
aqueous procedure, **148**, 228
nonaqueous procedure, **148**, 233
pea epicotyls, **148**, 172
rice endosperm, **148**, 172
soybean cell suspension cultures, **148**, 218
Anabaena
akinete formation in culture, **167**, 225
purification of
DNA-directed RNA polymerase, **167**, 593
ferredoxin, **167**, 409
phycobiliproteins, **167**, 295
Anacystis nidulans
isolation of
cytochromes *c*-549, *c*-552, and *c*-554, **167**, 393
DNA, **153**, 208
plasma membranes, **167**, 246
mutagenesis, **153**, 211
rapid permeabilization to electrolytes, **167**, 259
transformation, **153**, 209
Anaphylatoxin inactivator
isolation and assay, **162**, 223
properties, **162**, 224
Anaphylatoxins
generation and properties, **162**, 616
isolation, **162**, 617
Anemia
autoimmune hemolytic, treatment with platelet-delivered vinca alkaloids, **149**, 322
Angiogenesis
induction by glycyl-L-histidyl-L-lysine-Cu growth factor, **147**, 325

Angiotensin II
sequential hydrolysis with immobilized aminopeptidases, **136**, 175
Angiotensin-converting enzyme, *see* Dipeptidyl carboxypeptidase I
Aniline
assay with microsomal electrode, **137**, 155
1-Anilino-8-naphthalenesulfonate
in measurement of ionic permeability of sarcoplasmic reticulum vesicles, **157**, 455
Anthranilate phosphoribosyltransferase
–anthranilate synthase, *see* Anthranilate synthase–phosphoribosyltransferase
Anthranilate synthase
assay, **142**, 301
as coupling enzyme in *arom* multifunctional enzyme assay, preparation, **142**, 331
purification from *Pseudomonas putida*, **142**, 302
Anthranilate synthase–phosphoribosyltransferase
assays, **142**, 368
catalyzed reactions, **142**, 368
properties, **142**, 370
purification from *Salmonella typhimurium*
AS partial complex, **142**, 377
component I subunit, **142**, 380
component II subunit, **142**, 381
whole complex, **142**, 376
Anthrax toxin
assays, **165**, 113
production, **165**, 106
purification, **165**, 109
safety considerations, **165**, 105
Antibiotics
enrichment of cyanobacterial mutants, **167**, 725, 730
in purification of cyanobacteria, **167**, 26, 78
resistance mutations in *E. coli* rRNA genes, **164**, 673
Antibodies, *see also* Immunoglobulins
in analysis of hormonal activation of cAMP-dependent protein kinase isozymes, **159**, 97
–antigen interactions, in affinity immobilization, **137**, 653
antinuclear, assay with ligand-bearing liposomes, **149**, 210
antireceptor, in isolation of lymphocyte surface receptors, **150**, 393
antistreptococcal, assay with ligand-bearing liposomes, **149**, 211
auto-, *see* Autoantibodies
biotinylation for liposome preparation, **149**, 122
bovine serum albumin, in serum sickness, detection, **162**, 491, 496
to branched-chain α-keto acid dehydrogenase proteins
in analysis of mutations, **166**, 109
preparation, **166**, 108
in screening of expression libraries, **166**, 110
calcimedin, production, **139**, 27
calcineurin
applications, **159**, 647
characterization, **159**, 634
preparation, **159**, 628
calmodulin
produced during trypanosome infection, assays, **139**, 265
site-directed
assay, **139**, 402
preparation, **139**, 393
calmodulin-dependent phosphodiesterase
applications, **159**, 647
characterization, **159**, 634
preparation, **159**, 628
to cell-surface carbohydrates, antigen analysis, **138**, 195
to chloroplast envelope polypeptides
in immunochemical studies of
envelope membrane fractions, **148**, 211
intact plastids, **148**, 208
preparation, **148**, 207
in cloning of Na^+,K^+-ATPase α-subunit gene, **156**, 379
elastin
characterization, **144**, 247
production, **163**, 670
erythrocyte subgroup, assay with ligand-bearing liposomes, **149**, 212
exogenous, renal glomerular deposits, quantitation, **162**, 454
Fab fragments, in immunoelectron mi-

croscopy of ribosomal subunits, **164**, 519
Fab′ fragments, conjugation with horseradish peroxidase, **139**, 499
factor VIII, adsorption to matrix-bound protein A, **137**, 458
factor IX, adsorption to macrobead-linked factor IX, **137**, 458
fatty acylation for immunoliposome preparation, **149**, 89
fibronectin segment-specific, preparation, **144**, 462
glomerular basement membrane, preparation, **162**, 444
hapten-specific, in localization of ribosomal proteins, **164**, 517
to inserted polypeptide sequence of pMR100 fusion proteins, preparation, **154**, 144
insulin-like growth factor II, production, **146**, 250
in isolation of specific polysomes, **152**, 248
isovaleryl-CoA dehydrogenase, preparation, **166**, 157
kallikrein, immobilization, **163**, 164
kininogen
  application to ELISA, **163**, 255
  coupling to peroxidase, **163**, 254
  purification, **163**, 254
α_2-macroglobulin, preparation, **163**, 489
metanephrine
  catecholamine radioimmunoassay with, **142**, 559
  preparation, **142**, 556
to MHC-restricted T-cell receptors, production, **150**, 643
monoclonal, *see* Monoclonal antibodies
nerve growth factor
  coupling reactions for NGF assay, **147**, 171
  NGF enzyme immunoassay with, **147**, 178
  preparation, **147**, 169
polyclonal, *see* Polyclonal antibodies
as probes for screening DNA libraries, overview, **152**, 396
production in lymphocyte cultures
  human systems, **150**, 309
  murine systems, **150**, 304
as rRNA probes, production, characterization, and applications, **164**, 493
in screening of
  cDNA expression libraries, **152**, 451
  λgt11 libraries, for gene isolation, **152**, 458; **154**, 115
  phage library for fibronectin cDNA, **144**, 456
serum albumin, interactions with antigen, ellipsometric studies, **137**, 371
in splenic focus assay
  detection of producing clones, **150**, 201
  monoclonality and donor origin, **150**, 203
to spermatozoa, in postvasectomy autoimmune orchitis, detection, **162**, 466
to S-100 protein subunits, preparation, **139**, 497
to surface antigen receptors, stimulation of lymphocytes, **150**, 92
to synthetic T-cell receptor D_β/J_β peptides, production, **150**, 647
thyroglobulin, detection of circulating forms, **162**, 473
Antigens
  –antibody interactions, in affinity immobilization, **137**, 653
  carbohydrate, on cell surfaces, analysis, **138**, 195
  CD2
    antibody preparation, **150**, 538
    characterization, **150**, 539
    endo-F treatment, **150**, 545
    function, **150**, 537
    heterogeneity, **150**, 547
    immunoblotting, **150**, 543
    immunoprecipitation, **150**, 541
    neuraminidase treatment, **150**, 545
    radiolabeling, **150**, 539
    SDS–PAGE analysis, **150**, 543
  cell substrate attachment, laminin binding properties, **144**, 504
  HLA, L cells with, sorting, **151**, 161
  peptide, in cloning T cells with distinct functions, **150**, 342
  proliferating cell nuclear, *see* Proliferating cell nuclear antigen
  specific lymphocyte responses in basal serum-free medium, **150**, 131

Tac, *see* Interleukin-2 receptors
T-cell, –T-cell receptor complex, antibodies, activation of T cells, **150**, 103
viral, induction of delayed hypersensitivity in mice, **163**, 363
Anti-inflammatory drugs
assessment with adjuvant arthritis model, **162**, 351
Antimony
atomic absorption spectrometry, **158**, 130
Antisera
bone glycoprotein, production, **145**, 280
calbindin D, preparation, **139**, 538
calcineurin, preparation, **159**, 628
calmodulin-binding protein, preparation, **139**, 151
calmodulin-dependent phosphodiesterase, preparation, **159**, 628
cAMP-dependent protein kinase, preparation, **159**, 228
elastin
cross-reactivity, **144**, 248
production, **144**, 248
glomerular basement membrane, characteristics, **162**, 445
hetero-, *see* Heteroantisera
insulin-like growth factor I, preparation, **146**, 218
kallikrein, preparation, **163**, 163
kininogen, production, **163**, 252
Na^+,K^+-ATPase
preparation, **156**, 95
in screening of λgt11 cDNA library, **156**, 382
oncomodulin, preparation, **139**, 158
polyclonal, to transferrin receptor, preparation, **147**, 246
prostaglandin
characterization, **141**, 326
preparation, **141**, 323
Shiga toxin, production, **165**, 251
staphylococcal epidermolytic toxin, preparation, **165**, 338
T-cell receptor, listing, **150**, 631
transferrin receptor
evaluation, **147**, 257
preparation, **147**, 255
α_1-Antitrypsin
characterization in biological fluids, **163**, 401
genetic polymorphism, **163**, 409
purification from human plasma, **163**, 407
Aorta
elastin synthesis in culture, analysis, **144**, 218
Aphanizomenon flos-aquae
purification of
cytochromes *c*-550 and *c*-553, **167**, 396
ferredoxin, **167**, 408
Aphanocapsa
carotenoid synthesis *in vitro*, **167**, 329
thylakoid membrane preparation from, **167**, 333
Aplysia californica
dissection for analysis of calmodulin-encoding nucleic acids, **139**, 279
Apoenzymes
aminopeptidase, preparation, **158**, 27
carboxypeptidase A, preparation, **158**, 30
Aprotinin
^{125}I-labeled, in assay of tissue kallikrein, **163**, 160
APS kinase, *see* Adenylylsulfate kinase
L-Arabinan
preparation, **160**, 542
properties, **160**, 544
1,5-L-Arabinan
preparation, **160**, 543
properties, **160**, 544
α-L-Arabinofuranosidase
assay, **160**, 707, 712
properties, **160**, 711, 717
purification from
Aspergillus niger, **160**, 708
Scopolia japonica, **160**, 713
Arabinogalactanase
assay, **160**, 720
properties, **160**, 723
purification from *Bacillus subtilis*, **160**, 721
Arabinose
quantitative TLC, **160**, 172
Arachidonate
formation from linoleate in intact cells, **141**, 354

release by cultured cells, **141**, 354
uptake by platelets and cultured cells, **141**, 353
Arachidonate-CoA ligase
assay, **141**, 351
Arachidonoyl-CoA synthetase, *see* Arachidonate-CoA ligase
Archaebacteria
ribosomal proteins, HPLC for microsequence analysis, **164**, 542
Arginase
reactor, preparation from dialyzer, **137**, 485
Arginine carboxypeptidase
assay, **163**, 186
distribution, **163**, 190
inhibitors, **163**, 191
properties, **163**, 189
purification from human plasma, **163**, 188
quantification, **163**, 189
Arginine–glycine–aspartic acid receptors
multifunctional
isolation from platelets, **144**, 486
properties, **144**, 477
Arogenate
assay, **142**, 431
HPLC, **142**, 427
preparation from *Neurospora crassa*, **142**, 426, 515
Arogenate dehydratase
amino acid composition, **142**, 518
assay, **142**, 488, 495, 513
properties, **142**, 494, 502, 517
purification from
Phenylobacterium immobile, **142**, 516
Pseudomonas diminuta, **142**, 501
tobacco, **142**, 490
substrate preparation, **142**, 515
Aromatic amine dehydrogenase
assay, **142**, 650
properties, **142**, 653
purification from *Alcaligenes faecalis*, **142**, 651
Aromatic-amino-acid aminotransferase
assay, **142**, 256, 267
catalyzed reactions, **142**, 255
limited proteolysis, **142**, 264
properties, **142**, 261, 265, 270
purification from
Escherichia coli, **142**, 258
rat small intestine, **142**, 268
Aromatic-L-amino-acid decarboxylase
assay, **142**, 171, 180
properties, **142**, 177, 183
purification from porcine kidney, **142**, 176, 182
Aromatic-amino-acid–glyoxylate aminotransferase
assay, **142**, 274
properties, **142**, 275
purification from rat liver, **142**, 275
Arsenate
effect on Ca^{2+} release from sarcoplasmic reticulum vesicles, **157**, 226
Arsenazo III
in analysis of TRH-induced changes in cellular calcium metabolism, **141**, 43
Arsenic
As(III), voltammetric assay, **158**, 264
atomic absorption spectrometry, **158**, 130
Artemia salina, *see* Brine shrimp
Arteries
coronary, preparation for analysis of calmodulin–phosphodiesterase binding, **159**, 596
Arthritis
adjuvant
cell-mediated passive transfer, **162**, 353
clinical evaluation, **162**, 344
effects of pharmacologic agents, **162**, 351
genetics, **162**, 341
histological evaluation, **162**, 347
paw volume measurement of inflammatory response, **162**, 344
radiographic evaluation, **162**, 345
tolerance to, induction, **162**, 354
chronic, in rabbits
evaluation, **162**, 384
immunopathology, **162**, 382
induction, **162**, 380
overview, **162**, 379
pathogenesis, **162**, 383
pathology, **162**, 381

streptococcal cell wall-induced, in rats
induction and assessment, **162**, 376
overview, **162**, 373
relevance to human lesion, **162**, 378
type II collagen-induced
in mice
assessment, **162**, 365
associated immune responses, **162**, 368
in experimental immunotherapy, **162**, 372
overview, **162**, 361
passive transfer, **162**, 370
strain selection for, **162**, 362
in rats
assessment, **162**, 358
overview, **162**, 355
Arthrobacter
A. globiformis, polyacrylamide gel-entrapped
incubation in nutrient medium, **135**, 200
microcultural analysis, **135**, 202
preparation, **135**, 199
steroid-transforming activity, **135**, 201, 204
viability, **135**, 202, 204
A. simplex, prepolymer-entrapped
applications, **135**, 247
preparation, **135**, 243
entrapment by cellulose acetate beads, **135**, 287
Arthus reaction
overview, **162**, 478
reverse passive form
elicitation in rat skin, **162**, 479
hemorrhage measurement, **162**, 482
leukocyte infiltration measurement, **162**, 480
plasma leakage measurement, **162**, 481
Arylalkylamine *N*-acetyltransferase
assay, **142**, 584
properties, **142**, 588
purification from ovine pineal gland, **142**, 585
Arylformamidase
assay, **142**, 226
properties, **142**, 229
purification from *Streptomyces parvulus*, **142**, 227
Arylsulfatase
assays, **143**, 207, 361
properties, **143**, 365
purification from *Helix pomatia*, **143**, 363
Aryl sulfotransferase
chromatographic assay, **143**, 204
ion-pair extraction, **143**, 202
Ascitic fluid
extracellular matrix proteins, analysis, **145**, 396
human, purification of
C-reactive protein, **163**, 383
haptoglobin, **163**, 457
Ascorbate
assay with
enzyme reactor, **137**, 281
microsomal electrode, **137**, 157
effect on elastin synthesis, **144**, 240
L-Ascorbate oxidase reactor
ascorbic acid assay with, **137**, 281
preparation, **137**, 282
Asialoglycoprotein receptors
mediated endocytosis, analytical methods, **149**, 3
Asialoorosomucoid
receptor-mediated endocytosis, analysis
binding to cell surface, **149**, 4
degradation, **149**, 9
internalization, **149**, 4
intracellular dissociation from receptor, **149**, 5
intracellular segregation from receptor, **149**, 7
subcellular distribution, **149**, 7
Asolectin
concentration, effect on Na^+,K^+-ATPase reconstitution, **156**, 146
liposome preparation with, **156**, 142
Asparaginase
–albumin–IgG conjugate, preparation, **137**, 570
chemotherapeutic applications, **137**, 447
microparticle-entrapped
hemofilter with, **137**, 493
preparation, **137**, 492
reactor, preparation
from dialyzer, **137**, 483, 487
with fibrin, **137**, 482

Asparagine
  assay with enzyme transistor, **137**, 244
  glycosyl derivatives
    attachment to proteins, **138**, 409
    incorporation into glutamine residues, **138**, 413
    noncovalent coupling to proteins, **138**, 418
Asparagus
  purification of asparagusate reductase, **143**, 518
Asparagusate reductase (NADH)
  assay, **143**, 517
  properties, **143**, 519
  purification from asparagus, **143**, 518
Asparagusic acid
  synthesis, **143**, 516
Asparagus officinalis, *see* Asparagus
Aspartame
  production with immobilized thermoase
    batchwise system, **136**, 514
    continuous column system, **136**, 513
Aspartase, *see* Aspartate ammonia-lyase
Aspartate aminotransferase
  assays, **142**, 256
    with dry reagent chemistries, **137**, 414
  catalyzed reactions, **142**, 256
  complex formation with oligosaccharide, **138**, 421
  cysteinesulfinic acid assay with, **143**, 160
  immobilization and assay, **136**, 486, 488, 490
  as marker for chromosome detection in somatic cell hybrids, **151**, 183
  properties, **142**, 261, 265
  purification from *E. coli*, **142**, 258
Aspartate ammonia-lyase
  assay, **136**, 465
  industrial L-aspartic acid production with, **136**, 466
L-Aspartate β-decarboxylase
  activity stabilization by glutaraldehyde, **136**, 476
Aspartate transcarbamoylase
  conformational states, chemical stabilization, **135**, 569
  crosslinked subunits
    preparation, **135**, 574
    reassociation, **135**, 575
    separation, **135**, 573
  crosslinking with tartryl diazide, **135**, 571
L-Aspartic acid
  industrial production by immobilized bacteria, **136**, 463
Aspergillus
  A. aculeatus, purification of cellulase, **160**, 279
  A. fumigatus
    culture conditions, **160**, 267
    purification of cellulase and hemicellulase, **160**, 268
  A. niger
    immobilization, **137**, 688
    polyacrylamide gel-entrapped mycelium
      enzymatic activity, **135**, 203, 212, 215
      preparation, **135**, 202
      viability, **135**, 203
    production of α-amylase and glucoamylase, **137**, 689
    purification of
      α-L-arabinofuranosidase, **160**, 708
      cellulase, **160**, 261
      β-glucosidase, **160**, 577
      β-mannanase, **160**, 604
  A. parasiticus, purification of poly(*N*-acetylgalactosamine) deacetylase, **161**, 516
  A. saitoi, purification of α-mannosidases I and II, **138**, 782
Astroblasts
  in glia maturation factor assay, **147**, 234
  isolation, **147**, 232
Atherosclerosis
  premature, treatment by low-density lipoprotein pheresis, **137**, 466
Atomic absorption spectrometry
  aluminum assay, **158**, 293
  cobalt assay
    applications, **158**, 349
    contamination control, **158**, 345
    instrumentation, **158**, 347
    interferences, **158**, 346
    standards, **158**, 348
    technique selection, **158**, 345

copper assay
materials and instrumentation, **158**, 352
quality assurance, **158**, 355
sample preparation, **158**, 354
flame technique
calcium assay in biological fluids, **158**, 314
essential features
atomic vapor production, **158**, 312
optical design, **158**, 312
photodetection and readout, **158**, 313
radiation sources, **158**, 311
sample preparation and contamination, **158**, 127
theory, **158**, 310
trace metal determinations, **158**, 128
furnace technique
chromium assays, **158**, 334
overview, **158**, 118
sample preparation and contamination, **158**, 127
trace metal determinations, **158**, 128
magnesium assay
instrument parameters, **158**, 366
interferences, **158**, 367
reference standards, **158**, 368
sample preparation, **158**, 368
molybdenum assay
instrument setup and operating conditions, **158**, 378
interferences, **158**, 379
sample preparation, **158**, 377
standards, **158**, 378
multielement
analytical methods
high-accuracy technique, **158**, 154
microsurvey technique, **158**, 153
atomization parameters, **158**, 149
data handling, **158**, 152
detection limits, **158**, 148
equipment, **158**, 146
principles, **158**, 146
sample preparation, **158**, 151
nickel assay, electrothermal technique
computations, **158**, 389
equipment, **158**, 384
procedures, **158**, 387
reagents, **158**, 386
selenium assay, **158**, 399
vanadium assay, **158**, 402
zinc assay
buffers and solutions, **158**, 423
electrothermal technique, **158**, 432
flame technique, **158**, 430
method verification, **158**, 430
sample preparation, **158**, 425
zinc standard solutions, **158**, 429
Atomic emission spectrometry
aluminum assay, **158**, 291
flame technique
analytical applications, **158**, 188
detection limits, **158**, 189
instrumentation, **158**, 182
interferences, **158**, 186
sample preparation, **158**, 185
standard additions, **158**, 187
theory, **158**, 180
Atomic fluorescence spectrometry
hollow cathode lamp/inductively coupled plasma
detection limits, **158**, 227
instrumentation, **158**, 225
interferences
physicochemical, **158**, 232
spectral, **158**, 231
introduction of organic solvents, **158**, 233
simultaneous multielement analyses, **158**, 234
mercury assay, **158**, 235
theory, **158**, 224
ATP (adenosine 5′-triphosphate)
–ADP transport in mitochondria, analytical technique, **148**, 680
analogs as photoaffinity labels
application to Na^+,K^+-ATPase studies, **156**, 319
synthesis, **156**, 313
binding to
Ca^{2+}-ATPase
during ATP hydrolysis, assay, **157**, 238
fluorescence assay, **157**, 214
Na^+,K^+-ATPase
centrifugation assay, **156**, 200
kinetics, **156**, 239
rate of dialysis assay, **156**, 193
binding site in Na^+,K^+-ATPase, NMR studies, **156**, 358

in cytochemical localization of Na^+,K^+-ATPase, **156**, 417
dextran-bound
glucose 6-phosphate production with, **136**, 52
partition in aqueous two-phase system, **136**, 50
preparation, **136**, 47
effect on phosphoinositide hydrolysis, **141**, 120
enzymatic regeneration, **136**, 263
hydrolysis, associated ligand binding to ATP-driven pumps, assays, **157**, 233
modification in N-6 position, **136**, 3
phosphoryls, isolation, **159**, 68
–$^{32}P_i$ exchange in clathrin-coated vesicles, assay, **157**, 636
γ-^{32}P-labeled, purification, **157**, 636
postlabeling of membrane phosphoinositides, **141**, 125
reaction with Na^+,K^+-ATPase, **156**, 5
role in coupling chemical reactions and transport of Na^+ and K^+, **156**, 17
in sarcoplasmic reticulum vesicles
Ca^{2+}-ATPase phosphorylation with, **157**, 168
exchange reactions, **157**, 182, 197
hydrolysis, measurement, **157**, 200
synthesis, **157**, 180
in absence of Ca^{2+} gradient, **157**, 204
measurement, **157**, 203
net production, **157**, 201
synthesis from Ca^{2+} gradient by Ca^{2+}-ATPase, **157**, 220
ATPase
Ca^{2+}-
analysis by controlled proteolysis, **139**, 801
assay, **139**, 794
in sarcoplasmic reticulum fractions, **157**, 43
ATP binding during ATP hydrolysis, assay, **157**, 238
calcium binding during ATP hydrolysis, analysis, **157**, 236
cardiac sarcoplasmic reticulum-associated
assay, **157**, 117
calmodulin-dependent stimulation, **157**, 134
cAMP-dependent stimulation, **157**, 128
functional reconstitution, **157**, 314
role in Ca^{2+} transport, **157**, 120
crystals
freeze-fracture-etch electron microscopy, **157**, 279
induction in sarcoplasmic reticulum, **157**, 272
negative staining electron microscopy, **157**, 273
structural analysis by image reconstruction, **157**, 283
unstained, electron microscopic observation, **157**, 278
monomeric, preparation with
deoxycholate, **157**, 267
octaethylene glycol dodecyl monoether, **157**, 268
occluded Ca^{2+}, assay, **157**, 229
oligomeric, preparation with Tween 80, **157**, 265
phosphorylation, assay, **157**, 230
plasma membrane-associated
assays, **157**, 350, 353
isolation, **157**, 345
reconstitution into liposomes, **157**, 347
solubilization, purification, and reconstitution, **157**, 355
purification from human erythrocytes, **139**, 792
reconstitution, **139**, 795
sarcoplasmic reticulum-associated
active cation pumping, induction by electric field, **157**, 240
associated ATP or ITP hydrolysis, **157**, 200
associated exchange reactions, **157**, 182, 197
ATP binding, fluorescence assay, **157**, 214
ATP synthesis, **157**, 180, 201, 220
calcium binding, measurement, **157**, 163, 210
catalytic cycle, **157**, 191

magnesium-induced changes, fluorescence assay, **157**, 217
phosphoenzyme assay, **157**, 195
phosphoenzyme functional characterization, **157**, 174
phosphoenzyme hydrolysis and reversal, **157**, 176
phosphorylation, **157**, 168, 192, 215
role in Ca^{2+} transport, **157**, 156
soluble, inactivation, **157**, 196
Ca^{2+},Mg^{2+}-, labeling with maleimides and iodoacetamide derivative, **157**, 251
F_1-
assays, **167**, 351
electrophoretic and immunoblotting analyses, **167**, 353
extraction and purification, **167**, 357
properties, **167**, 360
reconstitution of photophosphorylation, **167**, 359
H^+-
chromaffin granule-associated
assay, **157**, 622
purification, **157**, 625
reconstitutively active, purification, **157**, 628
resolution and reconstitution, **157**, 622
in clathrin-coated vesicles
assays, **157**, 634
molecular weight determination, **157**, 642
purification, **157**, 639
reconstitution, **157**, 642
endoplasmic reticulum-associated
assays, **157**, 614
characteristics, **157**, 618
reconstitution, **157**, 617
Golgi membrane-associated
assays, **157**, 614
characteristics, **157**, 618
reconstitution, **157**, 617
lysosomal, properties, **157**, 600
mitochondrial and plasma and vacuolar membrane-associated, discrimination, **157**, 570
plasma membrane-associated
assay, **157**, 513, 528, 534
properties, **157**, 527, 542
purification, **157**, 521, 530, 537, 538, 574
reconstitution, **157**, 525, 540
solubilization, **157**, 520
tonoplast-associated
assay, **148**, 126; **157**, 584
properties, **148**, 128; **157**, 584
purification, **148**, 124; **157**, 580
subunit composition, **148**, 127
vacuolar membrane-associated
assay, **157**, 548
polypeptide composition, **157**, 558
properties, **157**, 549, 555
purification, **157**, 552
H^+,K^+-, preparation from
porcine stomach, **157**, 649
rabbit stomach, **157**, 651
rat stomach, **157**, 653
K^+-, from *E. coli*
genetic analysis, **157**, 655
phosphorylation, **157**, 677
properties, **157**, 676
solubilization and purification, **157**, 671
Mg^{2+}-, tonoplast-associated
assay, **148**, 110
detection and physiological role, **148**, 91
properties, **148**, 97, 112
purification, **148**, 96, 110
solubilization, **148**, 94
transport activity, **148**, 111
myosin-like, from intestinal brush border, purification and properties, **139**, 137
Na^+,K^+-
acidic pH gel electrophoresis, **156**, 124
active cation pumping, induction by electric field, **157**, 240
active site, **156**, 14
amino acid sequence, **156**, 15
antiserum preparation, **156**, 95
assay, **156**, 36, 51, 105
in reconstituted vesicles, **156**, 135
in sarcolemmal vesicles, **157**, 32
ATP and ADP binding
centrifugation assay, **156**, 200
rate of dialysis assay, **156**, 193
ATP binding during ATP hydrolysis, assay, **157**, 238

back-door phosphorylation, **156**, 121
assay, **156**, 121
data analysis, **156**, 122
cardiac glycoside binding, assay, **156**, 201
cation binding during ATP hydrolysis, assay, **157**, 233
$C_{12}E_8$-solubilized, incorporation into liposomes
determination of sidedness, **156**, 160
procedure, **156**, 157
conformational states
eosin-based fluorescence assay, **156**, 278
equilibrium fluorescence assay, **156**, 273
stopped-flow fluorescence assay, **156**, 275
conformational transition, **156**, 4
coupled assay, **156**, 116
coupling of chemical reactions and transport, **156**, 17
crystallization, **156**, 16
detergent-dispersed, preparation, **156**, 335
digitalis binding site
affinity labels, **156**, 323
covalent labeling, **156**, 331
mapping on α subunit, **156**, 332
effects of Na^+ and K^+, **156**, 2
in erythrocyte inside-out vesicles
enzyme assays, **156**, 177
transport assays, **156**, 175
ESR studies, **156**, 371
front-door phosphorylation, **156**, 121
identification as Na,K-pump, **156**, 2
immunoassays, **156**, 120
inactivation by detergents, protective agents, **156**, 76
incorporated into sealed vesicles, density gradient purification, **156**, 139
inhibitors, **156**, 9
isozyme $\alpha(+)$, preparation from mammalian axolemma, **156**, 65
kinetic studies
effects of inhibitors and modifiers, **156**, 248
general considerations, **156**, 237
monovalent cation interactions, **156**, 244
substrate interactions, **156**, 239
labeling with fluorescein 5′-isothiocyanate, **156**, 272
liberated phosphate, measurement, **156**, 107
lipid requirement for activity, **156**, 12
membrane-bound
crystallization in two dimensions, **156**, 80
electron microscopic analysis, **156**, 83, 430
image analysis of 2-D crystals, **156**, 85
preparation, **156**, 81
proteolysis for structural and conformational analyses, **156**, 291
subunit interactions, detection with crosslinking reagents, **156**, 345
microscopic localization
autoradiographic method for ouabain-binding sites, **156**, 428
cytochemical method, **156**, 417
immunocytochemical method, **156**, 429
molecular weight, **156**, 13
monoclonal antibodies
applications, **156**, 407
isotype identification, **156**, 98
preparation, **156**, 96, 98, 394
screening procedures, **156**, 399
NMR studies
nuclear relaxation with paramagnetic probes, **156**, 355
transferred nuclear Overhauser enhancement measurements, **156**, 367
ouabain binding
assay in intact cells, **156**, 219, 223
calcium effects, **156**, 223
equilibrium binding conditions, **156**, 219
equilibrium binding constant, **156**, 219, 221
potassium effects, **156**, 222
specificity, **156**, 215
phosphatase activities, **156**, 7
K^+-dependent, assay, **156**, 109
Na^+/K^+-activated, assay, **156**, 110

photoaffinity labeling
applications, **156**, 319
procedures, **156**, 317
physiological function, **156**, 1
polyclonal antibodies, preparation, **156**, 95
potassium, sodium, and rubidium binding
centrifugation assay, **156**, 230
characteristics, **156**, 235
criteria for specific binding, **156**, 233
preparative problems
inactivation by detergent, **156**, 31
latency in plasma membrane preparations, **156**, 30
purity and integrity criteria, **156**, 31
properties, **156**, 60
purification from
brine shrimp, **156**, 48
duck salt gland, **156**, 46
mammalian kidney, **156**, 29, 88
shark rectal gland, **156**, 43
purification, overview, **156**, 10, 16
purity criteria
molar activity, **156**, 112
specific activity, **156**, 110
subunit purity, **156**, 114
quaternary structure determination, **156**, 333
reaction with ATP, **156**, 5
reconstitution by
cholate dialysis, **156**, 127
freeze–thaw sonication, **156**, 141
release of occluded ions, measurement technique, **156**, 281
solubilization, **156**, 72
solubilized, purification and analysis, **156**, 77
stability after purification, **156**, 43
α subunit
folding through membrane and functional domain identification, **156**, 300
gene cloning with antibody probes, **156**, 379
orientation of proteolytic fragments, **156**, 298
subunits
isolation, **156**, 90
peptide preparation from, **156**, 91
preparation for immunization, **156**, 95
sulfhydryl group modification
assays, **156**, 309
procedures, **156**, 307
thioinosine triphosphate derivatives as affinity labels, **156**, 310
transport assays, **156**, 120, 136
116K, purification from clathrin-coated vesicles, **157**, 643
potassium-stimulated, as plant plasma membrane marker, assay, **148**, 543
in *Streptococcus faecalis*
assay, **157**, 683
properties, **157**, 688
purification, **157**, 684
vanadate-sensitive, as plant plasma membrane marker, assay, **148**, 544
ATPγS (adenosine 5′-*O*-(γ-thio)triphosphate)
in preparation of protein phosphatase-resistant substrates, **159**, 346
ATP synthase
genes in *Synechococcus*
cloning, **167**, 795
sequencing, **167**, 798
Autoantibodies
anti-DNA, in murine lupus
allotype analysis, **162**, 402
idiotypic analysis, **162**, 408
isoelectric focusing, **162**, 406
isotype analysis, **162**, 402
production by hybridomas, **162**, 401
specificity for nucleic acids and phospholipids, **162**, 406
tests, **162**, 393
anti-erythrocyte, in murine lupus, tests, **162**, 393
Autoimmunity
in murine lupus, analytical procedures, **162**, 390
Autophosphorylation
insulin receptors in intact cells, **150**, 718
role in regulation of calmodulin-dependent protein kinases, **139**, 690
Autoradiography
in analysis of tissue-specific gene expression, **152**, 656

calmodulin-dependent myosin heavy chain kinase, **139**, 106
cholera toxin ADP-ribosyltransferase, **165**, 238
chromosomes after *in situ* hybridization, **151**, 285
collagen peptides, **145**, 200
in colony screening of cosmid libraries, **151**, 421
cyanobacterial carbon metabolites on chromatograms, **167**, 548
in detection and cloning of amplified DNA, **151**, 351
in DNA dideoxy sequencing, **155**, 81
in DNA sequencing by chemical methods, **152**, 536
in DNA solid-phase sequencing, **155**, 319
glycosaminoglycans, **138**, 263
haptoglobin and encoding mRNA in tissues, **163**, 473
inositol phospholipids in superior cervical ganglion, **141**, 201
lipopolysaccharide monomers, **138**, 270
in localization of Na^+,K^+-ATPase ouabain-binding sites, **156**, 428
nucleic acids
with ^{32}P, **152**, 57
with ^{35}S, **152**, 55
in oligonucleotide solid-phase sequencing, **155**, 319
PDGF-like proteins, **147**, 33
plaque-forming cells on extra-thin agar layers, **150**, 215
proteins transported into isolated chloroplasts, **148**, 291
tRNA fragments, in tRNA sequencing, **155**, 384, 393
Avena sativa, *see* Oat
Avian myeloblastosis virus
reverse transcriptase, first-strand cDNA synthesis with, **152**, 321
Avicel
cellobiohydrolase assay with, **160**, 99
dyed
preparation, **160**, 24
total cellulase assay with, **160**, 97
Avidin
–biotin technology for liposome targeting, **149**, 119
catalase-labeled
in biotin assay with bioaffinity sensors, **137**, 120
preparation, **137**, 115
purification on 1,1′-carbonyldiimidazole-activated support, **135**, 114
Axolemma
brain stem
isolation, **156**, 66
purification of Na^+,K^+-ATPase, **156**, 69
Axonemes
spermatozoal, motility regulation by calcium, analysis, **139**, 808
Axons
enriched PC12 cell cultures, preparation, **147**, 215
squid
dialyzed, sodium efflux, vanadate effects, **156**, 252
isethionic acid, GLC analysis, **143**, 173
8-Azido-ATP
synthesis, **156**, 313
8-Azido-cyclic AMP
photoaffinity labeling of cAMP receptor, **159**, 269
8-Azido-cyclic IMP
^{32}P-labeled, synthesis, **159**, 733
5′-Azido-5′-deoxythymidine
synthesis, **155**, 270
3′-[*O*-3-(4-Azido-2-nitrophenyl)propionyl]-ATP
synthesis, **156**, 315

B

Bacillus
alkalophilic thermophilic species, purification of xylanases, **160**, 656
B. anthracis, purification of anthrax toxin, **165**, 109
B. cereus, diarrheal toxin, ileal loop test, **165**, 316
B. pumilus, purification of xylanase, **160**, 634
B. sphaericus, purification of *O*-acetylserine sulfhydrylase, **143**, 475

B. subtilis
entrapped, proinsulin production, **137**, 632
purification of
arabinogalactanase, **160**, 721
branched-chain 2-keto acid dehydrogenase–pyruvate dehydrogenase complex, **166**, 332
cAMP phosphodiesterase-inhibiting acylpeptides, **159**, 498
dihydrolipoamide dehydrogenase, **166**, 335
leucine dehydrogenase, **166**, 284
lichenase, **160**, 573
β-mannanase, **160**, 603
prephenate dehydratase, **142**, 509
purification of chitosanase, **161**, 502
Bacitracin
as inhibitor of lipid-linked saccharide pathway, **138**, 674
Bacteria, *see also specific bacteria*
alleviating underrepresentation of specific sequences in genomic cloning, overview, **152**, 173
colonies, screening with
oligonucleotide probes, **152**, 439
radioactive nucleic acid probes, **152**, 415
cyano-, *see* Cyanobacteria
detection of β-D-glucanase and β-D-xylanase, **160**, 70
glycolipid receptors
overlay analysis, **138**, 212
overlay and solid-phase analyses, **138**, 220
immobilization after permeabilization, **137**, 639
immobilized, isomaltulose production, **136**, 432
isolation of acetolactate synthase structural genes, **166**, 101
lipopolysaccharide preparation from gram-negative species, **138**, 268
in molecular cloning
handling, culturing, and storage, **152**, 145
isolation of plasmids, **152**, 164
preparation of competent cells, **152**, 151
nitrifying, hybrid biosensors with, **137**, 131
permeabilization, **137**, 638
permeabilized, acetolactate synthase assay, **166**, 230
plating and growth for colony screening of cosmid libraries, **151**, 418
protein secretion by gram-positive species, genetic engineering, **153**, 507
sulfometuron methyl-hypersensitive mutants, isolation, **166**, 105
sulfometuron methyl-resistant mutants, isolation, **166**, 103
susceptibility to sulfometuron methyl, assays, **166**, 99
taurine-decomposing, purification of taurine dehydrogenase, **143**, 497
tryptophan biosynthetic pathway
in gram-negative species, **142**, 293
in gram-positive species, **142**, 297
Bactericidal agents
in purification of cyanobacteria, **167**, 26
Bacteriophages
entrapment in erythrocytes, **149**, 228
filamentous
biology and replication, **153**, 14
in production of ssDNA, **153**, 24
λ
in construction of cyanobacterial DNA libraries, **167**, 780
DNA packaging *in vitro*, **153**, 95
DNA preparation based on chromatographic phage purification, **153**, 69
EMBL vectors, cloning in, **153**, 103
genomic library construction and characterization in, **152**, 190
growth and DNA preparation, **152**, 154
growth protocol, **153**, 76
in transduction of cDNA expression libraries into mammalian cells, **151**, 434
λgt10, cDNA cloning into, **152**, 360
λgt11
cDNA cloning into, **152**, 364
DNA libraries, immunoscreening for gene isolation, **152**, 458
epitope mapping on, **154**, 122

gene isolation by immunoscreening, **154**, 108
incorporation of fibronectin cDNA, **144**, 452
protein preparation from, **154**, 121
screening for fibronectin cDNA, **144**, 456
transposon mutagenesis, **154**, 125
M13
growth and DNA preparation, **152**, 154; **155**, 69
origin of replication, in preparation of DNA probes, **155**, 519
in production of single-stranded plasmid DNA, **153**, 3
ϕ, 92
colominic acid depolymerization, **138**, 789
propagation and purification, **138**, 786
plaques, screening with
oligonucleotide probes, **152**, 440
radioactive nucleic acid probes, **152**, 415
SP6, transcription system, in preparation of RNA probes, **155**, 519
Streptomyces, as cloning vectors
overview, **153**, 151
restriction maps, **153**, 154
shotgun cloning with, **153**, 152
T4, lysozyme structure and thermal stability, **154**, 511
T5, promoter, in transcription–translation system for protein analysis, **155**, 416
Baeocytes
cyanobacterial
development, **167**, 233
growth media and conditions, **167**, 233
motility, **167**, 234
nature of multiple fission, **167**, 234
Banding analysis
chromosomes in tumor cells
constitutive heterochromatin technique, **151**, 274
Giemsa technique, **151**, 269
nucleolus organizer regions technique, **151**, 274
quinacrine technique, **151**, 271
reverse technique, **151**, 273
Barium
atomic absorption spectrometry, **158**, 131
Barley
assay of (1→3),(1→4)-β-D-glucan, **160**, 548
Basement membranes
constituent proteins
biosynthesis, analysis in cell and tissue cultures, **145**, 386
properties, **145**, 364
purification, **145**, 366
electron immunohistochemistry, **145**, 95
glomerular, preparation from rat kidney, **162**, 443
sources and extractions, **145**, 365
Basophils
allergic reaction mediators, overview, **163**, 3
guinea pig, isolation, **162**, 535
human, isolation by
affinity chromatography, **162**, 525
centrifugal elutriation, **162**, 533
density gradient centrifugation, **162**, 529
with cell panning, **162**, 531
flow microfluorometry, **162**, 533
localization of lysophospholipase, **163**, 43
mediator release, assay, **162**, 538
overview, **162**, 522
Batch reactor
aspartic acid production by immobilized bacteria, **136**, 466
B-cell activators
polyclonal, listing, **150**, 17
B cells
antibody-producing
growth of clones in microcultures, **150**, 240
hapten-specific, individual culture, **150**, 265
antibody-secreting
hemolytic plaque assay, **150**, 209
protein A plaque assay, **150**, 217
antigen-activated, poly(L-lysine) plaque assay, **150**, 223
antigen-binding
characterization, **150**, 324
purification, **150**, 316

bone marrow precursors, long-term culture, **150**, 275
calcium mobilization, flow cytometric analysis, **141**, 53
double-layer agar cultures, **150**, 257
membrane immunoglobulin, isolation with lectins, **150**, 469
mitogen- and antigen-reactive, cloning on filter paper disks, **150**, 287
polyclonal hyperactivity in murine lupus, tests, **162**, 398
progenitors, adherent layer-dependent development in semisolid agar, **150**, 251
stimulation
with antibodies to surface antigen receptors, **150**, 92
in splenic focus assay, **150**, 196
tonsil, isolation of Epstein–Barr virus/C3d receptor, **150**, 550
Bead-Beater
in preparation of fungal mitochondria and plasma and vacuolar membranes, **157**, 567
Beads
positively charged, in labeling of plant plasma membranes, **148**, 574
Beer
assay of (1→3),(1→4)-β-D-glucan, **160**, 549
Beet
tonoplasts, isolation from
microsomes, **148**, 107
vacuoles, **148**, 106
tonoplast vesicles
applications to H^+-coupled transport, **148**, 118
characteristics, **148**, 117
preparation, **148**, 115
Benzofuroxan
in spectrophotometric assay of thiols, **143**, 58
N-[*p*-(2-Benzoxazolyl)phenyl]maleimide
derivatized penicillamine, HPLC, **143**, 188
S-Benzyl-DL-α-methylcysteine sulfone
synthesis, **143**, 276
Benzylpenicillin
enzymatic deacylation in aqueous two-phase system, **137**, 665
synthesis with immobilized penicillin amidase, **136**, 284
Beryllium
atomic absorption spectrometry, **158**, 131
Betaine–homocysteine methyltransferase
assay, **143**, 384
properties, **143**, 387
purification from human liver, **143**, 385
Beta vulgaris, *see* Beet
Beverages
alcoholic and nonalcoholic, glucose flow enthalpimetric assay, **137**, 230
Bile
extracellular matrix proteins, analysis, **145**, 396
Bile acids
assay with nylon tube-immobilized enzymes, **137**, 168
Bilins
in phycobiliproteins
spectroscopic properties, **167**, 302
structure, **167**, 302
Binding assay
in analysis of
glycosphingolipid biosynthesis, **138**, 580
ribosome structure and function, **164**, 404
asialoorosomucoid binding to cell surface receptors, **149**, 4
cGMP binding to phosphodiesterase, **159**, 524
competitive
calcium-binding proteins CAB-27 and CAB-48, **139**, 75
PDGF-like proteins, **147**, 35
peptide–calmodulin complexes, **139**, 473
CR3 receptors, **150**, 608
crystal-induced chemotactic factor, **162**, 197
DNA in yeast extracts, electrophoretic method, **155**, 528
EGF receptor antibodies, **146**, 70
EGF receptors, **146**, 62
factor H receptors, **150**, 601
Fc receptors for IgG, **150**, 525
hemopexin binding to heme and porphyrin, **163**, 562

insulin-like growth factor I receptors
IM-9 lymphoid cell, **146**, 238
mononuclear cell, **146**, 235
lactoferrin, **147**, 311
laminin receptors, **144**, 494
leukotriene B_4 receptors, **141**, 372
Na^+,K^+-ATPase interactions with
ATP and ADP, **156**, 191
cardiac glycosides, **156**, 191
K^+, Na^+, and Rb^+, **156**, 229
ouabain, **156**, 213
PDGF-binding proteins, **147**, 61
phorbol 12,13-dibutyrate
in intact cells, **141**, 278
by mixed micellar method, **141**, 287
in subcellular fractions, **141**, 282
ryanodine receptor from sarcoplasmic reticulum, **157**, 491
solid-phase, sulfatides, **138**, 477
transferrin, **147**, 311
transforming growth factor β, **146**, 177, 182
trifluoperazine and W-7 binding to calmodulin, **159**, 656
tRNA interactions with
ribosomal P and A sites, **164**, 597
ribosomal preparations, **164**, 665
Vicia villosa lectins, **138**, 538
Bioaffinity sensors
biotin assay with, **137**, 120
dethiobiotin assay with, **137**, 122
optical-fiber
component characteristics, **137**, 350
glucose-sensing system
performance tests, **137**, 364
prototype construction, **137**, 360
preparation, **137**, 114
typical response, **137**, 118
Bioassay
anthrax toxin, **165**, 115
bone morphogenetic protein and polypeptides, **146**, 308
botulinum neurotoxins, **165**, 84
BSC-1 cell growth inhibitor, **146**, 165
cartilage-derived factor, **146**, 313
clostridial enterotoxin, **165**, 101
connective tissue-activating peptides, **163**, 735
CR3 receptor, **150**, 610
cytosolic factor required for cholera toxin activity, **165**, 246
E. coli heat-stable enterotoxin, **165**, 128
endotoxin, with toxic shock syndrome toxin-1
in vitro method, **165**, 304
in vivo method, **165**, 303
erythropoietin, **147**, 348
fibroblast growth factor, **147**, 107, 121
glia maturation factor, **147**, 232
hemolytic toxins, **165**, 213
hepatocyte-stimulating factor, **163**, 600
kinin receptors, various methods, **163**, 211
lactoferrin, **147**, 306
lymphocyte and macrophage-derived fibroblast growth factors, **163**, 717
melanoma growth stimulatory activity, **146**, 113
pertussis toxin, **165**, 125
phospholipid asymmetry of loaded erythrocytes, **149**, 288
platelet-activating factor, **163**, 49
platelet-derived growth factor, **147**, 15
Shiga toxin, **165**, 158
skeletal growth factor, **146**, 283
staphylococcal epidermolytic toxin, **165**, 334
staphylococcal leukocidin, **165**, 28
streptococcal erythrogenic toxin, **165**, 66
tissue kallikrein, **163**, 129
toxic shock syndrome toxin-1, **165**, 344
transferrin, **147**, 306
transforming growth factor α, **146**, 142
Vibrio cytolytic toxins, **165**, 177
Biocatalysts
immobilized
activity and effectiveness, **135**, 12
chemical properties, **135**, 4, 7
guidelines for characterization, **135**, 21
operational stability, **135**, 17
physical properties, **135**, 4, 9
productivity, **135**, 17
trends in characterization, **135**, 18
Biocatalytic reactors
two-liquid phase
applications, **136**, 148
organic solvent effects, **136**, 145
reaction kinetics, **136**, 143

reaction types, **136**, 138
selection, **136**, 146
Biochemical oxygen demand
analysis with microbial sensor, **137**, 125
Biocytin hydrazide
preparation, **138**, 437
Biodegradation
lignin
assays with polymeric dyes, **161**, 74
measurement based on ethylene production, **161**, 79
Bioelectrocatalysis
conductive and semiconductive matrices for immobilized enzymes, **137**, 430
Biological fluids, *see also specific fluids*
ionized calcium, measurement, **158**, 320
methylmalonic acid levels, GC–MS assay, **166**, 47
total calcium, measurement, **158**, 302
Biological response modifiers
liposome-encapsulated, evaluation with macrophages *in vivo*, **149**, 150
Bioluminescent assay
with coimmobilized enzymes, **136**, 82
monoamine oxidase, **142**, 619
Bioluminescent enzymes
immobilized
metabolite analysis with, **137**, 176
preparation, **137**, 172
reactor construction, **137**, 175
nylon tube-immobilized
bile acid assay with, **137**, 168
continuous-flow assays with, **137**, 163
NAD(P)H assay with, **137**, 167
preparation, **137**, 163
properties, **137**, 165
reactors
NADH assay with, **137**, 284
preparation, **137**, 285
Biomass
cellulose availability in derived materials, enhancement, **160**, 3
Biomolecules
coupling to silicon surfaces
application to ellipsometry, **137**, 392
procedures, **137**, 389
surface immobilization techniques, **137**, 381
Biopsy
splenic, for retroviral and autoimmunity studies in murine lupus, **162**, 390
Biosensors, *see also specific types*
hybrid
for creatinine, **137**, 132
for urea, **137**, 135
microbial
for biochemical oxygen demand, **137**, 125
for glutamate, **137**, 129
overview, **137**, 3
solid-phase optoelectronic
construction, **137**, 342
dye membranes for, **137**, 341
enzyme–dye membranes for, **137**, 342
penicillin G assay, **137**, 346
pH response, **137**, 345
serum albumin assay, **137**, 343
Biotin
assay with bioaffinity sensor, **137**, 120
–avidin technology for liposome targeting, **149**, 119
labeled probes, in detection of transcription in transfected cells, **151**, 530
–streptavidin, in visualization of ribosomal crosslink sites, **164**, 396
Biotinyl-ε-aminocaproyl-calmodulin
preparation, **159**, 618
properties and applications, **159**, 622
Biotinylation
phospholipids, for liposome preparation, **149**, 121
4,4′-Bisdimethylaminodiphenylcarbinol
in spectrophotometric assay of thiols, **143**, 58
Bis(2-ethylhexyl)sodium sulfosuccinate, *see* Docusate sodium
Bismuth
atomic absorption spectrometry, **158**, 131
Bis(1,10-phenanthroline)copper(I)
in footprinting of protein–DNA contacts, **155**, 543
comparison with hydroxyl radical method, **155**, 553
Blood, *see also* Plasma, Serum
coagulation, and kinins, in inflammatory reactions, overview, **163**, 14
cobalt levels, measurement, **158**, 349
constituents, analysis with dry reagent chemistries, **137**, 410
deheparinization with immobilized heparin lyase, **137**, 515

extracellular matrix proteins, analysis, **145**, 396
glucose
needle-type sensor, **137**, 326
portable continuous analyzer, **137**, 319
glycosphingolipids, HPLC, **138**, 3
human, isolation of genomic DNA, **152**, 182
ionized calcium, measurement by ion-selective electrodes, **158**, 325
nerve growth factor assay, **147**, 183
substitutes, medical applications, **137**, 452
total calcium, flame atomic absorption spectrometry, **158**, 314
Blotting, *see specific techniques*
Blue dextran
–Sepharose, preparation and regeneration, **159**, 773
Bone
noncollagenous glycoproteins
biosynthesis, **145**, 282
characterization, **145**, 277
ELISA, **145**, 280
extraction, **145**, 270
gene cloning, **145**, 286
purification, **145**, 272
noncollagenous proteins, overview, **145**, 256
proteoglycans
characterization, **145**, 267
extraction, **145**, 263
purification, **145**, 265
radiolabeling *in vivo*, **145**, 268
skeletal growth factor purification, **146**, 287
Bone cells
characterization, **145**, 315
cultured, effects of hormones, **145**, 324
isolation and cultivation, **145**, 309
Bone marrow
B-cell precursors, long-term culture, **150**, 275
murine, transfection with hypoxanthine phosphoribosyltransferase gene, **149**, 24
TdT^+ lymphoid precursor cells, selective culture system, **150**, 368
Bone morphogenetic protein
amino acid composition, **146**, 301
bioassay, **146**, 308
proteolysis-derived polypeptides, **146**, 302
purification, **146**, 295
Bordetella pertussis
purification of pertussis toxin, **165**, 122
Borohydride
pyridoxal phosphate-tritiated, labeling of plant plasma membranes, **148**, 572
reduction of disulfides, **143**, 251
Boronate
in electrophoretic gels for analysis of RNA termini, **155**, 438
Botrytis cinerea
purification of
pectinesterase, **161**, 367
polygalacturonase, **161**, 372
Botulinum neurotoxins
assays, **165**, 84
heavy and light chains, separation, **165**, 84
isolation, **165**, 79
structure–activity relationships, **165**, 76
Boyden chamber assay
aortic endothelial cell, fibroblast, and macrophage migration, **147**, 145
capillary endothelial cell migration, **147**, 142
Bradykinin
analogs, synthesis, **163**, 263
HPLC
acid separation, **163**, 260
modifications, **163**, 262
neutral separation, **163**, 259
sequential hydrolysis with immobilized aminopeptidases, **136**, 175
Brain, *see also specific regions*
bovine
isolation of clathrin-coated vesicles, **157**, 638
purification of
adenylate cyclase, **139**, 777
calcium-binding proteins CAB-27 and CAB-48, **139**, 69
calmodulin, **157**, 343
calmodulin-dependent protein phosphatase subunits, **139**, 80
calmodulin-stimulated phosphodiesterase, **159**, 549
calmodulin-stimulated phosphodiesterase isozymes, **159**, 582
fibroblast growth factor, **147**, 123

glia maturation factor, **147**, 226
S-100 protein, **139**, 170
21-kDa calcium-binding protein, **139**, 89
mammalian, purification of
calmodulin-activated phosphodiesterase, **159**, 557
microsomal Na^+,K^+-ATPase, **156**, 71
murine, catecholamine metabolites, HPLC assay, **142**, 535
ovine, purification of dihydropteridine reductase, **142**, 128
polysialosyl glycoproteins, detection, **138**, 169
primate, immunocytochemical localization of monoamine oxidase, **142**, 649
rat
catecholamine metabolites, GC–MS assay, **142**, 577
isethionic acid, GLC analysis, **143**, 173
purification of
calmodulin-kinase II, **139**, 709
calmodulin-stimulated phosphodiesterase, **159**, 555
cAMP phosphodiesterase, **159**, 767
protein kinase C, **141**, 426
Brain stem
mammalian, axolemma
isolation, **156**, 66
purification of Na^+,K^+-ATPase, **156**, 69
rat, purification of tryptophan 5-monooxygenase, **142**, 84
Branched-chain amino acid aminotransferase
assay, **166**, 270, 275
properties, **166**, 277, 279, 280
purification from
canine pancreas, **166**, 280
human pancreas, **166**, 278
porcine heart, **166**, 272
rat pancreas, **166**, 276
Branched-chain α-keto acid decarboxylase
assay, **166**, 144, 147
properties, **166**, 150
Branched-chain keto acid dehydrogenase
from *Pseudomonas*
assay, **166**, 345
gene cloning, **166**, 350
purification, **166**, 346
Branched-chain α-keto acid dehydrogenase
activation by α-chloroisocaproate, **166**, 118
activator protein, assay, **166**, 183
antibodies to constituent proteins
in analysis of mutations, **166**, 109
preparation, **166**, 108
in screening of expression libraries, **166**, 110
assays, **166**, 175, 189, 298, 309, 313
with maple syrup urine disease fibroblasts, **166**, 135
potential interfering substances, **166**, 196
potential sources of error, **166**, 176
component E1
assay, **166**, 147
properties, **166**, 150
component E2
assay, **166**, 152
properties, **166**, 154
flux rates in perfused heart and liver, measurement, **166**, 484
in liver, assay, **166**, 204
in muscle, assay, **166**, 209
properties, **166**, 301, 312
purification for kinase assays, **166**, 174
purification from
bovine kidney, **166**, 310
bovine liver, **166**, 299
rabbit liver, **166**, 316
Branched-chain 2-keto acid dehydrogenase
assay, **166**, 303
phosphorylation sites, analysis, **166**, 10
purification from bovine kidney, **166**, 304
–pyruvate dehydrogenase complex
assay, **166**, 331
properties, **166**, 337
purification from *Bacillus subtilis*, **166**, 332
reconstitution of components, **166**, 308
resolution into active components, **166**, 306
Branched-chain α-keto acid dehydrogenase kinase
assays, **166**, 166, 314
potential sources of error, **166**, 167

complex
properties, **166**, 319
purification from
rabbit heart, **166**, 319
rabbit liver, **166**, 316
inhibition by α-chloroisocaproate, **166**, 114
Branched-chain α-keto acid dehydrogenase phosphatase
assay, **166**, 321
inhibitor protein
assay, **166**, 322
properties, **166**, 327
purification, **166**, 326
properties, **166**, 325
purification from bovine kidney, **166**, 323
Brassica oleracea, *see* Cabbage, Cauliflower
Brevibacterium
B. ammoniagenes
immobilization by radiation polymerization, **135**, 148
polyacrylamide-immobilized
L-malic acid production, **136**, 458
preparation, **136**, 457
B. flavum
κ-carrageenan-immobilized
L-malic acid production, **136**, 459, 462
preparation, **136**, 459, 460
purification of *O*-acetylhomoserine sulfhydrylase, **143**, 471
Brick
porous, yeast immobilization by adsorption, **135**, 308
Bridge visual assay
chemotaxis, attractant concentration gradients, **162**, 87
Brine shrimp
purification of Na^+,K^+-ATPase, **156**, 48
storage and rearing for enzyme purification, **156**, 49
Bromobimanes
biological properties, **143**, 79
chemical and photophysical properties, **143**, 77
thiol labeling
analytical applications, **143**, 82
procedures, **143**, 80
Bromoconduritol
as glycoprotein processing inhibitor, **138**, 693
Bromodeoxyuridine
antibodies, in assay for lymphocyte proliferation, **150**, 117
photolysis, in analysis of calmodulin role in DNA repair, **139**, 723
in suicide selection of CHO cell mutants, **151**, 127
Buffers
metal ion
associated analytical complications, **158**, 53
capacity, **158**, 39
free metal ion concentration, calculation, **158**, 36
sources of uncertainty, **158**, 40
future developments, **158**, 54
ligands, interaction with metals in enzyme active sites, resulting complications, **158**, 42
Buoyancy
planktonic cyanobacteria, analytical techniques, **167**, 521
1,2-Butanediol
racemic, enzymatic preparation, **136**, 133
Butanol
precipitation of DNA, **152**, 47
production from glucose in aqueous two-phase system, **137**, 663
racemic, enzymatic preparation, **136**, 131
Butylene glycol
qualitative TLC, **160**, 173

C

Cabbage
purification of cystine lyase, **143**, 440
Cachectin, *see* Tumor necrosis factor
Cadmium
atomic absorption spectrometry, **158**, 132
Cd(II), voltammetric assay, **158**, 263
Cadoxen
solvolysis of cellulose, **160**, 11
Calbindin D
antiserum preparation, **139**, 538
immunocytochemical localization, **139**, 541
properties, **139**, 537

purification from
bovine intestine, **139**, 586
rat kidney, **139**, 535
radioimmunoassay, **139**, 539
X-ray crystallography, **139**, 585
Calcimedins
antibody production, **139**, 27
purification from
bovine liver, **139**, 24
chicken gizzard, **139**, 21
rat liver, **139**, 25
Calcineurin
antibodies, preparation, characterization, and applications, **159**, 627
conformation, analysis with monoclonal antibodies, **139**, 526
Calcium
atomic absorption spectrometry, **158**, 132
binding to Ca^{2+}-ATPase
during ATP hydrolysis, assay, **157**, 236
fluorescence assay, **157**, 210
in sarcoplasmic reticulum vesicles, **157**, 163
cellular fluxes
effects of maitotoxin
fractional efflux analysis, **141**, 71
gross uptake analysis, **141**, 67
net flux analysis, **141**, 70
measurement, **146**, 394
cellular metabolism, TRH-induced changes, analysis with
arsenazo III, **141**, 43
chlortetracycline, **141**, 45
permeabilized cells, **141**, 50
quin 2, **141**, 47
radiocalcium, **141**, 37
cellular mobilization, flow cytometric analysis, **141**, 53
distribution in blood, **158**, 322
effect on
Ca^{2+} release from sarcoplasmic reticulum, **141**, 3
cyanobacterial photosystem II *in vivo*, **167**, 280
[^{3}H]ouabain binding to Na^+,K^+-ATPase, **156**, 223
phosphoinositide hydrolysis, **141**, 120
efflux from sarcoplasmic reticulum vesicles, measurement, **157**, 203, 422, 464
gradient, as driving force in ATP synthesis by Ca^{2+}-ATPase, **157**, 220
homeostasis, **158**, 321
induced conformational transition of troponin C, model, **139**, 627
intracellular levels
fluorescence assay
in neutrophils, **162**, 292
in oxidant-exposed cells, **163**, 337
glycogen phosphorylase as noninvasive probe, **141**, 18
measurement, **146**, 396
intracellular messenger system, effects of specific inhibitors, **139**, 570
loading of junctional terminal cisternae, assay, **157**, 44
measurement in biological fluids by ion-selective electrodes
accuracy and quality control, **158**, 332
electrode calibration, **158**, 330
instrumentation, **158**, 328
sample processing, **158**, 327
sampling procedures, **158**, 325
membrane transport, overview, **157**, 3
minielectrode, preparation and utilization, **141**, 25
national reference system for, **158**, 318
occluded, in Ca^{2+}-ATPase, assay, **157**, 228
physiological significance, **158**, 323
regulation of
prolactin gene expression, analysis, **139**, 655
sperm axonemal motility, analysis, **139**, 808
release from sarcoplasmic reticulum
calcium-induced, measurement, **141**, 10
chemical-quench assay, **157**, 471
inositol trisphosphate-induced, measurement, **141**, 15
measurement techniques, **141**, 5
models, **141**, 4
spectrophotometric assay, **157**, 470
T tubule-mediated, induction and measurement, **157**, 472

release in skinned skeletal muscle fibers, measurement, **157**, 12
–sodium exchange in plasma membrane vesicles, measurement, **157**, 505
total, in biological fluids
assay methods, **158**, 303
concentrations, **158**, 306
flame atomic absorption spectrometry, **158**, 314
transport in
cardiac sarcoplasmic reticulum
calmodulin-dependent, **157**, 134
cAMP-dependent, **157**, 128
oxalate-facilitated, **157**, 114
at presteady state, **157**, 115
role of Ca^{2+}-ATPase, **157**, 120
cyanobacteria, **167**, 571
isolated transverse tubules, **157**, 79
platelet membranes, **157**, 353
sarcolemmal vesicles, **157**, 36
skeletal sarcoplasmic reticulum, role of Ca^{2+}-ATPase, **157**, 156
tonoplasts, **148**, 112
video microscopic localization in cultured cells, **139**, 830
voltage-induced uptake by sarcoplasmic reticulum vesicles, **157**, 248
Calcium alginate
entrapment of chloroplasts, **135**, 443
gels
cell immobilization in, **135**, 180
entrapped microbial cells, applications, **135**, 187
hardening, **135**, 186
preparation, **135**, 179
problems associated with use, **135**, 185
properties, **135**, 183
solubilization, **135**, 183
immobilization of
Claviceps purpurea, **136**, 320
enoate reductase, **136**, 313
plant protoplasts, **135**, 436
immobilized yeast, in industrial ethanol production, **136**, 394
supported nylon netting, plant cell entrapment, **135**, 415
Calcium-binding proteins
binding to gap junctions, **139**, 452
CAB-27 and CAB-48
characterization, **139**, 74
Chelex assay, **139**, 75
physiological function, **139**, 78
purification from bovine brain, **139**, 69
enzymatic iodination, **139**, 450
9-kDa, cDNA cloning and characterization, **139**, 337
sarcoplasmic reticulum-associated
crystals, properties, **157**, 324
purification and crystallization, **157**, 321
Spec
electrophoretic analysis, **139**, 358
encoding DNA sequences, screening, **139**, 357
expression in *E. coli*, **139**, 362
purification for immunization, **139**, 360
in tumor cells, analysis, **139**, 207
21-kDa
functional studies, **139**, 104
isolation from bovine brain, **139**, 89
metal ion-induced conformational changes, **139**, 97
properties, **139**, 91
vitamin D-dependent, *see* Calbindin D
Calcium channels
cardiac sarcoplasmic reticulum, effect of ryanodine, **157**, 104
pituitary, maitotoxin effects, **141**, 63
sarcoplasmic reticulum
characteristics, **157**, 502
observation in planar lipid bilayers, **157**, 480
purification
cardiac preparations, **157**, 500
skeletal muscle preparations, **157**, 493
ryanodine binding assay, **157**, 491
Calcium phosphate
–DNA coprecipitation, for plasmid transfer into cultured cells, **151**, 387
Calcium pump, *see* ATPase
Calelectrin
binding to biomembranes and lipids, **139**, 34
purification from
bovine tissues, **139**, 32
Torpedo electric organ, **139**, 31

Calmodulin
adenylate cyclase reconstitution with, **139**, 787
affinity purification, **139**, 10
antibodies
produced during trypanosome infection, assays, **139**, 265
site-directed
assay, **139**, 402
preparation, **139**, 393
association with phosphodiesterase in intact cells, assay, **159**, 594
bacterial expression vectors
application, **139**, 314
construction, **139**, 307
overview, **139**, 303
binding
associated domain in myosin light chain kinase, preparation and properties, **139**, 115
associated sequences in peptides and proteins
characterization, **139**, 463
identification, **139**, 456
to gap junctions, **139**, 452
to microtubule-associated proteins, analytical methods, **139**, 836
biotinylated derivatives
preparation, **159**, 618
properties and applications, **159**, 622
bovine papillomavirus-based expression vectors, analytical methods, **139**, 642
cDNA, isolation, **139**, 236
chemically modified
chromatography, **139**, 421
functional analysis, **139**, 430
preparation with
N-hydroxysuccinimide-containing compounds, **139**, 426
nitrosoureas and isocyanates, **139**, 423
chemical studies, **139**, 56
crosslinking derivatives
preparation, **159**, 612
properties and applications, **159**, 614
dependent protein methylation, assay, **139**, 668
effect on actin polymerization, **139**, 852
encoding DNA, analysis with synthetic oligonucleotides, **139**, 299
enzymatic iodination, **139**, 450
fluorescent derivatives
preparation, **159**, 606
properties and applications, **159**, 609
fragment 1–77, in affinity purification of phosphodiesterase, **159**, 573
functional studies, **139**, 62
gene
in *Aplysia*, cloning, **139**, 287
in chicken
cloning, **139**, 217
structural analysis, **139**, 224
in fruit fly
cloning, **139**, 230
exon sequencing, **139**, 232, 239
structural comparisons with other calmodulin genes, **139**, 243
site-specific mutagenesis, **139**, 291
in trypanosome, organization, **139**, 268
in yeast, cloning, **139**, 248
HPLC analysis, **157**, 334
immobilized, preparation and use, **157**, 336
immunolocalization in
plants, **139**, 478
trypanosome, **139**, 262
immunological studies, **139**, 59
immunoperoxidase staining in cells and tissues, **139**, 495
interaction with
cyclosporine in T cells
flow cytometric analysis, **139**, 683
fluorometric analysis, **139**, 680
peptides, NMR analysis, **139**, 197
phosphofructokinase, assays, **139**, 745
iodinated, preparation and use in calmodulin-binding protein studies, **159**, 667
N-methylation, assay, **139**, 674
microinjection into cultured cells, **139**, 826
monoclonal antibodies
characterization, **139**, 509
production, **139**, 506
–phenothiazine adduct
preparation, **139**, 409

properties, **139**, 413
specificity for calmodulin-regulated proteins, **139**, 417
preparation from
Acanthamoeba, **139**, 50
bovine brain, **157**, 343
chicken gizzard, **157**, 329
porcine brain, **139**, 129
–protein complex
properties, **139**, 145
purification from chicken intestine, **139**, 137
radioiodination, **139**, 838
regulation of prolactin gene expression, analysis, **139**, 655
regulatory systems, analysis with monoclonal antibodies, **139**, 518
role in DNA repair, analytical methods, **139**, 715
stimulation of protein phosphatase subunit A, **139**, 84
video microscopic localization in cultured cells, **139**, 830
X-ray crystallography, **139**, 632
Calmodulin antagonists
effects on
CaM-dependent phosphodiesterase and other enzymes, **159**, 652
transmembrane Ca^{2+} signaling, **139**, 570
Calmodulin-binding proteins
analysis with
gel overlay technique, **139**, 433
iodinated calmodulin, **159**, 667
detection in GH_3 cell nuclear matrix, **139**, 664
macrophage-associated
immunopurification, **139**, 155
radioimmunoassay, **139**, 152
preparation from human platelets, **139**, 851
role in actin polymerization, **139**, 852
Calmodulin-kinase II
autophosphorylation, role in enzyme regulation, **139**, 701
purification from rat brain, **139**, 709
Calmodulin-lysine methyltransferase
calmodulin N-methylation assay with, **139**, 674
Calothrix
sheath isolation from, **167**, 184
Calpain
cDNA cloning, **139**, 372
evolution, **139**, 377
mRNA distribution and complexity, **139**, 373
structure, **139**, 374
Calregulin
immunocytochemical localization, **139**, 47
immunoprecipitation, **139**, 46
purification from bovine liver, **139**, 37
radioimmunoassay, **139**, 43
Calspermin
assay, **139**, 127
properties, **139**, 134
purification from porcine testis, **139**, 131
Calvatia cyathiformis
purification of chitinase and chitobiase, **161**, 494
Campylobacter
diarrheal toxin, ileal loop test, **165**, 322
Cancer therapy
with plasma perfusion system containing immobilized protein A, **137**, 496
Candida
C. tropicalis, immobilized, phenol degradation, **137**, 706
C. utilis
coupling to modified Separon H 1000, efficiency, **135**, 346
(*R*)-propanediol preparation with, **136**, 316
Capsaicin
production by immobilized pepper cells, **135**, 420
Capsicum annuum, *see* Red pepper
Capsicum frutescens, *see* Chili pepper
Carbachol
induced inositol phosphate accumulation in hippocampal slices, assay, **141**, 192
Carbohydrates
cell-surface, analysis with antibodies, **138**, 195
chemical ionization MS, **138**, 62
composition in hemopexins, **163**, 550

direct chemical ionization MS, **138**, 71
in factor XII, composition, **163**, 65
human monoclonal antibodies, overview, **138**, 492
interactions with *Erythrina* lectins, **138**, 549
–lignin complexes
fractionation, **161**, 14
isolation from
barley straw, **161**, 192
various plants, **161**, 13
properties, **161**, 14
size-exclusion chromatography
conventional method, **161**, 195
high-performance method, **161**, 197
monomers and oligomers, liquid chromatography, **160**, 145
murine and rat monoclonal antibodies, overview, **138**, 484
N-linked, content in CD2 antigens, **150**, 547
NMR analysis in cyanobacteria, **167**, 530, 550
nondegradable ligands, synthesis, **138**, 404
transport in cyanobacteria, assay, **167**, 551
Carbon
inorganic, cyanobacterial uptake, **167**, 534
metabolism in cyanobacteria, analysis with radioisotopes, **167**, 539
relative flux in leucine metabolic pathways, measurement, **166**, 127
Carbon dioxide
fixation during cyanobacterial anoxygenic photosynthesis, assay, **167**, 381
Carbon matrices
as conductors in bioelectrocatalysis, **137**, 432
1,1′-Carbonyldiimidazole
activated supports
ligand coupling, **135**, 110
overview, **135**, 102
preparation, **135**, 108
protein purification with, **135**, 111
Carboxylesterase
as marker for chromosome detection in somatic cell hybrids, **151**, 181
preparation of racemic primary alcohols with, **136**, 121
Carboxymethylamylose
dye-labeled, in assay for α-amylase, **160**, 81
Carboxymethylation
E. coli heat-stable enterotoxin, **165**, 133
Carboxymethylcellulase
assay, **160**, 377
properties, **160**, 381
purification from *Sclerotium rolfsii*, **160**, 378
viscosimetric assay, **160**, 130
Carboxymethylcellulose
dyed, preparation, **160**, 25
dye-labeled, in assay for endo-1,4-β-D-glucanase, **160**, 78
endo-1,4-β-glucanase assay with, **160**, 100
trinitrophenyl derivative, endo-1,4-β-glucanase assay with, **160**, 106
Carboxymethylchitin
characterization, **161**, 409
preparation, **161**, 408
Carboxymethyl-β-glucan
dye-labeled, in assay for endo-1,3(4)-β-D-glucanase, **160**, 82
Carboxypeptidase A
apoenzyme preparation, **158**, 30
coupling to 1,1′-carbonyldiimidazole-activated support, **135**, 116
immobilized
in analysis of pancreatic ribonuclease thermal transitions, **135**, 597
via monoclonal antibodies
assay, **135**, 163
preparation, **135**, 165
properties, **135**, 166
metal substitution
in crystalline state, **158**, 77
in solution, **158**, 72
Carboxypeptidase Y
coupling to 1,1′-carbonyldiimidazole-activated support, **135**, 116
ethylenediamine adducts
coupling to carboxyalkyl agarose, **135**, 145
preparation, **135**, 145
immobilized, deblocking in peptide synthesis, **136**, 157
oxidation, **135**, 144

1-(2-Carboxyphenylamino)-1-deoxyribulose 5-phosphate
  preparation, **142**, 387
Carboxysomes
  cyanobacterial, research methodology, **167**, 204
Carob
  purification of galactomannan, **160**, 524
Carotene oxygenase
  assay based on determination of
    crocetindial, **167**, 339
    β-cyclocitral, **167**, 337
  enzyme system features, **167**, 340
  occurrence and general features, **167**, 336
Carotenes
  plant membrane, pesticide effects, **148**, 673
Carotenoids
  $A^{1\%}_{1\,\mathrm{cm}}$ of sum, **148**, 371
  –chlorophyll proteins, isolation and pigment content, overview, **148**, 353
  chromatography, **148**, 357
  cyanobacterial, isolation, **167**, 182
  extraction, **148**, 357
  specific absorption coefficients, **148**, 368
  spectral characteristics, **148**, 366
  synthesis by *Aphanocapsa in vitro*, **167**, 329
  total content determination, **148**, 369, 378
Carrageenan
  characteristics, **135**, 190
  in foot edema test, **162**, 320
  induced pleurisy
    analytical methods, **162**, 335
    comparison with other exudative models, **162**, 338
  type κ
    bead-forming procedure, **135**, 225
    Brevibacterium flavum immobilization with, **136**, 459, 460
    cell immobilization with, **135**, 194, 227
    plant protoplast immobilization with, **135**, 436
Carrot
  protoplasts, electroporation-mediated uptake of
    plasmid DNA, **153**, 342
    RNA, **153**, 345
  purification of galacturan 1,4-α-galacturonidase, **161**, 374
Cartilage
  bovine articular, type II collagen preparation from, **162**, 364
  explant cultures
    preparation, **144**, 414
    proteoglycan turnover measurement in, **144**, 415
  growth factor purification, **146**, 320
  macromolecular constituents
    biosynthesis and turnover, **145**, 345
    extraction, **145**, 342
    interactions, **145**, 356
    localization methods, **145**, 350
    overview, **145**, 339
    purification, **145**, 342
Cartilage-derived factor
  amino acid composition, **146**, 319
  assay, **146**, 313
  properties, **146**, 320
  purification, **146**, 315
Casein
  structural changes, analysis with immobilized Pronase, **135**, 598
Castanospermine
  as glycoprotein processing inhibitor, **138**, 688
Castor bean
  endoplasmic reticulum
    isolation from endosperm, **148**, 576
    phosphoglyceride synthesis, **148**, 585
  glyoxysomal membranes
    isolation, **148**, 505
    protein and phospholipid composition, **148**, 526
  glyoxysomes, isolation, **148**, 505
  protoplasts, preparation from endosperm, **148**, 30
Catalase
  amino groups, assay, **137**, 603
  assay, **137**, 601
  avidin labeling with, **137**, 115
  crosslinked
    characterization, **137**, 609
    gel electrophoresis, **137**, 603
    immobilization, **137**, 604
    preparation, **137**, 602
    thermostability, **137**, 613

entrapment with photocrosslinkable resin prepolymers, **135**, 234
glucosone effects, **137**, 606
and oxygen electrode, sulfhydryl oxidase assay with, **143**, 120
prepolymer-entrapped, applications, **135**, 245
stabilization toward glucosone, **137**, 611
thermodenaturation, **137**, 605
Catecholamines
enzymatic radioimmunoassay in biological samples, **142**, 550
HPLC assay in plasma, **142**, 521
metabolites, HPLC assay in brain, **142**, 535
Catechol methyltransferase
catecholamine radioimmunoassay with, **142**, 559
preparation from rat liver, **142**, 554
Catharanthus roseus, *see* Periwinkle
Cathepsin C
in production of natural sequence growth hormone from derivatives, **153**, 399
Cathepsin G
assay, **163**, 324
properties, **163**, 314
purification from human neutrophils, **163**, 313
Cathepsin T
assay, **142**, 280
properties, **142**, 287
purification from rat liver, **142**, 285
Catheters
Evatate tube, urokinase-treated
clinical application, **137**, 540
fibrinolytic activity after application, **137**, 543
Cations
permeability of endoplasmic and sarcoplasmic reticulum vesicles to, **157**, 434
transport in cyanobacteria, **167**, 561
Cauliflower
plastids, isolation from buds, **148**, 234
CCS paper
in solid-phase nucleotide sequencing, **155**, 310
synthesis, **155**, 309
CDPdiacylglycerol–glycerol-3-phosphate 3-phosphatidyltransferase
in plant endoplasmic reticulum
assay, **148**, 593
properties, **148**, 593
CDPdiacylglycerol–inositol 3-phosphatidyltransferase
in plant endoplasmic reticulum
assay, **148**, 594
properties, **148**, 595
Celite, *see* Infusorial earth
Cell adhesion
fibroblasts and myofibroblasts, analysis by
haptotactic substratum and related methods, **163**, 631
interference reflection microscopy, **163**, 638
Cell adhesion molecules
neural, *see* Neural cell adhesion molecules
Cell aggregation
neutrophils, assay, **162**, 72
Cell–cell communication
inhibition by phorbol ester, fluorescence photobleaching assay, **141**, 439
Cell cultures
Aeromonas hydrophila, for purification of aerolysin, **165**, 195
applications of glycyl-L-histidyl-L-lysine, **147**, 318
Bacillus anthracis, for anthrax toxin production, **165**, 106
bacterial, equipment required for molecular cloning work, **152**, 7
B cells
in double-layer agar, **150**, 257
filter paper cloning of mitogen- and antigen-reactive cells, **150**, 292
hapten-specific antibody-producing cells in individual wells, **150**, 265
progenitors in semisolid agar, adherent layer-dependent development, **150**, 251
bone cells, methodology, **145**, 303
calcium and calmodulin video microscopic localization, **139**, 830
calmodulin microinjection, **139**, 826
calmodulin and S-100 protein immunoperoxidase staining, **139**, 495
chili pepper cells
column systems, **135**, 414
flat bed system, **135**, 412

CHO cells, for genetic analysis, **151**, 3
chondrocytes, for analysis of macromolecular metabolism, **145**, 346
Claviceps purpurea, **136**, 318
Clostridium acetobutylicum, immobilized cell stability, control, **137**, 626
Clostridium thermocellum, for macrocellulase complex isolation, **160**, 484
cyanobacteria
  conditions for akinete formation, **167**, 223
  conservation without growth, **167**, 92
  enrichment, **167**, 71
  growth and maintenance conditions, **167**, 90
  growth media, composition, **167**, 6, 79
  light regimes, **167**, 17
  major sources, listing, **167**, 67
  marine planktonic species
    growth media, **167**, 101
    maintenance, **167**, 104
  solid media, preparation, **167**, 15
  symbiotically associated species, **167**, 114
  temperature, pH, and CO_2 regimes, **167**, 16
  thermophilic species
    growth media, **167**, 99
    maintenance, **167**, 98
cyanobacterial baeocytes, **167**, 233
elastin biosynthesis in, **144**, 225; **163**, 669
elastin-producing cells, **144**, 233
Escherichia coli, for production of α-hemolysin, **165**, 139
Flavobacterium heparinum, **137**, 518
genomic DNA isolation from, **152**, 182
GH_3 cells in defined medium, **139**, 655
green plant cells, initiation from
  nonphotosynthetic cells, **148**, 18
  photosynthetic cells, **148**, 24
growth regulator-induced proteins in, assay, **147**, 427
HeLa cells, for genetic analysis, **151**, 38
hepatocytes
  establishment for fibrinogen biosynthesis studies, **163**, 480
  preparation from rat, **149**, 3
hybridoma and myeloma cells, for genetic analysis, **151**, 50
insulin-like growth factor mRNAs, isolation, **146**, 244
Leuconostoc mesenteroides, dextransucrase production, **136**, 242
liver-derived cells, for genetic analysis, **151**, 19
long-term, bone marrow B-cell precursors, **150**, 275
lymphocytes
  for analysis of mitogen effects on locomotion, **162**, 189
  for fibroblast growth factor production, **163**, 723
  mixed, establishment, **150**, 87
  production of antibodies
    human systems, **150**, 309
    murine systems, **150**, 304
  for stimulation with proteolytic enzymes, **150**, 111
lymphoid cells, in generation of TdT^+ precursors, **150**, 363
lymphoma S49 cells, for genetic analysis, **151**, 9
macrophages
  for fibroblast growth factor production, **163**, 723
  preparation from mice, **141**, 251
mammalian cells
  in flat-bed hollow-fiber reactor, **135**, 393
  important products, **135**, 388, 391
  RNA detection by *in situ* hybridization, **151**, 539
  systems for, **135**, 391
monolayer
  growth factor-induced pH_i changes, measurement, **147**, 415
  membrane transport assay, **146**, 382
myoblasts and myofibers, **139**, 327
osteoblast-rich, preparation from rat calvaria, **145**, 326
PC12 pheochromocytoma cells, preparation and maintenance, **147**, 208
pore-forming toxins in, radioisotopic assay, **165**, 278
Ruminococcus albus, for cellulolytic enzyme purification, **160**, 216, 393, 409
Schwann cells
  maintenance and purification, **147**, 218

preparation, **147**, 218
proliferation assay, **147**, 219
Shigella, for production of Shiga toxin, **165**, 153
soybean cell suspensions
isolation of
amyloplasts, **148**, 222
nuclei, **148**, 535
maintenance, **148**, 220
Sporocytophaga myxococcoides, for cellulase purification, **160**, 339
Sporotrichum thermophile, for cellobiose dehydrogenase purification, **160**, 444
Staphylococcus aureus, for γ-hemolysin production, **165**, 11
Streptomyces on lignocellulose, **161**, 37, 250
sycamore cell suspensions
cell density measurement and control, **148**, 15
culture apparatuses, **148**, 12
effects of deficient nutrient media, **148**, 9
growth parameters, **148**, 7
nutritional aspects, **148**, 4
preparation, **148**, 4
Synechococcus
axenic systems, **167**, 107
batch systems, **167**, 107
enrichment, **167**, 106
synchronous systems, **167**, 109
Synechocystis, **167**, 192
Talaromyces emersonii, for cellulolytic enzyme purification, **160**, 363, 438
T cells, for production of monoclonal lines, **150**, 326
teratocarcinoma cells, for genetic analysis, **151**, 66
Thermomonospora fusca, for cellulase purification, **160**, 316
Cell cycle
growth factor-modulated events, characterization, **147**, 400
Cell differentiation
adipocytes, effects of specific phosphodiesterase inhibitors, **159**, 517
Cell-directed inhibitor
chemotaxis-inhibiting
assay, **162**, 233
properties, **162**, 232
Cell envelopes
cyanobacterial
isolation, **167**, 174
monomeric constituents, analyses, **167**, 186
polysaccharides from external layers, isolation, **167**, 185
Cell fusion
macromolecule-loaded erythrocytes with recipient cells, **149**, 295
Cell hybridization
application to growth factor studies, **147**, 379
HeLa cells, **151**, 49
methodology, **147**, 376
Cell immobilization
by adsorption to solid supports, **135**, 307
on agarose beads, **135**, 402
analytical applications, overview, **137**, 3
in aqueous two-phase systems, **137**, 657
in beaded polymers, **135**, 222
in biofilm-fluidized sand beds, **135**, 300
in calcium alginate, **135**, 175
in κ-carrageenan, **135**, 194
on celite, **135**, 318
in cellulose acetate beads, **135**, 282
in cellulose gel, **135**, 268
in chitosan, **135**, 259
in column reactors, **135**, 414
combination with mass spectrometry, **137**, 421
by covalent linkage, **135**, 341
in crosslinked polyacrylamide–hydrazide gels, **135**, 216
by entrapment in
gels, **135**, 440
membrane reactors, **135**, 372
with enzymes, **137**, 641
in epoxy carrier system, **135**, 252
on fibrin beads, **135**, 401
in flat bed hollow fiber reactor, **135**, 393
in flat bed reactors, **135**, 412
in fundamental biochemical studies, overview, **135**, 475
in gelatin, **135**, 293
industrial applications, overview, **136**, 353
in mammalian cell cultivation, **135**, 387
medical applications
evaluations *in vivo*, **137**, 453

examples, **137**, 445
methodology, **137**, 454
overview, **137**, 443
in organic synthesis, overview, **136**, 117
overview, **135**, 173
in polyacrylamide gels, **135**, 198
by prepolymer methods, **135**, 230
techniques and aspects, overview, **137**, 583
on transition metal-activated supports, **135**, 357
on wood chips, **135**, 333
Cell lines
BRL-3A (rat liver)
insulin-like growth factor II purification, **146**, 259
MSA carrier protein purification, **146**, 273
BSC-1 (African green monkey kidney), growth inhibitor isolation, **146**, 163
CHO (Chinese hamster ovary), *see* CHO cells
eukaryotic, transfection
for functional analysis of cloned genes, **152**, 691
for identification of gene regulatory elements, **152**, 613
hepatic, description, **151**, 19
indicator, in transforming growth factor bioassays, **146**, 342
mammalian
in full-length cDNA cloning and screening, **154**, 23
phage-mediated transduction of cDNA expression libraries, **151**, 434
mutant, resistant to transferrin receptor antibodies, selection, **147**, 276
protein kinase C-deficient, preparation, **141**, 419
RL-PR-C (rat liver), description, **151**, 30
transfected, fluorescence-activated cell sorting, **151**, 150
variant
with altered responses to hormones and growth factors, isolation, **147**, 370
mitogen-specific nonproliferative, isolation, **147**, 355
resistant to growth factor–toxin conjugates, isolation, **147**, 386
WRK-1, phosphoinositide pools, measurement, **141**, 83
Cell locomotion
Boyden chamber assay, **147**, 142
lymphocytes, mitogens as activators, **162**, 187
phagokinetic assay, **147**, 136
spectrophotometric assay, **147**, 144
Cell morphology
lymphocytes after mitogen treatment, analysis, **150**, 69
Cellobiase, *see* β-Glucosidase
Cellobiohydrolase, *see* Cellulose 1,4-β-cellobiosidase
Cellobiose
enzyme thermistor monitoring during cellulose hydrolysis, **137**, 311
Cellobiose dehydrogenase
assay, **160**, 444, 448, 457
in assay for cellulase, **160**, 112
detection in gels, **160**, 144
properties, **160**, 452, 459
purification from
Monilia sp., **160**, 458
Sclerotium rolfsii, **160**, 450
Sporotrichum thermophile, **160**, 445
Cellobiose dehydrogenase (quinone)
assay, **160**, 464
properties, **160**, 466
purification from *Sporotrichum pulverulentum*, **160**, 464
Cellobiose phosphorylase
assay, **160**, 468
properties, **160**, 470
purification from *Cellvibrio gilvus*, **160**, 469
Cellobiosidase
assay, **160**, 392
properties, **160**, 395
purification from *Ruminococcus albus*, **160**, 393
Cellodextrins
glycosides
applications, **160**, 51
assays, **160**, 48
preparation, **160**, 46
identification and analysis, **160**, 34
preparation, **160**, 26, 38
properties, **160**, 27

Cell panning
with density gradient centrifugation, isolation of human basophils, **162**, 531
Cell proliferation
effects of transferrin receptor antibodies, **147**, 275
entrapped bacteria, inhibition for proinsulin production, **137**, 632
lymphocytes
assessment by enzyme immunoassay, **150**, 117
after mitogen treatment, analysis, **150**, 69
lymphoid cells
correlation with transferrin receptor expression, **150**, 734
triggering receptors for, overview, **150**, 384
osteoblast-like cells, hormonal effects, **145**, 334
regulation by growth factors, characterization, **147**, 400
role of growth factor–nutrient interactions, kinetic analysis, **147**, 392
T cells, assay, **150**, 332
transforming growth factor effects, bioassays, **146**, 341
Cells, *see also specific cell types*
reconstituted from cytoplasts and karyoplasts
genetic isolation, **151**, 231
physical isolation, **151**, 229
Cell sorting
fluorescence-activated
cybrid isolation by, **151**, 232
hybrid and transfected cells
isolation of HLA antigen-expressing cells, **151**, 161
labeling procedures, **151**, 155
live/dead cell discrimination, **151**, 157
parameters for analysis and sorting, **151**, 156
recovery of rare cells, **151**, 159
sterile methodology, **151**, 157
Cell synchronization
procedure, **151**, 277
Cellular regulators
related articles previously published in Methods in Enzymology, **141**, 469
Cellulase
amorphous, crystalline, and dyed substrates, preparation, **160**, 19
assays, **160**, 87
based on cellobiose dehydrogenase, **160**, 112
complexes in *Clostridium thermocellum*
assays, **160**, 486
fractionation, **160**, 487
properties, **160**, 490
–hemicellulase complex
properties, **160**, 335
purification from *Phoma hibernica*, **160**, 333
nephelometric and turbidometric assay, **160**, 117
as plant plasma membrane marker, assay, **148**, 545
production *in vitro*, effect of culture conditions, **160**, 207
properties
Aspergillus enzyme, **160**, 262, 293
Cellulomonas enzymes, **160**, 214
Eupenicillium enzymes, **160**, 256
Humicola enzymes, **160**, 330
kidney bean enzymes, **160**, 350
mutant *Trichoderma* enzymes, **160**, 248
Pseudomonas enzymes, **160**, 206
Ruminococcus enzyme, **160**, 221
Sporocytophaga enzymes, **160**, 341
Thermoascus enzymes, **160**, 304
Thermomonospora enzyme, **160**, 316
Trichoderma enzymes, **160**, 233, 241
purification from
Aspergillus aculeatus, **160**, 279
Aspergillus fumigatus, **160**, 268
Aspergillus niger, **160**, 261
Cellulomonas uda, **160**, 212
Eupenicillium javanicum, **160**, 254
Humicola, **160**, 325
kidney bean, **160**, 347
Pseudomonas fluorescens, **160**, 202
Ruminococcus albus, **160**, 219
Sporocytophaga myxococcoides, **160**, 341
Thermoascus aurantiacus, **160**, 303
Thermomonospora fusca, **160**, 316
Trichoderma koningii, **160**, 224
Trichoderma reesei, **160**, 237

Trichoderma viride QM 9414 mutant strain, **160**, 245
type B, *in vitro* conversion to type A, **160**, 210
Cellulomonas uda
purification of cellulases, **160**, 212
Cellulose
alkali-swollen, preparation, **160**, 22
amorphous
endo-1,4-β-glucanase assay with, **160**, 105
preparation, **160**, 24
availability in biomass materials, enhancement by
biological pretreatments, **160**, 9
chemical pretreatments, **160**, 4
physicochemical pretreatments, **160**, 9
biocatalyst-immobilizing gels
activity yield, **135**, 275
characterization, **135**, 272
operational stability, **135**, 278
preparation, **135**, 270, 280
material costs, **135**, 280
temperature effects, **135**, 271
degrading enzymes, detection in prokaryotes, **160**, 180
dissolution
in *N*-ethylpyridinium chloride, **135**, 269
and hydrolysis, for cellodextrin preparation, **160**, 29, 30
hydrolysis, enzyme thermistor monitoring, **137**, 311
hydrolyzates, TLC, **160**, 176
isonitrile-derivatized, coupling of urease, **135**, 101
microcrystalline form, *see* Avicel
phosphoric acid-swollen
cellobiohydrolase assay with, **160**, 99
preparation, **160**, 23
regenerated, preparation, **160**, 24
solvolysis with cadoxen, **160**, 11
α-Cellulose
preparation, **160**, 21
Cellulose acetate
asymmetric membrane with immobilized enzymes, preparation, **137**, 63
biocatalyst-immobilizing beads
cell aggregation by, **135**, 283
cell entrapment, **135**, 287, 291
preparation, **135**, 283, 284
properties, **135**, 285
Cellulose 1,4-β-cellobiosidase
assays, **160**, 98, 126
detection in gels, **160**, 142
function, **160**, 313
properties, **160**, 311, 402
purification from
Humicola, **160**, 326
Penicillium pinophilum, **160**, 188, 400
Sclerotium rolfsii, **160**, 310
Trichoderma koningii, **160**, 231
Trichoderma reesei, **160**, 188
Cellulose disks
oligonucleotide synthesis on, **154**, 234
Cellulose nitrate, *see* Pyroxylin
Cellulosomes
assays, **160**, 473
properties, **160**, 479
purification from *Clostridium thermocellum*, **160**, 477
Cellvibrio gilvus
purification of cellobiose phosphorylase, **160**, 469
Cell volume
measurement technique, **146**, 385
Cell wall
bacterial, induction of hepatic granuloma for inflammation and repair studies, **163**, 608
cyanobacterial
isolation, **167**, 175
outer membrane
isolation of associated polymers, **167**, 180
preparation, **167**, 176
rigid layer and associated polymers, isolation, **167**, 177
stability and function, **167**, 177
streptococcal, induction of arthritis in rats, **162**, 373
Centrifugation
in analysis of
asialoorosomucoid subcellular distribution, **149**, 7
ATP and ADP binding to Na^+,K^+-ATPase, **156**, 200
extracellular matrix proteins, principles, **145**, 11
K^+, Na^+, and Rb^+ binding to Na^+,K^+-ATPase, **156**, 230
polysomes, **152**, 246
bacteriophage purification, **152**, 159

branched-chain α-keto acid dehydrogenase kinase complex purification, **166**, 316
branched-chain 2-keto acid dehydrogenase purification, **166**, 304
branched-chain α-keto acid dehydrogenase purification, **166**, 300
cardiac sarcolemmal vesicle isolation, **157**, 87
cardiac sarcoplasmic reticulum vesicle isolation, **157**, 94
density gradient
amyloplast isolation, **148**, 222
basophil isolation, **162**, 529
with cell panning, **162**, 531
bone proteoglycan purification, **145**, 266
chloroplast purification, **148**, 179
cyanobacterial plasma membrane isolation, **167**, 246, 437
cytochrome b_6/f complex isolation, **167**, 342
endosomes and lysosomes, **157**, 605
F_1-ATPase isolation, **167**, 357
Na^+,K^+-ATPase purification, **156**, 44
oxygen-evolving photosystem II particle isolation, **167**, 274
plant mitochondria isolation, **148**, 409, 415
plastid isolation, **148**, 157
prolamellar bodies and prothylakoids, **148**, 249
proteoglycans, **144**, 324, 332
vesicle-incorporated Na,K-pump purification, **156**, 139
differential, crude renal membrane preparation, **156**, 36
equilibrium, *E. coli* α-hemolysin purification, **165**, 142
equipment required for molecular cloning work, **152**, 8
filtering technique in analysis of cyanobacterial C_i uptake, **167**, 535
granule-enriched subcellular fractions of neutrophil preparation, **162**, 543
H^+-ATPase purification, **157**, 521, 530, 537, 539, 552, 625, 640
junctional and longitudinal sarcoplasmic reticulum isolation, **157**, 101
monoamine oxidase B purification, **142**, 634
Na^+,K^+-ATPase purification, **156**, 46, 50, 69
by airfuge, **156**, 42
by angle rotor, **156**, 41
by zonal rotor, **156**, 39
neutrophil and subcellular fraction preparation, **162**, 272
nonglyoxysomal peroxisome isolation, **148**, 519
Phanerochaete mitochondrial, ribosomal, and chromosomal DNA separation, **161**, 218
phycobiliprotein purification, **167**, 295
phycobilisome preparation, **167**, 305
plant mitochondria isolation, **148**, 444
plant plasma membrane isolation, **148**, 548
plant submitochondrial particle isolation, **148**, 448
plasmid purification, **152**, 168
RNA isolation, **152**, 224
skeletal muscle triad isolation, **157**, 55
ultra-, *see* Ultracentrifugation
zonal rate, proteoglycans, **144**, 333
Centrifuge column technique
in assay for Ca^{2+}-ATPase calcium occlusion and phosphorylation state, **157**, 229
Cephalexin
synthesis with immobilized penicillin amidase, **136**, 284
Cephalosporins
assay with
enzyme photometer, **137**, 210
enzyme thermistor, **137**, 209
Cerebroside-sulfatase
diagnostic assay
metachromatic leukodystrophy, **138**, 742
multiple sulfatase deficiency, **138**, 745
Cerebrospinal fluid
catecholamine metabolites, GC–MS assay, **142**, 577
extracellular matrix proteins, analysis, **145**, 396
Cerenkov radiation
in assay of ^{32}P-labeled nucleic acids, **152**, 51

Ceruloplasmin
assays, **163**, 442
characterization, **163**, 448
genetic polymorphism, **163**, 450
purification from human plasma, **163**, 446
Cesium chloride
with guanidinium thiocyanate, in RNA isolation, **152**, 224
Cetrimonium bromide
preparation of reverse micelles with, **136**, 216
Cetyltrimethylammonium bromide, *see* Cetrimonium bromide
Chaetomium trilaterale
purification of β-xylosidase, **160**, 697
Charcoal
–collodion-immobilized protein A
in cancer treatment, **137**, 513
preparation, **137**, 502
Chelating agents
inhibition of metalloenzymes, **158**, 110
Chelex 100
calcium-binding protein assay with, **139**, 75
removal of adventitious metals from solutions, **158**, 10
Chemical-quench assay
calcium release from sarcoplasmic reticulum, **157**, 471
Chemoattractant receptors
leukocyte-associated, measurement, **162**, 236
Chemokinesis
description of types, **162**, 9
effects of various factors, **162**, 129
neutrophils, agarose microdroplet assay, **162**, 75
Chemostat
sycamore cell suspension cultures in, **148**, 12
Chemotactic factor inactivator
assay, **162**, 229
isolation, **162**, 225
properties, **162**, 224
Chemotactic factors, *see also specific types*
bacterial, purification, **162**, 198
concentration gradients in chemotaxis assays, **162**, 85
crystal-induced
assay, **162**, 196
binding assay, **162**, 197
induction and isolation, **162**, 194
radioiodination, **162**, 196
for fibroblasts, preparation, **163**, 701
induced biochemical changes in leukocytes, analysis, **162**, 279
induced membrane phenomena in phagocytes, analysis, **162**, 271
mast cell-derived, assay, **162**, 521
overview, **162**, 127
Chemotaxis
attractant concentration gradients, assessment for
bridge visual assay, **162**, 87
Millipore filter assay, **162**, 90
under agarose assay
cylindrical system, **162**, 94
linear system, **162**, 97
effect of attractant degradation, **162**, 98
eosinophils
assay, **162**, 166
natural mediators, **162**, 165
fibroblasts, assay, **163**, 697
leukocytes
agonist peptides, synthesis, **162**, 133, 135
antagonist peptides, synthesis, **162**, 134, 138
chromium-51 radioimmunoassay, **162**, 59
collagen and fibrin gel assays, **162**, 23
effects of various factors, **162**, 129, 131
measurement *in vivo*
air sac procedure, **162**, 107
associated problems, **162**, 114
histopathological approaches, **162**, 102
with radiolabeled cells, **162**, 104
skin window technique and modifications, **162**, 105
micropore filter assay, **162**, 38
natural mediators, **162**, 153
overview, **162**, 3
polarization assay, **162**, 17
time-lapse cinematography, **162**, 29
visual assays on plane 2-D substrata, **162**, 35

macrophages, natural mediators, **162**, 140
mononuclear cells toward fibronectin fragments, assay, **162**, 222
neutrophils
agarose gel assay, **162**, 50
inhibitors, **162**, 223
natural mediators, **162**, 161
orientation chamber assay, **162**, 65
semiautomated assay with image analyzer, **162**, 79
Chemotherapy
with immobilized enzymes, **137**, 447
Chili pepper
cell culture establishment and maintenance, **135**, 411
cell immobilization techniques, **135**, 412
immobilized cells
capsaicin production, **135**, 420
nutrient uptake, **135**, 418
viability, **135**, 417
Chitin
colloidal, preparation, **161**, 423
conformation and structure, determination by physical methods, **161**, 435
degree of acetylation, determination by
colorimetric assay, **161**, 445
first derivative UV spectroscopy, **161**, 444
gas chromatography, **161**, 446
IR spectroscopy, **161**, 444
pyrolysis gas chromatography, **161**, 445
thermal analysis, **161**, 446
titration, **161**, 443
degree of deacetylation, spectroscopic determination, **161**, 422
enzymatic assay, **161**, 457
hydrolyzates
preparation, **161**, 411
TLC, **160**, 176
isolation from crustaceans, **161**, 417
molecular weight determination, **161**, 422
oligomeric fragments, HPLC, **161**, 410
protein content, assay, **161**, 421
purification, **161**, 403
solvents and solutions, **161**, 404
Chitinase
assay, **161**, 430, 460, 467, 471, 474, 480, 485, 493, 496
properties
kidney bean enzyme, **161**, 482
Neurospora enzyme, **161**, 473
Pycnoporus enzyme, **161**, 470
Serratia enzyme, **161**, 461
tomato enzyme, **161**, 489
Verticillium enzyme, **161**, 477
purification from
kidney bean leaves, **161**, 481
Neurospora crassa, **161**, 472
puffballs, **161**, 494
Pycnoporus cinnabarinus, **161**, 468
Serratia marcescens, **161**, 460
soybean seeds, **161**, 492
tomato, **161**, 487
Verticillium albo-atrum, **161**, 475
radiochemical assay, **161**, 424
viscosimetric assay, **161**, 426
Chitin deacetylase
assay, **161**, 511, 520
biological implications, **161**, 523
properties, **161**, 513, 522
protein determination, **161**, 522
purification from
Colletotrichum lindemuthianum, **161**, 521
Mucor rouxii, **161**, 512
Chitin synthase
assay, **138**, 643
properties, **138**, 648
purification from yeast, **138**, 645
Chitobiase, *see* *N*-Acetyl-β-glucosaminidase
Chitooligosaccharides
and reduced forms
enzymatic hydrolysis, **161**, 456
HPLC, **161**, 453
Chitosan
biocatalyst-immobilizing
applications, **135**, 266
chemical structure, **135**, 259
preparation, **135**, 259
properties, **135**, 264
degree of acetylation, determination by
colorimetric assay, **161**, 445
first derivative UV spectroscopy, **161**, 444
gas chromatography, **161**, 446
IR spectroscopy, **161**, 444
pyrolysis gas chromatography, **161**, 445

thermal analysis, **161**, 446
titration, **161**, 443
molecular weight distribution, determination by HPLC, **161**, 447
Chitosanase
assay, **161**, 502, 506
properties, **161**, 503, 508
purification from
Bacillus, **161**, 502
Streptomyces griseus, **161**, 507
Chlamydomonas reinhardii
chloroplasts, isolation, **148**, 166
Chloramphenicol acetyltransferase
assay
in identification of gene regulatory elements, **152**, 717
in plant transformants, **153**, 303
encoding gene, in collagen gene expression analysis, **144**, 67
reporter gene
applications in cyanobacteria, **167**, 741
optimization of gene transfer into eukaryotic cells
CAT assay, **151**, 389
plasmid preparation, **151**, 383
plasmid transfer, **151**, 387
Chlorella pyrenoidosa
purification of adenylylsulfate–ammonia adenylyltransferase, **143**, 357
Chloride
uptake by clathrin-coated vesicles, assay, **157**, 637
Chlorine
in lignin quantitation, **161**, 99
Chloroacetic acids
as solvents for chitin, **161**, 407
Chlorocarbonates
activated supports
active groups, determination, **135**, 88
affinity chromatography with, **135**, 89
coupling of proteins and ligands, **135**, 88
preparation, **135**, 87
Chlorogloeopsis
sheath isolation from, **167**, 185
α-Chloroisocaproate
activation of branched-chain α-keto acid dehydrogenase, **166**, 118
inhibition of branched-chain α-keto acid dehydrogenase kinase, **166**, 114
synthesis, **166**, 115
2-Chloro-10-(3-isothiocyanatopropyl)-phenothiazine
CAPP$_1$–calmodulin preparation with, **139**, 409
synthesis and titration, **139**, 407
p-Chloromercuribenzenesulfonic acid
in spectrophotometric assay of thiols, **143**, 60
p-Chloromercuribenzoic acid
in spectrophotometric assay of thiols, **143**, 60
2-(*p*-Chlorophenoxy)propionic acid
racemic ester, enzymatic preparation, **136**, 135
Chlorophylls
–carotenoid proteins, isolation and pigment content, overview, **148**, 353
chromatography, **148**, 357
extraction, **148**, 357
measurements in cyanobacterial mats, **167**, 483
pheophytinization and assay, **148**, 374
quantitation, **148**, 364
specific absorption coefficients, **148**, 362
spectral characteristics, **148**, 359
Chloroplasts
cytochromes
EPR spectra, **148**, 264
estimation *in situ*, **148**, 269
gel electrophoretic behavior, **148**, 267
molecular weights, **148**, 267
optical spectra, **148**, 260
redox potentials, **148**, 265
cytochromes c_6, sequence comparisons with cyanobacteria, **167**, 151
DNA, ribosomal protein genes, cloning and identification, **164**, 748
entrapment in alginate with hydrogenases, **135**, 453
envelope membrane fractions
immunochemical characterization, **148**, 214
level of cross-contamination, estimation, **148**, 216
polypeptide analysis, **148**, 213
preparation, **148**, 211
envelope polypeptides
antibodies to, preparation, **148**, 207
electrophoretic analysis, **148**, 200
galactolipid synthesis, **148**, 346
characteristics, **148**, 348

immobilization in
agar, **135**, 446
alginate, **135**, 443
polyurethane, **135**, 447
intact, immunochemical studies, **148**, 208
integrity criteria, **148**, 154
isolated
analysis after protein transport, **148**, 288
fatty acid synthesis, **148**, 331
product analysis, **148**, 333
newly synthesized fatty acids, analysis, **148**, 335
uptake of translation products, **148**, 284
isolation from
Chlamydomonas reinhardii, **148**, 166
Euglena gracilis, **148**, 169
pea, **148**, 149, 195
protoplasts
by mechanical rupture, **148**, 191
by rapid fractionation, **148**, 192
spinach, **135**, 443; **148**, 149, 165, 195
membranes
analysis for tocopherol and plastoquinone, **148**, 391
associated lipids
class analysis, **148**, 307
fatty acid analysis, **148**, 303
isolation, **148**, 301
molecular species, analysis, **148**, 311
quantification, **148**, 315
isolation, **148**, 296
phosphatidylglycerol synthesis, **148**, 342
characteristics, **148**, 344
phosphatidylglycerol synthesis, **148**, 339
3-phosphoshikimate 1-carboxyvinyltransferase purification, **142**, 353
plastocyanins, sequence comparisons with cyanobacteria, **167**, 148
plastoquinone synthesis, **148**, 390
characteristics, **148**, 340
prenylquinone synthesis, localization, **148**, 390
purification with silica sols, **148**, 179
synthesis of tocopherol and intermediates
associated enzymes, purification, **148**, 391
enzymatic assay, **148**, 387
identification and purification of labeled products, **148**, 388
thermolysin treatment
applications, **148**, 202
procedure, **148**, 198
1-Chloro-2-propanol, *see sec*-Propylene chlorohydrin
Chlortetracycline
in analysis of TRH-induced changes in cellular calcium metabolism, **141**, 45
fluorescence assay in neutrophils, **162**, 294
in measurement of Ca^{2+} fluxes in sarcoplasmic reticulum vesicles, **157**, 464
CHO cells
growth, **151**, 4
history, **151**, 4
mutagenesis, **151**, 6
mutants
drug-resistant
genetic analysis, **151**, 119
problems associated with isolation, **151**, 117
suitable drug concentration for isolation, **151**, 114
endocytosis-defective
genotypic analysis, **138**, 467
isolation, **138**, 460
quantitative testing, **138**, 466
glycosylation-defective, biochemical characterization, **138**, 443
repetitive cloning of dihydrofolate reductase gene, **151**, 451
suicide selection procedure, **151**, 121
preparation of leucine–tRNA ligase from, **166**, 263
replica plating
application to detection of antigen variants, **151**, 108
miscellaneous applications, **151**, 111
procedure, **151**, 106
special techniques with, **151**, 8
storage, **151**, 7
Cholate
dialysis, in reconstitution of Na,K-pump, **156**, 127
Cholera toxin
ADP-ribosyltransferase activity
assays, **165**, 245

comparison with pertussis toxin enzymatic activity, **165**, 242
competing reactions and artifactual results, **165**, 243
in presence of high NADase activity, **165**, 241
procedures and analyses *in vitro*, **165**, 236
required cytosolic factor, assay and purification, **165**, 246
assays, **165**, 170
production, **165**, 171
properties, **165**, 175
purification, **165**, 171
Cholesterol
assay with enzyme thermistor, **137**, 221
Cholesterol esterase
enzyme replacement therapy with, **137**, 574
Cholesterol oleate
radiolabeled, in measurement of liposome uptake by macrophages, **149**, 130
Cholic acid methyl ester
enzymatic oxidation in two-phase system, **136**, 156
Choline
^{3}H-labeled monopositive, isotope spaces and flux rates in sarcoplasmic reticulum vesicles, **157**, 422
Choline oxidase
immobilization, **137**, 174, 179
Cholinephosphotransferase
in plant endoplasmic reticulum
assay, **148**, 586
properties, **148**, 587
Chondroblasts
cultivation for elastin synthesis studies, **144**, 235
Chondrocytes
cultivation for analysis of macromolecular metabolism, **145**, 346
Chorion
human, DNA extraction for analysis of procollagen gene RFLPs, **145**, 209
Chorismate
assay, **142**, 424
conversion to tryptophan
in bacteria, **142**, 293
in fungi and higher eukaryotes, **142**, 299
HPLC, **142**, 423
preparation, **142**, 423, 440
Chorismate mutase
assay, **142**, 451, 463
–prephenate dehydratase
assays, **142**, 432
properties, **142**, 435
purification from *E. coli*, **142**, 433
–prephenate dehydrogenase
assays, **142**, 442
properties, **142**, 448
purification from *E. coli*, **142**, 444
substrate preparation, **142**, 440
properties
mung bean isozymes, **142**, 457
sorghum isozymes, **142**, 462
Streptomyces enzyme, **142**, 470
purification from
mung bean, **142**, 455
sorghum, **142**, 459
Streptomyces aureofaciens, **142**, 466
Chorismate synthase
assay, **142**, 362
as coupling enzyme in *arom* multifunctional enzyme assay, preparation, **142**, 331
properties, **142**, 365
purification from *Neurospora crassa*, **142**, 363
Chromaffin granules
preparation from bovine adrenals, **157**, 620
Chromatic adaptation
complementary, in cyanobacteria
action spectra, **167**, 326
rapid screening, **167**, 319
Chromatofocusing
cellobiohydrolases, **160**, 400
complement component C5b–6, **162**, 559
insulin-like growth factor I, **146**, 202
lignin peroxidase, **161**, 269
manganese peroxidase, **161**, 269
Chromatography, *see also specific techniques*
acetolactate synthase II, **166**, 449
acetolactate synthase III, **166**, 456
β-N-acetylhexosaminidase, **161**, 465
O-acetylhomoserine sulfhydrylase, **143**, 467, 472

O-acetylserine–*O*-acetylhomoserine sulfhydrylase, **143**, 481
O-acetylserine sulfhydrylase, **143**, 476
acid proteases, **160**, 503
actin, **139**, 847
adenosylhomocysteinase, **143**, 380, 432
adenylylsulfate–ammonia adenylyltransferase, **143**, 357
adenylylsulfate kinase, **143**, 344
aerolysin, **165**, 195
Aeromonas hemolysin, **165**, 201
alanine sulfodisulfane, **143**, 224
amine *N*-methyltransferase A, **142**, 663
amine *N*-methyltransferase B, **142**, 665
ω-amino acid–pyruvate aminotransferase, **143**, 501
1-aminocyclopropane-1-carboxylate synthase, **143**, 428
in analysis of glycosphingolipid biosynthesis, **138**, 580
anthranilate synthase, **142**, 302
anthranilate synthase–phosphoribosyltransferase
  AS partial complex, **142**, 379
  component I subunit, **142**, 381
  component II subunit, **142**, 382
  whole complex, **142**, 376
anthrax toxin, **165**, 111
α-L-arabinofuranosidase, **160**, 709, 714
arabinogalactanase, **160**, 721
arginine carboxypeptidase, **163**, 188
arogenate dehydratase, **142**, 502
arogenate dehydrogenase, **142**, 491, 516
aromatic amine dehydrogenase, **142**, 652
aromatic-amino-acid aminotransferase, **142**, 259, 269
aromatic-L-amino acid decarboxylase, **142**, 183
aromatic-amino-acid–glyoxylate aminotransferase, **142**, 275
arom multifunctional enzyme, **142**, 337
arylalkylamine *N*-acetyltransferase, **142**, 585
arylformamidase, **142**, 227
asparagusate reductase, **143**, 519
aspartate aminotransferase, **142**, 259
ATP-sulfurylase, **143**, 344
bacteriophage λ, for phage DNA preparation, **153**, 69
betaine–homocysteine methyltransferase, **143**, 385
bone morphogenetic protein, **146**, 299
bone proteoglycans, **145**, 265
branched-chain-amino-acid aminotransferase, **166**, 272, 276, 278, 280
branched-chain keto acid dehydrogenase, **166**, 347
branched-chain α-keto acid dehydrogenase phosphatase, **166**, 324
  inhibitor protein, **166**, 326
BSC-1 cell growth inhibitor, **146**, 165
calbindin D, **139**, 586
calcimedins, **139**, 21
calcium-binding proteins CAB-27 and CAB-48, **139**, 73
calcium channels, **157**, 495, 502
calmodulin, **157**, 330
calmodulin-activated phosphodiesterase, **159**, 559
calmodulin-dependent myosin heavy chain kinase, **139**, 111
calmodulin-stimulated phosphodiesterase inhibitor, **159**, 662
calregulin, **139**, 37
cAMP-binding protein from *Trypanosoma*, **159**, 289
cAMP-dependent protein kinase
  holoenzyme, **159**, 206
  isozyme components, **159**, 108
cAMP phosphodiesterase, **159**, 741, 754, 764, 775
carboxymethylcellulase, **160**, 379
cathepsin T, **142**, 286
cellobiohydrolase, **160**, 326
cellobiose dehydrogenase, **160**, 447, 450
cellobiose dehydrogenase (quinone), **160**, 465
cellobiose phosphorylase, **160**, 469
cellobiosidase, **160**, 393
cellodextrins
  analytical procedures, **160**, 34
  preparative procedures, **160**, 31
cellulase–hemicellulase complex, **160**, 334
cellulases, **160**, 202, 212, 245, 254, 261, 268, 279, 303, 316, 341, 348
cellulosomes, **160**, 478
cGMP-binding phosphodiesterase, **159**, 715, 724
cGMP phosphodiesterase, **159**, 709
chitinase, **161**, 468, 473, 476, 482, 488
chitin deacetylase, **161**, 512

chitosanase, **161**, 503, 508
cholera toxin, **165**, 172
chorismate mutase, **142**, 469
mung bean-derived isozymes, **142**, 456
sorghum-derived isozymes, **142**, 460
chorismate mutase–prephenate dehydratase, **142**, 434
chorismate mutase–prephenate dehydrogenase, **142**, 446
chorismate synthase, **142**, 364
clostridial enterotoxin, **165**, 96
collagen type IV, **145**, 55
complement components
C3, **162**, 601
C4, **162**, 597
C5, **162**, 601, 607
coniferyl-alcohol dehydrogenase, **161**, 303
C-reactive protein, **163**, 385
cyanobacterial cytochromes, **167**, 389
cyanobacterial ferredoxins, **167**, 405
cyclic nucleotide analogs, **159**, 76
cystathionine β-lyase, **143**, 447, 484
cystathionine γ-lyase, **143**, 489
cystathionine β-synthase, **143**, 392
cysteamine dioxygenase, **143**, 412
S-cysteine conjugates, **143**, 233
D-cysteine desulfhydrase, **143**, 451
cysteine dioxygenase, **143**, 396
cysteinesulfinate decarboxylase, **143**, 407
L-cysteine sulfoxide lyase, S-substituted, **143**, 436
cystine lyase, **143**, 441
cytochrome-*c* oxidase, **148**, 498
cytosolic factor required for cholera toxin activity, **165**, 248
damselysin, **165**, 186
3-dehydroquinate dehydratase, **142**, 322
dehydroquinate synthase, **142**, 310
dextransucrase, **138**, 653
dihydropteridine reductase, **142**, 100, 108, 123, 128
dipeptidyl carboxypeptidase I, **163**, 200
diphtheria toxin, **165**, 70
diphtheria toxin fragment A, **165**, 205
disialyl oligosaccharides, **138**, 296
dopamine β-monooxygenase, **142**, 600, 605, 612
EGF receptor antibodies, **146**, 72
endo-β-*N*-acetylglucosaminidase, **138**, 773
endo-β-*N*-acetylglucosaminidase H, **138**, 766
endochitinase, **161**, 499
endo-1,4-β-glucanases, **160**, 327, 329, 365, 371
endopectate lyase, **161**, 382
endopolygalacturonase, **161**, 388
eosinophil chemotactic factors, **162**, 167, 172, 176
Escherichia heat-stable enterotoxin, **165**, 129
Escherichia α-hemolysin, **165**, 141
estradiol receptor-kinase, **139**, 737
exocellulase, **160**, 405
exo-1,4-β-mannanase, **160**, 585
exopolygalacturonase, **161**, 388
factor XII, **163**, 72
factor XIIa, **163**, 77
factor D, **162**, 650
factor P, **162**, 652
ferredoxin–thioredoxin reductase, **167**, 426
fibroblast-derived growth factor, **147**, 43
fibroblast growth factor, **147**, 123
fibronectin, **162**, 215
fibronectin chemotactic fragments, **162**, 219
β-galactosidase, **149**, 34
galacturan 1,4-α-galacturonidase, **161**, 375, 378
glial growth factor, **147**, 221
glia maturation factor, **147**, 227
β-glucanases, **160**, 386
1,4-β-D-glucan cellobiohydrolase, **160**, 311
glucose oxidase, **161**, 310
β-glucosidases, **160**, 329, 410, 420, 439, 577
glycolipids, **138**, 203, 212
G protein subunits, **159**, 704
haptoglobin, **163**, 457
hemicellulases, **160**, 268
heparin lyase, **137**, 521
hepatocyte-stimulating factor, **163**, 598
4-hydroxyphenylpyruvate dioxygenase, **142**, 135, 140, 146, 152
indoleamine 2,3-dioxygenase, **142**, 193
indoleamine *N*-methyltransferase, **142**, 670
inhibitor-1 and -2 of protein phosphatase, **159**, 430

L-*myo*-inositol-1-phosphate synthase, **141**, 132
inositol trisphosphate, **141**, 145
insulin-like growth factor I, **146**, 202
β-isopropylmalate dehydrogenase, **166**, 432
α-isopropylmalate synthase, **166**, 417
isovaleryl-CoA dehydrogenase, **166**, 378
Kdp-ATPase, **157**, 672
α-ketoisocaproate dioxygenase, **166**, 292
kininogens, **163**, 241
kynurenine-oxoglutarate aminotransferase, **142**, 220
laminin receptor, **144**, 497
latent kallikrein, **163**, 99
leucine-binding proteins, **166**, 249
leucine dehydrogenase, **166**, 285
lichenase, **160**, 573
ligninase, **161**, 243
lipoamide dehydrogenase-valine, **166**, 349
lipocortins, **163**, 27
lymphocyte chemotactic factors, **162**, 154
lysophospholipase, **163**, 36
macrophage chemotactic factors, **162**, 144
manganese peroxidase, **161**, 260, 268
β-mannanase, **160**, 604, 613, 622
β-mannosidase, **160**, 592, 617
α-mannosidases I and II, **138**, 782
methanol oxidase, **161**, 324
methionine γ-lyase, **143**, 461
2-methyl branched-chain acyl-CoA dehydrogenase, **166**, 363
methylmalonate-semialdehyde dehydrogenase, **166**, 391
D-methylmalonyl-CoA hydrolase, **166**, 397
L-methylmalonyl-CoA mutase, **166**, 411
DL-methylmalonyl-CoA racemase, **166**, 403
monoamine oxidase, **142**, 625
monoamine oxidase A, **142**, 632
monophenol monooxygenase, **142**, 156, 166
monosialyl oligosaccharides, **138**, 293
NADPH-specific dihydropteridine reductase, **142**, 112
neisserial IgA protease, **165**, 119
neutrophil chemotactic factor, **162**, 162
norepinephrine *N*-methyltransferase, **142**, 657
oligosaccharides, effective size data, **138**, 84
oligosaccharyl sulfatase, **138**, 820
parvalbumin, **139**, 764
pectinesterase, **161**, 357, 388
pectin lyase, **161**, 353, 388
phenylalanine ammonia-lyase, **142**, 246, 250
phenylalanine 4-monooxygenase, **142**, 10, 20, 30, 39, 47, 52
phosphatidylinositol kinases, **141**, 213
phosphoribosylanthranilate isomerase–indoleglycerol-phosphate synthase, **142**, 393
phosphorylase phosphatase, **159**, 383
3-phosphoshikimate 1-carboxyvinyltransferase, **142**, 345, 351, 353
plague murine toxin, **165**, 164
plasma kallikrein, **163**, 93
plasma prekallikrein, **163**, 89
platelet-derived growth factor, **147**, 6, 16
PNGase F, **138**, 773
poly(*N*-acetylgalactosamine) deacetylase, **161**, 516
poly(A)$^+$ RNA, **152**, 254
polygalacturonase, **161**, 363, 396
prephenate aminotransferase, **142**, 486
prephenate dehydratase, **142**, 510
prephenate dehydrogenase, **142**, 506
proliferating cell nuclear antigen/cyclin, **150**, 150
prolyl 4-hydroxylase, **144**, 98
prostaglandins, **141**, 333
protein A, **143**, 401
protein disulfide-isomerase, **144**, 113
protein phosphatase
  high-molecular-weight forms, **159**, 371
  Mg^{2+}-dependent form, **159**, 441
  type 1, **159**, 397
  type 2A, **159**, 403
  type 2B, **159**, 411
  type 2C, **159**, 423
  types C-I and C-II, **159**, 362
protein S, **139**, 381
proteoglycans, **144**, 331
protopectinase, **161**, 338, 342, 344
Pseudomonas exotoxin A, **165**, 149

pyranose 2-oxidase, **161**, 318
restriction enzymes
*Hgi*CI, **155**, 4
*Not*I, **155**, 19
*Sfi*I, **155**, 20
sarcoplasmic reticulum protein cDNA, **157**, 295
selenocysteine β-lyase, **143**, 416, 494
Shiga toxin, **165**, 156, 250, 401
shikimate dehydrogenase, **142**, 317
shikimate kinases, **142**, 358, 360
sialyl oligosaccharide phosphates, **138**, 297
skeletal growth factor, **146**, 287
S-100 protein, **139**, 170
staphylococcal alpha toxin, **165**, 5
staphylococcal enterotoxins
type A, **165**, 46, 50
type B, **165**, 47
type C_1, **165**, 48, 51
type C_2, **165**, 48, 50
type D, **165**, 49, 51
type E, **165**, 50
staphylococcal leukocidin, **165**, 25
Streptococcus ATPase, **157**, 685
streptolysin O, **165**, 55
streptolysin S, **165**, 62
subunit-exchange, in purification of α-chymotrypsin, **135**, 488
succinate dehydrogenase (ubiquinone), **148**, 494
sugars, in characterization of radiolabeled lignins, **161**, 24
sulfatases, **143**, 364
sulfhydryl oxidase, **143**, 513
sulfotransferases, **143**, 204
taurine dehydrogenase, **143**, 498
thermostable direct hemolysin, **165**, 191
thioredoxin *f*, **167**, 422
thioredoxin *m*, **167**, 413, 419
thiosulfate reductase, **143**, 352
thrombospondin, **144**, 439
thylakoid lipids, **148**, 321
tissue kallikrein, **163**, 135
T-kininogen, **163**, 278
tonoplast H^+-ATPase, **148**, 124
transforming growth factor α, **146**, 140
tryptophanase, **142**, 417
tryptophan *C*-methyltransferase, **142**, 239
tryptophan 5-monooxygenase, **142**, 84, 94
tryptophan side chain oxidases I and II, **142**, 203
tryptophan synthase, **142**, 404
tyrosine 3-monooxygenase, **142**, 60, 67, 77
vitronectin, **144**, 436
vulnificolysin, **165**, 179
xylanase, **160**, 634, 639, 649, 656, 666, 673, 676
xylobiose and xylooligomers, **160**, 531
β-xylosidase, **160**, 666, 686, 690, 693, 697
zinc-containing phosphodiesterases, **159**, 780, 782
Chromatophores
Rhodopseudomonas capsulata
immobilized in crosslinked proteins
activity assays, **135**, 457
activity yields, **135**, 460
application of reactors, **135**, 463
preparation, **135**, 457
spectroscopy, **135**, 467
isolation, **135**, 456
Chromium
assay by atomic absorption spectrometry
in foods and tissues, **158**, 342
in serum, **158**, 339
in urine, **158**, 337
atomic absorption spectrometry, **158**, 133
^{51}Cr in radioimmunoassay for leukocyte chemotaxis, **162**, 59
as exchange-inert metal
applications to enzymes, **158**, 98
incorporation into enzymes, **158**, 102
as paramagnetic probe for Na^+,K^+-ATPase, **156**, 357
Chromobacterium violaceum
purification of phenylalanine 4-monooxygenase, **142**, 46, 52
Chromophores
selection for photosome preparation, **149,** 100
Chromoplasts
carotenogenic enzymes from membranes, solubilization and reconstitution, **148**, 392

isolated
enzymatic properties, **148**, 245
purity criteria, **148**, 244
isolation from
daffodil flowers, **148**, 242
red pepper fruits, **148**, 176
Chromosome hopping
in mapping and cloning of mammalian genomes, **151**, 475
Chromosomes
banding analysis in tumor cells
constitutive heterochromatin technique, **151**, 274
Giemsa technique, **151**, 269
nucleolus organizer regions technique, **151**, 274
quinacrine technique, **151**, 271
reverse technique, **151**, 273
cyanobacterial, genetic engineering, **153**, 215
in interspecific somatic cell hybrids, isozyme-based identification, **151**, 169
metaphase, mammalian, microcloning, **151**, 503
metaphase and prometaphase, *in situ* hybridization
principle, **151**, 279
problems encountered, diagnosis and correction, **151**, 289
procedures, **151**, 282
required materials, **151**, 280
microcell-mediated transfer
general strategy, **151**, 315
overview, **151**, 313
procedures, **151**, 317
required materials and reagents, **151**, 316
preparation for
banding analysis, **151**, 276
flow cytometric sorting, **151**, 253
in situ hybridization, **151**, 282
sorted
disease locus mapping, **151**, 311
identification, **151**, 299
miniaturized restriction enzyme analysis, **151**, 311
primary physical gene mapping, **151**, 312
spot-blot hybridization, **151**, 307
sorting, **151**, 296
by flow cytometry, **151**, 259
suspensions, preparation from
fibroblasts, **151**, 295
lymphocytes, **151**, 293
in T-cell clones and hybrids, analysis, **150**, 352
walking, *see* Genomic walking
yeast
introduction of *ura3*$^-$ deletions, **154**, 172
ploidy, visual assays, **155**, 351
site-directed rearrangements, **153**, 243
Chromozym PK
in assay of plasma prekallikrein, **163**, 171
Chroococcus minutus
sheath isolation from, **167**, 183
Chymase
assay, **163**, 327
properties, **163**, 314
purification from human mast cells, **163**, 323
Chymotrypsin
α-form
immobilized, peptide synthesis with, **136**, 284
thermostabilization, **137**, 619
coupling to tresyl-agarose, **135**, 69
gel-entrapped
catalytic activity, **135**, 593
preparation, **135**, 593
reactivation, **135**, 594, 595
thermoinactivation, **135**, 593
glass-immobilized
denaturation studies, **135**, 544
kinetics, **135**, 538
mechanistic studies in aprotic solvents, **135**, 540
preparation, **135**, 537
immobilized
in analysis of protein digestibility changes, **135**, 601
peptide synthesis with, **136**, 182, 184
purification by subunit-exchange chromatography, **135**, 488
Sepharose-immobilized
EPR studies, **135**, 504, 510
preparation, **135**, 504
spin labeling, **135**, 504

Cibacron Blue F3GA
in isolation of kinetic intermediates in ribosomal assembly *in vitro*, **164**, 270
Cinematography
time-lapse, leukocyte chemotaxis
analysis of cell tracks, **162**, 30
assessment of directional responses, **162**, 33
equipment, **162**, 29
C$\bar{1}$ inhibitor
assays, **163**, 180
properties, **163**, 184
purification from human plasma, **163**, 182
Cinnamic acid
carbon isotope-labeled, preparation, **161**, 51
Cinnamyl alcohol
carbon isotope-labeled, preparation, **161**, 51
Circular dichroism
calmodulin-binding sequences in peptides and proteins, **139**, 468
fibronectin, **145**, 50
laminin, **145**, 39
S-100 protein, **139**, 177
in titration of initiation factor 3 binding to RNA, **164**, 247
Cisternae
junctional terminal
characteristics of cardiac preparations, **157**, 102
isolation from
cardiac muscle, **157**, 101
skeletal muscle, **157**, 36
terminal, in isolated triads, enzymatic characteristics, **157**, 64
Citric acid
in isolation of plague murine toxin subunits, **165**, 168
Citrobacter freundii
purification of selenocysteine β-lyase, **143**, 493
Citronellol
racemic, enzymatic preparation, **136**, 126
Clathrin
coated vesicles, preparation from bovine brain, **157**, 638
Claviceps purpurea
alkaloid production after immobilization, **136**, 320
cultivation, **136**, 318
immobilization, **136**, 319
Cloacin
generated 3′ terminal fragment of 16S rRNA, preparation, **164**, 244
Cloning
alloreactive T cells from humans, **150**, 333
amplified genes, application of in-gel DNA renaturation, **151**, 366
antigen-specific T cells
with distinct functions, **150**, 342
from mice, **150**, 325
ATP synthase genes in *Synechococcus*, **167**, 795
bone glycoprotein genes, **145**, 286
branched-chain keto acid dehydrogenase genes from *Pseudomonas putida*, **166**, 350
calcium protease cDNA, **139**, 372
calmodulin gene from
Aplysia, **139**, 287
chicken, **139**, 217
fruit fly, **139**, 230
yeast, **139**, 248
cAMP-dependent protein kinase cDNA
catalytic subunit, **159**, 311
regulatory subunit II, **159**, 318
cDNA
by differential colony hybridization, **147**, 64
directional, in plasmid vectors, **152**, 349
high-efficiency protocol, **154**, 3
and library generation, overview, **152**, 307
into phages λgt10 and λgt11, **152**, 359
in plasmid vectors, **154**, 28
with primer–restriction end adapters, **154**, 64
vector–primer method, **152**, 371; **154**, 41
cosmids, simplified technique, **153**, 82
CR3 receptor cDNA, **150**, 614
cyanobacterial genes, with host–vector systems, **153**, 199

DNA fragments into M13 vectors, for dideoxy sequencing, **155**, 63
elastase mRNA sequences, **144**, 304
elastin cDNA, **144**, 265
in EMBL λ phage vectors, **153**, 103
epidermal growth factor gene, **146**, 39
in eukaryotes, with mitochondrial vectors, **153**, 366
fibronectin cDNA, **144**, 449
genes, with long oligonucleotide probes, **152**, 443
genomic DNA
with cosmid vectors, **151**, 397; **152**, 199
underrepresentation of specific sequences in, alleviation, **152**, 173
HeLa cells, **151**, 44
interleukin-2 receptor cDNA, **150**, 692
α-isopropylmalate synthase gene, **166**, 422
Kdp-ATPase genes, **157**, 663
kininogen cDNAs and genomic DNAs, **163**, 230
laminin cDNA, **144**, 464
macroregions of mammalian genomes, strategies, **151**, 461
micro-, *see* Microcloning
mismatch, application of restriction enzyme *Hgi*CI, **155**, 9
mitogen- and antigen-reactive B cells on filter paper disks, **150**, 287
mitogen-nonresponsive 3T3 cell variants, **147**, 364
molecular
laboratory equipment for, **152**, 3
partial DNA digestion for, **152**, 183
phage and plasmid DNA preparation for, **152**, 145
RNA preparation for, prevention of RNase contamination, **152**, 20
use of ^{32}P- and ^{35}S-labeled compounds, safety considerations, **152**, 25
Na^+,K^+-ATPase α subunit gene with antibody probes, **156**, 379
9-kDa calcium-binding protein cDNA from rat, **139**, 343, 351
parvalbumin gene from rat, **139**, 317
in plant cells, vectors for, **153**, 277
protein S gene from *Myxococcus*, **139**, 383
rare mRNAs, subtraction technique, **152**, 425
repetitive, dihydrofolate reductase gene from mutant CHO cells
construction of
*Bgl*I-free cosmid vector, **151**, 451
cosmid cloning vectors, **151**, 453
ligation of genomic and vector DNA fragments, **151**, 453
packaging in λ phage heads, **151**, 455
screening drug-resistant colonies by hybridization, **151**, 456
ribosomal protein genes
in chloroplast DNA, **164**, 748
L24 missense mutants, strategy, **164**, 707
sarcoplasmic reticulum protein cDNAs, **157**, 289
shotgun
prokaryotic promoters, **153**, 61
with *Streptomyces* phage vector, **153**, 152
with *Streptomyces* plasmid vector, **153**, 119
single hapten-specific B cells, **150**, 271
slow/cardiac troponin C cDNA from quail, **139**, 328, 334
Streptomyces DNA, with cosmid shuttle vectors, **153**, 166
sub-, *see* Subcloning
T-cell receptor genes, **150**, 655
transferrin receptor gene, **147**, 280; **150**, 739
yeast genes, by complementation, **152**, 481
Clostridium
C. acetobutylicum, immobilized cells, long-term stability, control, **137**, 626
C. botulinum, isolation of neurotoxins, **165**, 79
C. butyricum, hydrogen production in fuel cell, **137**, 708
C. perfringens
diarrheal toxin, ileal loop test, **165**, 315
purification of
alpha toxin, **165**, 91
enterotoxin, **165**, 97
C. tetani, purification of tetanus toxin, **165**, 86

C. thermocellum
cultivation, **160**, 484
isolation of
macrocellulase complexes, **160**, 487
yellow affinity substance, **160**, 495
purification of
cellulosomes, **160**, 477
endoglucanase, **160**, 352
C. tyrobutyricum, strain La 1, (*E*)-2-methylbutenoate reduction, **136**, 315
Clostridium histolyticum aminopeptidase
assay, **136**, 171, 173
immobilization and properties, **136**, 172
sequential peptide hydrolysis with, **136**, 174
Clotting assay
factor XII, **163**, 57, 71
plasma prekallikrein, **163**, 87
CMP (cytidine 5′-monophosphate)
intracellular, manipulation for glycerophospholipid synthesis alteration, **141**, 224
Cobalt
assay by atomic absorption spectrometry
in animal and plant tissues, **158**, 350
in blood, **158**, 349
in serum, **158**, 349
atomic absorption spectrometry, **158**, 134
as exchange-inert metal
applications to enzymes, **158**, 98
incorporation into enzymes, **158**, 102
Cobalt chloride, *see* Cobaltous chloride
Cobaltous chloride
in cytochemical localization of Na^+,K^+-ATPase, **156**, 422
Coenzyme A
short-chain compounds, HPLC assay, **166**, 70
Coenzymes
immobilized active
applications, **136**, 7
preparation, **136**, 3
properties, **136**, 6
Colchicine
in isolation of mitogen-specific nonproliferative cells, **147**, 356
Colicin
generated 16S rRNA fragments, isolation and characterization, **164**, 188
Collagen
antibodies, characterization and isolation, **145**, 148
assay in cell cultures, **147**, 409
biosynthesis, analytical methods, **163**, 642, 675
borohydride-reducible and mature crosslinking residues, assay, **144**, 120
α chains
electrophoretic isolation, **145**, 173
peptide mapping
with cyanogen bromide, **145**, 175
by limited proteolysis, **145**, 177
crosslinked polypeptides
chromatographic assay, **144**, 135
fluorescent detection in gels, **144**, 137
crosslinking amino acids, overview, **144**, 115
degradation, analytical methods, **163**, 689
enzymatic hydrolysis, specificity, **144**, 153
genes
expression, **144**, 65; **163**, 688
structural analysis, **144**, 62
group 1 molecules
common structural features, **144**, 13
HPLC, **144**, 59
hydroxypyridinium crosslinks, assay, **144**, 127
mutations
assessment by R-loop analysis of procollagen mRNA, **145**, 235
detection by 2-D gel electrophoresis, **145**, 204
peptides generated by cyanogen bromide digestion
detection in gels, **145**, 199
preparation, **145**, 188
radiolabeling, **145**, 188
2-D gel electrophoresis, **145**, 191
–polymer composite material, with bound enzymes
characterization and applications, **137**, 549
preparation, **137**, 546
synthesis in bone cell cultures, analysis, **145**, 318
type I
assay in solution, **144**, 145

chemical labeling, **144**, 143
electron immunohistochemistry, **145**, 80
enzymatic hydrolysis, assay, **144**, 144
extraction, **144**, 142
fibrillar, assays, **144**, 147
as fibroblast chemoattractant, **163**, 701
gene transcription assay, **144**, 75
mRNA steady-state levels, assay, **144**, 79
mRNAs, translation *in vitro*, **144**, 84
in osteogenesis imperfecta fibroblasts, mRNA analysis, **145**, 223
overview, **144**, 6
pro-α2 gene, mutation analysis, **145**, 213
purification, **144**, 143
type II
antibody response to, **162**, 368
cellular response to, **162**, 369
electron immunohistochemistry, **145**, 81
as fibroblast chemoattractant, **163**, 701
induced arthritis
assessment, **162**, 358, 365
in experimental immunotherapy, **162**, 372
passive transfer in mice, **162**, 370
mRNAs, *in vitro* translation, **144**, 84
overview, **144**, 7
preparation from
bovine articular cartilage or chicken sternum, **162**, 364
bovine nasal septum, **162**, 358
type III
electron immunohistochemistry, **145**, 82
fibrillar, assays, **144**, 150
as fibroblast chemoattractant, **163**, 701
overview, **144**, 8
preparation, **144**, 148
type IV
assays, **144**, 152
assembly *in vitro*, **145**, 62
basement membrane-associated
properties, **145**, 364
purification, **145**, 368
derived antigens, purification, **145**, 409
electron immunohistochemistry, **145**, 96
electron microscopy, **145**, 58, 65
isolation
and characterization, **144**, 45, 47, 57
from tumor tissue, **145**, 54
overview, **144**, 20; **145**, 53
preparation and labeling, **144**, 150
type V
assays, **144**, 153
electron immunohistochemistry, **145**, 84
isolation and characterization, **144**, 44, 54
overview, **144**, 9
preparation and labeling, **144**, 152
type VI
electron immunohistochemistry, **145**, 85
electron microscopy, **145**, 67
isolation and characterization, **144**, 45, 48
overview, **144**, 26
type VII
electron immunohistochemistry, **145**, 85
isolation and characterization, **144**, 56
overview, **144**, 31
type VIII
isolation and characterization, **144**, 49
overview, **144**, 32
type IX
isolation and characterization, **144**, 44, 50, 56
overview, **144**, 33
type X
isolation and characterization, **144**, 52
overview, **144**, 36
type K
isolation and characterization, **144**, 47, 54
overview, **144**, 11
Collagenase
in analysis of fibroblast collagen production, **163**, 646
antibodies, production, **144**, 166
in assay for
collagen degradation, **163**, 690
collagen production, **163**, 677
competition ELISA, **144**, 169
detection and measurement, **144**, 141

hydrolysis of collagen, specificity, **144**, 153
immunoprecipitation, **144**, 170
immunoreactivity on blots, **144**, 168
isolation from
fibroblasts, **144**, 162
neutrophils, **144**, 164
metallo-, *see* Metallocollagenase
mRNA isolation and *in vitro* translation, **144**, 170
type IV, isolation, **144**, 165
type V, isolation, **144**, 166
Collagen gel assay
leukocyte chemotaxis
assessment of cell distribution in gels, **162**, 27
direct observation in 3-D gels, **162**, 28
gel preparation, **162**, 23
mitogen-activated lymphocyte locomotion, **162**, 190
Collagen membranes
enzyme electrode preparation with, **137**, 23
Colletotrichum lindemuthianum
culture conditions, **161**, 519
purification of chitin deacetylase, **161**, 521
Collodion
–charcoal-immobilized protein A
in cancer treatment, **137**, 513
preparation, **137**, 502
Colominic acid
depolymerization by polysialic acid depolymerase, **138**, 789
Colony hybridization
differential, cDNA cloning by, **147**, 64
in screening of cosmid libraries, **151**, 420
Colony-stimulating factors
granulocyte–macrophage, gene isolation, **150**, 752
Colorimetric assay
acetolactate synthase I, **166**, 436
O-acetylhomoserine sulfhydrylase, **143**, 465, 470
α-L-arabinofuranosidase, **160**, 707, 712
arylsulfatases, **143**, 207
ATP-sulfurylase, **143**, 335
branched-chain-amino-acid aminotransferase, **166**, 275
chitinase, **161**, 474, 485
comparison with other assays, **161**, 434
interferences, **161**, 434
modifications, **161**, 433
procedure, **161**, 431
chitin and chitosan acetylation degree, **161**, 445
chitin deacetylase, **161**, 520
complexed iron in biological samples
procedure, **158**, 359, 362
reagents, **158**, 358, 360
cystathionine β-synthase, **143**, 390
cysteamine, **143**, 152
cysteinesulfinic acid, fuchsin method, **143**, 160
L-cysteine sulfoxide lyase, S-substituted, **143**, 435
cystine lyase, **143**, 439
ferredoxin–thioredoxin reductase, **167**, 423
α-galactosidase, **160**, 628
galacturan 1,4-α-galacturonidase, **161**, 373
β-glucosidase, **160**, 576
H^+-ATPase, **157**, 514, 529, 635
heparin lyase, **137**, 516
β-mannosidase, **160**, 616
molybdenum, **158**, 372
Na^+,K^+-ATPase, **156**, 36
penicillamine, **143**, 190
phenylalanine 4-monooxygenase, **142**, 5
polygalacturonase, **161**, 362
sialic acids, **138**, 138
sulfite, fuchsin method, **143**, 15
thiocyanate, **143**, 23
thioredoxin *f*, **167**, 421
thiosulfate, **143**, 24
tungsten, **158**, 373
in presence of molybdenum, **158**, 374
Colostrum
bovine
growth factor purification, **146**, 45
isolation of sialyl oligosaccharides and phosphates, **138**, 289
Compactin, *see* Mevastatin
Complement
alternative pathway
assays, **162**, 641

C3
assays, **162**, 645
purification from human plasma, **162**, 648
factor B, assay, **162**, 648
factor D, purification from human plasma, **162**, 650
factor H, assay, **162**, 651
factor P, purification from human plasma, **162**, 652
C1
functional analysis, **162**, 628
immunochemical analysis, **162**, 631
inhibitor purification, **162**, 636
purification from human plasma, **162**, 632
C2
functional analysis, **162**, 628
immunochemical analysis, **162**, 631
purification from human plasma, **162**, 634
C3
functional analysis, **162**, 589
immunochemical analysis, **162**, 583
purification from human plasma, **162**, 596
structure and function, **162**, 612
C3a
generation and properties, **162**, 616
immunochemical analysis, **162**, 583
isolation from human serum, **162**, 617
C3b
metastable binding site, **162**, 620
physiological fragmentation, **162**, 621
C3c, significance, **162**, 622
C3d-containing fragments, generation and properties, **162**, 622
C4
functional analysis, **162**, 628
immunochemical analysis, **162**, 631
purification from human plasma, **162**, 596, 633
C4a
generation, **162**, 616
immunochemical analysis, **162**, 583
isolation from human serum, **162**, 617
C4b, binding protein purification, **162**, 637
C4b/C3b, inactivator purification, **162**, 638
C5
derived fibroblast chemoattractant, **163**, 705
functional analysis, **162**, 589
immunochemical analysis, **162**, 583
purification from human plasma, **162**, 596
structure and function, **162**, 612
C5a
assay, **150**, 620
bacterially produced
biological characterization, **162**, 665
chemical characterization, **162**, 663
identification, **162**, 659
purification, **162**, 662
reconstitution from extracts, **162**, 662
chemotactic and other biological activities, **162**, 619
gene
nucleotide sequence, **162**, 654
synthesis, **162**, 653
generation and properties, **162**, 616
immunochemical analysis, **162**, 583
isolation from human serum, **150**, 617; **162**, 617
radioiodination, **150**, 619
C5b, metastable binding site, **162**, 620
C5b–6
assay, **162**, 557
purification from human plasma, **162**, 557
C7
assay, **162**, 561
purification from human plasma, **162**, 561
C8
assays, **162**, 565
properties, **162**, 572
purification from human plasma, **162**, 564
C9
assays, **162**, 573
properties, **162**, 576
purification from human plasma, **162**, 573
component proteins, overview, **162**, 551, 579, 626
dependent liposome uptake by macrophages, **149**, 124

and immune complexes, as mediators of inflammation, overview, **163**, 7
liposome opsonization by, **141**, 248
selective depletion in animals, **162**, 453
Complementation
in isolation of yeast genes
background, **152**, 481
limitations, **152**, 486
method extensions, **152**, 487
plasmid library construction, **152**, 492
proving clone identity, **152**, 500
mutant cells by mRNA after microinjection
applications, **151**, 379
overview, **151**, 371
procedure, **151**, 373
mutations, in isolation of cyanobacterial photosynthetic apparatus genes, **167**, 764
Complementation analysis
Kdp-ATPase gene mutations, **157**, 664
Tn*5* insertions in *Rhizobium* nitrogen fixation genes, **154**, 208
Complement receptors
C1q
assays, **150**, 560
properties, **150**, 578
purification, **150**, 571
C5a
binding assay on neutrophils, **162**, 241
crosslinking studies, **150**, 621
detection on neutrophils, **150**, 624
type 1
assays, **150**, 582
properties, **150**, 585
purification from human erythrocytes, **150**, 583
type 2, assays, **150**, 589
type 3
assays, **150**, 607
biosynthesis, **150**, 613
cloning, **150**, 614
purification from human granulocytes, **150**, 605
Computer applications
data acquisition system for neutrophil chemotaxis assay, **162**, 80
enzyme electrodes, **137**, 50
enzyme thermistor-based process monitoring and control, **137**, 307
EPR studies of immobilized chymotrypsin, **135**, 507
image analysis of
ribosomal electron micrographs, **164**, 3, 53
ribosomal subunit electron micrographs, **164**, 35
nucleic acid sequence analysis, **164**, 765
programs for
calculating ionic concentrations in aqueous solutions of metals and ligands
data entering instructions, **157**, 397
overview and availability, **157**, 378
SPECS program source code, **157**, 383
DNA and protein sequencing, **155**, 582
simulation of DNA melting, application to gel electrophoresis, **155**, 482
simultaneous synthesis of oligonucleotides, **154**, 226, 229
Concanavalin A
affinity immobilization with, **137**, 650
crosslinked derivatives
biological activities, **150**, 27
characterization, **150**, 25
preparation, **150**, 18
in erythrocyte-mediated microinjection, **149**, 298
glycopeptide chromatography with, **138**, 244
labeling of plant plasma membranes, **148**, 572
liposome opsonization by, **141**, 251
plant protoplast immobilization with, **135**, 424, 426
T-cell proliferative responses in basal serum-free medium, **150**, 130
Condensation
hydrolase-catalyzed, factors influencing product yield, **136**, 286
Congo Red
complex formation with polysaccharides, in assay of β-D-glucanases, **160**, 59
Coniferyl alcohol
^{14}C-labeled
polymerization, **161**, 71
synthesis, **161**, 66
Coniferyl-alcohol dehydrogenase
assay, **161**, 302

properties, **161**, 304
purification from *Rhodococcus erythropolis*, **161**, 303
Conjugation
bacterial, mediation of plasmid transfer to cyanobacteria, **153**, 232; **167**, 747
proteins to liposomes, **149**, 111
Connective tissue-activating peptide
analytical methods, **163**, 735
characteristics, **163**, 733, 748
granulocyte-derived, isolation, **163**, 745
overview, **163**, 731
type I, amino acid composition, **163**, 747
type Ib
amino acid composition, **163**, 747
isolation, **163**, 744
type III
amino acid sequence, **163**, 746
isolation, **163**, 743
type V, isolation, **163**, 743
Contact activation
in human plasma, fluorometric assay, **163**, 80
Contact sensitivity
induction in mice, **163**, 361
Continuous-flow analysis
with enzyme reactors, **137**, 263
Copper
assay by atomic absorption spectrometry, **158**, 351
atomic absorption spectrometry, **158**, 135
binding to chromatographic media, analysis, **158**, 19
cold cyanolysis of sulfane in presence of, **143**, 28
in cyanobacterial cytochrome-*c* oxidase, EPR detection, **167**, 450
–*o*-phenanthroline, in detection of Na^+,K^+-ATPase subunit interactions, **156**, 347
in spectrophotometric assay of thiols, **143**, 60
Corn
biosensors based on kernel tissues, **137**, 150
Corn syrup
high-fructose, industrial production with immobilized glucose isomerase, **136**, 356
Corticium rolfsii
purification of polygalacturonase, **161**, 363
Cortisone
enzymatic reduction in two-phase system, **136**, 154
Corynebacterium
C. diphtheriae, purification of diphtheria toxin, **165**, 69
entrapment with photocrosslinkable resin prepolymers, **135**, 236
prepolymer-entrapped, applications, **135**, 247
Cosmid libraries
colony screening
analytical cosmid DNA extraction, **151**, 425
autoradiography, **151**, 421
colony selection and subsequent screening, **151**, 423
plasmid blotting on nitrocellulose, **151**, 424
plating and growth of bacteria, **151**, 418
prehybridization and hybridization, **151**, 420
preparation of filters, **151**, 419
required materials and solutions, **151**, 417
construction
ligation, **151**, 413
linearization and phosphatase treatment of vector DNA, **151**, 411
packaging, infection, and plating, **151**, 413
preparation of
high-molecular-weight DNA, **151**, 407
partial restriction fragments, **151**, 410
fractionation for gene isolation by sib selection
serial procedure, **151**, 447
simple procedure, **151**, 446
Cosmids
analytical extraction, **151**, 425
in cloning large segments of genomic DNA, **152**, 199
for genomic walking and restriction mapping, **152**, 604

λSV2*cos*
construction, **151**, 398
integration and excision, **151**, 398, 402
in vivo packaging, **151**, 400
pSV13, construction, **151**, 403
as shuttle vectors
for cloning and analysis of *Streptomyces* DNA, **153**, 166
principle, **151**, 426
procedures, **151**, 427
simplified cloning technique, **153**, 82
Cotton fibers
in assays for
endo-1,4-β-glucanase, **160**, 108
total cellulase, **160**, 93
dewaxed, preparation as cellulase substrate, **160**, 19
p-Coumaryl alcohol
preparation, **161**, 73
Covalent coupling
enzyme immobilization by, **135**, 37
C-reactive protein
assays, **163**, 392
chemical characterization, **163**, 387
purification, **163**, 383
purity criteria, **163**, 392
Creatine kinase
as marker for chromosome detection in somatic cell hybrids, **151**, 180
Creatinine
assay with
biosensor, **137**, 67
dry reagent chemistries, **137**, 413
enzyme reactor, **137**, 267
enzyme transistor, **137**, 245
hybrid biosensor, **137**, 132
Creatinine deiminase
as creatinine sensor component, **137**, 132
Crocetindial
determination in carotene oxygenase assay, **167**, 339
Crosslinking
in analysis of transferrin receptor, **150**, 736
aspartate transcarbamoylase with tartryl diazide, **135**, 571, 574
catalase with diimido esters, **137**, 602
complement C5a to neutrophils or derived membranes, **150**, 621
dCMP deaminase with glutaraldehyde
in activated R conformation, **135**, 579
in inhibited T conformation, **135**, 581
elastin, **144**, 182
α-glucosidase to albumin, **137**, 569
[^{125}I]calmodulin to MAP_2, **139**, 841
IgG to asparaginase–albumin, **137**, 570
insulin to α-glucosidase–albumin, **137**, 569
intramolecular, enzyme stabilization by, **137**, 615
intra-RNA, in *E. coli* ribosomes, **164**, 290
Na^+,K^+-ATPase with
glutaraldehyde, for analysis of quaternary structure, **156**, 333
various reagents, for detection of subunit interactions, **156**, 345
protein–protein, identification in *E. coli* ribosomes, **164**, 64
pyranose oxidase with diimido esters, **137**, 602
RNA–protein
in *E. coli* ribosomes, **164**, 298
with soluble carbodiimide, **164**, 311
rRNA by psoralen, **164**, 330
5S RNA, inter- and intramolecular, **164**, 730
Crosslinking reagents
membrane-impermeant, in analysis of Fc receptor for IgE, **150**, 503
Crown ethers
as metal ion buffer ligands, **158**, 54
Cryopreservation
cyanobacteria, **167**, 123
Cryptands
as metal ion buffer ligands, **158**, 54
Cryptococcus albidus
purification of xylanase, **160**, 639
Crystallization
Ca^{2+}-ATPase in sarcoplasmic reticulum, **157**, 272
calbindin D, **139**, 588
calcium-binding protein from skeletal muscle sarcoplasmic reticulum, **157**, 321
calmodulin, **139**, 633
catabolite gene activator protein with cAMP, for structural analysis, **159**, 278

endoglucanase D, **160**, 360
isomaltulose, **136**, 448
leukocidin F and S components, **165**, 26
Na^+,K^+-ATPase, **156**, 16
troponin C, **139**, 612
two-dimensional
membrane-bound Na^+,K^+-ATPase, **156**, 80
Na^+,K^+-ATPase in large membrane fragments, **156**, 42
tetanus toxin, **165**, 389
Crystallography
5S RNA and fragments, sample preparation for, **164**, 158
X-ray, *see* X-ray crystallography
Cucumber
biosensors based on foliar tissues, **137**, 151
Culture media
basal serum-free, for analysis of T-cell responses
applications, **150**, 130
composition, **150**, 130
properties, **150**, 133
chemically defined, for lymphocyte growth
applications, **150**, 145
cell cultivation in, **150**, 140
formulation, **150**, 135
variations, **150**, 142
for cyanobacteria
composition, **167**, 6, 79
solid systems, preparation, **167**, 15
liquid, single B-cell culture in, **150**, 265
for marine planktonic cyanobacteria, **167**, 101
for production of
aerolysin, **165**, 194
anthrax toxin, **165**, 107
bacterial streptolysin O, **165**, 54
Escherichia α-hemolysin, **165**, 139
Shiga toxin, **165**, 154
staphylococcal γ-hemolysin, **165**, 12
for thermophilic cyanobacteria, **167**, 99
for transforming growth factor bioassays
plasma-supplemented, **146**, 348
serum-free, **146**, 349
serum-supplemented, **146**, 347
Culture substrata
elastic rubber, for analysis of myofibroblast contractile forces, **163**, 624
haptotactic, for analysis of fibroblast and myofibroblast adhesion, **163**, 631
Cyamopsis tetragonolobus, *see* Guar
Cyanidium caldarium
purification of phycobilisomes, **167**, 308
Cyanobacteria, *see also specific cyanobacteria*
akinetes, identification, production in culture, isolation, and properties, **167**, 222
alternate strain designations and numbers in culture collections, **167**, 811
ammonium transport, **167**, 557
application of reporter genes, **167**, 736
baeocytes, description, **167**, 233
bioenergetics, ESR and NMR analyses, **167**, 509
carbohydrate transport, **167**, 551
carbon metabolism, analysis with radioisotopes, **167**, 539
carboxysomes, research methodology, **167**, 204
cation transport, **167**, 561
cell envelopes
isolation, **167**, 174
monomeric constituents, analyses, **167**, 186
polysaccharides from external layers, isolation, **167**, 185
cell numbers, estimation methods, **167**, 237
cell walls
isolation, **167**, 175
outer membrane
isolation of associated polymers, **167**, 180
preparation, **167**, 176
rigid layer and associated polymers, isolation, **167**, 177
stability and function, **167**, 177
chromatic adaptation, physiological conditions and action spectra, **167**, 318
cloned genes, listing, **167**, 831
cloning vectors, listing, **167**, 824
coflocculation with clay particles, **167**, 621
communities, analysis with microsensors, **167**, 639
conjugal transfer of plasmids, **153**, 232; **167**, 747

culture conservation without growth, **167**, 92
cyanophycin inclusions, research methodology, **167**, 207
cyclic nucleotides, extraction and analysis, **167**, 584
cytochrome b_6/f complex purification, **167**, 342
cytochrome-*c* oxidase, characterization, **167**, 437
cytochromes c_6, sequence comparisons with chloroplasts, **167**, 151
cytoplasmic inclusions, ultrastructural analysis, **167**, 163
ecological distribution, **167**, 30
elective cultures, **167**, 71
energetic status, NMR analysis, **167**, 667
ESR oximetry, **167**, 670
extracellular proteins, analysis, **167**, 599
fimbriae
  electron microscopic analysis, **167**, 192
  isolation, **167**, 194
  overview, **167**, 189
  ultrastructural analysis, **167**, 167
fine structure
  analysis by
    freeze-drying, **167**, 159
    freeze-etching, **167**, 158
    metal shadowing, **167**, 159
    negative staining, **167**, 160
  observation in thin sections, **167**, 160
flocculants, **167**, 616
gas vesicles, chemical and physical properties, **167**, 213
gene cloning with host–vector systems, **153**, 199
genera not represented in Pasteur Collection, taxonomic position, **167**, 58
geosmin production, **167**, 605
growth
  conditions, **167**, 16
  and harvest, **167**, 438
  and maintenance conditions, **167**, 90
growth medium composition, **167**, 6, 79
heat shock response, **167**, 622
heterocysts
  electron donation to nitrogenase, **167**, 496
  isolation, **167**, 490
hormogonia, description, **167**, 235, 851
identification, **167**, 33
immobilization in
  agar, **135**, 446
  polyurethane, **135**, 448
  solid matrices, **167**, 629
inorganic carbon uptake, **167**, 534
iron–sulfur proteins, EPR characterization, **167**, 427
isolation
  procedures, **167**, 21, 72
  tools, **167**, 18
major sources of cultures, listing, **167**, 67
major subgroups in axenic culture, description, **167**, 62
marine planktonic
  culture media, **167**, 101
  overview, **167**, 100
  purification, **167**, 104
  stock cultures, **167**, 104
mats
  chlorophyll measurements, **167**, 483
  nitrogen fixation, **167**, 474
membrane lipids, analysis, **167**, 251
membrane vesicles, size and surface charge analysis, **167**, 691
mutagenesis, **167**, 714
mutants in energetic pathways, enrichment, **167**, 728
nitrogen fixation genes, in DNA hybridizations, **167**, 803
nitrogen-fixing filamentous
  purification, **167**, 93
  storage, **167**, 95
nitrogen and hydrogen metabolism, **167**, 459, 467
oxygen exchange, MS analysis, **167**, 686
pH gradients in intact cells, ESR assay, **167**, 677
photosynthetic apparatus components, gene isolation, **167**, 755
photosystem II, calcium effects *in vivo*, **167**, 280
phycobilisomes
  preparation, **167**, 305
  stability, **167**, 313
phylogenetic studies by 16S rRNA sequencing, **167**, 132, 138
planktonic, buoyancy and suspension, analytical techniques, **167**, 521

plasma membranes
energization, measurement, **167**, 371
isolation, **167**, 245
plasmids, listing, **167**, 816
plastocyanins, sequence comparisons with chloroplasts, **167**, 148
polyglucose granules, research methodology, **167**, 196
poly-β-hydroxybutyrate granules, research methodology, **167**, 202
polyphosphate granules, research methodology, **167**, 199
pores, ultrastructural analysis, **167**, 169
prenylquinones, HPLC assay, **167**, 270
preparation for EPR spectroscopy, **167**, 428
preservation
in liquid nitrogen, **167**, 123
by lyophilization, **167**, 124
purification
hydrogenases, **167**, 501
methods, **167**, 25
recognition in natural habitats, **167**, 29
respiratory proton extrusion, measurement, **167**, 363
salt-stressed, osmotic adjustment, **167**, 528
sample transportation, **167**, 69
sampling, **167**, 4
sequence-specific endonucleases, listing, **167**, 821
sheaths, isolation, **167**, 183
storage, **167**, 5
sulfur metabolism, **167**, 572
symbiotic associations
cultivation, **167**, 114
isolation of bacteria, **167**, 119
reconstitution, **167**, 117
taxonomy, application of DNA/DNA hybridization, **167**, 125
thermophilic
collection, culture, and storage, **167**, 97
culture maintenance, **167**, 98
culture media, **167**, 99
isolation, **167**, 97
overview, **167**, 96
thioredoxin *m* purification, **167**, 413
transformation, **167**, 703, 712
by chromosomal recombination, **153**, 216
turgor pressure measurements, **167**, 660
unicellular, synchronously growing, nitrogen fixation, **167**, 484
volatile organic compounds, analysis, **167**, 609
Cyanogen bromide
activation of agarose, **135**, 50
cleavage of ribosomal proteins, **164**, 264
in peptide mapping of collagens, **145**, 175, 188
Cyanolysis
disulfides, **143**, 255
in sulfane sulfur analysis
cold method, **143**, 26
with cupric ion, **143**, 28
hot method, **143**, 27
Cyanophycin
cyanobacterial
isolation and purification, **167**, 209
proteolysis, assay, **167**, 212
quantitative analysis, **167**, 210
synthesis induction, **167**, 209
Cybrids
isolation by
fluorescence-activated cell sorting, **151**, 232
toxin–antitoxin protocols, **151**, 233
Cyclic AMP
activated lipolysis, analysis in permeabilized adipocytes, **159**, 197
analogs for analysis of low-K_m hormone-sensitive phosphodiesterase, **159**, 531
and analogs, in cells treated with cyclic nucleotide analogs
assay, **159**, 79
purification, **159**, 75
automated radioimmunoassay, **159**, 45
cascade system
regulatory steps, theoretical considerations, **159**, 19
yeast mutants, isolation and characterization, **159**, 27
crystallized with catabolite gene activator protein, for CAP structural analysis, **159**, 278
cyanobacterial
extraction, **167**, 584
radioimmunoassay, **167**, 587
TLC analysis, **167**, 586
typical concentrations, **167**, 587

effect on cAMP receptor electrophoretic mobility, **159**, 275
as elastin synthesis modulator, **144**, 245
export, analysis and regulation by prostaglandin A_1, **159**, 83
intracellular antagonist (*Rp*)-cAMPS
inhibition of cAMP-dependent protein kinase, **159**, 163
synthesis, **159**, 160
levels in leukocytes, effects of chemoattractants, **162**, 281
radiolabeled, preparation, **159**, 459
role in cyclic cascade regulation, **159**, 3
site-selective analogs, synergistic pairs, use in intact cells, **159**, 118
turnover in intact cells, **159**, 50
Cyclic AMP phosphodiesterase
acylpeptide inhibitors from *Bacillus subtilis*, **159**, 497
hormone-sensitive low-K_m
analysis with cAMP analogs, **159**, 531
assay, **159**, 737
hormone effects, observation conditions, **159**, 738
particulate, solubilization, **159**, 740
properties, **159**, 744
purification, **159**, 740
subcellular distribution, **159**, 742
insulin-sensitive
assay, **159**, 745, 752
properties, **159**, 750, 757
purification from rat liver, **159**, 753
stimulation by insulin, **159**, 748
low-K_m
assay, **159**, 767
properties, **159**, 770
purification from
human platelets, **159**, 772
rat brain, **159**, 767
subcellular distribution and regulation, **159**, 771
purification from canine kidney, **159**, 760
Cyclic AMP receptors
in *Dictyostelium*, identification and ligand-induced modification, **159**, 267
Cyclic AMP transporter
characterization, **159**, 86
Cyclic GMP
and analogs, in cells treated with cyclic nucleotide analogs
assay, **159**, 81
purification, **159**, 75
automated radioimmunoassay, **159**, 45
binding to phosphodiesterase, assay, **159**, 524
cyanobacterial
extraction, **167**, 584
radioimmunoassay, **167**, 587
TLC analysis, **167**, 586
typical concentrations, **167**, 587
as elastin synthesis modulator, **144**, 245
noncatalytic binding sites on cGMP phosphodiesterase, photoaffinity labeling, **159**, 730
radiolabeled, preparation, **159**, 459
Cyclic GMP phosphodiesterase
noncatalytic binding sites, photoaffinity labeling, **159**, 730
purification from rod outer segments, **159**, 707
stimulation by calmodulin in intact tissue, assay, **159**, 601
Cyclic nucleotide phosphodiesterase
activation by *Acanthamoeba* calmodulin, assay, **139**, 62
assay
in cyanobacteria, **167**, 589
with radiolabeled and fluorescent substrates, **159**, 457
calmodulin-sensitive
catalytic site, analysis, **159**, 526
purification, **159**, 522
calmodulin-stimulated
analysis with calmodulin derivatives, **159**, 605
antibodies, preparation, characterization, and applications, **159**, 627
assay, **159**, 548, 664
association with calmodulin in intact cells, assay, **159**, 594
heat-sensitive inhibitor
assay, **159**, 665
preparation, **159**, 662
properties, **159**, 666
high-affinity testicular isoforms
age effects and distribution, **159**, 679
assay, **159**, 678
isozymes
properties, **159**, 593
purification from bovine brain, **159**, 588

properties, **159**, 566
purification
from bovine brain, **159**, 573
from mammalian brain, **159**, 557
from mammalian tissues, **159**, 543
selective inhibitors, description, **159**, 652
calspermin assay with, **139**, 127
cGMP-stimulated
allosteric site, analysis, **159**, 528
binding of cyclic nucleotide analogs, **159**, 529
catalytic site, analysis, **159**, 526
purification, **159**, 523
histo- and cytochemical assays, **159**, 477
methylxanthine inhibitors, **159**, 489
mutational analysis in *Drosophila*, **159**, 786
preparation from porcine brain, **139**, 128
reaction mixtures, HPLC analysis, **159**, 471
relationship to cAMP turnover, **159**, 52
specific inhibitors, effects on adipocytes, **159**, 504
in yeast cAMP cascade system mutants, assay, **159**, 38
zinc-containing, purification from yeast, **159**, 777

Cyclic nucleotides
analogs for analysis of phosphodiesterase catalytic and allosteric sites, **159**, 520
binding to proteins, precipitation assay, **159**, 147
cascade system
covalent modification and allosteric control, **159**, 17
energy consumption, **159**, 14
properties, **159**, 9
regulatory role of cAMP, **159**, 3
theoretical considerations of regulatory steps, **159**, 19
yeast mutants, isolation and characterization, **159**, 27
metabolism in intact cells, monitoring, **159**, 60

Cyclin, *see* Proliferating cell nuclear antigen

β-Cyclocitral
determination in carotene oxygenase assay, **167**, 337

Cycloheximide
as glycosylation inhibitor, **138**, 704

1-Cyclohexyl-3-(2-mopholinoethyl)carbodiimide metho-*p*-toluene sulfonate
in probing of
RNA structure, monitoring by primer extension, **164**, 482
rRNA–protein interactions, **164**, 464

Cyclosporin A
–calmodulin interaction in T cells
flow cytometric analysis, **139**, 683
fluorometric analysis, **139**, 680

Cyclosporine, *see* Cyclosporin A

Cystamine
enzymatic assay, **143**, 150

Cystathionine β-lyase
assay, **143**, 439, 444, 484
properties
cabbage enzyme, **143**, 441
E. coli enzyme, **143**, 486
spinach enzyme, **143**, 447
purification from
cabbage, **143**, 440
E. coli, **143**, 484
spinach, **143**, 446

Cystathionine γ-lyase
assay, **143**, 487
properties, **143**, 491
purification from *Streptomyces phaeochromogenes*, **143**, 488

Cystathionine β-synthase
assay, **143**, 389
properties, **143**, 393
purification from human liver, **143**, 391

Cysteamine
colorimetric assay, **143**, 152
enzymatic assay, **143**, 150

Cysteamine dioxygenase
assay, **143**, 411
cystamine and cysteamine assay with, **143**, 150
properties, **143**, 413
purification from porcine liver, **143**, 411

Cysteic acid
HPLC assay, **143**, 164

Cysteine
and derivatives, metabolism in microorganisms, **143**, 454
dietary, quantification, **143**, 300

enantiomers
analytical resolution, **143**, 167
preparative resolution, **143**, 169
intracellular delivery, **143**, 313
metabolism in
mammals, **143**, 372
plants, **143**, 420
o-phthalaldehyde derivatives, HPLC, **143**, 141
S-substituted forms, preparation, **143**, 292
S-Cysteine conjugates
purification, **143**, 233
synthesis, **143**, 230
D-Cysteine desulfhydrase
assay, **143**, 450
properties, **143**, 451
purification from spinach, **143**, 450
Cysteine dioxygenase
assay, **143**, 397, 403
properties, **143**, 400
purification from rat liver, **143**, 396
Cysteinesulfinate decarboxylase, *see* Sulfinoalanine decarboxylase
Cysteinesulfinic acid
colorimetric assay, fuchsin method, **143**, 160
HPLC assay, **143**, 155, 164
^{35}S-labeled, synthesis, **143**, 272
L-Cysteine sulfoxide lyase
S-substituted
assay, **143**, 435
properties, **143**, 437
purification from shiitake mushroom, **143**, 436
Cystine
assay with cystine-binding protein, **143**, 146
dietary, quantification, **143**, 300
o-phthalaldehyde derivatives, HPLC, **143**, 141
Cystine lyase, *see* Cystathionine β-lyase
Cytarabine
in analysis of calmodulin role in DNA repair, **139**, 726
entrapment in liposomes
drug capture efficiency, **149**, 68
drug permeability, **149**, 70
Cytochemistry
localization of
cAMP-dependent protein kinase
free catalytic subunit, **159**, 236
regulatory subunit, **159**, 255
cyclic nucleotide phosphodiesterase, **159**, 477
Na^+,K^+-ATPase
with ATP, **156**, 417
cobalt chloride method, **156**, 422
lead citrate method, **156**, 423
strontium–lead method, **156**, 418
Cytochrome b_6/f
assays, **167**, 348
genes, isolation and characterization from *Nostoc*, **167**, 779
properties, **167**, 344
purification from cyanobacteria, **167**, 342
Cytochrome *c*
^{125}I-labeled, glycosaminoglycan detection with, **138**, 260
Cytochrome c_6
in cyanobacteria and chloroplasts, sequence comparisons, **167**, 151
Cytochrome *c*-549
purification from *Anacystis nidulans*, **167**, 393
Cytochrome *c*-550
purification from
Aphanizomenon flos-aquae, **167**, 396
Synechococcus vulcanus, **167**, 395
Cytochrome *c*-552
purification from *Anacystis nidulans*, **167**, 393
Cytochrome *c*-553
purification from
Aphanizomenon flos-aquae, **167**, 396
Synechococcus vulcanus, **167**, 395
Cytochrome *c*-554
purification from
Anacystis nidulans, **167**, 393
Spirulina platensis, **167**, 394
Cytochrome-*c* oxidase
in cyanobacterial membranes
associated Cu^{2+}, EPR detection, **167**, 450
identification, **167**, 443
oxidation rates, **167**, 441
mitochondrial
properties, **148**, 501

purification from sweet potato, **148**, 497
Cytochrome P450
assay, **137**, 678
immobilization and properties, **137**, 682
preparation from yeast, **137**, 679
Cytochromes
chloroplast-associated
EPR spectra, **148**, 264
estimation *in situ*, **148**, 269
gel electrophoretic behavior, **148**, 267
molecular weights, **148**, 267
optical spectra, **148**, 260
oxidation–reduction potentials, **148**, 265
Cytodex
immobilization of plant protoplasts, **135**, 424, 426
Cytofluorometry
in detection of transferrin receptors, **150**, 726
Cytokines
role in inflammation and repair, hepatic granuloma model, **163**, 611
Cytoplasm
preparation by rapid fractionation of protoplasts, **148**, 192
Cytoplasmic inclusions
cyanobacterial, ultrastructural analysis, **167**, 163
Cytosine arabinoside, *see* Cytarabine
Cytoskeleton
associated actin
role in leukocyte locomotion, **162**, 246
triton-insoluble, analysis, **162**, 265
erythrocyte, erythrosome preparation with, **149**, 218
Cytotoxicity
diphtheria toxin fragment A–hormone hybrid, assay, **165**, 209
phospholipid-containing liposomes toward tumor cells, assays, **141**, 464
Pseudomonas exotoxin A, assay, **165**, 227
Shiga toxin, **165**, 232
T cells, assay, **150**, 333
Cytotoxic T cells
alloreactive, responses in basal serum-free medium, **150**, 132

D

Dabsyl chloride
activation of agarose, **135**, 80
Daffodil
chromoplasts
carotenogenic enzymes, solubilization and reconstitution, **148**, 392
isolation from flowers, **148**, 241
Damselysin
bioassay, **165**, 177
properties, **165**, 187
purification, **165**, 186
Dansyl chloride
activation of agarose, **135**, 81
Darkroom
equipment required for molecular cloning work, **152**, 9
Daucus carota, *see* Carrot
dCMP deaminase
activated R conformation
crosslinked, properties, **135**, 580
crosslinking, **135**, 579
conformational states, chemical stabilization, **135**, 577
inhibited T conformation
crosslinked, properties, **135**, 583
crosslinking, **135**, 581
30^{B}-Dealanine insulin
porcine
human insulin semisynthesis with, **136**, 168
preparation, **136**, 167
Deglutamyllentinic acid
preparation, **143**, 438
Deglycosylation
glycoproteins
chemical procedure, **138**, 343
enzymatic procedure, **138**, 351, 354
Deheparinization
blood, with immobilized heparin lyase, **137**, 515
3-Dehydroquinase
as coupling enzyme in *arom* multifunctional enzyme assay, preparation, **142**, 330
3-Dehydroquinate
preparation for *arom* multifunctional enzyme assay, **142**, 328

3-Dehydroquinate dehydratase
assay, **142**, 321
properties, **142**, 323
purification from *E. coli*, **142**, 322
Dehydroquinate synthase
assay, **142**, 308
properties, **142**, 313
purification from *E. coli*, **142**, 309
substrate isolation from bacterial growth medium, **142**, 312
3-Dehydroshikimate
preparation for *arom* multifunctional enzyme assay, **142**, 328
Demethoxylation
lignin by *Phanerochaete chrysosporium*, assays, **161**, 83
Denaturation
partial, 5S RNA fragment 1, **164**, 166
thermo-, *see* Thermodenaturation
Dendritic cells
as accessory cells in T-cell oxidative mitogenesis, **150**, 108
Denitrification
liquid membrane-entrapped cell system for, **137**, 707
Densitometric analysis
cAMP-dependent protein kinase regulatory subunit phosphorylation, **159**, 145
phospholipid asymmetry of loaded erythrocytes, **149**, 286
phycobilisome stability, **167**, 317
thrombin-sensitive membrane proteins, **147**, 161
Dentin
noncollagenous proteins
overview, **145**, 260
preparation and extraction, **145**, 290
phosphophoryns
phosphoserine analysis, **145**, 299
purification and characterization, **145**, 296
2-Deoxyadenosine 5′-(α-thio)triphosphate
^{35}S-labeled, in DNA sequencing, **155**, 97
Deoxycholate
solubilization of Ca^{2+}-ATPase in monomeric form, **157**, 267
2-Deoxy-D-glucose
tritiated, in selection of glucose transport- and glycolysis-defective mutants, **151**, 132
3-Deoxy-D-*arabino*-heptulosonic acid 7-phosphate
as dehydroquinate synthase substrate, **142**, 306
isolation from bacterial growth medium, **142**, 312
Deoxyinosine triphosphate
in resolution of compressions during dideoxy DNA sequencing, **152**, 551
Deoxymannojirimycin
as glycoprotein processing inhibitor, **138**, 693
3-Deoxy-D-manno-2-octulosonic acid
derivatives, chemical synthesis, **138**, 402
Deoxynojirimycin
as glycoprotein processing inhibitor, **138**, 690
Deoxynucleoside 3′-phosphoramidites
preparation, **154**, 299
in synthesis of oligonucleotides, **154**, 287, 314
Deoxyribonuclease I
in assay for
actin, **162**, 261, 270
protein–DNA interactions *in vitro*, **152**, 731
DNA modification with, **152**, 110
in DNA sequencing, **155**, 93
in footprinting of protein–DNA contacts
comparison with hydroxyl radical method, **155**, 551
procedure, **155**, 542
thiol–disulfide interchange in, thiol pK_a values, **143**, 134
DE 81 paper
in solid-phase nucleotide sequencing, **155**, 326
Dephosphorylation
lipocortins, **163**, 28
Derivatization
chemical, Ca^{2+},Mg^{2+}-ATPase, **157**, 251
prostaglandins for GC and GC–MS analyses, **141**, 348
Desferal
entrapment in erythrocytes, **149**, 247

Desmosine
ELISA, **144**, 255
polyclonal antibodies, production, **144**, 250
radioimmunoassay, **144**, 254
Detergents
inactivation of Na^+,K^+-ATPase, **156**, 31
Dethiobiotin
assay with bioaffinity sensor, **137**, 122
Detoxification
with immobilized enzymes, **137**, 447
Deuteration
ribosomal components, **164**, 136
Dextran
bound ATP, preparation, **136**, 47
bound NADH, multienzymatic recycling in microcapsules, **136**, 74
–dimethylfluorescein
cell loading method, **147**, 416
pH_i measurements with, **147**, 422
synthesis, **147**, 425
in fluorescence photobleaching assay of nucleocytoplasmic transport, **141**, 451
synthesis with immobilized dextransucrase, **136**, 239
Dextransucrase
assay, **136**, 241; **138**, 650
dextran synthesis with, **136**, 251
immobilization, **136**, 245
production in *Leuconostoc* fed-batch culture, **136**, 242
properties, **138**, 655
purification from
Leuconostoc mesenteroides, **136**, 242
streptococcus, **138**, 652
Diabetes
treatment with immobilized cells, **137**, 451
Diabsyl chloride
activation of agarose, **135**, 82
synthesis, **135**, 81
N,*N′*-Diacetylchitobiase, *see* *N*-Acetyl-β-glucosaminidase
Diacylglycerol hydrolase
hydrolysis of
dioleoylglycerol, assay, **141**, 310
palmitoylacetylglycerol, assay, **141**, 311
preparation from various sources, **141**, 303
Diacylglycerols
in analysis of protein kinase C-mediated phosphorylation in intact cells, **141**, 414
quantitative assay, **141**, 294
Diagnosis
enzymatic, sphingolipidoses, **138**, 727
Dialysis
cholate, in reconstitution of Na,K-pump, **156**, 127
in control of lipid/protein ratio during sarcoplasmic reticulum membrane reconstitution, **157**, 308
equilibrium
in preparation of metallocarboxypeptidases, **158**, 72, 75
21-kDa calcium-binding protein, **139**, 93
in erythrocyte entrapment of
drugs, **149**, 242
enzymes, **149**, 229
proteins, **149**, 235
flow, in measurement of plasma membrane energization in cyanobacteria, **167**, 373
rate, in assay of ATP and ADP binding to Na^+,K^+-ATPase, **156**, 193
tubing, metal-free
handling, **158**, 13
preparation, **158**, 13
vacuum, fibronectin, in preparation of chemotactic fragments, **162**, 218
Dialyzer
enzyme reactor preparation from, **137**, 483
for erythrocyte entrapment of proteins
application, **149**, 238
description, **149**, 235
Diamide
disulfide formation with, **143**, 267
properties, **143**, 266
Diaminopolyethylene glycol
synthesis, **136**, 494
Diaphorase, *see* Dihydrolipoamide dehydrogenase
Diatomaceous earth, *see* Infusorial earth
Diazomethane
ethereal, preparation, **161**, 135

Dichlorophenolindophenol
in spectrophotometric assay of thiols, **143**, 58
2,3-Dichloropropanol
racemic, enzymatic preparation, **136**, 133
Dictyostelium discoideum
cAMP receptor identification and ligand-induced modification, **159**, 267
1,4-Dideoxy-1,4-imino-D-mannitol
as glycoprotein processing inhibitor, **138**, 696
Dideoxyribonucleotides
as chain terminators in large-scale DNA sequencing, **152**, 538
Diet
methionine and cyst(e)ine content, analysis, **143**, 300
selenium-deficient, for rats, **143**, 309
for sulfur amino acid studies, preparation, **143**, 301
Diethyl pyrocarbonate, *see* Pyrocarbonic acid diethyl ester
1,5-Difluoro-2,4-dinitrobenzene
in detection of Na^+,K^+-ATPase subunit interactions, **156**, 347
Digestion
DNA
for detection and cloning of amplified fragments, **151**, 339
high-molecular-weight genomic, for cosmid library construction, **151**, 410
size fractionation, **152**, 183
Digitalis
binding site in Na^+,K^+-ATPase, affinity labeling, **156**, 323
Digitonin
permeabilized adipocytes, in cAMP studies, **159**, 193
Diglycosylceramides
perbenzoylated, HPLC, **138**, 117
Dihydrofolate reductase
gene from mutant CHO cells, repetitive cloning, **151**, 451
Dihydrolipoamide dehydrogenase
assay, **166**, 332
coimmobilized with NAD
application to ethanol analyzer, **136**, 31
assay, **136**, 28
preparation, **136**, 26
luciferase-coimmobilized
bioluminescent assays with, **136**, 88
preparation, **136**, 85
as marker for chromosome detection in somatic cell hybrids, visualization, **151**, 181
purification from *Bacillus subtilis*, **166**, 335
Dihydrolipoyl acyltransferase
assay, **166**, 144, 152
properties, **166**, 154
Dihydrolipoyl dehydrogenase
assay, **166**, 144
Dihydropteridine reductase
assay, **142**, 97, 104, 118, 127
NADPH-specific
assay, **142**, 111
properties, **142**, 115
purification from bovine liver, **142**, 112
staining, **142**, 113
properties, **142**, 101, 110, 125, 131
purification from
bovine liver, **142**, 107
human liver, **142**, 120
ovine brain, **142**, 128
ovine liver, **142**, 99
Dihydroxyacetone phosphate
enzymatic synthesis with ATP regeneration, **136**, 277
2,5-Dihydroxymethyl-3,4-dihydroxypyrrolidine
as glycoprotein processing inhibitor, **138**, 694
3,4-Dihydroxyphenylacetic acid
GC–MS assay in brain and body fluids, **142**, 577
HPLC assay in brain, **142**, 539, 547
3,4-Dihydroxyphenylethylene glycol
GC–MS assay in brain and body fluids, **142**, 577
1,25-Dihydroxyvitamin D_3 receptors
in bone cells, binding studies, **145**, 332
3,5-Diiodotyrosine hydrazide
radioiodinated, preparation, **138**, 433
4-(Dimethylamino)-4′-azobenzene
sulfinate and thiosulfonate anion

sulfurtransferase assays with, **143**, 238
synthesis, **143**, 236
5-Dimethylamino-1-naphthalene
sulfinate and thiosulfonate anion
sulfurtransferase assays with, **143**, 239
synthesis, **143**, 239
5-Dimethylaminonaphthalene-1-sulfonyl-calmodulin
preparation, **159**, 606
properties and applications, **159**, 609
Dimethylcitraconate
properties, **166**, 95
synthesis, **166**, 94
4′,5′-Dimethylfluorescein
–dextran
cell loading method, **147**, 416
pH_i measurements with, **147**, 422
synthesis, **147**, 425
3,7-Dimethyl-1-octanol
racemic, enzymatic preparation, **136**, 126
Dimethyl sulfate
modification of 5S RNA, **164**, 727
in probing of
large RNA molecules, **164**, 476
RNA structure, monitoring by primer extension, **164**, 482
rRNA–protein interactions, **164**, 464
Dimethyl sulfoxide
and hydrogen bromide, cleavage of ribosomal protein tryptophanyl peptide bond, **164**, 262
preparation of isonitrile-derivatized polymers with, **135**, 95, 96
2,4-Dinitrophenylethylenediamine
labeled 30S ribosomal subunits at 5′-end of 16S RNA, preparation, **164**, 85
2,4-Dinitrophenyl groups
and anti-DNP, in visualization of ribosomal crosslink sites, **164**, 395
derivatized thiols and disulfides, preparation for HPLC, **143**, 103
Dioctanoylglycerol
and A23187, synergistic effects in pituitary cells, **141**, 429
analogs for protein kinase C structure–function analysis
analogs with altered ester linkages, **141**, 318
3-hydroxyl analogs, **141**, 316
positional analogs, **141**, 317
Dioleoylglycerol
enzymatic hydrolysis, assays, **141**, 310
preparation, **141**, 305
Dioxane
linin fractions, preparation by acidolysis, **161**, 31
mechanistic sudies of immobilized chymotrypsin in, **135**, 540
Dipeptidyl carboxypeptidase I
assays, **163**, 196
distribution, **163**, 209
inhibitors
assays, **163**, 206
overview, **163**, 205
properties, **163**, 203
purification from
human lung, **163**, 201
human plasma, **163**, 202
Diphenylpicrylphenylhydrazine
in spectrophotometric assay of thiols, **143**, 57
Diphtheria toxin
ADP-ribosyltransferase activity, assay, **165**, 222
fragment A
gonadotropin-coupled
characterization, **165**, 208
preparation, **165**, 204
growth factor-conjugated
cell variant isolation with, **147**, 386
preparation, **147**, 384
purification, **147**, 382
NAD^+ glycohydrolase activity, assay, **165**, 224
properties, **165**, 74
purification from *Corynebacterium diphtheriae*, **165**, 69
toxicity assays, **165**, 219
Dipsyl chloride
activation of agarose, **135**, 81
1,2-Diretinoyl-*sn*-glycero-3-phosphocholine
synthesis, **149**, 102
Dissection
Aplysia, for analysis of calmodulin-encoding nucleic acids, **139**, 279
Distillation
in preparation of metal-free water, **158**, 4
Disulfides
aromatic, in spectrophotometric assay of thiols, **143**, 45, 51

assay with 2-nitro-5-thiosulfobenzoate, **143**, 115
L-[^{35}S]cysteinesulfinate synthesis from, **143**, 272
dinitrophenol derivatives, HPLC, **143**, 101
formation with diamide, **143**, 267
interchange with thiols
reaction rates, **143**, 130
Brønsted plots, **143**, 132
thiol equilibrium constants, **143**, 138
thiol pK_a values, **143**, 132
mixed, in spectrophotometric assay of thiols, **143**, 50
protein-mixed, assay in cells
liberation of protein-bound thiols, **143**, 127
sample preparation, **143**, 125
reduction
direct method, **143**, 251
electrochemical method, **143**, 257
with substitution, **143**, 253
by thiols, **143**, 246
sulfhydryl oxidase-catalyzed formation, assay, **143**, 119
Dithiobis(succinimidyl propionate)
reversible protein immobilization with, **135**, 140
3,3′-Dithiobis(sulfosuccinimidyl propionate)
active esters, in analysis of Fc receptor for IgE, **150**, 503
reversible protein immobilization with, **135**, 140
Dithioerythritol
reduction of disulfides, **143**, 246
1,3-Dithiopropan-2-ol
thiol–disulfide interchange in, thiol pK_a values, **143**, 132
Dithiothreitol
reduction of disulfides, **143**, 246
thiol–disulfide interchange in, thiol pK_a values, **143**, 132
Dithizone
extraction of adventitious metals from solutions, **158**, 7
DNA
amplified fragments
cloning by in-gel renaturation, **151**, 366
detection by in-gel renaturation, **151**, 359
Anacystis nidulans, isolation, **153**, 208
Aplysia
analysis with calmodulin cDNA, **139**, 287
isolation, **139**, 286
binding domains in catabolite gene activator protein, analysis, **159**, 281
binding proteins, gel electrophoretic assay, **151**, 551
–calcium phosphate coprecipitation, for plasmid transfer into cultured cells, **151**, 387
calmodulin-encoding, analysis with synthetic oligonucleotides, **139**, 299
changes in mitogen-treated lymphocytes, **150**, 58
chloroplast, ribosomal protein genes, cloning and identification, **164**, 748
clones
corresponding genomic sequences, identification, **152**, 588
identification and characterization, **152**, 507
isolated by complementation in yeast, identification, **152**, 500
positive, selection and purification methods, **152**, 422
selection from libraries, overview, **152**, 393
complementary
adaptor ligation, **152**, 348
calcium protease, cloning, **139**, 372
calmodulin from fruit fly, isolation, **139**, 236
cAMP-dependent protein kinase
catalytic subunit, cloning, **159**, 311
regulatory subunit II, cloning, **159**, 318
cloning
by differential colony hybridization, **147**, 64
directional, in plasmid vectors, **152**, 349
high-efficiency protocol, **154**, 3
into phages λgt10 and λgt11, **152**, 359
in plasmid vectors, **154**, 28

with primer–restriction end adapters, **154**, 64
vector–primer method, **154**, 41
double-stranded
annealing to vector, **152**, 341
homopolymeric tailing with dG, **152**, 339
large fragments, purification, **152**, 335
elastin, cloning, **144**, 265
epidermal growth factor, isolation, **147**, 194
fibronectin and receptor
incorporation into λgt11 phage, **144**, 452
screening in λgt11 libraries, **144**, 456
synthesis, **144**, 449
haptoglobin, isolation, **163**, 461
insulin-like growth factors
isolation, **146**, 245
mRNA analysis with, **146**, 247
interleukin-2 receptor, cloning, **150**, 692
kininogens, cloning, **163**, 230
laminin, cloning, **144**, 464
libraries, *see* DNA libraries, complementary
ligninase, identification with synthetic oligonucleotides, **161**, 228
linker ligation, **152**, 344
Na^+,K^+-ATPase α subunit, cloning, **156**, 380
nerve growth factor, isolation, **147**, 194
9-kDa calcium-binding protein, cloning and characterization, **139**, 337
^{32}P-labeled, synthesis, **152**, 324
preparation, overview, **152**, 307
primer extension, in mapping 5′ termini of gene transcripts, **152**, 629
in probing ribosome structure and function, **164**, 401
prolyl 4-hydroxylase, preparation, **144**, 103
sarcoplasmic reticulum proteins, cloning, **157**, 289
serum amyloid A protein, cloning, **163**, 513
slow/cardiac troponin C
cloning, **139**, 328, 334
codon sequence, **139**, 331, 335
transferrin receptor, cloning, **150**, 741
consensus sequences, determination with random-sequence oligonucleotides, **155**, 568
cosmid, *see* Cosmids
cyanobacterial
hybridizations with nitrogen fixation gene probes, **167**, 803
isolation from transformants, **153**, 225
transformation, **167**, 703
defined regions, mutagenesis with degenerate oligonucleotides, **155**, 558
dot blotting, **152**, 582
double strands, construction with synthetic oligonucleotides, **154**, 245
3′-end labeling with [α-^{32}P]ddATP, **152**, 104
5′-end labeling with ^{32}P, **152**, 100
extraction for analysis of procollagen gene RFLPs, **145**, 206
gapped duplex, in oligonucleotide-directed mutagenesis, **154**, 350
gel electrophoresis
in agarose, **152**, 75
in polyacrylamide, **152**, 64
genomic
cloning with cosmid vector systems, **151**, 397
high-molecular-weight, for cosmid library construction
preparation, **151**, 407
restriction enzyme digestion, **151**, 410
hybridization to AT-rich oligonucleotide probes, **152**, 447
isolation from
cultured cells, **152**, 182
human blood, **152**, 182
whole organs, **152**, 181
for kininogens, cloning, **163**, 230
large segments, cloning with cosmid vectors, **152**, 199
libraries, *see* Genomic libraries
mammalian, cloning and mapping strategies, **151**, 461
partial digestion, **152**, 183

restriction digests, hybridization to oligonucleotides, **154**, 98, 104
sequences corresponding to cDNA clones, identification, **152**, 588
specific sequences, underrepresentation in cloning, alleviation, **152**, 173
high-molecular-weight
gel electrophoresis
orthogonal-field-alternation technique, **155**, 468
pulsed-field technique, **155**, 453
purification, **155**, 449
specific fragmentation, **155**, 463
homopeptide leader sequence-encoding, in proinsulin production by *E. coli*, **153**, 385
hybridization to
DNA, in taxonomic classification of cyanobacteria, **167**, 125
RNA, in chemical probing of large RNA molecules, **164**, 475
immobilization for hybrid selection of mRNA, **152**, 568
λ phage
preparation
based on chromatographic phage purification, **153**, 69
for cloning applications, **152**, 154
in vitro packaging technique, **153**, 95
ligation, **152**, 108
from lignin-degrading fungus
dot-blot hybridization for genome size estimation, **161**, 215
fractionation into mitochondrial, ribosomal, and chromosomal forms, **161**, 218
preparation and construction of genomic libraries, **161**, 213
restriction fragment length polymorphisms, use for genetic mapping, **161**, 219
sequence variation between strains, **161**, 219
linker, preparation for cDNA cloning, **154**, 11, 45
liposome-encapsulated
preparation, **149**, 158
transport *in vivo*, efficiency, **149**, 160
melting, computer simulation, application to gel electrophoresis, **155**, 482
microinjection into *Xenopus* oocytes, cell preparation for, **152**, 284
minute amounts, detection via polymerase chain reaction *in vitro*, **155**, 347
mitochondrial
cloning of fragments, **153**, 372
isolation from various fungi, **153**, 370
M13 phage, preparation for cloning applications, **152**, 154
mutations, obtainment of large number in defined region, **154**, 415
nick translation, methodology, **152**, 91
open reading frames
cloning strategy, **154**, 129
expression vectors, application, **154**, 156
oxidant-induced strand breaks in inflammation, *in vitro* analysis, **163**, 331
plasmid, *see* Plasmids
precipitation with
butanol, **152**, 47
ethanol, **152**, 42
isopropanol, **152**, 46
spermine, **152**, 46
protein-coding regions, identification by open reading frame cloning strategy, **154**, 129
–protein complexes, hydroxyl radical footprinting, **155**, 537
–protein interactions
in vitro assays, **152**, 721
in vivo footprinting, **152**, 735
in yeast extracts, gel electrophoretic assay, **155**, 528
radiolabeling for *in situ* hybridization to chromosomes, **151**, 283
recombinant technology, role in gene amplification, **151**, 99
removal of
3′-phosphates, **152**, 102
5′-phosphates, **152**, 99
renaturation, in-gel technique for detection and cloning of amplified genes, **151**, 332
repair, role of calmodulin, analytical methods, **139**, 715

restriction enzyme *Hgi*CI cleavage sites, determination, **155**, 4
restriction and modification by *E. coli* K12, **152**, 130
segments cloned into multicopy plasmids, Tn*5* mutagenesis, **154**, 178
sequences encoding Spec proteins, screening, **139**, 357
single base changes, detection by gel electrophoresis, **155**, 501
single-stranded, production with pKUN plasmids, **153**, 12
site-specific cleavage at 8-, 9-, and 10-bp sequences, **155**, 22
small-scale phenol extraction, **152**, 40
Streptomyces, cloning and analysis with cosmid shuttle vectors, **153**, 166
subcloning, **152**, 512
templates
  single-stranded, in oligonucleotide-directed mutagenesis, **154**, 329
  transcription for preparation of specific rRNA fragments, **164**, 226
Ti plasmid-derived
  microscale preparation from tobacco, **153**, 290
  molecular analysis, **153**, 290
transcription regulatory sequences, analysis with expression vectors, **153**, 452
transfer
  from *E. coli* to *Agrobacterium*, **153**, 299
  electroporation-mediated, into plant protoplasts, **153**, 336, 351
  into embryonal carcinoma cells, **151**, 81
in transformed plant protoplasts, molecular analysis, **153**, 333
treatment with
  DNA polymerase I Klenow fragment, **152**, 94
  DNase I, **152**, 110
  exonucleases, **152**, 104
  restriction enzymes, overview, **152**, 113
from vaccinia virus recombinants, analysis, **153**, 556
vector primer, preparation for cDNA cloning, **154**, 11, 45

DNA-directed RNA polymerase
  cyanobacterial
    assay, **167**, 592
    properties, **167**, 596
    purification, **167**, 593
    σ factor, purification, **167**, 595
  metal-hybrid, preparation, **158**, 89
DNA libraries
  amplification, storage, and replication, **152**, 407
  complementary
    construction, **139**, 364; **154**, 5
    generation, overview, **152**, 307
    phage-mediated transduction into mammalian cells, **151**, 434
    prepared in plasmid vectors, immunoscreening, **152**, 451
    screening for
      branched-chain α-keto acid dehydrogenase genes, **166**, 110
      cyanobacterial photosynthetic apparatus genes, **167**, 762
      transient expression of clones, **154**, 25
  cyanobacterial, construction in bacteriophage λ vectors, **167**, 780
  genomic, *see* Genomic libraries
  plasmid, construction for cloning in yeast, **152**, 492
  recombinant
    λgt11, gene isolation by immunoscreening, **152**, 458; **154**, 108
    screening with oligonucleotide probes, **152**, 432
  selection of clones from, overview, **152**, 393
DNA ligase
  T4, DNA modification with, **152**, 108
DNA methylase
  in alteration of apparent recognition specificities of restriction enzymes, **155**, 41
  type II
    assays, **155**, 35
    purification, **155**, 40
DNA polymerase
  assay in mitogen-treated lymphocytes, **150**, 63

Klenow fragment
in direct sequencing of denatured plasmid DNA, **152**, 558
modification of DNA, **152**, 94
RNA-directed, *see* RNA-directed DNA polymerase
T4
blunt ending of cDNA, **152**, 333
modification of DNA, **152**, 107
DNA probes
nick-translated
size optimization, **152**, 93
sizing, **152**, 93
preparation with phage M13 origin of replication, **155**, 519
single-stranded, in mapping gene transcripts, **152**, 612
DNase I, *see* Deoxyribonuclease I
DNA sequenator
fluorescence-based, operation, **155**, 289
DNA sequencing
antibiotic resistance mutations in *E. coli* rRNA genes, **164**, 680
application of
direct transfer electrophoresis, **155**, 250
transposon-promoted deletions, **155**, 177
unidirectional digestion with exonuclease III, **155**, 156
with bifunctional pAA-PZ plasmids and *E. coli* strain AA102F′, **155**, 216
computer programs for, **155**, 582
denatured plasmid DNA, direct method, **152**, 556
dideoxy method, **154**, 283
strategies for large-scale projects, **152**, 538
directed, application of Tn*5* mutagenesis, **154**, 187
with DNase I, buffer gradient gel, and ^{35}S label, **155**, 93
fluorescence-based method, **155**, 260
four-lane methods, **152**, 522
M13/dideoxy chain termination method
with ordered deletions, **155**, 119
with sonication, **155**, 51
one-lane methods, **152**, 527
with pBR322 plasmids, **155**, 139
phosphoramidite method, **152**, 287
via RNA intermediates generated *in vitro*, **152**, 563
segmented paper method, **152**, 250
simultaneous synthesis of fragments, **152**, 221
with single-stranded pEMBL plasmids, **155**, 111
single-stranded strategy, **155**, 204
solid-phase methods, **155**, 301
specific, via polymerase chain reaction *in vitro*, **155**, 335
by subcloning in pGV451 and derived vectors, **155**, 231
with α-thiodeoxynucleotides, **155**, 166
DNA synthesis
complementary
first-strand, with oligo(dT) as primer, **152**, 316
second-strand
classical method, **152**, 325
with mRNA fragments as primers, **152**, 330
vector-primed technique, **152**, 323, 371
estradiol 17β and tamoxifen effects
de novo pathway analysis, **146**, 337
scavenger pathway analysis, **146**, 334
manual protocol, **146**, 24
transferrin effects, myotrophic assay, **147**, 295
DNA synthesizer
automated, employing deoxynucleoside 3′-phosphoramidites
design, **154**, 316
operation, **154**, 319
DNA topoisomerase
assay in mitogen-treated lymphocytes, **150**, 64
DNP-ethylenediamine, *see* 2,4-Dinitrophenylethylenediamine
Docusate sodium
reverse micelle preparation with, **136**, 189
Dopamine
HPLC assay
in brain, **142**, 539, 547
in plasma, **142**, 521
Dopamine β-monooxygenase
assay, **142**, 597, 603, 608
properties, **142**, 600, 606, 613

purification from
bovine adrenal gland, **142**, 599
chicken adrenal gland, **142**, 610
human plasma, **142**, 604
Dot blotting
DNA and RNA from cells, various techniques, **152**, 582
Doxorubicin
entrapment in liposomes, **149**, 74
Drainage tubes
polyvinyl chloride, urokinase-treated
clinical application, **137**, 541
fibrinolytic activity after application, **137**, 543
Drosophila melanogaster, *see* Fruit fly
Drugs
delivery by multiple emulsions, **149**, 52
entrapment in erythrocytes, **149**, 229, 242, 256
induced endocytosis in
erythrocyte ghosts, **149**, 267
erythrocytes, **149**, 260
targeting by
multiple emulsions, **149**, 54
temperature- and pH-sensitive liposomes, **149**, 77
Dry reagent chemistries
analytical applications
drugs and hormones, **137**, 416
electrolytes, **137**, 416
enzymes, **137**, 413
metabolites, **137**, 410
basic features, **137**, 395
element construction
by hybrid methods, **137**, 408
by multilayers, **137**, 406
by multisaturation, **137**, 404
layer construction
by component entrapment in preformed matrices, **137**, 403
by simultaneous component entrapment and matrix formation, **137**, 401
monitoring, **137**, 396
performance, **137**, 420
Duimycin
as inhibitor of lipid-linked saccharide pathway, **138**, 677

E

Edetic acid
^{51}Cr-labeled, in assay for uptake of liposome-encapsulated agents, **149**, 143
in preparation of metal-free dialysis tubing, **158**, 14
EDTA, *see* Edetic acid
EGF, *see* Epidermal growth factor
Eicosanoids
precursor fatty acids
release by cultured cells, **141**, 354
uptake by
cultured cells, **141**, 353
platelets, **141**, 353
Elastase
assays, **144**, 289
classification, **144**, 299
leukocyte, *see* Leukocyte elastase
mRNA sequences, cloning, **144**, 304
pancreatic, *see* Pancreatic elastase
Elastic modulus
volumetric, in cyanobacteria, measurement with gas vesicles, **167**, 663
Elastin
amino acid composition, **144**, 175
antibodies
characterization of classes, **144**, 247
production, **163**, 670
antisera
cross-reactivity, **144**, 248
production, **144**, 248
assay in tissues, **163**, 660
biosynthesis, **144**, 186; **163**, 668
cDNA cloning, **144**, 265
crosslinking, **144**, 182
degradation, **144**, 191
analysis with tissue models, **144**, 302
assays, **144**, 289
derived peptides, as fibroblast chemoattractants, **163**, 702
desmosine
ELISA, **144**, 255
polyclonal antibodies, production, **144**, 250
radioimmunoassay, **144**, 254
electron immunohistochemistry, **145**, 91
fibers, structure and formation, **144**, 189

gene structure, **144**, 184, 261
genomic libraries, screening, **144**, 287
immunofluorescence, **144**, 257
immunologic studies, **144**, 192
immunoprecipitation, **144**, 220
insoluble
extraction, **144**, 196
isolation, **163**, 664
isolation, **144**, 173
microfibrillar component, **144**, 188
monoclonal antibodies, production, **144**, 249
mRNA levels in tissues, assay, **144**, 223
phylogenetic distribution, **144**, 175
precursor, immunoprecipitation, **144**, 256
primary structure, **144**, 177
producing cells, culture techniques, **144**, 233
radioimmunoassay, **144**, 251
soluble
isolation, **163**, 665
purification, **144**, 205
synthesis
in cell cultures, analysis, **144**, 225
modulation, *in vitro* studies
culture condition effects, **144**, 237
extracellular matrix effects, **144**, 241
soluble modulators, **144**, 243
in organ cultures, analysis, **144**, 216
quantification in explant and tissue cultures, **163**, 671
tropoelastin precursor, isolation, **163**, 665
visualization in tissues, **163**, 659
Western blotting, **163**, 671
Electric energy
production from hydrogen in bacterial fuel cell, **137**, 671, 708
Electric field
induction of active cation pumping by Ca^{2+}- and Na^+,K^+-ATPase, **157**, 240
locomotory response of granulocytes, analysis, **162**, 115
Electric organ
Torpedo, purification of calelectrin, **139**, 31
Electroblotting
platelet-derived growth factor-like proteins, **147**, 26
Electrochemical detection
with HPLC
catecholamine metabolites
one-column technique, **142**, 536
two-column switching technique, **142**, 544
catecholamines, **142**, 521
penicillamine, **143**, 186
thiols, **143**, 110
Electrochemical reduction cell
disulfide reduction in, **143**, 258
for preparation of chiral compounds
design, determining factors, **136**, 308
practical aspects, **136**, 314
Electrodes, *see also specific electrodes*
animal tissue-based, preparation, **137**, 141
ion-selective, Ca^{2+} measurement in biological fluids, **158**, 325
measurement of respiratory proton extrusion from cyanobacterial cells, **167**, 363
microsome-based
activity yield and recovery, **137**, 155
aniline assay, **137**, 155
ascorbate assay, **137**, 157
diffusion control, **137**, 160
endogenous enzyme coupling in, **137**, 157
hybrid systems, **137**, 159
NAD(P)H assay, **137**, 156
operation, **137**, 154
preparation, **137**, 153
plant tissue-based, preparation, **137**, 149
sulfide-specific, sulfide assay with, **143**, 29
Electroencephalogram
in neurotoxicity assays, **165**, 270
Electrofusion
breakdown parameters, adjustment, **151**, 196
electronics and pulse chambers, **151**, 216
media constraints, **151**, 215
requirements for membrane contact, **151**, 206
resealing parameters, adjustment, **151**, 203

Electron immunohistochemistry
cartilage macromolecules, **145**, 353
collagens
type I, **145**, 80
type II, **145**, 81
type III, **145**, 82
type IV, **145**, 96
type V, **145**, 84
type VI, **145**, 85
type VII, **145**, 85
elastin, **145**, 91
entactin, **145**, 99
fibronectin, **145**, 88
laminin, **145**, 97
methodology for extracellular matrix studies
embedding, **145**, 113, 144
fixation, **145**, 104, 135
prestaining treatments, **145**, 114, 139
sectioning, **145**, 112, 138
staining, **145**, 118, 140
proteoglycans, **145**, 92, 99
reticulin, **145**, 87
Electron microscopy
actin filaments in crab sperm acrosomal bundles, **162**, 264
Ca^{2+}-ATPase crystals
freeze-fracture-etch technique, **157**, 279
negative staining technique, **157**, 273
unstained preparations, **157**, 278
chitin, **161**, 435
collagens
type IV, **145**, 58, 65
type VI, **145**, 67
crosslinked RNA, **164**, 324
cyanobacterial fimbriae, **167**, 192
cyanobacterial gas vesicles, **167**, 221
extracellular matrix proteins, principles, **145**, 4
fibronectin, **145**, 47
freeze-fracture and freeze-etching technique for microbial toxin assays
chemical fixation of cells and membranes, **165**, 377
cleaning of replica, **165**, 389
cryoprotection, **165**, 378
freezing, **165**, 380
fracture, **165**, 383
replication, **165**, 386
specimen mounting for freezing, **165**, 380
vacuum, contamination, and etching conditions, **165**, 384
immuno-, *see* Immunoelectron microscopy
laminin, **145**, 28
liposome–cell interactions, **149**, 138
membrane-bound Na^+,K^+-ATPase
freeze-etching, **156**, 437
freeze-fracture, **156**, 435
negative staining, **156**, 432
thin sectioning, **156**, 431
in monitoring plant protoplast coupling to microcarriers, **135**, 425
Na^+,K^+-ATPase crystals, **156**, 83
negative staining technique for microbial toxin assays
instrument operation, **165**, 373
preparation of
required materials, **165**, 360
specimen supports, **165**, 365
staining procedure, **165**, 368
organelles immobilized in crosslinked proteins, **135**, 468
plant cell membranes
dehydration and embedding, **148**, 615
fixation, **148**, 600, 611
instrument operation and photographic procedures, **148**, 622
postfixative treatments, **148**, 603
sectioning, **148**, 617
staining of sections, **148**, 618
proteoglycans, **144**, 371; **145**, 72
ribosomal subunits, **164**, 35
ribosomes, **164**, 3
in R-loop analysis of procollagen mRNA, **145**, 248
rRNA, **164**, 76
scanning transmission, ribosomes, **164**, 49
thylakoid lipids, **148**, 322
tissue kallikrein, immunostaining for, **163**, 155
transmission, tetanus toxin 2-D crystals, **165**, 393
vesicles reconstituted with Na^+,K^+-ATPase
freeze-fracture, **156**, 439
thin sectioning, **156**, 438
wood degradation, **160**, 194

Electron paramagnetic resonance
copper detection in cyanobacterial cytochrome-*c* oxidase, **167**, 450
immobilized chymotrypsin, **135**, 502
computer analysis of data, **135**, 507
deactivation studies, **135**, 510
integration of data, **135**, 506
iron–sulfur proteins in cyanobacteria, **167**, 427
vanadium assay, **158**, 420
Electron spin resonance
in analysis of
cyanobacterial bioenergetics, **167**, 509
lymphoid cell receptors
applications, **150**, 432
methodology, **150**, 420
Na^+,K^+-ATPase, **156**, 371
in assay for pH gradients in intact cyanobacteria, **167**, 677
Electron spin resonance oximetry
cyanobacteria
calibration, **167**, 673
internal O_2 measurements, **167**, 675
overview, **167**, 670
spin probe
concentration, **167**, 672
selection, **167**, 671
total O_2 measurements, **167**, 673
viscosity, **167**, 673
Electron transport
photosynthetic, in *Oscillatoria limnetica*, assays for reactions, **167**, 380
Electropermeabilization
breakdown parameters, adjustment, **151**, 196
electronics and pulse chambers, **151**, 216
media constraints, **151**, 215
resealing parameters, adjustment, **151**, 203
Electrophoresis, *see also specific techniques*
cAMP receptor, mobility shift induced by cAMP binding, **159**, 275
with chromatography, cyanobacterial carbon metabolites, **167**, 548
collagenase, **144**, 146
glycosaminoglycans, **138**, 260
recombinant human C5a, **162**, 664
sucrose gradient, chorismate mutase, **142**, 470
Electroporation
erythrocyte membranes for drug loading, **149**, 251
plant protoplasts
for direct gene transfer, **153**, 313
for DNA and RNA introduction, **153**, 336, 351
ELISA, *see* Enzyme-linked immunosorbent assay
Ellipsometry
in biomedical analysis
applications, **137**, 371
instrumentation, **137**, 369
principles, **137**, 368
biomolecular interactions on silicon surfaces, **137**, 392
IgG interaction with immobilized protein A, **137**, 385
Ellman's reagent
sulfhydryl oxidase assay with, **143**, 511
Elongation factors
effect on structure of ribosomal proteins L7/L12, **164**, 154
EF-G
and GTP, binding to ribosomal pretranslocation complex, kinetics, **164**, 594
titrations, in analysis of translational kinetics, **164**, 616
EF-Ts, role in *in vitro* translation, **164**, 620
EF-Tu, titrations, in analysis of translational kinetics, **164**, 616
Elutriation
centrifugal, isolation of human basophils, **162**, 533
Embedding
tissue for kallikrein immunohistochemical staining, **163**, 149
Embryo
chicken, purification of protein disulfide-isomerase, **144**, 113
tissue cultures, in analysis of basement membrane biosynthesis, **145**, 389
Emericella nidulans
purification of β-xylosidase, **160**, 692
Emulsions
multiple
characterization, **149**, 62
drug delivery by, **149**, 52

drug release from, **149**, 64
drug targeting by, **149**, 54
formulation
added stabilizing components, **149**, 62
emulsifying equipment, **149**, 57
nature of entrapped materials, **149**, 61
nature of oil phase, **149**, 59
nature and quantity of emulsifying agents, **149**, 60
volumes of dispersed phases, **149**, 59
Encephalomyelitis
experimental allergic
active induction, **162**, 417
adoptive transfer, **162**, 418
overview, **162**, 413
Endo-β-*N*-acetylglucosaminidase, *see* Mannosyl-glycoprotein endo-β-*N*-acetylglucosaminidase
Endo-*N*-acetylneuraminidase
bacteriophage-derived
assay, **138**, 171
polysialosyl glycoprotein detection with, **138**, 183
properties, **138**, 177
purification, **138**, 174
effects on
cerebral membranes, **138**, 634
polysialic acid electrophoresis, **138**, 631
Endocellulase, *see* Endo-1,4-β-glucanase
Endochitinase
assay, **161**, 498
properties, **161**, 500
purification from wheat germ, **161**, 498
Endocytosis
defective CHO cell mutants, isolation, **138**, 458
drug-induced
in erythrocyte ghosts, **149**, 267
in erythrocytes, **149**, 260
erythrocyte loading by, **149**, 226
fluid-phase, growth factor-stimulated
assay, **146**, 409
markers for, **146**, 402
receptor-mediated, analytical methods
intracellular ligand–receptor dissociation, **149**, 5
intracellular ligand and receptor segregation, **149**, 7
ligand binding to cell surface, **149**, 4
ligand degradation, **149**, 9
ligand internalization, **149**, 4
ligand subcellular distribution, **149**, 7
Endo-1,3(4)-β-D-glucanase
assay with dye-labeled CM-β-glucan, **160**, 82
Endo-1,4-β-glucanase
activation by acid proteases, **160**, 507
assays, **160**, 100, 352, 364
detection in
gels, **160**, 139
prokaryotes, **160**, 182
properties, **160**, 354, 367, 375
purification from
Clostridium thermocellum, **160**, 352
Humicola, **160**, 327, 329
Sporotrichum pulverulentum, **160**, 370
Talaromyces emersonii, **160**, 365
Trichoderma koningii, **160**, 227
type C, purification from *E. coli*, **160**, 192
type D from *Clostridium thermocellum*
assay, **160**, 356
crystallization, **160**, 360
properties, **160**, 360
purification from *E. coli*, **160**, 357
Endoglycosidases
deglycosylation of glycoproteins, **138**, 354
Endonucleases
sequence-specific, from cyanobacteria, listing, **167**, 821
Endopectate lyase
assay, **161**, 381
biological function, **161**, 384
properties, **161**, 383
purification from *Erwinia aroideae*, **161**, 382
Endoplasmic reticulum
calcium-transporting systems, properties, **157**, 7
castor bean
bound polysomes, isolation and characterization, **148**, 584
electron transport enzymes and cytochromes, **148**, 578
isolation from endosperm, **148**, 577

monosaccharide lipid synthesis, **148**, 580
phosphoglyceride synthesis, **148**, 585
phospholipid biosynthesis, **148**, 578
purity of isolated membranes, **148**, 581
preparation from rat liver, **157**, 613
rough, vesicles, sugar nucleotide transport assay, **138**, 715
vesicles
ionic permeability
light-scattering measurements, **157**, 427
membrane potential–dye measurements, **157**, 430
radioisotope flux–Millipore filtration measurements, **157**, 419
isolation, **157**, 418
Endopolygalacturonase
chromatography with Spheron ion exchangers, **161**, 388
β-Endorphin
model peptide, design, **154**, 478
structural analysis, **154**, 476
Endosomes
acidification, fluorescence assay, **157**, 607
density gradient centrifugation, **157**, 605
fluorescent labeling, **157**, 604
Endothelial cell growth factors
assays, **147**, 98
properties, **147**, 156
purification, **147**, 96, 153
storage, **147**, 98
Endothelial cells
aortic
migration, spectrophotometric assay, **147**, 145
purification of plasminogen activator inhibitor, **163**, 301
capillary, migration
Boyden chamber assay, **147**, 142
phagokinetic assay, **147**, 136
corneal, cultivation for elastin synthesis studies, **144**, 236
Endotoxins
bacterial, detection with toxic shock syndrome toxin-1
in vitro method, **165**, 304
in vivo method, **165**, 303
detection in assay for endogenous pyrogens, **163**, 502
2-Enoate reductase
assay, **136**, 312
immobilization, **136**, 313
purification from *Clostridium* La 1, **136**, 309
repeated use after immobilization, **136**, 315
substrate specificity and kinetic data, **136**, 305
Enolase
as marker for chromosome detection in somatic cell hybrids, **151**, 181
5-Enolpyruvylshikimate 3-phosphate
preparation for *arom* multifunctional enzyme assay, **142**, 330
Enoyl-CoA hydratase
peroxisomal, assay, **148**, 523
Entactin
electron immunohistochemistry, **145**, 99
Enterotoxins
clostridial
bioassay, **165**, 101
production, **165**, 96
purification, **165**, 97
in commercial protein A, radioimmunoassay, **137**, 514
Escherichia, heat-stable
bioassay, **165**, 128
biochemical properties, **165**, 136
carboxymethylation, **165**, 133
purification, **165**, 126
sequence analysis, **165**, 135
staphylococcal
monkey feeding test, **165**, 324
overview, **165**, 43
properties, **165**, 44
purification
type A, **165**, 46, 50
type B, **165**, 47
type C_1, **165**, 48, 51
type C_2, **165**, 48, 50
type D, **165**, 49, 51
type E, **165**, 50
Enucleation
biology of, **151**, 225
gradient techniques, **151**, 223
monolayer techniques, **151**, 222

in production of
cybrids, **151**, 232
nuclear hybrids, **151**, 235
reconstituted cells, **151**, 229
Enzymatic assay
branched-chain amino acids, **166**, 3
branched-chain keto acids, **166**, 6
chitin, **161**, 457
collagen degradation, **163**, 691
coupled, Na^+,K^+-ATPase, **156**, 116
cystamine and cysteamine, **143**, 150
glutathione, in analysis of oxidant injury, **163**, 334
inositol 1-phosphate, **141**, 129
Enzymatic radioimmunoassay
catecholamines
critical reagents, **142**, 552
general principles, **142**, 552
procedures, **142**, 559
troubleshooting, **142**, 568
Enzyme channeling immunoassay
coated-surface, **136**, 98
soluble-reagent
microtiter plate protocol, **136**, 99
rapid protocol, **136**, 102
Enzyme electrodes
α-amylase assay, **137**, 64
commercial availability, **137**, 28
creatinine assay, **137**, 67
for fermentation monitoring
externally buffered systems, **137**, 304
external systems, **137**, 298
in situ systems, **137**, 301
glucose-measuring, long-term peritoneal implantation, **137**, 68
with immobilized flavins, **137**, 103
for industrial applications
analytical characteristics, **137**, 56
apparatus description, **137**, 45
calibration, **137**, 51
computerization, **137**, 50
measurements, **137**, 49
preparation, **137**, 48
instrumentation, **137**, 15
mediated amperometric
development, **137**, 102
electrodes, **137**, 96
enzymes, **137**, 94
immobilization procedures, **137**, 99
instrumentation, **137**, 91
mediators, **137**, 95
method of use, **137**, 28
multienzymatic
amplification principle, **137**, 39
antiinterference principle, **137**, 42
competition principle, **137**, 40
enzyme membrane characterization, **137**, 31
preparation, **137**, 30
sequence principle, **137**, 36
preparation, **137**, 20
required materials, **137**, 20
types and characteristics, **137**, 16
Enzyme immobilization
adsorption methods, **135**, 58
analytical applications, overview, **137**, 3
in aqueous two-phase systems, **137**, 657
in bioelectrocatalysis, **137**, 430
via carbohydrate chains, **135**, 141
on 1,1′-carbonyldiimidazole-activated matrices, **135**, 102
with cells, **137**, 641
in cellulose acetate beads, **135**, 282
on chlorocarbonate-activated supports, **135**, 84
combination with mass spectrometry, **137**, 421
covalent coupling methods, **135**, 37
on derivatized transition metal-activated supports, **135**, 127
by entrapment in gels, **135**, 440
entrapment methods, **135**, 61
in fundamental biochemical studies, overview, **135**, 475
in gelatin, **135**, 293
on hydrous transition metal oxides, **135**, 123
industrial applications, overview, **136**, 353
on isonitrile-derivatized supports, **135**, 90
matrices for, *see* Immobilization matrices
medical applications
evaluations *in vivo*, **137**, 453
examples, **137**, 445
methodology, **137**, 454
overview, **137**, 443

by microencapsulation, **136**, 68
with monoclonal antibodies, **135**, 160
in organic synthesis, overview, **136**, 117
on polyacrylamide gels, **135**, 593
on porous glass, **135**, 537
by prepolymer methods, **135**, 230
on pyridoxal phosphate-bound Sepharose, **135**, 157
by radiation polymerization, **135**, 146
reversible
with *N*-hydroxysuccinimide esters containing cleavable crosslinks, **135**, 136
on succinamidopropyl-glass, **135**, 131
on succinylthiopropyl-glass, **135**, 133
on Sepharose 4B, **135**, 592
techniques and aspects, overview, **137**, 583
on transition metal-activated supports
chemistry, **135**, 117
metal-link method, **135**, 119
on tresyl chloride-activated supports, **135**, 65
Enzyme immunoassay
for assessment of lymphocyte proliferation, **150**, 117
with ^{125}I-labeled aprotinin, tissue kallikrein, **163**, 160
two-site, for nerve growth factor
antibody preparation, **147**, 169
first antibody coupling, **147**, 171
marker enzyme selection, **147**, 171
optimization, **147**, 177
second antibody coupling, **147**, 176
sensitivity, **147**, 183
specificity, **147**, 183
standard procedure, **147**, 181
Enzyme-linked immunosorbent assay
antibody production in lymphocyte cultures
human systems, **150**, 314
murine systems, **150**, 304
bone glycoproteins, **145**, 280
calcineurin antibodies, **159**, 634
cholera toxin, **165**, 170
collagen, **163**, 681
competitive, collagenase, **144**, 169
C1q receptor, **150**, 567
C-reactive protein, **163**, 398
cyclic nucleotide phosphodiesterase antibodies, **159**, 634
desmosine, **144**, 255
elastin, **144**, 226
in tissue cultures, **163**, 672
Epstein–Barr virus/C3d receptor, **150**, 552
extracellular matrix components, **147**, 412
fibrinogen, **163**, 479
kininogens, **163**, 252
α_2-macroglobulin, **163**, 492
melanoma growth stimulatory activity, **146**, 114
oncomodulin antibodies, **139**, 159
parvalbumin, **139**, 766
proteoglycans, **144**, 370
in screening of Na^+,K^+-ATPase antibodies, **156**, 400
serum amyloid A protein, **163**, 519
comparison with radioimmunoassay, **163**, 521
Shiga toxin, **165**, 161
staphylococcal epidermolytic toxin, **165**, 337
T-cell receptors, **150**, 679
thermometric, for process monitoring
instrumentation, **137**, 335
procedure, **137**, 337
reagent preparation, **137**, 336
tissue kallikrein, **163**, 106, 134
toxic shock syndrome toxin-1, **165**, 343
Enzyme–membrane reactor
coenzyme retention, **136**, 13
experimental setup, **136**, 15
performance, **136**, 19
Enzyme photometers
in β-lactam flow–injection analysis
applications, **137**, 210
data evaluation, **137**, 206
instrumentation, **137**, 202
procedures, **137**, 203
reagents and chemicals, **137**, 202
theory, **137**, 199
Enzyme reactors
for amino acid removal from plasma
associated problems, **137**, 479
preparation methods, **137**, 481

construction with bioluminescent enzymes, **137**, 172
continuous-flow analysis with, **137**, 263
with dextran-bound ATP
application to repeated enzyme reaction, **136**, 52
design, **136**, 46
optimal partition of reagents, **136**, 50
flow–injection analysis with, **137**, 268, 271
heparin lyase
construction, **137**, 525
testing *in vivo*, **137**, 525
nylon tube
applications, **137**, 295
operation principle, **137**, 288
performance characteristics, **137**, 292
preparation, **137**, 289
routine analysis with, **137**, 289
testing and optimization, **137**, 289
with polyethylene glycol-bound NAD
application, **136**, 42
theoretical analysis, **136**, 40
Enzyme replacement therapy
with enzyme–albumin conjugates, **137**, 566
Enzymes, *see also specific enzymes*
albumin-conjugated, for enzyme replacement therapy, **137**, 566
calmodulin-dependent, effects of CaM antagonists, **159**, 658
calmodulin-regulated, $CAPP_1$–calmodulin specificity for, **139**, 417
carotenogenic, solubilization and reconstitution from chromoplasts, **148**, 392
cellulolytic
assay, **137**, 692
glycosides as substrates and ligands, **160**, 45
production by *Penicillium funiculosum*, **137**, 690, 694
high-efficiency entrapment in erythrocytes, **149**, 229
immobilized, industrial applications, **136**, 371
introduction of exchange-inert metal ions, **158**, 95
involved in lymphocyte stimulation, analysis, **150**, 47
metal-containing, inhibition by chelating agents, **158**, 110
metal exchange in, selectivity, **158**, 83
metal-free, preparation, **158**, 21
metal-hybrid, assignment of metal ions, **158**, 86
in reverse micelles
activity and conformation, **136**, 188
synthesis of apolar compounds, **136**, 216
sulfur metabolism-associated, additional articles dealing with, **143**, 523
thrombolytic, therapeutic applications, **137**, 552
Enzyme stabilization
by chemical modification, **137**, 599
by immobilization, **137**, 585
by immobilization and solute addition
cofactor addition effects, **137**, 594
polymer addition effects, **137**, 595
substrate addition effects, **137**, 593
by intramolecular crosslinking, **137**, 615
Enzyme thermistors
in flow enthalpimetric assay of glucose, **137**, 225
in flow–injection analysis of β-lactams, **137**, 197
instrumentation, **137**, 182
for lipid assays in body fluids, **137**, 217
operation, **137**, 186
for process monitoring and control, **137**, 307
various applications, **137**, 188
Enzyme transistors
ammonia measurements, **137**, 241
background, **137**, 233
field effect
glucose-sensitive system, **137**, 250, 259
overview, **137**, 247
urea-sensitive system, **137**, 258
gas measurements in aqueous solution, **137**, 236
hydrogen gas measurements, **137**, 238
instrumentation, **137**, 236
response characteristics, **137**, 235

Eosin
as fluorescence probe for Na^+,K^+-ATPase conformational analysis, **156**, 278
Eosinophil chemotactic factors
ECF-A
assay, **162**, 172
peptide synthesis, **162**, 173
purification from human lung, **162**, 171
sequence analysis, **162**, 173
high-molecular-weight
assay, **162**, 175
properties, **162**, 179
purification from *Schistosoma* eggs, **162**, 176
purification and characterization, **162**, 167
Eosinophils
chemotaxis, assay, **162**, 166
human, purification of lysophospholipase, **163**, 36
Epidermal growth factor
cDNA isolation, **147**, 194
gene
cloning and expression in yeast, **146**, 39
construction and purification, **146**, 34
immunoaffinity purification, **146**, 3
α_2-macroglobulin-conjugated
preparation, **147**, 389
transmembrane delivery, **147**, 390
monoclonal antibodies
production, **146**, 13
properties, **146**, 17
radioiodination, **146**, 13
role in tissue repair processes, **163**, 714
toxin-conjugated
cell variant isolation with, **147**, 386
preparation, **147**, 385
Epidermal growth factor receptors
affinity labeling
in intact cells, **146**, 147
in isolated membranes, **146**, 149
biosynthesis, kinetic analysis, **146**, 54
in bone cells, binding studies, **145**, 329
cell-secreted, identification, **146**, 61
glycosylation analysis, **146**, 57
immunoaffinity purification, **146**, 82
isolation and identification, **146**, 50
monoclonal antibodies
biological effects, **146**, 73
characterization, **146**, 68
production, **146**, 67
purification, **146**, 72
Epinephrine
enzymatic radioimmunoassay
in plasma, **142**, 559, 568
in tissues, **142**, 565
in urine, **142**, 565
HPLC assay in plasma, **142**, 521
Epitopes
allogeneic, recognized by alloreactive T cells, characterization, **150**, 340
on λgt 11 clones, mapping, **154**, 122
Epoxy resins
biocatalyst-immobilizing
applications, **135**, 258
chemical structure, **135**, 253
physical properties, **135**, 256
preparation, **135**, 253
Epstein–Barr virus/C3d receptors
assays, **150**, 552
purification from
Raji cells, **150**, 549
tonsil B cells, **150**, 550
72-kDa fragment isolation, **150**, 551
Equilibrium constants
nomenclature for metal-buffered systems, **158**, 35
E-rosette receptors, *see* Antigens, CD2
Erwinia
E. aroideae, purification of endopectate lyase, **161**, 382
E. rhapontici, immobilized
activity regeneration, **136**, 447
isomaltulose production in reactor, **136**, 443
preparation, **136**, 437, 441
stabilization, **136**, 445
storage stability, **136**, 449
Erythroblasts
erythropoietin bioassay with, **147**, 348
isolation from spleen, **147**, 342
responsiveness to erythropoietin, **147**, 347
Erythrocyte ghosts
drug-induced endocytosis, **149**, 267

ferrofluid-containing
characterization, **149**, 307
preparation, **149**, 303
white resealable
preparation by column method, **149**, 272
properties, **149**, 274
Erythrocytes
disulfide formation after diamide treatment, **143**, 267
drug-induced endocytosis, **149**, 260
entrapment of
bacteriophages, **149**, 228
drugs
by dialysis, **149**, 242
by electric cell fusion, **149**, 258
by electric membrane perforation, **149**, 256
by hemolysis or endocytosis, **149**, 229
enzymes, **149**, 229
proteins
by dialysis, **149**, 235
by endocytosis, **149**, 226
by hypotonic hemolysis, **149**, 223
human, purification of
Ca^{2+}-ATPase, **139**, 792
CR1 receptor, **150**, 583
inside-out vesicles
Na,K-transport, assays, **156**, 175
preparation and adjustment of ionic composition, **156**, 173
loaded, phospholipid asymmetry
bioassay, **149**, 288
dye-removal assay, **149**, 286
maintenance, **149**, 292
staining assay, **149**, 282
loading by
endocytosis, **149**, 226
hypotonic hemolysis, **149**, 225
macromolecule-loaded
fusion with recipient cells, **149**, 295
preparation, **149**, 293
plasma membranes
electric perforation, **149**, 249
preparation, **157**, 341
resealing after electric perforation, **149**, 253
potassium efflux, vanadate effects, **156**, 261
preparation for loading, **149**, 224
resealed ghosts, sodium efflux, vanadate effects, **156**, 263
resealing after loading, **149**, 225
resialylated
in analysis of sialyloligosaccharide-binding protein specificity, **138**, 166
preparation, **138**, 163
rubidium influx, vanadate effects, **156**, 255
sodium efflux, vanadate effects, **156**, 259
subgroups, antibodies to, assay with ligand-bearing liposomes, **149**, 212
Erythropoietin
assay, **147**, 330
bioassay, **147**, 348
characterization, **147**, 337
immunoaffinity purification, **147**, 333
Erythrosomes
preparation, **149**, 218
properties, **149**, 219
Escherichia coli
acetolactate synthase isozymes I and III, assay, **166**, 241
κ-carrageenan-immobilized
L-alanine production in reactor, **136**, 477
preparation, **135**, 195; **136**, 474
conjugal transfer of plasmids to cyanobacteria, **153**, 236; **167**, 751
cosmid vectors, for cloning and analysis of *Streptomyces* DNA, **153**, 166
coupling to hydrous transition metal oxides, **135**, 358
detergent removal of *N,N'*-diacetylchitobiase, **161**, 526
diarrheal toxin, ileal loop test, **165**, 314
entrapment by
cellulose acetate beads, **135**, 289
applications, **135**, 291
photocrosslinkable resin prepolymers, **135**, 239
entrapped, proinsulin production, **137**, 632
expression of
bovine growth hormone derivatives, **153**, 390
calmodulin, **139**, 303

cAMP-dependent protein kinase regulatory subunit, **159**, 325
eukaryotic genes with synthetic two-cistron system, **153**, 401
foreign proteins, **153**, 492
heterologous unfused protein, **153**, 416
human C5a gene, **162**, 657
Spec 1 protein, **139**, 362
as glutamate sensor component, **137**, 129
in immunoscreening of cDNA expression libraries, **152**, 451
polyazetidine-immobilized
phenylalanine production, **136**, 498, 501
preparation, **136**, 500
polyurethane-immobilized
L-aspartic acid production in reactor, **136**, 466
preparation, **136**, 464
properties, **136**, 467
prepolymer-entrapped, applications, **135**, 247
production of human proinsulin, enhancement technique, **153**, 385
purification of
acetolactate synthase I, **166**, 438
acetolactate synthase III, **166**, 456
aromatic-amino-acid aminotransferase, **142**, 258
aspartate aminotransferase, **142**, 258
chemotactic factor, **162**, 200
chorismate mutase–prephenate dehydratase, **142**, 433
chorismate mutase–prephenate dehydrogenase, **142**, 444
Clostridium endocellulase C, **160**, 192
Clostridium endoglucanase D, **160**, 357
cystathionine β-lyase, **143**, 484
cystine-binding protein, **143**, 144
3-dehydroquinate dehydratase, **142**, 322
dehydroquinate synthase, **142**, 309
heat-stable enterotoxin, **165**, 126
α-hemolysin, **165**, 140
K^+-ATPase, **157**, 671
phosphoribosylanthranilate isomerase–indoleglycerol-phosphate synthase, **142**, 392
3-phosphoshikimate 1-carboxyvinyltransferase, **142**, 344
shikimate dehydrogenase, **142**, 317
shikimate kinases, **142**, 357, 359
tryptophanase, **142**, 415
tryptophan synthase, **142**, 402
recombinant gene expression and product analysis, **152**, 661
recovery of human C5a, **162**, 662
restriction and modification of DNA, **152**, 130
restriction site bank vectors, construction, **153**, 40
ribosomes and subunits active in tRNA binding, preparation, **164**, 658
secretion of foreign proteins, **153**, 499
strain AA102F′, construction for DNA sequencing, **155**, 221
strain HB101, growth for 5S RNA production, **164**, 159
synthesis of eukaryotic proteins via pLcII(nic$^-$) vector, **153**, 464
transformation with double-cut vectors, **152**, 354
translation of single messenger, effects of rRNA mutations, **153**, 432
transport of branched-chain amino acids, assay, **166**, 245
tryptophan biosynthetic pathway, **142**, 293
Esterase
assay in immobilized pepper cells, **135**, 418
Esterification
stereoselective, *dl*-menthol by immobilized lipase, **136**, 299
Esters
racemic, enzymatic preparation, **136**, 134
Estradiol 17β
effect on cellular DNA synthesis
de novo pathway analysis, **146**, 337
scavenger pathway analysis, **146**, 334
Estradiol receptor-kinase
assay, **139**, 735
properties, **139**, 739
purification from bovine uterus, **139**, 737
Estradiol receptor-phosphatase
purification from bovine uterus, **139**, 733
Estradiol receptors
gel electrophoresis, **139**, 734
purification from bovine uterus, **139**, 733

Ethacridine
in purification of hemopexin, **163**, 543
Ethanol
assay with
biosensor, **137**, 53
enzyme reactor, **137**, 276
enzyme thermistor, **137**, 190
fermentation with yeast cells immobilized on wood chips, **135**, 340
flow-injection analysis with immobilized enzymes, **137**, 176
industrial production with immobilized yeast, **136**, 380, 394
oxidation by enzyme-coimmobilized NAD, **136**, 61
precipitation of
DNA, **152**, 42
RNA, **152**, 44
Ethanolaminephosphotransferase
in plant endoplasmic reticulum
assay, **148**, 588
properties, **148**, 588
Ethanol analyzer
dehydrogenase–NAD gel application to, **136**, 31
1-Ethyl-3-(3-dimethylaminopropyl)carbodiimide
crosslinking of RNA and protein, **164**, 311
Ethylene
production, measurement in lignin biodegradation assay, **161**, 79
Ethylenediamine
–glycoenzyme adducts
coupling to carboxyalkyl agarose, **135**, 145
preparation, **135**, 145
Ethylene glycol
qualitative TLC, **160**, 173
Ethylene glycol bis(succinimidyl succinate)
reversible protein immobilization with, **135**, 137
N-Ethylmaleimide
in spectrophotometric assay of thiols, **143**, 64
Ethyl methanesulfonate
mutagenesis with, **147**, 371
α-Ethylmethionine sulfoximine
synthesis, **143**, 288
N-Ethylpyridinium chloride
dissolution of cellulose, **135**, 270
preparation, **135**, 269
Etioplasts
isolation, **148**, 248
prolamellar bodies
characterization, **148**, 251
isolation, **148**, 250
separation from prothylakoids, **148**, 249
prothylakoids
characterization, **148**, 251
separation from prolamellar bodies, **148**, 249
Euglena gracilis
isolation of chloroplasts and proplastids, **148**, 169
Eukaryotes
higher, heterologous gene expression, functional analysis, **152**, 684
Eupenicillium javanicum
purification of cellulase, **160**, 254
Eupergit C
antibody-mediated enzyme immobilization with, **135**, 165, 169
5′-phosphodiesterase immobilization with, **136**, 518
Evatate
immobilized urokinase
clinical applications, **137**, 540
preparation, **137**, 530
Evolution
calcium protease, **139**, 377
cyanobacteria, analysis by 16 S rRNA sequencing, **167**, 132, 138
RNA-based analysis, **164**, 793
Exo-cellobiohydrolase, *see* Cellulose 1,4-β-cellobiosidase
Exocellulase
assay, **160**, 403
properties, **160**, 406
purification from *Irpex lacteus*, **160**, 404
Exocytosis
neutrophil granules, description, **162**, 549
Exo-1,4-β-glucanase, *see* Cellulose 1,4-β-cellobiosidase
Exoglycosidases
deglycosylation of glycoproteins, **138**, 351

Exo-β-D-mannanase
β-D-mannopyranoside substrates, synthesis, **160**, 515
Exo-1,4-β-mannanase
assay, **160**, 583
properties, **160**, 587
purification from *Aeromonas hydrophila*, **160**, 585
Exonucleases
DNA modification with, **152**, 104
in mapping gene transcripts, **152**, 625
unidirectional DNA digestion, in DNA sequencing, **155**, 156
Exopolygalacturonase
chromatography with Spheron ion exchangers, **161**, 388
Exotoxins
Pseudomonas aeruginosa, immunotoxin construction with, **151**, 139
Extension inhibition analysis
translation initiation complexes, **164**, 419
Extracellular matrix
constituent proteins, analysis in biological fluids, **145**, 391
effect on elastin synthesis, **144**, 241
electron immunohistochemical studies
overview, **145**, 78
specific methodology, **145**, 103, 133
growth factor-stimulated synthesis, assays, **147**, 407
preparation, **144**, 241
release of radioactivity, in assay for collagen degradation, **163**, 693
Extraction
adventitious metals from solutions, **158**, 7
cartilage-derived growth factor, **146**, 321
collagens
and procollagens, **144**, 42
type I, **144**, 142
type IV from EHS tumor, **145**, 370
cyanobacterial membrane lipids, **167**, 252
cyclic nucleotides from cyanobacteria, **167**, 584
elastin
autoclaving method, **144**, 200
hot alkali method, **144**, 199
Rasmussen method, **144**, 202
Ross–Bornstein method, **144**, 201
Starcher method, **144**, 203
inositol phospholipids from rat pituitary, **141**, 165
inositol trisphosphate from human platelets, **141**, 144
insulin-like growth factor I from tissues, **146**, 227
ion-pair, sulfotransferases, **143**, 202
leukotriene B_4 and ω-oxidation products, **141**, 357
lignin–carbohydrate complexes, **161**, 13
matrix proteins from bovine bone, **146**, 286
mRNA from lignin-degrading fungus, **161**, 222
nuclear RNA
guanidinium isothiocyanate method, **152**, 237
hot phenol method, **152**, 236
phenol-based, nucleic acids, large- and small-scale procedures, **152**, 33
phospholamban from cardiac SR vesicles, **157**, 362
placental fibronectin, **144**, 422
prostaglandins from biological samples, **141**, 332, 346
proteoglycans, **144**, 320; **145**, 263
RNA from free ribosomal and microsomal fractions, **152**, 246
soluble elastin, **144**, 267
thylakoid lipids, **148**, 321

F

Fabry's disease
enzymatic diagnosis, **138**, 751
Factor VII
activation by factor XIIa, **163**, 68
Factor VIII
antibodies, adsorption to matrix-bound protein A, **137**, 458
Factor IX
antibodies, adsorption to macrobead-linked factor IX, **137**, 458
purification from plasma, **137**, 463
Factor X
activated, digestion of fusion proteins produced in *E. coli*, **153**, 475

Factor XI
activation by factor XII or XIIa, **163**, 67
Factor XII
activated
enzymatic properties, **163**, 79
factor VII activation, **163**, 68
factor XI and plasma prekallikrein activation, **163**, 67
purification, **163**, 77
activation, **163**, 75
by plasma kallikrein, **163**, 66
assays, **163**, 57, 71
with synthetic substrates, **163**, 170
factor XI and plasma prekallikrein activation, **163**, 67
properties, **163**, 62, 75
purification from
bovine plasma, **163**, 55
human plasma, **163**, 72
structure, **163**, 75
Factor H receptors
assays, **150**, 599
Farber disease
enzymatic diagnosis, **138**, 734
Fast protein liquid chromatography
calmodulin, **139**, 51
ligninase, **161**, 243
110-kDa protein–calmodulin complex, **139**, 141
phosphorylase phosphatase subunits, **159**, 388
restriction enzymes *Rsr*I and *Rsr*II, **155**, 12
vitronectin, **144**, 436
Fatty acid methyl esters
phosphatidylinositol-derived
gas–liquid chromatography, **141**, 241
preparation, **141**, 240
Fatty acids
composition in mitochondrial phosphoglycerides, **148**, 481
in cyanobacterial membrane lipids
gas chromatography, **167**, 254
positional distribution, **167**, 255
effects on
branched-chain α-keto acid metabolism, **166**, 492
cGMP-binding phosphodiesterase, **159**, 717
leucine metabolism to branched-chain α-keto acids, **166**, 495
incorporation into plant membrane lipids, pesticide effects, **148**, 670
newly synthesized in isolated chloroplasts, analysis, **148**, 335
peroxisomal degradation, assay methods, **148**, 520
synthesis by isolated chloroplasts, **148**, 331
product analysis, **148**, 333
Fc receptors
IgE
analysis with membrane-impermeant crosslinking reagents, **150**, 503
characterization, **150**, 520
detection methods, **150**, 513
induction, **150**, 518
IgG
binding assays, **150**, 525
fluorescence flow cytometric study, **150**, 478
purification, **150**, 532
Feces
aluminum levels, measurement, **158**, 296
Fermentation
in alkaloid production by immobilized *Claviceps purpurea*, **136**, 323
with celite-immobilized mycelial cells, **135**, 326, 330, 332
ethanol, with yeast cells immobilized on wood chips, **135**, 340
industrial, with immobilized yeast, **136**, 380, 394
in isomaltulose production, **136**, 433
monitoring with enzyme sensors, **137**, 298
Fermenter
air-lift, alkaloid production by immobilized *Claviceps purpurea*, **136**, 327
draft tube, for cultivation of bead-entrapped animal cells, **135**, 406
Ferredoxins
cyanobacterial
assays, **167**, 403
overview, **167**, 397
properties, **167**, 400
purification
from *Anabaena variabilis*, **167**, 409

from *Aphanizomenon flos-aquae*, **167**, 408
general procedure, **167**, 405
Ferredoxin–thioredoxin reductase
assays, **167**, 423
properties, **167**, 427
purification from *Nostoc muscorum*, **167**, 426
Ferric sodium tartrate
pretreatment of biomass materials for enhancing cellulose availability, **160**, 8
Ferricyanide
in spectrophotometric assay of thiols, **143**, 59
Ferrofluids
entrapment in erythrocyte ghosts, **149**, 303
Fetuin
chemical deglycosylation, **138**, 343
Fever
produced by endogenous pyrogens, assays, **163**, 502, 504
Fibrin
–asparaginase reactor, preparation, **137**, 482
entrapment of animal cells, **135**, 401
Fibrin autography
plasminogen activator, **163**, 296
reverse, plasminogen activator inhibitors, **163**, 299
Fibrin gel assay
leukocyte chemotaxis
assessment of cell distribution in gels, **162**, 27
direct observation in 3-D gels, **162**, 28
gel preparation, **162**, 27
Fibrinogen
assay, **163**, 479
biosynthesis, primary hepatocyte culture for analysis, **163**, 480
polyclonal antibodies, production, **163**, 479
purification from rat plasma, **163**, 476
subunit isolation, **163**, 476
Fibrin plates
^{125}I-labeled, in assay for plasminogen activator, **163**, 294
plasminogen activator inhibitors, **163**, 298
Fibroblast-activating factor
mRNA isolation and translation, **163**, 729
Fibroblast-derived growth factor
assay, **147**, 41
properties, **147**, 41
purification, **147**, 42
Fibroblast growth factor
acidic
amino acid sequence, **147**, 132
assays, **147**, 107, 121
properties, **147**, 114, 129
purification, **147**, 110, 123
tissue distribution, **147**, 117
basic
assays, **147**, 107
properties, **147**, 114
purification, **147**, 110
tissue distribution, **147**, 117
lymphocyte- and macrophage-derived forms
assays, **163**, 716
production, **163**, 723
purification, **163**, 726
role in tissue repair processes, **163**, 714
Fibroblasts
BALB/c 3T3, PDGF-modulated proteins, identification, **147**, 85
bovine cardiac, calregulin localization, **139**, 47
chemoattractants, preparation, **163**, 701
chemotaxis assay, **163**, 697
collagen production, analytical methods, **163**, 642
cultivation for elastin synthesis studies, **144**, 233, 235
EGF receptor internalization, antibody effects, **146**, 77
elastin synthesis *in vitro*, analysis, **144**, 229
growth
effects of EGF receptor antibodies, **146**, 74
in flat bed hollow fiber reactor, **135**, 398
growth factor-stimulated glucose uptake, assay, **146**, 400

human
  chromosome suspension preparation
    from, **151**, 295
  DNA extraction for analysis of procol-
    lagen gene RFLPs, **145**, 208
  EGF receptor biosynthesis, **146**, 49
  ion fluxes and levels, measurement, **146**,
    384
  in isovaleric acidemia, mutant isova-
    leryl-CoA dehydrogenase assay and
    properties, **166**, 155
  L, sorting for expression of HLA sur-
    face antigen, **151**, 161
  LM, protein synthesis, inhibition by
    Pseudomonas exotoxin A, **165**, 228
  in maple syrup urine disease, branched-
    chain α-keto acid dehydrogenase
    assays, **166**, 135
  mechanical and adhesive properties,
    analysis
    elastic rubber substratum method, **163**,
      624
    haptotactic substratum and related
      methods, **163**, 631
    interference reflection microscopy,
      163, 638
  migration, spectrophotometric assay,
    147, 145
  mixed vesicle preparation from cultures,
    146, 376
  NR6, mitogen-induced pH$_i$ changes,
    assay, **147**, 424
  oligosaccharides, HPLC, **138**, 110
  in osteogenesis imperfecta, analysis of
    collagen I mRNAs, **145**, 223
  Swiss 3T3, mitogen-specific nonprolif-
    erative variants, isolation, **147**,
    355
Fibronectin
  cDNA
    incorporation into λgt11 phage, **144**,
      452
    screening in λgt11 libraries, **144**, 456
    synthesis, **144**, 449
  CD spectroscopy, **145**, 50
  chemotactic fragments, preparation, **162**,
    214
  electron immunohistochemistry, **145**, 88
  electron microscopy, **145**, 47
  as fibroblast chemoattractant, **163**, 702

fusion proteins
  preparation, **144**, 459
  purification from λgt11 lysogens, **144**,
    461
  segment-specific antibodies, **144**, 462
genomic libraries, preparation, **144**, 462
glycopeptides, preparation, **144**, 428
oligosaccharides, preparation, **144**, 429
overview, **145**, 46
purification on 1,1'-carbonyldiimidazole-
  activated support, **135**, 113
purification from human placenta, **144**,
  421
role in tissue repair processes, **163**, 710
solution studies, **145**, 48
structure–function relationships, **145**, 52
Fibronectin receptors
  cDNA and genomic clones, isolation and
    analysis, **144**, 447
  functional analysis with synthetic pep-
    tides, **144**, 479
  incorporation into liposomes, **144**, 489
  isolation from
    cultured cells, **144**, 483
    placenta, **144**, 485
  properties, **144**, 476, 478
Filter assay
  DNA–protein interactions *in vitro*, **152**,
    723
  micropore
    eosinophil chemotactic factors, **162**,
      172
    leukocyte chemotaxis, **162**, 41
      checkerboard method, **162**, 46
  Millipore
    chemotaxis, attractant concentration
      gradients, **162**, 90
    eosinophil chemotactic factor from
      Schistosoma eggs, **162**, 175
    ^{32}P-labeled nucleic acids, **152**, 51
    ribosomal protein–RNA complexes, **164**,
      203
Filter paper
  disks, in cloning of mitogen- and anti-
    gen-reactive B cells, **150**, 287
  total cellulase assay with, **160**, 94
Filters
  in preparation of metal-free water, **158**, 5
Filtration
  in analysis of ligand binding to ATP-

driven pumps during ATP hydrolysis, **157**, 233
in purification of cyanobacteria, **167**, 25, 77
rapid technique for analysis of metabolite fluxes across organelles
examples, **148**, 680
experimental conditions, **148**, 678
instrument description, **148**, 678
overview, **148**, 675
technique description, **148**, 676
ultra-, *see* Ultrafiltration
Fimbriae
cyanobacterial
electron microscopic analysis, **167**, 192
isolation, **167**, 194
overview, **167**, 189
ultrastructural analysis, **167**, 167
Fingerprinting
in intra-RNA crosslink analysis, **164**, 297
in RNA–protein crosslink analysis, **164**, 308
Fixation
tissue for kallikrein immunohistochemical localization, **163**, 145
Flat bed reactor
for cultivation of plant cells, **135**, 412
hollow-fiber, for cultivation of animal cells
application, **135**, 397
design and construction, **135**, 394
operation, **135**, 396
Flavin coenzymes
immobilized, enzyme electrodes with, **137**, 103
Flavobacterium
F. heparinum, purification of heparin lyase, **137**, 521
F. meningosepticum, purification of Endo F and PNGase F, **138**, 772
Flavomycin
as inhibitor of lipid-linked saccharide pathway, **138**, 678
Flocculants
cyanobacterial
assay, **167**, 618
preparation and properties, **167**, 617
Flour
barley, purification of (1→3),(1→4)-β-D-glucan, **160**, 511
Flow cytometry
in analysis of
cellular calcium mobilization, **141**, 53
cyclosporine–calmodulin interaction in T cells, **139**, 677
liposome targeting to hepatocytes, **149**, 170
application to gene amplification in mammalian cells, **151**, 97
chromosome sorting by, **151**, 252
fluorescence-based, in analysis of lymphoid cell receptors, **150**, 478
Flow enthalpimetric assay
glucose, **137**, 225
Flow injection analysis
ceruloplasmin, **163**, 444
with enzyme reactors, **137**, 268, 271
with immobilized chemiluminescent and bioluminescent columns, **137**, 171
β-lactams, with immobilized enzyme reactors, **137**, 197
Flow microfluorometry
in isolation of human basophils, **162**, 533
5′-*N*-(Fluorenylmethyloxycarbonyl)-5′-amino-5′-deoxythymidine
synthesis, **155**, 271
Fluorescein–β-alanine hydrazide
synthesis, **138**, 434
Fluorescein 5′-isothiocyanate
labeling of Na^+,K^+-ATPase, **156**, 272
Fluorescence
activated cell sorting, hybrid and transfected cells, **151**, 150
in analysis of actin in presence of modulating proteins, **162**, 263
ANS, in measurement of ionic permeability of sarcoplasmic reticulum vesicles, **157**, 455
in assessment of inhibitory potency of calmodulin antagonists, **159**, 654
in conformational analysis of Na^+,K^+-ATPase, **156**, 273, 278
in DNA sequencing, **155**, 286
in liposome–cell interaction studies, **149**, 137
in monitoring plant protoplast coupling to microcarriers, **135**, 432
stopped-flow technique in kinetic studies of ribosomal translocation, **164**, 582
X-ray, *see* X-ray fluorescence

Fluorescence microscopy
calcium release in skinned skeletal muscle fibers, **157**, 23
in detection of transferrin receptors, **150**, 726
phospholipid asymmetry of loaded erythrocytes, **149**, 282
Fluorescence photobleaching assay
nucleocytoplasmic transport
associated photodamage, **141**, 457
fluorescent dextran influx assay, **141**, 451
future directions, **141**, 457
instrumentation, **141**, 448
transport analysis *in vivo*, **141**, 454
phorbol ester inhibition of cell–cell communication
applications and future directions, **141**, 447
experimental protocol and data analysis, **141**, 442
instrumentation, **141**, 440
Fluorescence quenching
in measurement of ionic permeability of sarcoplasmic reticulum vesicles, **157**, 455
Fluorescence spectroscopy
in assessment of phycobilisome stability, **167**, 314
calmodulin-binding sequences in peptides and proteins, **139**, 464
S-100 protein, **139**, 180
Fluorescent labeling
endosomes and lysosomes, **157**, 604
gangliosides, **138**, 314
Na^+,K^+-ATPase, **156**, 272
RNA and ribosomal proteins, **164**, 174
thiols
with bromobimanes
analytical applications, **143**, 82
procedures, **143**, 80
with monobromobimane, **143**, 86
1-Fluoro-2,4-dinitrobenzene
hydrogen selenide and methylselenol assays with, **143**, 32
Fluorography
cell-free translation products, **152**, 301
collagen peptides in gels, **145**, 200
nucleic acids
with ^{32}P, **152**, 57
with ^{35}S, **152**, 55
PDGF-modulated proteins, **147**, 91
proteins methylated in presence of calmodulin, **139**, 669
Fluorometric assay
anthranilate phosphoribosyltransferase, **142**, 371
anthranilate synthase, **142**, 301, 368
arogenate dehydratase, **142**, 496
arogenate dehydrogenase, **142**, 488
arylsulfatases, **143**, 207
Ca^{2+}-ATPase conformational changes, **157**, 206
chorismate synthase, **142**, 362
complement component C3, **162**, 645
contact activation in human plasma, **163**, 80
cyclic nucleotide phosphodiesterase, **159**, 457
cyclosporine–calmodulin interaction in T cells, **139**, 677
cytochrome P450, **137**, 678
endosomal and lysosomal acidification, **157**, 607
with HPLC, collagen hydroxypyridinium crosslinks, **144**, 127
intracellular Ca^{2+} in oxidant-exposed cells, **163**, 337
ionic events in neutrophils, **162**, 292
monophenol monooxygenase, **142**, 163
nucleic acids, **152**, 50
oligopeptide receptor binding in neutrophils, **162**, 244
phenylalanine 4-monooxygenase, **142**, 35, 44, 50
3-phosphoshikimate 1-carboxyvinyltransferase, **142**, 349
prephenate aminotransferase, **142**, 481
prephenate dehydrogenase, **142**, 503
selenium, **158**, 397
thiols, with fluorobenzoxadiazoles, **143**, 67
tryptophan 5-monooxygenase, **142**, 83, 93
tyrosine 3-monooxygenase, **142**, 64
Fluorometry
crosslinked collagen polypeptides in gels, **144**, 137

growth factor-induced pH_i changes, **147**, 422
immobilized alcohol dehydrogenase, **135**, 552
5-Fluoroorotic acid
as selective agent in yeast molecular genetics, **154**, 164
Fluorophores
in flow cytometric study of lymphoid cell receptors
preparation, **150**, 478
receptor labeling with, **150**, 481
p-Fluorosulfonylbenzoyl adenosine
affinity labeling of myosin light chain kinase, **139**, 188
^{14}C-labeled, synthesis, **139**, 190
Foods
glucose content, flow enthalpimetric assay, **137**, 230
Foot edema test
with carrageenan
applications, **162**, 326
methodology, **162**, 321
Footprinting
equimolar, in analysis of initiation factor 3 binding to RNA cloacin fragment, **164**, 246
hydroxyl radical-based, protein–DNA complexes, **155**, 537
comparison with other methods, **155**, 551
protein–DNA interactions *in vivo*, **152**, 735
Formaldehyde
identification as product of vanillate *O*-demethylase activity, **161**, 299
Formate dehydrogenase
NAD-coimmobilized
assay, **136**, 28
L-malate production with, **136**, 29
preparation, **136**, 26
Formic acid
as solvent for chitin, **161**, 407
Formycin nucleotides
preparation for Na^+,K^+-ATPase conformational analysis, **156**, 272
N-Formylaminopropanol
1-tosyloxy-3-isocyanopropane synthesis with, **135**, 94
Forskolin
effect on cGMP-binding phosphodiesterase, **159**, 721
Fractionation
alcohol-mediated, elastin, **144**, 210
cosmid library, for gene isolation by sib selection
serial procedure, **151**, 447
simple procedure, **151**, 446
hapten–gelatin, splenic B cells, **150**, 266
membrane proteins sensitive to thrombin, **147**, 158
poly(A)$^+$ RNA, **152**, 259
subcellular, rat hepatocytes, **141**, 97
Fractogel HW65
1,1′-carbonyldiimidazole-activated, in avidin purification, **135**, 114
Freeze-drying
in analysis of cyanobacterial fine structure, **167**, 159
Freeze-etching
in analysis of cyanobacterial fine structure, **167**, 158
with freeze-fracture, in electron microscopic assay for microbial toxins, **165**, 374
Freeze-fracture
with freeze-etching, in electron microscopic assay for microbial toxins, **165**, 374
Fremyella diplosiphon
purification of
phycobiliproteins, **167**, 298
phycobilisomes, **167**, 307
D-Fructans
reductive cleavage, **138**, 30
β-Fructofuranosidase
entrapment with urethane prepolymers, **135**, 241
gelatin-immobilized, properties, **135**, 297
–glucose oxidase electrode, description, **137**, 38, 42
immobilization by radiation polymerization, **135**, 150
immobilized
preparation, **137**, 585, 593
stability, **137**, 593

in industrial saccharification
    immobilized enzyme system, **136**, 377
    liquid enzyme system, **136**, 374
    prepolymer-entrapped, applications, **135**,
        245
Fruit fly
    mutational analysis of cyclic nucleotide
        phosphodiesterase, **159**, 786
    purification of cAMP-dependent protein
        kinase, **159**, 215
Fuchsin
    in assay for
        cysteinesulfinic acid, **143**, 160
        sulfite, **143**, 15
Fucosyltransferases
    glycosphingolipid-synthesizing, assays,
        138, 599
Fuel cell
    bacterial, electric energy production
        from hydrogen, **137**, 671, 708
Fumarase, *see* Fumarate hydratase
Fumarate hydratase
    assay, **136**, 456
    in L-malic acid production by immobi-
        lized bacteria, **136**, 458, 459
Fungi, *see also specific fungi*
    cell immobilization after permeabiliza-
        tion, **137**, 639
    cell permeabilization, **137**, 638
    purification of 1,3-β-glucan synthase,
        138, 638
    tryptophan biosynthetic pathway, **142**,
        299

G

G_{M2} activator
    assays, **138**, 805
    purification, **138**, 813
Gaitonde's reagent
    selenocysteine assay with, **143**, 148
Galactokinase
    as marker for chromosome detection in
        somatic cell hybrids, **151**, 182
Galactolipids
    plant membrane, pesticide effects, **148**,
        672
    synthesis by chloroplast membranes,
        148, 346
        characteristics, **148**, 348

Galactomannan
    from carob
        properties, **160**, 525
        purification, **160**, 524
        dye-labeled, in assay for β-D-man-
            nanase, **160**, 76
    from guar
        properties, **160**, 527
        purification, **160**, 526
Galactomyces reessii
    purification of protopectinase, **161**, 340
Galactose
    cell-surface, labeling with hydrazides,
        138, 441
    removal from glucocerebrosidase before
        cell targeting, **149**, 32
Galactose oxidase
    and neuraminidase, stimulation of T
        cells, **150**, 105
    and tritiated borohydride, in labeling of
        lymphocyte receptors, **150**, 412
Galactosialidosis
    enzymatic diagnosis, **138**, 760
α-Galactosidase
    assay, **160**, 628
    properties, **160**, 629, 631
    purification from
        guar seeds, **160**, 628
        lucerne seeds, **160**, 630
β-Galactosidase
    antibody coupling for nerve growth
        factor immunoassay, **147**, 176
    assay, in identification of gene regula-
        tory elements, **152**, 717
    coupled to cellulose acetate beads
        applications, **135**, 292
        assay, **135**, 291
    diagnostic assay
        galactosialidosis, **138**, 760
        G_{M1}-gangliosidosis, **138**, 752
    encapsulation of yeast cells with, **137**,
        646
    –fibronectin fusion proteins
        preparation, **144**, 459
        purification from λgt11 lysogens, **144**,
            461
        segment-specific antibodies, **144**, 462
    fusion proteins
        analysis in bacterial transformants,

in production of antibody probes, **154**, 144
glucocerebrosidase digestion before cell targeting, **149**, 32
immobilized composites
industrial applications
developments, **136**, 422
engineering considerations, **136**, 414
microbial contamination problems, **136**, 417
operating strategy, **136**, 419
preparation, **136**, 413
properties, **136**, 413
immobilized, galactosyl-*N*-acetylgalactosamine synthesis with, **136**, 230
in industrial saccharification
immobilized enzyme system, **136**, 377
liquid enzyme system, **136**, 374
purification from *Streptococcus pneumoniae*, **149**, 34
reporter gene, application in cyanobacteria, **167**, 744
Galactosyl-*N*-acetylgalactosamine
synthesis with immobilized β-galactosidase, **136**, 230
Galactosylceramidase
diagnostic assay for Krabbe disease, **138**, 739
Galactosylgalactosylglucosylceramidase
diagnostic assay for Fabry disease, **138**, 751
Galactosyltransferases
glycosphingolipid-synthesizing, assays, **138**, 581
Galacturan 1,4-α-galacturonidase
assay, **161**, 373
physiological function, **161**, 380
properties, **161**, 376, 379
purification from
carrot, **161**, 374
liverwort, **161**, 377
Galvanotaxis
granulocytes, assay, **162**, 115
Galvanotaxis chamber
construction, **162**, 120
Ganglia
superior cervical, autoradiography of inositol phospholipids, **141**, 201
Gangliosides
continuous-development TLC, **138**, 127
fluorescent, preparation, **138**, 314
isolation and purification from human plasma, **138**, 300
labeled
characterization, **138**, 340
synthesis, **138**, 334
labeling with hydrazides, **138**, 442
Gangliosidosis
G_{M1}, enzymatic diagnosis, **138**, 752
G_{M2}, enzymatic diagnosis, **138**, 755
Gap junctions
interactions with Ca^{2+}-modulated proteins, **139**, 452
isolation from rat liver, **139**, 447
Gas chromatography
in assay for
acetylation degree of chitin and chitosan, **161**, 445, 446
1-aminocyclopropane-1-carboxylate synthase, **143**, 426
carotene oxygenase, **167**, 337
cyanobacterial hydrogenases, **167**, 507
cyanobacterial nitrogenase, **167**, 460
capillary, inositol trisphosphate, **141**, 148
cyanobacterial volatile organic compounds, **167**, 614
fatty acids in cyanobacterial membrane lipids, **167**, 254
leucine 2,3-aminomutase, **166**, 132
lignin oxidation products, **161**, 120, 128
mercaptoacetate, **143**, 178
3-mercaptolactate, **143**, 178
3-mercaptopyruvate, **143**, 178
prostaglandins, **141**, 348
Gas chromatography–mass spectrometry
bacterial chemotactic factor, **162**, 210
branched-chain 2-hydroxy and 2-keto acids, **166**, 27
glycolipids, methylation analysis, **138**, 13
lignin degradation intermediates, **161**, 205, 208
lignin oxidation products, **161**, 121, 129
methylmalonic acid, **166**, 47
prostaglandins, **141**, 349
with selected ion monitoring, catecholamine metabolites
assay procedure, **142**, 577
principles, **142**, 571
selenium assay
chromatographic conditions, **158**, 395

instrumentation, **158**, 395
overview, **158**, 393
reagents, **158**, 395
sample digestion, **158**, 396
tuning, **158**, 396
Gas exchange systems
in analysis of cyanobacterial C_i uptake, **167**, 536
Gas–liquid chromatography
in assay for
acetolactate synthase, **166**, 235
inositol 1-phosphate, **141**, 130
α-4-*O*-methyl-D-glucuronidase, **160**, 565
selenium, **158**, 401
vanillate hydroxylase, **161**, 276
branched-chain α-keto acids, **166**, 39
fatty acids in chloroplast membrane lipids, **148**, 303
isethionic acid, **143**, 173
lignin aromatic fragments, **161**, 175
lipid molecular species in chloroplast membranes, **148**, 312
–mass spectrometry, sialic acids, **138**, 150
phosphatidylinositol fatty acid methyl esters, **141**, 241
sialic acids, **138**, 149
Gas vesicles
cyanobacterial
concentration measurements
by gas volume assessment, **167**, 216
optical technique, **167**, 217
constituent proteins
electrophoresis, **167**, 218
encoding genes, isolation, **167**, 220
enzymatic digestion, **167**, 219
sequence analysis, **167**, 218
description, **167**, 213
elastic compressibility, **167**, 221
electron microscopy, **167**, 221
isolation and purification, **167**, 215
in measurement of cyanobacterial turgor pressure, **167**, 660
Gaucher's disease
enzymatic diagnosis, **138**, 747
GC–MS, *see* Gas chromatography–mass spectrometry
GDP (guanosine 5′-diphosphate)
phosphoryls, isolation, **159**, 70
Gelatin
crosslinked, immobilization of organelles, **135**, 457, 471
immobilized biocatalysts
assay and use, **135**, 295
preparation, **135**, 294
properties, **135**, 296
Gelatinase
metallo-, *see* Metallogelatinase
Gel diffusion assay
β-D-glucanases, **160**, 63
Gel electrophoresis
α_1-acid glycoprotein variants, **163**, 427
aggregating proteoglycans, **144**, 391
in analysis of
calmodulin–peptide complex stoichiometry, **139**, 471
cAMP-dependent protein kinase regulatory subunit phosphorylation, **159**, 143
DNA–protein binding in yeast extracts, **155**, 528
restriction enzymes *Not*I and *Sfi*I, **155**, 17
RNA 3′ and 5′ ends, **155**, 433
rRNA–protein interactions, **164**, 456
bone glycoproteins, **145**, 278
bone morphogenetic protein, **146**, 299
branched-chain α-keto acid dehydrogenase, **166**, 301
calbindin D, **139**, 535
calcium-binding protein, 21-kDa form, **139**, 91
calmodulin-dependent protein phosphatase subunits, **139**, 82
cAMP phosphodiesterase, **159**, 767
CD2 antigens, **150**, 543
cell-free translation products, **152**, 296
cellobiose dehydrogenase, **160**, 451, 458
cellulases, **160**, 255, 267, 302, 304, 319
ceruloplasmin variants, **163**, 450
chloroplast envelope polypeptides, **148**, 200, 213
cholera toxin, **165**, 174
chorismate mutase, **142**, 469
clostridial enterotoxin, **165**, 96
collagen α chains, **145**, 173
collagen peptides, **145**, 179
collagen type I mRNA from osteogenesis imperfecta fibroblasts, **145**, 225

complement components C3, C5, C3a, and C5a, **162**, 585
connective tissue-activating peptides, **163**, 740
C-reactive protein, **163**, 390
crosslinked catalase, **137**, 603
crosslinked collagen polypeptides, **144**, 137
crosslinked pyranose oxidase, **137**, 604
crosslinked RNA, **164**, 320
cytochrome-*c* oxidase from cyanobacteria, **167**, 443
dentin proteins, **145**, 292
dihydropteridine reductase, **142**, 128
diphtheria toxin fragment A–hormone hybrid, **165**, 209
direct transfer method for DNA sequencing, **155**, 250
DNA
  in agarose, **152**, 75
  in denaturing gradient gels
    application of melting behavior simulation, **155**, 482
    detection of single base changes, **155**, 501
  for detection and cloning of amplified fragments, **151**, 343
  before dideoxy sequencing, **155**, 80
  in polyacrylamide, **152**, 64
  in solid-phase sequencing, **155**, 319
DNA-binding proteins, **151**, 551
in DNA footprinting *in vivo*, **152**, 749
DNA–protein complexes, mobility shift assay, **152**, 725
in DNA sequencing by chemical methods, **152**, 536
elastin, **144**, 268
endo-1,4-β-glucanases, **160**, 353, 366
epidermal growth factor gene, **146**, 37
epidermal growth factor receptors, **146**, 54
equipment required for molecular cloning work, **152**, 14
estradiol receptor, **139**, 734
extracellular matrix components, **147**, 412
extracellular proteins from cyanobacteria, **167**, 602
F_1-ATPase, **167**, 353
fibroblast-derived growth factor, **147**, 46
fibroblast growth factor, **147**, 114
β-galactosidase fusion proteins, **154**, 143
galacturan 1,4-α-galacturonidase, **161**, 375
gas vesicle proteins from cyanobacteria, **167**, 218
β-glucosidase, **160**, 440
growth regulator-induced proteins, **147**, 438
haptoglobin phenotypes, **163**, 453
with immunoblotting, crosslinked ribosomal proteins
  diagonal technique, **164**, 70
  one-dimensional technique, **164**, 65
  two-dimensional technique, **164**, 73
interrupted, collagens, **163**, 685
isovaleryl-CoA dehydrogenase, **166**, 158, 161, 383
isozyme markers for chromosome identification, **151**, 171
lipopolysaccharide monomers, **138**, 267
lymphocyte surface receptors, **150**, 394
melanoma growth stimulatory activity, **146**, 119, 126
membrane proteins sensitive to thrombin, **147**, 160
monophenol monooxygenase, **142**, 156
Na^+,K^+-ATPase at acidic pH, **156**, 124
oligonucleotides, **154**, 243, 281, 310
  in solid-phase sequencing, **155**, 319
orthogonal-field-alternation, large DNA molecules, **155**, 468
parvalbumin, **139**, 765
pectic enzymes, **161**, 330
pertussis toxin, **165**, 125
phosphopeptides derived from Na^+,K^+-ATPase, **156**, 296
photosystem I and II reaction center complexes from cyanobacteria, **167**, 264
photosystem II polypeptides from cyanobacteria, **167**, 279
phycobiliproteins, **167**, 325
plague murine toxin, **165**, 165
  subunits, **165**, 168
polysialic acid, endoneuraminidase effects, **138**, 631
prephenate aminotransferase, **142**, 483
proteins methylated in presence of calmodulin, **139**, 669

proteoglycans, **144**, 335
pulsed-field
intact genomic DNA, **155**, 453
in mapping and cloning of mammalian genomes, **151**, 464
ribosomal protein–RNA complexes, **164**, 215
ribosomal 30S reconstitution intermediates, **164**, 275
RNA
in agarose, **152**, 75
in polyacrylamide, **152**, 64
RNA polymerase σ subunit, **167**, 595
sarcoplasmic reticulum protein cDNA, **157**, 293
Shiga toxin, **165**, 157
16S rRNA colicin fragments, **164**, 196
skeletal growth factor, **146**, 291
staphylococcal δ-lysin, **165**, 20
T-cell receptors, **150**, 678
tetanus toxin, **165**, 87
transferrin receptor antibodies, **147**, 267
transforming growth factor α, **146**, 111
transforming growth factor β receptors, **146**, 189
tRNA, **155**, 380
two-dimensional
acute-phase proteins, **163**, 581
collagen peptides, **145**, 183
crosslinked RNA fragments, **164**, 294
[^{32}P]phosphoproteins in reactivated sperm, **139**, 820
proliferating cell nuclear antigen/cyclin, **150**, 156
radiolabeled ribosomal proteins, **164**, 431
ribosomal proteins, **164**, 576
ribosomal 50S reconstitution intermediates, **164**, 276
Spec proteins, **139**, 358, 360
S6 multiple phosphorylated forms, **146**, 365
S6 tryptic phosphopeptides, **146**, 373
thylakoid polypeptides, **148**, 623
Gel entrapment
enzyme immobilization by, **135**, 61
Gel filtration
in assay for lysosomal acidification, **157**, 596
bacterial chemotactic factor, **162**, 201
bone glycoproteins, **145**, 272
branched-chain 2-keto acid dehydrogenase, **166**, 306
branched-chain 2-keto acid dehydrogenase multienzyme complex, **166**, 334
calbindin D, **139**, 535
calcium-binding protein, 21-kDa form, **139**, 96
calelectrins, **139**, 31
calmodulin-dependent protein phosphatase subunits, **139**, 80
cDNA large double-stranded fragments, **152**, 335
cellobiose dehydrogenase, **160**, 458
cellulase, **160**, 220
clostridial alpha toxin, **165**, 93
collagen crosslinking amino acids, **144**, 121
complement proteins
C5b–6, **162**, 560
C7, **162**, 562
elastin, **144**, 268
endo-1,4-β-glucanase, **160**, 227
gangliosides, **138**, 302
Hummel–Dryer method for analysis of
calmodulin–microtubule protein binding, **139**, 840
calmodulin–phosphofructokinase binding, **139**, 749
insulin-like growth factor I, **146**, 202
insulin-like growth factor II, **146**, 265
lignin–carbohydrate complexes, **161**, 14
α_2-macroglobulin, **163**, 487
metallocarboxypeptidases, **158**, 74
myelin basic protein, **162**, 414
Na^+,K^+-ATPase, **156**, 77
nonequilibrium, in assay of oligonucleotide–rRNA binding, **164**, 449
110-kDa protein–calmodulin complex, **139**, 141
phospholipid transfer protein, **148**, 665
phosphopeptides derived from Na^+,K^+-ATPase, **156**, 298
plasmid DNA, **152**, 167
platelet-derived growth factor-binding proteins, **147**, 56

in preparation of white resealable erythrocyte ghosts, **149**, 272
proteoglycans, **138**, 281
recombinant human C5a, **162**, 663
sialyl oligosaccharides, **138**, 291
Synechocystis fimbriae, **167**, 194
toxic shock syndrome toxin-1, **165**, 40
transforming growth factor β, **146**, 157
xylobiose and xylooligomers, **160**, 534
β-xylosidase, **160**, 681
Gel filtration media
contamination problems, **158**, 15
interactions with metals
effect of eluent ionic strength, **158**, 18
elimination by alkaline reduction, **158**, 19
quantitation, **158**, 16
Gel overlay analysis
calmodulin-binding proteins, **139**, 433
gap junction interactions with Ca^{2+}-modulated proteins, **139**, 452
pectic enzymes, **161**, 330
Gels, *see also specific gels*
analytical, for preparation of 5S RNA-related materials, **164**, 170
boronate-containing, for electrophoretic analysis of RNA termini, **155**, 438
denaturing gradient, preparation for DNA electrophoresis, **155**, 512
detection of cellulolytic enzymes in, **160**, 135
Gelsolin
preparation from macrophages, **162**, 256
Gene amplification
in somatic mammalian cells
applications of flow cytometry, **151**, 97
assay methods, **151**, 102
associated drug resistance
multiple mechanisms, **151**, 88
selection procedures, **151**, 88
critical drug levels for enhancement, microtiter analysis, **151**, 94
documentation methods, **151**, 97
mechanism, **151**, 87
role of recombinant DNA technology, **151**, 99
Gene mapping
application of Tn*5* mutagenesis, **154**, 175
Kdp-ATPase mutant genes, **157**, 664
open reading frame strategy, **154**, 129
with sorted chromosomes, **151**, 292
Genes
acetolactate synthase, isolation from microbes and plants, **166**, 101
amplified, detection and cloning by in-gel DNA renaturation, **151**, 332
ATP synthase, in *Synechococcus*
cloning, **167**, 795
sequencing, **167**, 798
bone glycoprotein, cloning, **145**, 286
branched-chain keto acid dehydrogenase, cloning from *Pseudomonas putida*, **166**, 350
branched-chain α-keto acid dehydrogenase, isolation, **166**, 110
C5a
expression in *E. coli*, **162**, 657
nucleotide sequence, **162**, 654
synthesis, **162**, 653
calmodulin
in *Aplysia*, cloning, **139**, 287
in chicken
cloning, **139**, 217
structural analysis, **139**, 224
in fruit fly
cloning, **139**, 230
exon sequencing, **139**, 232, 239
structural comparisons with other calmodulin genes, **139**, 243
site-specific mutagenesis, **139**, 291
in trypanosome, organization, **139**, 268
in yeast, cloning, **139**, 248
chimeric, expression in plant cells, **153**, 283
chloramphenicol acetyltransferase
in collagen gene expression analysis, **144**, 67
in optimization of gene transfer into eukaryotic cells, **151**, 382
cloned
functional analysis in higher eukaryotes, **152**, 684
regulatory elements, identification with functional assays, **152**, 704
cloning with long oligonucleotide probes, **152**, 443
collagen
expression analysis, **163**, 688

expression techniques, **144**, 65
mutations, assessment by R-loop analysis of procollagen mRNA, **145**, 235
structural analysis, **144**, 62
type I, transcription assay, **144**, 75
colony-stimulating factor, isolation, **150**, 752
copy number, estimation, **152**, 594
cyanobacterial
cloned, listing, **167**, 831
cloning with host–vector systems, **153**, 199
for cyanobacterial gas vesicle proteins, isolation, **167**, 220
for cyanobacterial photosynthetic apparatus components, isolation, **167**, 755
cytochrome b_6/f, isolation and characterization from *Nostoc*, **167**, 779
differentially expressed, isolation by
+/− screening, **152**, 424
subtraction hybridization, **152**, 425
dihydrofolate reductase, from mutant CHO cells, repetitive cloning, **151**, 451
direct transfer to plant protoplasts, **153**, 313
elastin, structure, **144**, 184, 261
endogenous, inhibition by antisense RNA, **151**, 519
epidermal growth factor
cloning and expression in yeast, **146**, 39
construction and purification, **146**, 34
essential, conditional mutations, isolation by plasmid shuffling, **154**, 167
eukaryotic, expression in *E. coli* with synthetic two-cistron system, **153**, 401
foreign, vaccinia virus as expression vector, **153**, 545
full-length, isolation by genomic walking, **152**, 598
fusions, application of subcloning, **152**, 515
heterologous, expression in yeast, **152**, 673
hypoxanthine phosphoribosyltransferase, retrovirus-mediated transfection, **149**, 13
insulin, liposome-encapsulated
expression in liver and spleen, **149**, 161
targeting to hepatic cells, **149**, 163
interleukin-1, isolation, **150**, 748
interleukin-2, isolation, **150**, 749
interleukin-3, isolation, **150**, 751
isolation by
immunoscreening λgt11 libraries, **152**, 458
retroviral tagging, **152**, 469
isolation from
λgt 11 expression libraries by immunoscreening, **154**, 108
mammalian cells by sib selection, **151**, 445
α-isopropylmalate synthase, cloning and sequencing, **166**, 422
Kdp-ATPase, analysis, **157**, 655
lymphotoxin, isolation, **150**, 753
mast cell growth factor, isolation, **150**, 751
members of multigene families
detection and recovery, **152**, 590
discrimination, **152**, 591
mutant
isolation by insertional mutagenesis, **151**, 489
isolation strategies, **151**, 329
Na^+,K^+-ATPase α subunit, cloning with antibody probes, **156**, 379
nitrogen fixation
in cyanobacterial DNA hybridizations, **167**, 803
identification in *Rhizobium* by Tn*5* mutagenesis, **154**, 197
parvalbumin, cloning from rat, **139**, 317
photosystem II, deletion and replacement in *Synechocystis*, **167**, 768
pro-α2(I) collagen, mutation, characterization, **145**, 213
procollagen, restriction fragment length polymorphisms, analysis, **145**, 205
product overexpression, application of subcloning, **152**, 515

prolactin, regulation by Ca^{2+}/calmodulin, analysis, **139**, 655
protein S
cloning, **139**, 383
expression, **139**, 387
sequencing, **139**, 385
recombinant, expression and product analysis in *E. coli*, **152**, 661
replacement, application of Tn*5* mutagenesis, **154**, 188
reporter, application in cyanobacteria, **167**, 736
repressor, λ phage, in construction of pHE6 expression plasmid, **153**, 482
ribosomal protein
in chloroplast DNA, cloning and identification, **164**, 748
L24, mutations, analysis, **164**, 706
rRNA, in *E. coli*, antibiotic resistance mutations, **164**, 673
shotgun synthesis by segmented paper method, **154**, 253
T-cell receptor, cloning, **150**, 655
tissue-specific expression, *in situ* hybridization assay, **152**, 649
transcription in transfected cells, *in situ* detection with biotin-labeled probes, **151**, 530
transcripts
mapping by nuclease protection assays and cDNA primer extension, **152**, 611
prevalence levels in RNA populations, assay, **152**, 633
transfer into eukaryotic cells, optimization with CAT reporter gene, **151**, 382
transferred from cosmid libraries, rescue, **151**, 426
transferrin receptor, cloning, **150**, 739
by transfection, **147**, 281
troponin C, isolation, **139**, 330
tumor necrosis factor
isolation, **150**, 753
modification by site-specific mutagenesis, **154**, 403
yeast, isolation by complementation, **152**, 481

Genomic libraries
construction and characterization in phage λ, **152**, 190
elastin gene screening, **144**, 287
fibronectin and receptor, preparation, **144**, 462
for genomic walking, requirements, **152**, 599
jumping-type, generation by chromosome hopping, **151**, 475
linking-type, construction, **151**, 484
Phanerochaete chrysosporium, construction, **161**, 213
Streptomyces ambofaciens, generation in shuttle cosmid vector, **153**, 179
underrepresentation of specific sequences, alleviation, **152**, 173
Genomic walking
cosmid vectors for, **152**, 604
in isolation of full-length genes, **152**, 598
Genotyping
haptoglobin, **163**, 466
Geosmin
production by cyanobacteria, **167**, 605
Gizzard
chicken, purification of
calcimedins, **139**, 21
calmodulin, **157**, 330
turkey, purification of myosin phosphatase, **159**, 446
Glass
cysteinylsuccinamidopropyl-controlled-pore, preparation, **143**, 505
glycophase, activation with tresyl chloride, **135**, 67
porous
aminotransferase immobilization with, **136**, 486
enzyme immobilization with, **135**, 537
succinamidopropyl-derivatized
activation with thionyl chloride, **135**, 132
aqueous preparation method, **135**, 131
nonaqueous preparation method, **135**, 132
reversible protein coupling, **135**, 132
succinylthiopropyl-derivatized
preparation, **135**, 135
reversible protein coupling, **135**, 135

GLC, *see* Gas–liquid chromatography
Glial growth factor
  role in tissue repair processes, **163**, 713
  purification from bovine pituitary, **147**, 220
Glia maturation factor
  bioassay, **147**, 232
  purification from bovine brain, **147**, 226
β-Globin
  produced in *E. coli*, properties, **153**, 480
Gloeothece
  sheath isolation from, **167**, 183
Gloeotrichia echinulata
  akinete formation in culture, **167**, 226
Glomerulonephritis
  experimental, production and study of three models, **162**, 421
  induced with planted antigens
    characteristics, **162**, 451
    overview, **162**, 449
    production, **162**, 450
(1→3),(1→4)-β-D-Glucan
  assay in
    barley, **160**, 548
    beer, **160**, 549
    malt, **160**, 549
    oats, **160**, 548
    wort, **160**, 549
  purification from barley flour, **160**, 511
β-Glucanase
  assays, **160**, 59
  properties, **160**, 387
  purification from pea, **160**, 384
1,4-β-D-Glucan 4-glucanohydrolase, *see* Endo-1,4-β-glucanase
Glucan 1,4-α-glucosidase
  assay, **137**, 691
  coupling to derivatized titanium(IV)-activated supports, **135**, 127
  enlarged, preparation for coentrapment, **137**, 646
  ethylenediamine adducts, coupling to carboxyalkyl agarose, **135**, 145
  –glucose oxidase electrode, description, **137**, 36
  immobilization by radiation polymerization, **135**, 151
  in industrial saccharification
    immobilized enzyme system, **136**, 377
    liquid enzyme system, **136**, 374
  production by *Aspergillus niger*, **137**, 689, 693
  transition metal-chelated, glutaraldehyde crosslinking, **135**, 126
Glucan 1,4-β-glucosidase
  detection in gels, **160**, 143
D-Glucans
  reductive cleavage, **138**, 35
1,3-β-Glucan synthase
  assay, **138**, 637
  preparation from fungi, **138**, 638
  properties, **138**, 641
Glucan synthase II
  as plant plasma membrane marker, assay, **148**, 545
Glucoamylase, *see* Glucan 1,4-α-glucosidase
Glucocerebrosidase, *see* Glucosylceramidase
Glucomannan
  –AH-Sepharose, preparation for β-mannanase purification, **160**, 597
β-1,4-Glucomannooligosaccharides
  preparation from konjak glucomannan, **160**, 520
Glucosamine
  as inhibitor of lipid-linked saccharide pathway, **138**, 679
Glucose
  analogs as inhibitors of lipid-linked saccharide pathway, **138**, 680
  assay with
    biosensor, **137**, 52
    dry reagent chemistries, **137**, 412
    enzyme reactor, **137**, 279, 295
      continuous-flow system, **137**, 263
      flow–injection system, **137**, 268
    enzyme thermistor, **137**, 188
    enzyme transistor, **137**, 250, 259
  biosensors, long-term peritoneal implantation, **137**, 68
  cellular transport, defective mutants, isolation, **151**, 132
  conversion to acetone and butanol, **137**, 663
  enzyme thermistor monitoring and control during sucrose hydrolysis, **137**, 308
  flow enthalpimetric assay, **137**, 225

flow-injection analysis with immobilized enzymes, **137**, 176
growth factor-stimulated cellular uptake, assay, **146**, 399
oxidative pathway, defective mutants, isolation, **151**, 135
production from starch, **137**, 661
starvation, effect on protein glycosylation, **138**, 682
tritiated, in selection of glucose oxidative pathway-defective mutants, **151**, 135
Glucose analyzer
portable continuous
characteristics *in vitro*, **137**, 323
clinical results, **137**, 324
description, **137**, 320
general principles, **137**, 319
Glucose dehydrogenase
–glucose oxidase electrode, description, **137**, 39, 40
in multienzymatic NADH recycling, **136**, 71
Glucose isomerase
coupled to cellulose acetate beads, assay, **135**, 287
immobilized, industrial production and application, **136**, 356
Glucose oxidase
assay, **161**, 308
electrodes
long-term peritoneal implantation, **137**, 68
preparation, **137**, 25
enzyme channeling immunoassays with, **136**, 98
ethylenediamine adducts, coupling to carboxyalkyl agarose, **135**, 145
and horseradish peroxidase, cellobiase assay with, **160**, 110
multienzyme electrodes
amplification principle, **137**, 39
antiinterference principle, **137**, 42
competition principle, **137**, 40
sequence principle, **137**, 36
in needle-type glucose sensor, **137**, 326
in portable continuous blood glucose analyzer, **137**, 319
properties, **161**, 311
purification from *Phanerochaete chrysosporium*, **161**, 309
reactors
glucose assay with, **137**, 263, 279, 295
preparation, **137**, 261, 280, 289
transistor
glucose assay with, **137**, 259
preparation, **137**, 257
Glucose 6-phosphate
enzymatic synthesis with ATP regeneration, **136**, 279
inositol biosynthesis from, measurement, **141**, 127
production in enzyme reactor based on aqueous two-phase system, **136**, 52
Glucose-6-phosphate dehydrogenase
cell permeabilization test with, **137**, 640
as marker for chromosome detection in somatic cell hybrids, **151**, 182
in multienzymatic NADPH recycling, **136**, 72
thioredoxin *m* assay with, **167**, 411
Glucose-6-phosphate isomerase
as marker for chromosome detection in somatic cell hybrids, **151**, 182
Glucose sensors
needle-type
characteristics *in vitro*, **137**, 329
characteristics *in vivo*, **137**, 331
clinical applications, **137**, 334
construction, **137**, 327
optical-fiber
performance tests, **137**, 364
prototype construction, **137**, 360
α-Glucosidase
–albumin conjugate, preparation, **137**, 569
–albumin–insulin conjugate, preparation, **137**, 569
gelatin-immobilized, properties, **135**, 297
as marker for chromosome detection in somatic cell hybrids, **151**, 183
β-Glucosidase
assays, **160**, 109, 409, 416, 424, 432, 438, 576
detection in
gels, **160**, 138
prokaryotes, **160**, 184
enlarged, coentrapment with yeast cells, **137**, 642

function, **160**, 423
properties
Aspergillus enzyme, **160**, 579
Ruminococcus enzyme, **160**, 412
Sclerotium enzymes, **160**, 428
Sporotrichum enzymes, **160**, 421
Talaromyces enzymes, **160**, 442
purification from
Aspergillus niger, **160**, 577
Humicola, **160**, 328
Ruminococcus albus, **160**, 410
Schizophyllum commune, **160**, 435
Sclerotium rolfsii, **160**, 426
Sporotrichum pulverulentum, **160**, 419
Talaromyces emersonii, **160**, 439
Trichoderma koningii, **160**, 224
β-D-Glucoside glucohydrolase, *see* β-Glucosidase
Glucosinolate sulfatase
assay, **143**, 361
properties, **143**, 365
purification from *Helix pomatia*, **143**, 363
Glucosone
effects on
catalase, **137**, 606
pyranose oxidase, **137**, 608, 611
Glucosylceramidase
deglycosylated, targeting to hepatic cells, **149**, 38
diagnostic assay for Gaucher disease, **138**, 747
digestion before cell targeting
N-acetylglucosamine removal, **149**, 35
galactose removal, **149**, 32
sialic acid removal, **149**, 31
purification from human placenta, **149**, 29
Glucosyltransferase
glycosphingolipid-synthesizing, assay, **138**, 578
β-Glucuronidase
as marker for chromosome detection in somatic cell hybrids, **151**, 183
neutrophil granule-associated, assay, **162**, 546
in preparation of fungal mitochondria and plasma and vacuolar membranes, **157**, 563
D-Glucuronidase
assays, **160**, 561
Glutamate
assay with microbial sensor, **137**, 129
flow-injection analysis with immobilized enzymes, **137**, 176
Glutamate decarboxylase
entrapment with photocrosslinkable resin prepolymers, **135**, 239
prepolymer-entrapped, applications, **135**, 245
Glutamate dehydrogenase
in multienzymatic NAD(P)H recycling, **136**, 71, 77
Glutamate-oxaloacetate transaminase, *see* Aspartate aminotransferase
Glutamate–pyruvate transaminase
in multienzymatic NADH recycling, **136**, 71
Glutamic-oxaloacetic aminotransferase, *see* Aspartate aminotransferase
Glutamic-pyruvic aminotransferase, *see* Alanine aminotransferase
Glutamine
residues, incorporation of glycosyl-asparagine derivatives, **138**, 413
γ-Glutamyltransferase
assay with dry reagent chemistries, **137**, 415
Glutaraldehyde
activation of
modified Separon H 1000, **135**, 345, 349, 352
Sorfix, **135**, 354
crosslinking of
dCMP deaminase
in activated R conformation, **135**, 579
in inhibited T conformation, **135**, 581
transition metal-chelated enzymes, **135**, 126
enzyme electrode preparation with, **137**, 23
in quaternary structure determination of Na^+,K^+-ATPase, **156**, 333
stabilization of L-aspartate β-decarboxylase activity, **136**, 476
Glutathione
enzymatic assay, in analysis of oxidant injury, **163**, 334
HPLC, in analysis of oxidant injury, **163**, 335

Glutathione peroxidase
as marker for chromosome detection in somatic cell hybrids, **151**, 184
Glutathione reductase
as marker for chromosome detection in somatic cell hybrids, **151**, 184
Glycans
high-mannose, HPLC, **138**, 114
Glyceraldehyde-3-phosphate dehydrogenase
inhibition by H_2O_2, assay, **163**, 330
as marker for chromosome detection in somatic cell hybrids, **151**, 184
thermostabilization, **137**, 623
Glycerol
quantitative TLC, **160**, 173
Glycerol kinase
immobilized
dihydroxyacetone phosphate synthesis with, **136**, 277
sn-glycerol 3-phosphate synthesis with, **136**, 276
Glycerol 3-phosphate
enzymatic synthesis with ATP regeneration, **136**, 276
formation from nucleoside monophosphates, **159**, 71
Glyceryl-ether monooxygenase
assay, **141**, 395
properties, **141**, 396
Glycerylpropylsilica
1,1′-carbonyldiimidazole-activated
IgG purification with, **135**, 114
preparation, **135**, 109
tresyl chloride-activated
alcohol dehydrogenase coupling to, **135**, 69
preparation, **135**, 67
trypsin inhibitor coupling to, **135**, 70
Glycine max, *see* Soybean
Glycoconjugates
labeling with hydrazides, **138**, 429
Glycogen phosphorylase
as noninvasive indicator of cytoplasmic calcium, **141**, 18
Glycol chitin
characterization, **161**, 409
preparation, **161**, 408
Glycolipid receptors
for bacteria and viruses
overlay assay, **138**, 212, 221
solid-phase assay, **138**, 227
Glycolipids
analysis with antibodies to cell-surface carbohydrates, **138**, 195
continuous-development TLC, **138**, 127, 128
methylation analysis with GC–MS, **138**, 13
plant-derived, molecular species, HPLC, **148**, 650
sulfated, laminin binding properties, **144**, 505
total nonacid, preparation, **138**, 212
Glycolysis
defective cellular mutants, isolation, **151**, 132
Glycopeptides
fibronectin, preparation, **144**, 428
lectin affinity chromatography, **138**, 232
Glycophosphosphingolipids
biosynthetic pathway, **138**, 195
carbohydrate analysis, **138**, 193
isolation from plants, **138**, 190
oligosaccharides
purification, **138**, 192
release, **138**, 191
overview, **138**, 186
sequence analysis, **138**, 194
Glycoproteins
analysis with antibodies to cell-surface carbohydrates, **138**, 195
bone-associated, noncollagenous
biosynthesis, **145**, 282
characterization, **145**, 277
ELISA, **145**, 280
extraction, **145**, 270
gene cloning, **145**, 286
purification, **145**, 272
chemical deglycosylation, **138**, 341
enzymatic deglycosylation, **138**, 350
N-linked, glycosylation inhibitors, **138**, 661
polysialosyl, detection in brain, **138**, 169
in solution, labeling with hydrazides, **138**, 441
synthesis with glycosyl-asparagine derivatives, **138**, 409, 413, 418
Thy-1, T-cell-associated, isolation with lectins, **150**, 473

Glycosaminoglycans
assay in cell cultures, **147**, 411
bone-associated, characterization, **145**, 267
conformational studies, **145**, 77
detection with ^{125}I-cytochrome *c*, **138**, 260
proteoglycan
degree of epimerization, **144**, 351
extraction, **144**, 342
fractionation, **144**, 344
identification, **144**, 345
isolation, **144**, 343
size distribution, **144**, 352
sulfation pattern, **144**, 349
Glycosides
cardiac, binding to Na^+,K^+-ATPase, assay, **156**, 201
chromogenic and fluorogenic, as carbohydrase substrates and ligands
applications, **160**, 51
assays and analyses, **160**, 48
preparation, **160**, 46
cluster, preparation
from peptide backbone, **138**, 428
with tromethamine, **138**, 425
inhibition of Na^+,K^+-ATPase, **156**, 9
Glycosphingolipids
acetylated, HPLC, **138**, 9
biosynthesis
assay methods, **138**, 575
TLC analysis, **138**, 567
continuous-development TLC, **138**, 125
direct chemical ionization MS, **138**, 79
perbenzoylated, HPLC, **138**, 117
underivatized, HPLC, **138**, 4
Glycosylating agents
for oligosaccharide synthesis, **138**, 363
Glycosylation
animal cell mutants, biochemical characterization, **138**, 443
EGF receptor, analysis, **146**, 57
solvent effects, **138**, 376
Glycosylation inhibitors
for N-linked glycoproteins, **138**, 661
Glycosyltransferase
analysis in mitogen-treated lymphocytes, **150**, 50
cell-free isomaltulose-forming, characterization, **136**, 439
TLC-based assay
applications, **138**, 571
procedure, **138**, 568
Glycyl-L-histidyl-L-lysine
copper-binding
applications, **147**, 318, 325
growth-inhibitory analogs, **147**, 327
in growth-promoting mixtures, **147**, 324
preparation, **147**, 317
properties, **147**, 314
Glycyl-L-leucinamide
coupling to isonitrile-derivatized polymers, **135**, 100
Glyoxal
as crosslinking agent for polyacrylamide–hydrazide gels, **135**, 218
Glyoxalase I
as marker for chromosome detection in somatic cell hybrids, **151**, 185
Glyoxysomes
isolated
criteria for purity and intactness, **148**, 515
enzyme activities, **148**, 511
morphologic analysis, **148**, 513
isolation from castor bean endosperm, **148**, 507
membranes
isolation, **148**, 511
lipid analysis, **148**, 527
polypeptide analysis, **148**, 528
GMP (guanosine 5′-monophosphate)
endogenous, α-phosphoryl isolation, **159**, 71
Gold
atomic absorption spectrometry, **158**, 136
colloidal, in immunoelectron microscopy of cAMP-dependent protein kinase, **159**, 225
Golgi apparatus
vesicles, sugar nucleotide transport assay, **138**, 711
G proteins
role in
chemoattractant-induced phagocyte activation, **162**, 276
polyphosphoinositide breakdown in neutrophils, **141**, 261

subunit purification from rod outer
   segments, **159**, 703
Gradients, *see also specific gradients*
   concentration, chemotactic factors in
      chemotaxis assays, **162**, 85
Granules
   neutrophil
      isolation, **163**, 310
      primary and secondary
         constituents, assay, **162**, 546
      enriched subcellular fractions,
         preparation, **162**, 543
      exocytosis, **162**, 549
Granulocytes
   guinea pig, preparation of actin, **162**,
      252
   human
      preparation of cytoplasmic extracts,
         162, 248
      purification of
         connective tissue-activating peptide,
            163, 745
         CR3 receptor, **150**, 605
      necrotaxis and galvanotaxis assays, **162**,
         115
Granuloma
   hepatic, as model of inflammation and
      repair, overview, **163**, 605
Granulomatous reactions
   analysis in mice, **163**, 364
Granulosa cells
   rat, isolation and purification of pro-
      teoglycans, **138**, 279
Grape
   protoplasts, preparation from leaves,
      148, 28
Growth factors, *see also specific growth
      factors*
   analysis with cell hybrids, **147**, 379
   cartilage-derived
      extraction, **146**, 321
      purification, **146**, 323
   chondrosarcoma-derived, purification,
      147, 100
   colostrum-derived, purification, **146**, 45
   effects on membrane transport, assays,
      146, 376
   hepatoma-derived, purification, **147**,
      102
   hypothalamus-derived, purification, **147**,
      104

induced changes in pH$_i$, measurement,
      147, 415
induced proteins, assay in cell cultures,
      147, 427
interactions with cell nutrients, kinetic
      analysis, **147**, 392
milk-derived, purification, **146**, 43
modulated cell cycle events, character-
      ization, **147**, 400
role in tissue repair processes, overview,
      163, 707
stimulated cellular glucose uptake,
      assay, **146**, 399
stimulated extracellular matrix synthesis,
      assays, **147**, 407
stimulated fluid-phase endocytosis,
      assay, **146**, 402
stimulated kinases, assay, **146**, 355
Growth hormone
   bovine
      derivatives
         expression in *E. coli*, **153**, 397
         in production of natural sequence
            bGH, **153**, 399
         two-cistron expression vector, con-
            struction, **153**, 409
Growth hormone receptors
   lymphoid cell
      binding studies, **150**, 701
      labeling techniques, **150**, 710
      regulation plus competition assay, **150**,
         708
Growth inhibitors
   BSC-1 cell-derived
      bioassay, **146**, 165
      isolation, **146**, 163
GTP (guanosine 5'-triphosphate)
   effect on phosphoinositide hydrolysis,
      141, 120
   and EF-G, binding to ribosomal pre-
      translocation complex, kinetics, **164**,
      594
   phosphoryls, isolation, **159**, 70
   regulation of membrane phosphoino-
      sitide turnover, **141**, 122
Guaiacylglycerol-β-guaiacyl ether
   synthesis, **161**, 58
Guanidine hydrochloride
   in isolation of RNA from
      cultured cells, **152**, 226
      tissues, **152**, 221

Guanidinium thiocyanate
with cesium chloride, in RNA isolation, **152**, 224
in RNA isolation, **152**, 223
Guar
purification of
exo-1,4-β-D-mannanase, **160**, 591
galactomannan, **160**, 526
α-galactosidase, **160**, 628
β-mannanase, **160**, 601

H

Hageman factor, *see* Factor XII
Hair roots
extraction of leucine 2,3-aminomutase, **166**, 133
Halides
in spectrophotometric assay of thiols, **143**, 66
Halobacterium halobium
purification of thiols, **143**, 97
Haptens
labeled ribosomal subunits at RNA 3′-end, preparation, **164**, 82
labeled 30S ribosomal subunits at 5′-end of 16S RNA, preparation, **164**, 85
labeled 5S rRNA at internal position, preparation, **164**, 87
Haptoglobin
cDNA isolation, **163**, 461
and encoding mRNA, tissue localization, **163**, 468
genotyping and DNA polymorphisms, **163**, 466
phenotyping, **163**, 453
polymer isolation, **163**, 460
purification, **163**, 457
subunit isolation, **163**, 459
Heart
bovine, purification of
calmodulin-stimulated phosphodiesterase, **159**, 555
cAMP-dependent protein kinase
holoenzyme, **159**, 202
regulatory subunits, **159**, 208
cGMP-regulated phosphodiesterases, **159**, 685
isethionic acid, GLC analysis, **143**, 173
perfused, branched-chain α-keto acid dehydrogenase flux rates, measurement, **166**, 484
porcine, purification of branched-chain-amino-acid aminotransferase, **166**, 272
rabbit, purification of branched-chain α-keto acid dehydrogenase kinase complex, **166**, 319
sarcolemmal vesicles, preparation by sucrose flotation, **157**, 85
sarcoplasmic reticulum
junctional and longitudinal fractions
characteristics, **157**, 102
preparation, **157**, 101
preparation and structural characteristics, **157**, 109
purification of
calcium channel, **157**, 500
phospholamban, **157**, 360
vesicles, preparation, **157**, 92
Heat shock response
cyanobacterial, characterization, **167**, 622
HeLa cells
characteristics, **151**, 39
cloning, **151**, 44
cytotoxicity of Shiga toxin, **165**, 232
establishment, **151**, 38
growth conditions and nutritional requirements, **151**, 40
hybridization, **151**, 49
mutagenesis, **151**, 45
storage, **151**, 44
variant forms, **151**, 50
Helianthin
hemolytic assay, **165**, 217
Helianthus tuberosus, *see* Jerusalem artichoke
Helix pomatia, *see* Snail
Hemagglutination assay
Amphicarpaea bracteata lectin, **138**, 561
Cancer antennarius lectin, **138**, 520
concanavalin A derivatives, **150**, 26
Erythrina lectins, **138**, 547
Evonymus europaea lectin, **138**, 553
Limax flavus lectin, **138**, 528
thrombospondin, **144**, 443
Vicia villosa lectins, **138**, 537

Hematin
  effect on cGMP-binding phosphodiesterase, **159**, 717
Heme
  binding to hemopexin, assay, **163**, 562
Hemicellulase
  –cellulase complex
    properties, **160**, 335
    purification from *Phoma hibernica*, **160**, 333
  purification from *Aspergillus fumigatus*, **160**, 268
Hemicellulose
  degrading enzymes, detection in prokaryotes, **160**, 180
Hemofilter
  with immobilized asparaginase
    applications, **137**, 493
    preparation, **137**, 493
Hemoglobin
  allosteric effector, entrapment in erythrocytes, **149**, 246
  as chemotactic agent, preparation, **162**, 183
  liposome-encapsulated
    preparation, **149**, 189
    properties, **149**, 196
  solutions, preparation, **149**, 186
Hemolymph
  crab, purification of lectin, **138**, 521
Hemolysin
  bioassay, **165**, 200
  interaction with cell membranes, **165**, 203
  precursor detection in culture, **165**, 203
  production, **165**, 201
  properties, **165**, 202
  purification from *Aeromonas hydrophila*, **165**, 201
  thermostable direct
    bioassay, **165**, 190
    properties, **165**, 191
    purification, **165**, 190
α-Hemolysin
  Escherichia coli
    bioassay, **165**, 138
    production, **165**, 139
    purification, **165**, 140, 399
γ-Hemolysin
  staphylococcal
    overview, **165**, 8
    purification, **165**, 11
    titration, **165**, 10
Hemolysis
  hypotonic, methods for entrapment of agents in erythrocytes, **149**, 221
Hemolytic assay
  aerolysin, **165**, 193
  Aeromonas hemolysin, **165**, 200
  clostridial alpha toxin, **165**, 294
  complement alternative pathway activity, **162**, 641
  complement components
    C1, C2, and C4, **162**, 628
    C3 and C5, **162**, 589
  C1q receptor, **150**, 566
  E. coli α-hemolysin, **165**, 138
  staphylococcal δ-lysin, **165**, 21
  streptolysin O, **165**, 58
  streptolysin S, **165**, 63
  thermostable direct hemolysin, **165**, 190
Hemolytic plaque assay
  for antibody-secreting cells
    autoradiography, **150**, 215
    Cunningham's modification, **150**, 213
    direct and indirect assay, **150**, 213
    plaquing procedures, **150**, 211
    preparation of cell suspensions, **150**, 210
    principle, **150**, 209
  B cells cloned on filter paper disks, **150**, 289
  poly(L-lysine)-based, for antigen-activated B cells
    assay sensitivity and secretory rates of plaque-forming cells, **150**, 235
    cellular requirements for plaque formation, **150**, 238
    detection and nature of pseudoplaques, **150**, 232
    pharmacologic modulation of plaque-forming cells, **150**, 236
    procedures, **150**, 226
    theoretical considerations, **150**, 224
  protein A-based, for Ig-secreting cells, **150**, 217
  reverse, antibody production in human lymphocyte cultures, **150**, 315
Hemopexin
  amino acid content, **163**, 550

assays, **163**, 548
carbohydrate content, **163**, 550
heme and porphyrin binding, assay, **163**, 562
molecular and genetic polymorphism, **163**, 554
monoclonal and polyclonal antibodies, preparation, **163**, 545
physicochemical and optical properties, **163**, 555
purification, **163**, 539
purity criteria, **163**, 544
structural features, **163**, 550
Hemophilia
antibody adsorption in, extracorporeal systems, **137**, 458
Hemorrhage
in Arthus reaction, measurement, **162**, 482
during inflammation, quantitation, **162**, 314
Heparin
binding domain in thrombospondin, isolation, **144**, 442
Heparin lyase
assays, **137**, 516
immobilization, **137**, 523
production, **137**, 518
purification, **137**, 521
reactor
construction, **137**, 525
testing *in vivo*, **137**, 525
Hepatocytes, *see also* Liver
cAMP-dependent protein kinase activation, inhibition by cAMP antagonist, **159**, 165
coenzyme A content, **166**, 78
disulfides and thiols, HPLC analysis, **143**, 107
isolation, **166**, 255
liposome targeting, flow cytometric analysis, **149**, 170
phosphoinositide metabolism, assay of subcellular sites, **141**, 92
primary culture, establishment for fibrinogen biosynthesis studies, **163**, 480
protein kinase regulatory subunit phosphorylation, analysis, **159**, 143
rat, isolation of low-K_m cAMP phosphodiesterase, **159**, 534
receptor-mediated endocytosis, analytical methods, **149**, 3
regulated expression of acute-phase proteins, analysis, **163**, 584
subcellular fractionation, **141**, 97
SV40-transformed, description, **151**, 31
synergistic responses to site-selective cAMP analogs, **159**, 123
targeting of glucocerebrosidase, **149**, 38
transport of branched-chain amino or 2-keto acids, assay in
monolayer cultures, **166**, 258
suspension cultures, **166**, 259
Hepatocyte-stimulating factor
bioassay, **163**, 600
biological activities, **163**, 602
production, **163**, 595
purification, **163**, 598
Herpetosiphon giganteus
purification of restriction enzyme *Hgi*CI, **155**, 4
Heteroantisera
to T-cell receptors, production, **150**, 679
Heterocysts
cyanobacterial
electron donation to nitrogenase, **167**, 496
gas permeability, **167**, 665
isolation, **167**, 490
Heterokaryons
isolation by toxin–antitoxin technique, **151**, 237
Hevea brasiliensis, *see* Rubber tree
Hexafluoroacetone sesquihydrate
as solvent for chitin, **161**, 406
Hexafluoroisopropanol
as solvent for chitin, **161**, 406
Hexokinase
and acetate kinase, in glucose 6-phosphate production, **136**, 52
cell permeabilization test with, **137**, 640
immobilized, glucose 6-phosphate synthesis with, **136**, 279
Hexosaminidase, *see* β-*N*-Acetylhexosaminidase
Hexose monophosphate shunt
activation in oxidant-exposed cells, assay, **163**, 336
Heymann nephritis
active form

characteristics, **162**, 429
induction, **162**, 426
overview, **162**, 422
passive form
autologous phase
characteristics, **162**, 439
induction, **162**, 437
characteristics, **162**, 431
induction, **162**, 430
High-performance liquid chromatography
acyl-CoA esters in biological samples, **166**, 55
adenine nucleotides in plant vacuoles, **148**, 132
S-adenosylmethionine and sulfur metabolites, **143**, 192
amino acid-specific tRNA, **155**, 373
arogenate, **142**, 427
in assay for
arogenate dehydratase, **142**, 498
arogenate dehydrogenase, **142**, 489
collagen biosynthesis, **163**, 675
cyanobacterial prenylquinones, **167**, 270
leukotriene B_4 20-hydroxylase, **163**, 351
leukotriene C_4 synthase, **163**, 354
3-methylglutaconyl-CoA hydratase, **166**, 214
prephenate aminotransferase, **142**, 480
tryptophan 5-monooxygenase, **142**, 89
tyrosine 3-monooxygenase, **142**, 57
bacterial chemotactic factor, **162**, 202
bone glycoproteins, **145**, 275
branched-chain 2-keto acids, **166**, 20
BSC-1 cell growth inhibitor, **146**, 166
calmodulin, **157**, 334
calmodulin derivatives, **139**, 422
cartilage-derived factor, **146**, 316
cellodextrin glycosides, **160**, 47
cellulase, **160**, 239
chitin oligomeric fragments, **161**, 410
chitooligosaccharides and reduced forms, **161**, 453
chorismate, **142**, 423
collagen group 1 molecules, **144**, 59
collagen hydroxypyridinium crosslinks, **144**, 127
colostrum-derived growth factor, **146**, 47
crosslinked collagen peptides, **144**, 136
cysteic acid, **143**, 164
cysteine and cystine as *o*-phthalaldehyde derivatives, **143**, 141
cysteine enantiomers, **143**, 167
cysteinesulfinic acid, **143**, 155, 164
in determination of chitosan molecular weight distribution, **161**, 447
N,*N*′-diacetylchitobiase, **161**, 527
elastin, **144**, 212
with electrochemical detection
catecholamine metabolites
one-column technique, **142**, 536
two-column switching technique, **142**, 544
catecholamines, **142**, 521
thiols, **143**, 110
endothelial cell growth factors, **147**, 153
Escherichia heat-stable enterotoxin, **165**, 132
with fluorometry, thiols, **143**, 70, 73, 85
glucocerebrosidase, **149**, 29
glutathione, in analysis of oxidant injury, **163**, 335
glycosphingolipids, **138**, 3
as perbenzoylated derivatives, **138**, 117
hypotaurine, **143**, 155
inositol phosphates, **141**, 191
insulin-like growth factor II, **146**, 267
kinin analogs, **163**, 269
kinins
acid separation, **163**, 260
modifications, **163**, 262
neutral separation, **163**, 259
α-leucine, **166**, 17
β-leucine, **166**, 17
(3*R*) and (3*S*) enantiomers, **166**, 14
leukotriene B_4 and ω-oxidation products, **141**, 360
lignin aromatic fragments, **161**, 183
lignin oxidation products, **161**, 118
lipid classes in chloroplast membranes, **148**, 309
lipid molecular species in chloroplast membranes, **148**, 313
–mass spectrometry, sialic acids, **138**, 153
melanoma growth stimulatory activity, **146**, 119

methionine enantiomers, **143**, 167
milk-derived growth factor, **146**, 44
oligonucleotides, **154**, 279, 309
oligosaccharides, as perbenzoylated derivatives, **138**, 94
pectic enzymes, **161**, 385
penicillamine, **143**, 186
phosphodiesterase reaction mixtures, **159**, 471
plant glyco- and phospholipid molecular species, **148**, 650
prephenate, **142**, 427
prostaglandin A_1 metabolite, in analysis of cAMP export, **159**, 91
prostaglandins, **141**, 336, 347
protein kinase C, **141**, 424
proteoglycan link protein peptides, **144**, 410
radiochemical, in analysis of branched-chain amino acid metabolism, **166**, 79
recombinant human C5a, **162**, 663
ribosomal protein S6, **164**, 577
ribosomal proteins, **164**, 523
for microsequence analysis, **164**, 542
16S rRNA colicin fragments, **164**, 195
short-chain CoA compounds, **166**, 70
sialic acids, **138**, 144
skeletal growth factor, **146**, 289
taurine, **143**, 155
thiols, **143**, 97
T-kinin, **163**, 274
transforming growth factor α, **146**, 108
transforming growth factor β, **146**, 160
trimethylselenonium ion, **143**, 198
with UV detection, disulfides and thiols as dinitrophenol derivatives, **143**, 101
Hindlimb
rat, preparations for branched-chain amino acid metabolism studies, **166**, 476
Hippocampus
slices, agonist-induced inositol phosphate accumulation, assay, **141**, 192
Histamine
mast cell-derived, assay, **162**, 517
Histochemistry
horseradish peroxidase assay in pinosome osmotic lysis evaluation, **149**, 47
localization of cyclic nucleotide phosphodiesterase, **159**, 477
sialic acids, analysis, **138**, 158
visualization of isozyme markers for chromosome identification, **151**, 176
Histology
assessment of
adjuvant arthritis, **162**, 347
streptococcal cell wall-induced arthritis in rats, **162**, 376
type II collagen-induced arthritis
in mice, **162**, 365
in rats, **162**, 360
Histones
^{35}S-labeled
preparation, **159**, 350
as substrates for protein phosphatases, **159**, 352
Histopathology
in assay of leukocyte chemotaxis *in vivo*, **162**, 102
thyroid, in analysis of spontaneous autoimmune thyroiditis, **162**, 473
Homocysteine
metabolism in microorganisms, **143**, 454
S-substituted forms, preparation, **143**, 292
L-Homocysteinesulfonic acid
synthesis, **143**, 280
Homogenizer
Dounce, in isolation of brain stem axolemma, **156**, 67
milk, in isolation of soybean nuclei, **148**, 539
Homopeptides
template-free ribosomal synthesis from aminoacyl-tRNAs, **164**, 641
Homovanillic acid
GC–MS assay in brain and body fluids, **142**, 577
HPLC assay in brain, **142**, 539, 547
Hormogonia
cyanobacterial
cellular macromolecular composition, **167**, 238

differentiation, induction, **167**, 235, 851
dispersal, contributing properties, **167**, 239
isolation, **167**, 237

Hormone receptors
and other supportive receptors on lymphoid cells, overview, **150**, 385

Horseradish peroxidase
conjugation with Fab′ fragments, **139**, 499
enzyme channeling immunoassays with, **136**, 98
in evaluation of pinosome osmotic lysis, **149**, 46
as fluid-phase marker
endocytotic uptake, **146**, 413
overview, **146**, 405
and glucose oxidase, cellobiase assay with, **160**, 110
–glucose oxidase electrode, description, **137**, 37
immobilization, **137**, 173, 179
sulfhydryl oxidase assay with, **143**, 122

Host–vector systems
for gene cloning in cyanobacteria, **153**, 199

Hot tritium bombardment
in analysis of ribosome surface topography, **164**, 426

HPLC, *see* High-performance liquid chromatography

α_2-HS glycoprotein
assay, **163**, 432
chromosomal location and molecular biology, **163**, 440
genetic polymorphism, **163**, 436
properties, **163**, 436
purification from human plasma, **163**, 433

Human chorionic gonadotropin
chemical deglycosylation, **138**, 343
coupled to diphtheria toxin fragment A
characterization, **165**, 208
preparation, **165**, 204
enzymatic deglycosylation, **138**, 351

Human immunodeficiency virus receptors
functional analysis, **150**, 555

Humicola
culture conditions, **160**, 325
purification of cellulases, **160**, 325

Hyaluronate
binding to
newly synthesized proteoglycans, analysis, **144**, 381
proteoglycan link protein, analysis, **144**, 411
proteoglycans, assays, **144**, 361

Hybridization
in analysis of rRNA protein binding sites, **164**, 463
cell, *see* Cell hybridization
colony, *see* Colony hybridization
cyanobacterial DNA with nitrogen fixation gene probes, **167**, 803
differential, in isolation of ligninase cDNA clones, **161**, 232
DNA–DNA, in taxonomic classification of cyanobacteria, **167**, 125
DNA–RNA, in chemical probing of large RNA molecules, **164**, 475
dot-blot, *Phanerochaete* DNA for genome size estimation, **161**, 215
genomic DNA to AT-rich oligonucleotide probes, **152**, 447
high-specific-activity RNA as probe, synthesis, **155**, 408
in identification of genomic sequences corresponding to cDNA clones, **152**, 588
immobilized nucleic acids, theory and practice, **152**, 399
in isolation of cyanobacterial photosynthetic apparatus genes, **167**, 758
in mapping of gene transcripts, **152**, 620, 631
in measurement of transcript prevalence levels in RNA populations, **152**, 640
in mRNA selection and translation arrest, **152**, 567
Na^+,K^+-ATPase α subunit cDNA
Northern blot technique, **156**, 386
Southern blot technique, **156**, 389
with oligonucleotide probes, **154**, 94
conditions, **152**, 437
procedures, **152**, 438

RNA, in analysis of fibroblast collagen production, **163**, 648
in screening of
colonies and plaques with radioactive nucleic acid probes, **152**, 415
Southern and Northern blots, **152**, 576
selection scheme, in preparation of specific rRNA fragments, **164**, 221
in situ
in analysis of fibroblast collagen production, **163**, 651
biotinylated probes, in detection of gene transcription, **151**, 530
cDNA probes to rRNA, assays, **164**, 403
in demonstration of tissue-specific gene expression, **152**, 649
equipment required for molecular cloning work, **152**, 12
to metaphase and prometaphase chromosomes
principle, **151**, 279
problems encountered, diagnosis and correction, **151**, 289
procedures, **151**, 282
required materials, **151**, 280
radiolabeled RNA probes, in detection of RNA in culture, **151**, 539
spot-blot, sorted chromosomes, **151**, 307
subtraction technique for cloning rare mRNAs, **152**, 425
Hybridomas
producing lupus-associated anti-DNA antibodies, derivation, **162**, 401
Hybrids
asymmetric somatic, generation by plant protoplast fusion, **148**, 46
verification, **148**, 50
cellular
fluorescence-activated cell sorting, **151**, 150
isolation by toxin–antitoxin technique, **151**, 237
somatic, chromosome identification with isozyme markers, **151**, 169
cytoplasmic, *see* Cybrids
microcell, containing single chromosomes, preparation, **151**, 317
nuclear, genetic isolation, **151**, 235
Hydraulic conductivity
in cyanobacteria, measurement with gas vesicles, **167**, 665
Hydrazides
for glycoconjugate labeling
applications, **138**, 441
fluorescent reagents, **138**, 433
polymeric reagents, **138**, 438
radioactive reagents, **138**, 430
target reagents, **138**, 435
Hydrocarbons
aromatic, metabolism in mitogen-treated lymphocytes, **150**, 50
Hydrocellulose
preparation, **160**, 20
Hydrocortisone
transformation by gel-entrapped *Arthrobacter globiformis*, **135**, 201, 204
Hydrogen
assay with enzyme transistor, **137**, 238
–cation antiport, measurement in tonoplast vesicles, **148**, 118
conversion to electric energy in bacterial fuel cell, **137**, 671, 708
evolution during cyanobacterial anoxygenic photosynthesis, assay, **167**, 382
exchange in cyanobacteria, assays, **167**, 459, 467
ionic, permeability of sarcoplasmic reticulum vesicles, **157**, 433
Hydrogenase
cyanobacterial
assay, **167**, 506
induction, **167**, 463
isolation, **167**, 502
types and distribution, **167**, 501
entrapment in alginate with chloroplasts, **135**, 453
Hydrogen bromide
and dimethyl sulfoxide, cleavage of ribosomal protein tryptophanyl peptide bond, **164**, 262
Hydrogen fluoride
glycoprotein deglycosylation with, **138**, 343
Hydrogen peroxide
effect on cGMP-binding phosphodiesterase, **159**, 717

induced DNA strand breaks, assay, **163**, 331
inhibition of glyceraldehyde-3-phosphate dehydrogenase, assay, **163**, 330
pretreatment of biomass materials for enhancing cellulose availability, **160**, 4

Hydrogen selenide
assay with 1-fluoro-2,4-dinitrobenzene, **143**, 32

Hydrolases
condensation reactions
factors influencing product yield, **136**, 286
mechanism, **136**, 281
conformation and activity in reverse micelles, **136**, 206

Hydrolysis
ATP
associated ligand binding to ATP-driven pumps, assay, **157**, 233
in sarcoplasmic reticulum vesicles, measurement, **157**, 200
cellulose
for cellodextrin preparation, **160**, 29, 30
enzyme thermistor monitoring, **137**, 311
disulfides, **143**, 256
enzymatic
chitooligosaccharides and reduced forms, **161**, 456
collagen
measurement, **144**, 141
specificity, **144**, 153
diacylglycerols, assays, **141**, 310
phorbol diesters, assays, **141**, 308
in industrial saccharification
immobilized enzyme system, **136**, 377
liquid enzyme system, **136**, 374
inositol trisphosphate, **141**, 146
ITP in sarcoplasmic reticulum vesicles, measurement, **157**, 200
Klason procedure for characterization of radiolabeled lignins, **161**, 23
lactose, industrial applications, **136**, 412
phosphorylated Ca^{2+}-ATPase in sarcoplasmic reticulum vesicles, **157**, 176
polyphosphoinositides in neutrophils, N protein role, **141**, 261
site-specific, rRNA–oligonucleotide complexes, **164**, 440
stereoselective, *dl*-menthyl succinate by immobilized yeast cells, **136**, 293
sucrose, enzyme thermistor monitoring and control, **137**, 308

Hydrophobic chromatography
calcium-binding protein, 21-kDa form, **139**, 89
complement components
C3, **162**, 603
C5, **162**, 607
isopropylmalate dehydratase, **166**, 426
lignin–carbohydrate complexes, **161**, 14
proteoglycans, **138**, 284; **144**, 398
yellow affinity substance, **160**, 498

Hydrophobic moment
helical, calculation for calmodulin-binding sequences in peptides and proteins, **139**, 458

3-Hydroxyacyl-CoA dehydrogenase
peroxisomal, assay, **148**, 523

4-Hydroxybenzaldehyde
carbon isotope-labeled, preparation, **161**, 52

(2*R*)-Hydroxycarboxylate-viologen oxidoreductase
assay, **136**, 312
2-oxo-4-methylpentanoate reduction, **136**, 316
purification from *Proteus vulgaris*, **136**, 311
substrate specificity and kinetic data, **136**, 306

25-Hydroxycholesterol
as dolichol synthesis inhibitor, **138**, 698

1-(4-Hydroxy-3,5-dimethoxyphenyl)-2-(4-hydroxy-3-methoxyphenyl)propane-1,3 diol
synthesis, **161**, 61

5-Hydroxyindole-3-acetic acid
HPLC assay in brain, **142**, 539, 547

Hydroxyindole *O*-methyltransferase, *see* Acetylserotonin methyltransferase

2-Hydroxyisocaproic acid
GC–MS assay, **166**, 27

2-Hydroxyisovaleric acid
GC–MS assay, **166**, 27
Hydroxylamine
cleavage of ribosomal protein asparaginylglycyl peptide bond, **164**, 263
Hydroxylapatite
in purification of
staphylococcal γ-hemolysin, **165**, 14
staphylococcal δ-lysin, **165**, 17
Hydroxylapatite chromatography
complement protein C9, **162**, 575
Hydroxyl radicals
in footprinting of protein–DNA complexes, **155**, 537
comparison with other reagents, **155**, 551
p-Hydroxymercuribenzoic acid
in spectrophotometric assay of thiols, **143**, 60
Hydroxymethylation
polyacrylamide beads, **135**, 96
3-Hydroxy-3-methylglutaryl-CoA lyase
assays, **166**, 219
2-Hydroxy-3-methylvaleric acid
GC–MS assay, **166**, 27
β-Hydroxynorvaline
as glycosylation inhibitor, **138**, 704
4-Hydroxyphenylpyruvate dioxygenase
assay, **142**, 133, 139, 143, 148
properties, **142**, 136, 142, 147, 153
purification from
chicken liver, **142**, 150
human liver, **142**, 139
porcine liver, **142**, 135
Pseudomonas sp., **142**, 145
Hydroxyprolines
assay in cell cultures, **147**, 410
dialyzable, in assay for collagen degradation, **163**, 690
Hydroxypyridinium
collagen crosslinks, direct assay, **144**, 127
3*α*-Hydroxysteroid dehydrogenase
in multienzymatic NADH recycling, **136**, 73
12*α*-Hydroxysteroid dehydrogenase
immobilization, **136**, 152
steroid transformation in two-phase systems, **136**, 154
20*β*-Hydroxysteroid dehydrogenase
entrapped in reverse micelles
assay, **136**, 221
intrinsic rate parameters, **136**, 223
preparation, **136**, 221
immobilization, **136**, 152
immobilized subunits
preparation, **135**, 476
properties, **135**, 478
steroid transformation in two-phase systems, **136**, 154
N-Hydroxysuccinimide chlorocarbonate
activated supports
active groups, determination, **135**, 88
affinity chromatography with, **135**, 89
bound trypsin, coupling yield and activity, **135**, 89
coupling of proteins and ligands, **135**, 88
preparation, **135**, 87
preparation, **135**, 86
N-Hydroxysuccinimide compounds
calmodulin modification with, **139**, 426
N-Hydroxysulfosuccinimide
active esters, in analysis of Fc receptor for IgE, **150**, 503
5-Hydroxytryptamine, *see* Serotonin
Hygromycin
markers for plant transformation
application, **153**, 270
construction, **153**, 266
Hyperemia
during inflammation, quantitation, **162**, 306
Hypersensitivity
delayed
correlates *in vitro*, **163**, 367
elicitation *in vivo*, **163**, 359
granulomatous reactions, **163**, 364
overview, **163**, 357
passive transfer of immunity, **163**, 365
to tumors, **163**, 363
to viral antigens, **163**, 363
Hypotaurine
HPLC assay, **143**, 155
tritiated
hypotaurine aminotransferase assay with, **143**, 184
synthesis, **143**, 183

Hypotaurine aminotransferase
assay, **143**, 183
Hypothalamus
derived growth factor, purification, **147**, 104
Hypoxanthine phosphoribosyltransferase
gene
expression, inhibition by antisense RNA, **151**, 519
retrovirus-mediated transfection, **149**, 13
mRNA, detection by microinjection and complementation of mutant cells, **151**, 377

I

Idiotypic analysis
anti-DNA antibodies in murine lupus, **162**, 408
IgA proteases
neisserial, production and isolation, **165**, 117
Ileal loop test
for diarrheal toxins, **165**, 306
Image analysis
computerized
ribosomal electron micrographs, **164**, 3, 53
ribosomal subunit electron micrographs, **164**, 35
Na^+,K^+-ATPase crystals, **156**, 85
Image analyzer
for neutrophil chemotaxis assay, **162**, 80
Image reconstruction
in analysis of Ca^{2+}-ATPase crystalline structure, **157**, 283
three-dimensional, bacterial ribosomes, **164**, 95
Immobilization
affinity, *see* Affinity immobilization
biocatalysts
activity and effectiveness, **135**, 12
chemical parameters, **135**, 4, 7
physical parameters, **135**, 4, 9
by prepolymer methods, **135**, 230
biomolecules
on silanized silica, **137**, 383
on silicon wafers, **137**, 389
calmodulin, **157**, 337
cells, *see* Cell immobilization
cyanobacteria in solid matrices, **167**, 629
DNA
from cells, fast blot techniques, **152**, 582
for hybrid selection of mRNA, **152**, 568
enzymes, *see* Enzyme immobilization
microorganisms, by radiation polymerization, **135**, 146
organelles
in crosslinked proteins, **135**, 454
overview, **135**, 173
by prepolymer methods, **135**, 230
plant protoplasts
by entrapment in gels, **135**, 434, 440
on microcarriers, **135**, 421
proteins
on 1,1′-carbonyldiimidazole-activated matrices, **135**, 102
and ligands, on carbonate-activated carriers, **135**, 84
by techniques permitting subsequent release, **135**, 130
RNA from cells, fast blot techniques, **152**, 582
stabilization of enzymes by, **137**, 584
Immobilization matrices
amide- and ester-containing, **135**, 33
1,1′-carbonyldiimidazole-activated
ligand coupling, **135**, 110
overview, **135**, 102
preparation, **135**, 108
protein purification with, **135**, 111
commercially available, **135**, 32, 42
commercially unavailable, **135**, 35
configuration, **135**, 63
covalent enzyme coupling, **135**, 37
heterogeneous, **135**, 35
hydroxyl functions, activating reagents, **135**, 54
hydroxylic, **135**, 32
activation with
chlorocarbonates, **135**, 84
colored sulfonyl chloride, **135**, 79
magnetic types, **135**, 64
noncovalent enzyme coupling, **135**, 57

polyhydroxylic, isonitrile derivatives
coupling reactions, **135**, 91
ligand coupling, **135**, 97
preparation, **135**, 98
polymeric, activating reagents, **135**, 55
proteinaceous, activation with transition metals, **135**, 122
synthetic, **135**, 36
transition metal-activated, enzyme coupling, **135**, 117
tresyl chloride-activated
enzyme coupling, **135**, 68
preparation, **135**, 67
Immune complexes
and complement, as mediators of inflammation, overview, **163**, 7
Immune responses
cellular
in inflammation, overview, **163**, 21
to type II collagen, **162**, 369
humoral, to type II collagen, **162**, 368
lymphocytes *in vivo*, role of Zn^{2+}, **150**, 116
nonspecific defects in murine lupus, analytical methods, **162**, 396
Immunization
for induction of chronic arthritis, **162**, 380
in production of collagen antibodies, **145**, 156
with type II collagen, **162**, 358
Immunoadsorbents
in analysis of complement components C3, C5, C3a, and C5a, **162**, 585
medical applications, **137**, 451
Immunoadsorption
in analysis of hormonal activation of cAMP-dependent protein kinase isozymes, **159**, 97
in characterization of phosphodiesterase and calcineurin antibodies, **159**, 643
in purification of IgG Fc receptors, **150**, 534
Immunoaffinity chromatography
calmodulin-binding protein, **139**, 155
calmodulin and dependent enzymes, **139**, 513
calmodulin-stimulated phosphodiesterase, **159**, 543
isozymes, **159**, 588
complement protein C7, **162**, 561
connective tissue-activating peptides, **163**, 740
C1q receptor, **150**, 575
epidermal growth factor, **146**, 3
epidermal growth factor receptor, **146**, 82
erythropoietin, **147**, 334
Escherichia α-hemolysin, **165**, 146
interleukin-2 receptor, **150**, 690
melanoma growth stimulatory activity, **146**, 124
monoamine oxidase B, **142**, 645
oligosaccharides, **138**, 307
T-cell receptors, **150**, 650
thioredoxin *m*, **167**, 414
tissue kallikrein, **163**, 112
tissue prokallikrein A, **163**, 125
tissue prokallikrein B, **163**, 120
Immunoassay
α_1-antitrypsin, **163**, 401
bioluminescent, tissue kallikrein, **163**, 109
botulinum neurotoxins, **165**, 85
branched-chain α-keto acid dehydrogenase complex, **166**, 185
ceruloplasmin, **163**, 445
$C\bar{1}$ inhibitor, **163**, 180
endogenous pyrogens, **163**, 509
Na^+,K^+-ATPase, **156**, 120
Immunoblotting
calcineurin antibodies, **159**, 637
CD2 antigens, **150**, 543
chloroplast envelope membrane fractions, **148**, 214
cyclic nucleotide phosphodiesterase antibodies, **159**, 637
in detection of B cells cloned on filter paper disks, **150**, 290
dot, staphylococcal epidermolytic toxin, **165**, 336
F_1-ATPase, **167**, 353
with isoelectric focusing
α_1-acid glycoprotein variants, **163**, 428
α_2-HS glycoprotein variants, **163**, 437
proliferating cell nuclear antigen/cyclin, **150**, 154
protein–protein crosslinks in *E. coli* ribosomes, **164**, 64
radio-, *see* Radioimmunoblotting

in screening of Na^+,K^+-ATPase antibodies, **156**, 404
Immunochemical assay
anthrax toxin, **165**, 113
CR3 receptor, **150**, 607
lysophospholipase, **163**, 41
phospholamban, **157**, 144
toxic shock syndrome toxin-1, **165**, 339
Immunocytochemistry
Ca^{2+}-binding protein in tumor cells, **139**, 211
calbindin D in animal tissues, **139**, 541
calmodulin in plants, **139**, 478
calregulin in fibroblasts, **139**, 47
cAMP-dependent protein kinase RII subunit-binding proteins in animal tissues, **159**, 187
monoamine oxidase in brain, **142**, 649
Na^+,K^+-ATPase in kidney, **156**, 429
transferrin in muscle and nonmuscular tissues, **147**, 299
Immunodiffusion
complement components C3, C5, C3a, and C5a, **162**, 584
connective tissue-activating peptides, **163**, 739
double, toxic shock syndrome toxin-1, **165**, 341
Ouchterlony, staphylococcal epidermolytic toxin, **165**, 334
radial
C-reactive protein, **163**, 392
α_2-HS glycoprotein, **163**, 432
staphylococcal epidermolytic toxin, **165**, 337
Immunoelectron microscopy
calmodulin localization in plants
postembedding method, **139**, 493
preembedding method, **139**, 486
colloidal gold, cAMP-dependent protein kinase cellular compartmentalization, **159**, 225
ribosomal protein localization on *E. coli* ribosomal subunits, **164**, 503
rRNA topography, **164**, 80
Immunoelectrophoresis
amyloid P component, **163**, 534
complement components
C1, C2, and C4, **162**, 631
C3, C5, C3a, and C5a, **162**, 584
crossed
α_1-acid glycoprotein variants, **163**, 424
acute-phase plasma proteins, **163**, 566
cellulase, **160**, 235
α_2-macroglobulin, **163**, 490
Immunofluorescence
in analysis of heterologous gene expression in eukaryotic cells, **152**, 702
in assay of
CR2 receptor, **150**, 595
factor H receptor, **150**, 600
in detection of
anti-spermatozoal antibodies in postvasectomy autoimmune orchitis, **162**, 466
proliferating cell nuclear antigen/cyclin, **150**, 153
TdT^+ lymphoid precursor cells, **150**, 366
elastin, **144**, 257
in localization of calmodulin in trypanosome, **139**, 262
Immunogens
peptide, calmodulin antibody preparation with, **139**, 394
Immunoglobulins, *see also* Antibodies
B-cell membrane, isolation with lectins, **150**, 469
IgE, radiolabeled, in binding studies of Fc_ε receptors, **150**, 517
IgG
anti-brush border membrane fraction Fx1A
characterization, **162**, 435
isolation, **162**, 432
anti-oncomodulin, monitoring by ELISA, **139**, 159
crosslinking to asparaginase–albumin conjugate, **137**, 570
–eupergit C, in antibody-mediated lactate dehydrogenase immobilization, **135**, 169
interaction with immobilized protein A, ellipsometric study, **137**, 385
purification on 1,1′-carbonyldiimidazole-activated supports, **135**, 114
tryptic digestion, reflectometric study, **137**, 376
IgM
to liposomal antigens

conjugation with liposomes, **149**, 128
preparation, **149**, 126
polysialic acid-specific
polysialosyl glycoprotein detection, **138**, 182
preparation, **138**, 178
Immunohistochemistry
cartilage macromolecules, **145**, 351
collagen antibodies for, characterization and isolation, **145**, 148
electron, *see* Electron immunohistochemistry
glandular kallikreins, **163**, 157
oncomodulin, **139**, 162
parvalbumin, **139**, 552
Immunohistopathology
testis and kidney, in analysis of post-vasectomy autoimmune orchitis, **162**, 467
Immunoliposomes
pH-sensitive
acid-induced fusion, **149**, 94
cytoplasmic delivery of contents, mechanism, **149**, 96
interaction with target cells, **149**, 95
leakiness, **149**, 95
preparation, **149**, 90
Immunonephelometry
α_1-acid glycoprotein, **163**, 420
Immunopathology
chronic arthritis in rabbits, **162**, 382
Immunoperoxidase staining
calmodulin and S-100 protein in cells and tissues, **139**, 495
collagenase on Western blots, **144**, 168
TdT^+ lymphoid precursor cells, **150**, 367
Immunoplate assay
in detection of hybridoma and myeloma cell mutants, **151**, 54
Immunoprecipitation
in analysis of heterologous gene expression in eukaryotic cells, **152**, 703
bone proteins, **145**, 321
calregulin, **139**, 46
CD2 antigens, **150**, 541
cell-free translation products, **152**, 300
cGMP-regulated phosphodiesterases, **159**, 696
collagenase, **144**, 170
EGF receptors, **146**, 52, 70
elastin, **144**, 220
extracellular matrix components, **147**, 412
isovaleryl-CoA dehydrogenase, **166**, 158
α_2-macroglobulin, **163**, 493
parvalbumin translated *in vitro*, **139**, 773
platelet-derived growth factor-like proteins, **147**, 24
platelet-derived growth factor-modulated proteins, **147**, 90
proteoglycan core protein precursor and link protein, **144**, 382
radio-, *see* Radioimmunoprecipitation
in screening of Na^+,K^+-ATPase antibodies, **156**, 402
Shiga toxin, **165**, 161
transferrin receptors, **150**, 730
tropoelastin, **144**, 256, 269
Immunoscreening
cDNA expression libraries, **152**, 451
phage λgt11 libraries, for gene isolation, **152**, 458; **154**, 108
Immunostaining
cAMP-dependent protein kinase regulatory subunit II, **159**, 144
extracellular matrix components for electron immunohistochemistry, **145**, 118, 140
glycolipids on TLC plates, **138**, 129, 197
oligosaccharides on TLC plates, **138**, 208
phospholamban and sarcoplasmic reticulum, **157**, 145
Immunotherapy
experimental, collagen arthritis as model, **162**, 372
with transferrin receptor antibodies, **147**, 279
Immunotoxins
containing *Pseudomonas* exotoxin A
construction, **151**, 140
buffers for, **151**, 142
suggestions and alternative methodologies, **151**, 144
evaluation, **151**, 143
Immunoturbidometry
hemopexin, **163**, 548
Implantation
long-term, peritoneal glucose sensors, **137**, 68

Implants
sponge, as models of inflammation
N-acetylglucosaminidase levels, **162**, 330
associated fluid, assay of mediators, **162**, 332
leukocyte infiltration, estimation, **162**, 329
preparation and implantation, **162**, 327
Inbreeding
mice, theory and practice, overview, **150**, 163
Incubation vessel
thermostatted, for Ca^{2+}-selective minielectrode, description, **141**, 29
Indo-1
cell loading for calcium mobilization analysis, **141**, 58
Indoleamine 2,3-dioxygenase
assay, **142**, 192
catalyzed reaction, **142**, 189
properties, **142**, 194
purification from rabbit intestine, **142**, 193
Indoleamine *N*-methyltransferase
assay, **142**, 669
properties, **142**, 671
purification from rabbit lung, **142**, 670
Indoleglycerol-phosphate synthase
–phosphoribosylanthranilate isomerase
assays, **142**, 389
properties, **142**, 394
purification from *E. coli*, **142**, 392
substrate preparation, **142**, 387
Indolyl-3-acetic acid
transformation by gel-entrapped *Aspergillus niger*, **135**, 201, 212, 215
Industry
applications of immobilized cells and enzymes, **136**, 353
Inflammation
acute-phase response
in liver, assay of mediating factors, **163**, 587
overview, **163**, 373
assay in fluid of sponge implants, **162**, 332
associated oxidant-induced cell injury, *in vitro* analysis, **163**, 328
associated responses, measurement techniques
general considerations, **162**, 301
hemorrhage, **162**, 314
hyperemia, **162**, 306
lymphocyte accumulation, **162**, 315
microvascular platelet deposition and thrombosis, **162**, 312
monocyte infiltration, **162**, 311
neutrophil accumulation, **162**, 307
plasma exudation, **162**, 304
carrageenan pleurisy model
analytical methods, **162**, 335
comparison with other exudative models, **162**, 338
foot edema test, **162**, 320
implicated mast cell and neutrophil proteases, purification and assay, **163**, 309
induced changes in acute-phase proteins in serum, analysis, **163**, 567
mediators, overview, **163**, 3
and repair, hepatic granuloma as model, overview, **163**, 605
sponge implants as models, **162**, 327
Influenza virus
sialyloligosaccharide-binding protein specificity, analysis, **138**, 166
Infrared spectroscopy
in determination of chitin and chitosan acetylation degree, **161**, 444
Infusorial earth
adsorbed lipase, triglyceride interesterification with, **136**, 405
fungal spore entrapment, **135**, 322
immobilized mycelial cells, applications, **135**, 323
porous, aminotransferase immobilization with, **136**, 490
properties, **135**, 320
scanning electron microscopy, **135**, 322
Initiation factors
IF3, interaction with rRNA, physical studies, **164**, 238
Inorganic pyrophosphatase
immobilized, UDP-glucose synthesis with, **136**, 279
Inosine triphosphatase
as marker for chromosome detection in somatic cell hybrids, **151**, 186

Inositol
    activation of phosphoinositide break-
        down in cell-free system, **141**, 255
    availability, manipulation for gly-
        cerophospholipid synthesis altera-
        tion, **141**, 229
    biosynthesis from glucose 6-phosphate,
        measurement, **141**, 127
    ^{3}H-labeled, in analysis of cellular phos-
        phoinositide pools, **141**, 89
    myo-Inositol, *see* Inositol
Inositol hexaphosphate
    entrapment in erythrocytes, **149**, 246
myo-Inositol-1-phosphatase
    purification from rat testis, **141**, 140
myo-Inositol 1-phosphate
    assay methods, **141**, 129
Inositol phosphates
    agonist-induced accumulation in hippo-
        campal slices, assay, **141**, 192
    HPLC, **141**, 191
    ion-exchange chromatography, **141**, 182
    paper chromatography, **141**, 183
L-*myo*-Inositol-1-phosphate synthase
    purification from rat testis, **141**, 131
Inositol phospholipids
    autoradiographic localization in superior
        cervical ganglion, **141**, 201
    extraction from rat pituitary, **141**, 165
    metabolism in
        PC12 cells, measurement, **141**, 169
        platelets, measurement, **141**, 176
    radiolabeling, **141**, 158
    TLC analysis, **141**, 166
    turnover in pituitary cells, assay with
        Li$^+$, **141**, 149
Inositol 1,4,5-trisphosphate
    chromatographic resolution, **141**, 145
    effect on Ca^{2+} release from sarcoplasmic
        reticulum, **141**, 3
    extraction from cells, **141**, 144
    gas chromatographic assay, **141**, 148
    hydrolysis, **141**, 146
    role in TRH-induced changes in cellular
        calcium metabolism, **141**, 50
Insulin
    action in adipocytes, effects of specific
        phosphodiesterase inhibitors, **159**,
        514

    crosslinking to α-glucosidase–albumin
        conjugate, **137**, 569
    gene, liposome-encapsulated
        expression in liver and spleen, **149**,
            161
        targeting to hepatic cells, **149**, 163
    human
        tert-butyl ester, synthesis with immo-
            bilized trypsin, **136**, 186
        semisynthesis with immobilized
            Achromobacter protease I, **136**,
            162
    α_2-macroglobulin-conjugated
        preparation, **147**, 388
        transmembrane delivery, **147**, 390
    stimulation of cAMP phosphodiesterase,
        159, 748
    toxin-conjugated
        cell variant isolation with, **147**, 386
        preparation, **147**, 384
Insulin-like growth factor I
    antiserum preparation, **146**, 218
    extraction from tissues, **146**, 227
        procedure validation, **146**, 231
    monoclonal antibodies
        affinity and specificity, **146**, 211
        assay, **146**, 210
        characterization, **146**, 213
        production, **146**, 207
    mRNA
        biological sources, **146**, 240
        expression analysis, **146**, 247
        isolation from
            cultured cells, **146**, 244
            tissues, **146**, 242
    radioimmunoassay, **146**, 220, 229
    radiolabeling, **146**, 217
    serum levels, regulation, **146**, 225
    synthetic
        biological activities, **146**, 206
        characterization, **146**, 202
        preparation, **146**, 199
        purification, **146**, 202
    tissue levels, calculation, **146**, 230
Insulin-like growth factor II
    antibody production, **146**, 250
    mRNA
        biological sources, **146**, 240
        expression analysis, **146**, 247

isolation from
cultured cells, **146**, 244
tissues, **146**, 242
purification, **146**, 263
radioiodination, **146**, 249
radioligand assays, **146**, 248
standards for, **146**, 259
radioreceptor assay, **146**, 261
Insulin-like growth factor receptors
type I
on IM-9 lymphoid cells, assay, **146**, 238
on mononuclear cells, assay, **146**, 235
Insulin receptors
lymphoid cell
binding studies, **150**, 701
labeling techniques, **150**, 710
phosphorylation, **150**, 718
Interesterification
triglyceride with celite-adsorbed lipase, **136**, 405
Interference reflection microscopy
analysis of fibroblast and myofibroblast adhesion, **163**, 638
Interferon-α
characterization as pyrogen, **163**, 500
Interferon-β_2
characterization as pyrogen, **163**, 501
Interferon-γ
human, expression plasmid, construction, **153**, 423
Interleukin-1
characterization as pyrogen, **163**, 497
gene isolation, **150**, 748
Interleukin-2
in cloning of
alloreactive T cells, **150**, 336
antigen-specific T cells, **150**, 327
T cells with distinct functions, **150**, 342
gene isolation, **150**, 749
human, expression plasmid, construction, **153**, 427
Interleukin-3
gene isolation, **150**, 751
induced stabilization of protein kinase C–membrane complex, assay, **141**, 405
Interleukin-6, *see* Interferon-β_2
Interleukin-2 receptors
amino acid sequence, **150**, 699
biochemical properties, **150**, 685
biosynthesis, **150**, 687
cDNA cloning, **150**, 692
detection on T cells, **150**, 682
mRNA purification, **150**, 693
purification, **150**, 690
Intestine
bovine, purification of calbindin D, **139**, 586
chicken, purification of
calmodulin-dependent myosin heavy chain kinase, **139**, 110
lactose-binding lectins, **138**, 513
110-kDa protein–calmodulin complex, **139**, 137
murine, biosensors based on derived tissues, **137**, 145
rabbit, purification of indoleamine 2,3-dioxygenase, **142**, 193
rat, purification of aromatic-amino-acid aminotransferase, **142**, 268
Invertase, *see* β-Fructofuranosidase
Iodination
enzymatic
Ca^{2+}-modulated proteins, **139**, 450
plant plasma membranes, **148**, 570
radio-, *see* Radioiodination
5-(2-Iodoacetamidoethyl)aminonaphthalene 1-sulfonate
labeling of Ca^{2+},Mg^{2+}-ATPase, **157**, 251
Iodoazidobenzylpindolol
^{125}I-labeled
photoaffinity labeling of β-adrenergic receptors, **150**, 495
synthesis, **150**, 493
Iodosobenzoic acid
in spectrophotometric assay of thiols, **143**, 51
Ion-exchange chromatography
acetylxylan esterase, **160**, 556, 705
adenosylhomocysteinase, **143**, 378
adenylate cyclase, **139**, 777
alanine sulfodisulfane, **143**, 225
in assay for sulfate, **143**, 7
bacterial chemotactic factor, **162**, 201
bone glycoproteins, **145**, 274
botulinum neurotoxins, **165**, 82

calcium-binding protein, 21-kDa form, **139**, 91
calmodulin derivatives, **139**, 421
cAMP phosphodiesterase, **159**, 769
cartilage-derived growth factor, **146**, 323
cellobiohydrolases, **160**, 231, 400
cellulase, **160**, 237
ceruloplasmin, **163**, 447
chitin deacetylase, **161**, 521
collagen crosslinking amino acids, **144**, 123
collagens, **163**, 686
complement proteins
C5b–6, **162**, 558
C7, **162**, 562
C8, **162**, 567
C9, **162**, 574
elastin, **144**, 211
endo-1,4-β-glucanases, **160**, 229, 231, 353
β-glucosidases, **160**, 426, 435
inositol phosphates, **141**, 182
α_2-macroglobulin, **163**, 486
5′-nucleotides, **159**, 66
phospholipid transfer protein, **148**, 665
phycobiliproteins, **167**, 295
proteoglycans, **138**, 282; **144**, 327, 333, 397
ribosomal proteins, **164**, 532
transferrin, **147**, 292
toxic shock syndrome toxin-1, **165**, 40
Vicia villosa lectins, **138**, 540
Ion-exchange technique
measurement of occluded ion release from Na^+,K^+-ATPase, **156**, 281
Ionic transport
in neutrophils, measurement, **162**, 283
Ion microscopy
biological and medical
coupling with other microscopic techniques, **158**, 176
future developments, **158**, 179
image quantification, **158**, 175
instrumentation, **158**, 158
sample preparation, **158**, 167
Ionophores
A23187, and protein kinase C activators, synergistic effects in pituitary cells, **141**, 429
in assay for Na^+,K^+-ATPase sidedness in liposomes, **156**, 160
as glycoprotein transport inhibitors, **138**, 700
Ions
concentrations in aqueous solutions of metals and ligands, computer calculations, **157**, 378
Iron
atomic absorption spectrometry, **158**, 136
complexed, in biological samples, colorimetric assay, **158**, 357
transferrin-mediated uptake, assay, **147**, 274
Irpex lacteus
purification of
exocellulase, **160**, 404
β-mannanase, **160**, 609
Isethionic acid
gas–liquid chromatography, **143**, 173
Islet cells
pancreatic, microencapsulation, **137**, 575
1-Isobutyl-6-aminouracil
synthesis, **159**, 494
1-Isobutyl-5-nitroso-6-aminouracil
3-substituted compounds, synthesis, **159**, 494
Isobutylurea
synthesis, **159**, 493
Isocitrate dehydrogenase
cell permeabilization test with, **137**, 641
as marker for chromosome detection in somatic cell hybrids, **151**, 186
Isocyanates
calmodulin modification with, **139**, 423
Isoelectric focusing
acid proteases, **160**, 504
anti-DNA antibodies in murine lupus, **162**, 406
α-L-arabinofuranosidase, **160**, 714
aromatic-amino-acid aminotransferase, **142**, 259, 269
aromatic-amino-acid–glyoxylate aminotransferase, **142**, 275
aspartate aminotransferase, **142**, 259
bone morphogenetic protein, **146**, 299
cAMP phosphodiesterase, **159**, 768
cellobiohydrolases, **160**, 232, 400

cellobiose dehydrogenase, **160**, 458
cellobiosidase, **160**, 394
cellulase, **160**, 220
clostridial alpha toxin, **165**, 93
connective tissue-activating peptides, **163**, 741
dopamine β-monooxygenase, **142**, 606
endo-1,4-β-glucanases, **160**, 229, 230
Escherichia heat-stable enterotoxin, **165**, 130
fibroblast growth factor, **147**, 114, 126
flat-bed, acid proteases, **160**, 503
β-glucosidases, **160**, 224, 411, 419, 426
with immunoblotting
α_1-acid glycoprotein variants, **163**, 428
α_2-HS glycoprotein variants, **163**, 437
α-mannanase, **160**, 622
pectin lyase, **161**, 353
with print immunofixation, α_1-antitrypsin, **163**, 412
staphylococcal epidermolytic toxin, **165**, 34
streptococcal erythrogenic toxin, **165**, 66
streptolysin O, **165**, 57
thin-layer, α_1-antitrypsin, **163**, 410
toxic shock syndrome toxin-1, **165**, 38
vulnificolysin, **165**, 180
xylanase, **160**, 650
β-xylosidase, **160**, 681, 698
Isoleucine
biosynthetic intermediates, analysis, **166**, 106
metabolites
gradient system for resolution, **166**, 84
preparation for HPLC analysis, **166**, 88
Isomaltulose
crystallization, **136**, 448
production by fermentation, **136**, 433
production in immobilized cell reactors, **136**, 443
properties, **136**, 432
Isomerization
with immobilized glucose isomerase, industrial application, **136**, 356
Isonitrile
introduction into polyhydroxylic polymers, **135**, 96
Isopropanol
precipitation of DNA, **152**, 46
α-Isopropylmalate
properties, **166**, 95
synthesis, **166**, 93
β-Isopropylmalate
^{14}C-labeled, preparation, **166**, 228
properties, **166**, 95
synthesis, **166**, 94
Isopropylmalate dehydratase
assay, **166**, 424
properties, **166**, 428
purification from yeast, **166**, 425
β-Isopropylmalate dehydrogenase
assay, **166**, 225, 430
properties, **166**, 433
purification from yeast, **166**, 431
α-Isopropylmalate synthase
assay, **166**, 415
properties, **166**, 419
purification from yeast, **166**, 417
Isotachophoresis
lymphocyte chemotactic factor, **162**, 159
Isovaleryl-CoA dehydrogenase
assay, **166**, 375
mutant, in isovaleric acidemia fibroblasts, assay and properties, **166**, 155
properties, **166**, 384
purification from rat liver, **166**, 377
ITP (inosine 5′-triphosphate)
hydrolysis in sarcoplasmic reticulum vesicles, measurement, **157**, 200

J

Jerusalem artichoke
isolation of submitochondrial particles with different polarities, **148**, 442

K

Kallidin
HPLC
acid separation, **163**, 260
modifications, **163**, 262
neutral separation, **163**, 259
Kallikrein
plasma-type

activation of factor XII, **163**, 66
assay, **163**, 87
C$\bar{1}$ inhibitor, purification and properties, **163**, 179
properties, **163**, 91, 93
purification, **163**, 93
structure, **163**, 91
tissue-type
assays, **163**, 104, 129
with ^{125}I-aprotinin, **163**, 160
immunohistochemical localization, **163**, 143
latent form
assay, **163**, 96
purification from human urine, **163**, 99
properties, **163**, 138
purification from
human urine, **163**, 112, 162
rat submandibular gland, **163**, 150
rat urine, **163**, 135
purity and stability criteria, **163**, 114
Western blot analysis, **163**, 140
Kangaroo apple
cell coupling to modified Sorfix, **135**, 355
Karyotype analysis
T-cell clones and hybrids, **150**, 352
Kethoxal
in probing of
RNA structure, monitoring by primer extension, **164**, 482
rRNA–protein interactions, **164**, 464
Keto acids
branched-chain
enzymatic assay, **166**, 6
metabolism, analysis in perfused heart and liver, **166**, 484
transport in mammalian cells, assay, **166**, 252
α-Ketoisocaproate dioxygenase
assay, **166**, 289
properties, **166**, 294
purification from rat liver, **166**, 291
α-Ketoisocaproic acid
GLC assay, **166**, 39
2-Ketoisocaproic acid
GC–MS assay, **166**, 27
α-Ketoisovaleric acid
GLC assay, **166**, 39
2-Ketoisovaleric acid
GC–MS assay, **166**, 27
α-Keto-γ-methylthiolbutyric acid
ethylene production from, measurement in lignin biodegradation assay, **161**, 79
α-Keto-β-methylvaleric acid
GLC assay, **166**, 39
2-Keto-3-methylvaleric acid
GC–MS assay, **166**, 27
Keyhole limpet
purification of oligosaccharyl sulfatase, **138**, 819
Kidney
bovine, purification of
branched-chain α-keto acid dehydrogenase, **166**, 310
branched-chain 2-keto acid dehydrogenase, **166**, 304
branched-chain α-keto acid dehydrogenase phosphatase, **166**, 323
inhibitor protein, **166**, 326
canine, purification of cAMP phosphodiesterase, **159**, 760
chronic failure, treatment with immobilized enzymes, **137**, 448
glomerular deposits of exogenous antibody, quantitation, **162**, 454
human, purification of
G_{M2} activator, **138**, 813
sulfatide/G_{M1} activator, **138**, 812
mammalian, purification of Na^+,K^+-ATPase, **156**, 29
murine, isolation of genomic DNA, **152**, 181
ovine, purification of Na^+,K^+-ATPase, **156**, 88
porcine
biosensors based on derived tissues, **137**, 141
purification of aromatic-L-amino acid decarboxylase, **142**, 176, 182
rabbit, preparation for evaluation of serum sickness pathology, **162**, 487
rat
preparation of glomerular basement membranes, **162**, 443
purification of
calbindin D, **139**, 535

kynurenine-oxoglutarate aminotransferase, **142**, 219
Kidney bean
purification of
cellulase, **160**, 347
chitinase from leaves, **161**, 481
Kinetics
cell growth factor–nutrient interactions, **147**, 392
EGF receptor biosynthesis, **146**, 54
energy-transducing enzymes and partial reactions, **157**, 154
Na^+,K^+-ATPase
effects of inhibitors and modifiers, **156**, 248
general considerations, **156**, 237
monovalent cation interactions, **156**, 244
substrate interactions, **156**, 239
pre-steady-state, ribosomal translocation, **164**, 581
ribosomal translation activity, **164**, 611
Kininogens
high- and low-molecular-weight
assay, **163**, 277
cDNA and genomic DNA cloning, **163**, 230
ELISA, **163**, 252
heavy chain domains, isolation, **163**, 250
light and heavy chains, isolation, **163**, 247
purification from human plasma, **163**, 241
T-
assay, **163**, 277
properties, **163**, 281
purification from rat plasma, **163**, 277
radioimmunoassay, **163**, 276
role in inflammation, **163**, 282
Kinin receptors
antagonists, description, **163**, 287
bioassays, **163**, 211
characterization, **163**, 227
physiological roles, **163**, 229
radioreceptor assays, **163**, 224
Kinins
antagonists, description, **163**, 282
and blood coagulation, in inflammatory reactions, overview, **163**, 14
HPLC, **163**, 257
radioimmunoassay, **163**, 275
solid-phase synthesis, **163**, 263
T-
assay, **163**, 274
pharmacoligical properties, **163**, 273
role in inflammation, **163**, 282
Kluyveromyces
K. fragilis, purification of protopectinase, **161**, 338
K. lactis, coupling to modified Sorfix, **135**, 354
Krabbe's disease
enzymatic diagnosis, **138**, 739
Kupffer cells
culture for production of hepatocyte-stimulating factor, **163**, 596
targeting of glucocerebrosidase, **149**, 38
Kynurenine-oxoglutarate aminotransferase
assay, **142**, 218
properties, **142**, 221
purification from rat kidney, **142**, 219

L

Labeling
affinity, *see* Affinity labeling
avidin with catalase, **137**, 115
collagen
type I, **144**, 143
type IV, **144**, 150
type V, **144**, 152
cyclic nucleotides with ^{18}O, for metabolic studies, **159**, 60
DNA, for detection and cloning of amplified fragments, **151**, 343
fluorescent, *see* Fluorescent labeling
glycoconjugates with hydrazides, **138**, 429
photo-, *see* Photolabeling
plant plasma membranes
with concanavalin A, **148**, 572
with diazotized sulfanilic acid, **148**, 571
by enzymatic iodination, **148**, 570
with lanthanum, **148**, 573

with positively charged beads, **148**, 574
with pyridoxal phosphate-tritiated borohydride, **148**, 572
with UDPglucose, **148**, 574
by various approaches, **148**, 575
platelets
with [^{3}H]inositol, **141**, 178
with ^{32}P, **141**, 176
radioactive, *see* Radiolabeling
5S RNA with ^{15}N, **164**, 159
rRNA, for analysis of protein binding sites, **164**, 459
Laboratory equipment
for molecular cloning, overview, **152**, 3
Laccase
metal-hybrid, preparation, **158**, 92
β-Lactamase
in optoelectronic sensor for penicillin G, **137**, 346
β-Lactams
flow–injection analysis with enzyme reactors, **137**, 197
Lactase, *see* β-Galactosidase
L-Lactate
assay with
biosensor, **137**, 53
enzyme thermistor, **137**, 189
Lactate dehydrogenase
–alcohol dehydrogenase site-to-site complex
assays, **136**, 108
characterization, **136**, 110
immobilized system, **136**, 104
soluble system, **136**, 106
antibody-mediated immobilization, **135**, 169
coupling to 1,1′-carbonyldiimidazole-activated support, **135**, 115
as marker for chromosome detection in somatic cell hybrids, **151**, 187
Na^+,K^+-ATPase assay with, **156**, 116
Lactate dehydrogenase virus receptors
murine, characterization, **150**, 556
Lactoferrin
cell culture assay, **147**, 310
growth-inhibitory activity, assay, **147**, 309
growth-stimulatory activity, assay, **147**, 306
iron-saturated, cell binding assay, **147**, 311
neutrophil granule-associated, assay, **162**, 548
purification from milk, **147**, 304
β-Lactoglobulin
digestibility changes, analysis with immobilized enzymes, **135**, 601
structural changes, analysis with immobilized Pronase, **135**, 598
Lactoperoxidase
in radiolabeling of lymphocyte receptors, **150**, 408
Lactose
assay with biosensor, **137**, 52
binding lectins
properties, **138**, 514
purification, **138**, 511
hydrolysis, industrial applications, **136**, 412
Laminin
basemement membrane-associated
properties, **145**, 364
purification, **145**, 375
cDNA cloning, **144**, 464
CD spectroscopy, **145**, 39
derived antigens, purification, **145**, 410
electron immunohistochemistry, **145**, 97
electron microscopy, **145**, 28
overview, **145**, 25
self-assembly *in vitro*, **145**, 45
solution studies, **145**, 37
structural model, **145**, 41
structure–function relationships, **145**, 52
Laminin receptors
identification, **144**, 492
67–70K form
characterization, **144**, 500
isolation, **144**, 496
Lanthanum
labeling of plant plasma membranes, **148**, 573
Laser techniques
in analysis of cyanobacterial membrane vesicles, **167**, 691
Latex particles
preparation for liposome-enhanced agglutination assays, **149**, 207
Lavage fluid
pulmonary, extracellular matrix proteins, analysis, **145**, 396

Lavandula vera, *see* Lavender
Lavender
cell entrapment with urethane prepolymers, **135**, 244
Lead
atomic absorption spectrometry, **158**, 136
and strontium, in cytochemical localization of Na^+,K^+-ATPase, **156**, 418
Lead citrate
in cytochemical localization of Na^+,K^+-ATPase, **156**, 423
Lectins
Amphicarpaea bracteata
assay, **138**, 561
properties, **138**, 562
purification, **138**, 562
Cancer antennarius
assay, **138**, 520
properties, **138**, 525
purification, **138**, 521
Erythrina
biological activities, **138**, 551
interactions with carbohydrates, **138**, 549
properties, **138**, 547
purification, **138**, 544
Evonymus europaea
assay, **138**, 552
properties, **138**, 556
purification, **138**, 553
specificity, **138**, 558
in isolation of lymphocyte receptors, **150**, 463
lactose-binding
properties, **138**, 514
purification from
chicken tissues, **138**, 513
rat lung, **138**, 511
Lens culinaris, glycopeptide chromatography with, **138**, 248
Limax flavus
assay, **138**, 528
properties, **138**, 532
purification, **138**, 529
Sepharose-coupled
affinity chromatography with, **138**, 534
preparation, **138**, 534
listing, **150**, 6
mitogenic activity, assay, **150**, 4
parameters affecting lymphocyte transformation, **150**, 4
Pisum sativum, glycopeptide chromatography with, **138**, 246
Vicia villosa
assay, **138**, 537
properties, **138**, 541
purification, **138**, 538
Xenopus laevis
properties, **138**, 519
purification, **138**, 516
Lentinus edodes
deglutamyllentinic acid preparation, **143**, 438
S-substituted L-cysteine sulfoxide lyase purification, **143**, 436
Leucine
metabolic pathways, relative carbon flux measurement, **166**, 127
transport in *E. coli*, assay, **166**, 245
α-Leucine
HPLC, **166**, 17
β-Leucine
enantiomers (3*R*) and (3*S*), HPLC, **166**, 14
HPLC, **166**, 17
Leucine 2,3-aminomutase
assay, **166**, 130
clinical associations, **166**, 135
occurrence and distribution, **166**, 134
preparation, **166**, 133
Leucine dehydrogenase
assay, **166**, 283
in assay of branched-chain amino and keto acids, **166**, 3
production, **166**, 284
properties, **166**, 286
purification from *Bacillus subtilis*, **166**, 284
Leucine–tRNA ligase
assay, **166**, 262
preparation from CHO cells, **166**, 263
properties, **166**, 263
Leuconostoc mesenteroides
purification of dextransucrase, **136**, 242
Leukocidin
staphylococcal
bioassay, **165**, 28

crystallization of F and S components, **165**, 26
purification, **165**, 23
purity criteria, **165**, 31
storage, **165**, 32
Leukocyte elastase
affinity chromatography at subzero temperatures, **135**, 535
assay, **163**, 324
properties, **163**, 314
purification from human neutrophils, **163**, 313
Leukocytes
accumulation in sponge implants, estimation, **162**, 329
chemoattractant receptors, measurement, **162**, 236
chemotaxis
chromium-51 radioimmunoassay, **162**, 59
collagen and fibrin gel assays, **162**, 23
measurement *in vivo*, **162**, 101
micropore filter assay, **162**, 38
natural mediators, **162**, 140
overview, **162**, 3
polarization assay, **162**, 17
time-lapse cinematography, **162**, 29
visual assays on plane 2-D substrata, **162**, 35
contractile proteins, analysis, **162**, 246
eosinophil-enriched, preparation, **163**, 32
human, DNA extraction for analysis of procollagen gene RFLPs, **145**, 206
mononuclear, chemoattractant-induced biochemical changes, analysis, **162**, 279
radiolabeling for *in vivo* chemotaxis assays, **162**, 104
selective depletion in rabbits and rats, **162**, 453
Leukodystrophy
globoid cell, enzymatic diagnosis, **138**, 739
metachromatic, enzymatic diagnosis, **138**, 742
Leukotriene B_4
assay in fluid of sponge implants, **162**, 332
as fibroblast chemoattractant, **163**, 703
and ω-oxidation products
extraction, **141**, 357
HPLC assay, **141**, 360, 370
stability, **141**, 357
Leukotriene B_4 20-hydroxylase
assays, **163**, 350
properties, **163**, 352
Leukotriene B_4 receptors
neutrophil
binding assays, **141**, 372; **162**, 241; **163**, 340
solubilization **141**, 377
Leukotriene C_4 synthase
assay, **163**, 354
properties, **163**, 356
purification from rat basophilic leukemia cells, **163**, 355
Leukotrienes
mast cell-derived, assay, **162**, 519
Lichenase
assay, **160**, 573
with dye-labeled CM-cellulose, **160**, 78
properties, **160**, 574
purification from *Bacillus subtilis*, **160**, 573
Ligament
nuchal, elastin synthesis in culture, analysis, **144**, 216
Ligand displacement assay
asialoorosomucoid internalization by hepatocytes, **149**, 4
Ligation
adaptors to cDNA, **152**, 348
cDNA to plasmid vector, in directional cloning, **152**, 354
DNA restriction fragments, overview, **152**, 124
linkers to cDNA, **152**, 344, 351
Light
effects on cyanobacterial pH gradients, **167**, 682
spectral gradients in cyanobacterial communities, analysis, **167**, 650
ultraviolet, cyanobacterial mutagenesis, **167**, 718
Light microscopy
calmodulin localization in plants
immunofluorescent method, **139**, 485
immunohistochemical methods, **139**, 481

in monitoring plant protoplast coupling to microcarriers, **135**, 425
Light scattering
in analysis of actin in presence of modulating proteins, **162**, 262
laser, in analysis of cyanobacterial membrane vesicles, **167**, 691
in measurement of ionic permeability of
endoplasmic reticulum vesicles, **157**, 427
sarcoplasmic reticulum vesicles, **157**, 427, 446
Lignin
acid-precipitable polymeric
chemistry of, characterization, **161**, 43
compositional analysis, **161**, 40
production, **161**, 35
aromatic fragments generated by degradation
GLC analysis, **161**, 175
HPLC analysis, **161**, 183
biodegradation
assays with polymeric dyes, **161**, 74
measurement based on ethylene production, **161**, 79
–carbohydrate complexes
fractionation, **161**, 14
isolation from
barley straw, **161**, 192
various plants, **161**, 13
properties, **161**, 14
size-exclusion chromatography
conventional method, **161**, 195
high-performance method, **161**, 197
characterization by
acidolysis, **161**, 103
^{13}C NMR spectroscopy, **161**, 158
^{1}H NMR spectroscopy, **161**, 152
nitrobenzene oxidation, **161**, 106, 117
permanganate oxidation, **161**, 107
permanganate–periodate oxidation, **161**, 124
thioacidolysis, **161**, 104
^{14}C-labeled, *in vitro* synthesis, **161**, 65
degradation intermediates
analytical TLC, **161**, 200, 207
GC–MS analysis, **161**, 205, 208
preparative TLC, **161**, 204, 208
demethoxylation by *Phanerochaete chrysosporium*, assays, **161**, 83
depolymerization by *Streptomyces*
acidolytic assay, **161**, 256
turbidometric assay, **161**, 252
dioxane fractions, preparation by acidolysis, **161**, 31
isolation methods
Brauns' native preparation, **161**, 9
brown rot preparation, **161**, 10
general considerations, **161**, 3
kraft preparation, **161**, 10
lignosulfonate preparation, **161**, 11
milled wood enzyme preparation, **161**, 8
milled wood preparation, **161**, 6
milled wood
^{14}C-labeled
applications, **161**, 28
characterization, **161**, 23
labeling with ^{14}C, **161**, 18
preparation, **161**, 115
model compounds labeled with carbon isotopes
chemical synthesis, **161**, 47
NMR spectra, **161**, 54
oxidation products
GC analysis, **161**, 120, 128
GC–MS analysis, **161**, 121, 129
HPLC analysis, **161**, 118
qualitative determination
acidolysis method, **161**, 93
methylation–oxidation method, **161**, 91
nitrobenzene oxidation method, **161**, 89
thioacidolysis method, **161**, 95
quantitative determination
chlorine consumption method, **161**, 99
Goering–Van Soest method, **161**, 97
kappa method, **161**, 99
Klason method, **161**, 96
physical methods, **161**, 100
thioglycolic acid method, **161**, 99
total content in plant tissues, assay, **161**, 113
Ligninase, *see* Lignin peroxidase
Lignin peroxidase
assay, **137**, 693, **161**, 242
cDNA clones, identification with synthetic oligonucleotide probes, **161**, 228

chromatofocusing, **161**, 269
production by *Phanerochaete chrysosporium*, **137**, 690, 695
properties, **161**, 246
purification from *Phanerochaete chrysosporium*, **161**, 243
storage, **161**, 246
Lignocellulose
actinomycete cultivation on, **161**, 37
[^{14}C]lignin-labeled
applications, **161**, 28
characterization, **161**, 23
preparation, **161**, 18
compositional analysis, **161**, 40
degradation by actinomycetes, **161**, 39
preparation for actinomycete cultivation, **161**, 36
Lignols
and related compounds, synthesis, **161**, 57
Linked atom least-squares method
in chitin structural analysis, **161**, 437
Linkers
in dideoxy DNA sequencing, **152**, 547
in directional cDNA cloning, **152**, 349
ligation to cDNA, **152**, 344, 351
in vector-primed cDNA cloning, **152**, 371
Linoleate
conversion to arachidonate in intact cells, **141**, 354
Lipase, *see* Triacylglycerol lipase
Lipids
calelectrin binding, **139**, 34
cellular turnover stimulated by TRH, measurement, **141**, 100
chloroplast membrane
class analysis, **148**, 307
fatty acid analysis, **148**, 303
isolation, **148**, 301
molecular species, analysis, **148**, 311
quantification, **148**, 315
cyanobacterial, isolation, **167**, 182
cyanobacterial membrane
constituent fatty acids
gas chromatography, **167**, 254
positional distribution, **167**, 255
extraction, **167**, 252
thin-layer chromatography
class separation, **167**, 253
molecular species separation, **167**, 257
enzyme thermistor-based assays, **137**, 217
glyoxysomal membrane, analysis, **148**, 527
metabolism in mitogen-treated lymphocytes, **150**, 70
monosaccharide, synthesis in plant endoplasmic reticulum, **148**, 580
photoreactive, in analysis of membrane-penetrating toxins, **165**, 347
planar bilayers, observation of sarcoplasmic reticulum calcium channels in, **157**, 480
plant mitochondrial membrane
analysis, **148**, 475
temperature effects, **148**, 485
and proteins, ratio in sarcoplasmic reticulum membranes, control during reconstitution, **157**, 308
removal from crustacean chitin, **161**, 419
requirement for Na^+,K^+-ATPase activity, **156**, 12
selection for liposome syntheses, **149**, 81
synthesis in plant cell membranes, pesticide effects
general analytical methods, **148**, 667
specific analytical methods, **148**, 671
thylakoid
electron microscopy, **148**, 322
extraction, **148**, 321
purification, **148**, 321
reconstitution, **148**, 324
TLC analysis, **148**, 322
Lipoamide dehydrogenase
as scavenger enzyme in assay for site-to-site enzyme complex, **136**, 109, 112
Lipoamide dehydrogenase-valine
assay, **166**, 345
purification from *Pseudomonas aeruginosa*, **166**, 349
Lipocortins
assay, **163**, 28
biochemical aspects, **163**, 24
dephosphorylation, **163**, 28
induction, **163**, 26
nomenclature, **163**, 25
purification, **163**, 27

Lipolysis
  cAMP-activated, analysis in permeabilized adipocytes, **159**, 197
Lipopolysaccharides
  cyanobacterial, isolation, **167**, 180
  monomers, electrophoretic separation, **138**, 267
  T-cell proliferative responses in basal serum-free medium, **150**, 130
Lipoproteins
  low-density, pheresis, **137**, 466
Liposomes
  antibody-coated, *see* Immunoliposomes
  –antibody–complement complexes
    preparation, **149**, 128
    uptake by macrophages, **149**, 131
  association with plant protoplasts, assessment, **148**, 37
  Ca^{2+}-ATPase reconstitution into, **157**, 347
  content delivery to cells, analysis, **149**, 139, 142
  covalent attachment of proteins, **149**, 111
  DNA transport *in vivo*, efficiency, **149**, 160
  encapsulated agents
    cellular availability assay, **149**, 143
    evaluation with macrophages *in vivo*, **149**, 150
  encapsulated hemoglobin, properties, **149**, 196
  encapsulation of
    DNA, **149**, 158
    hemoglobin, **149**, 189
    nucleic acids, **149**, 177
    plasmid DNA, **148**, 42
    tobacco mosaic virus RNA, **148**, 41
  entrapment of adriamycin, **149**, 74
  incorporation of
    cell surface receptors, **144**, 489
    $C_{12}E_8$-solubilized Na^+,K^+-ATPase
      determination of sidedness, **156**, 160
      procedure, **156**, 157
    laminin receptor, **144**, 501
    toxin pores, **165**, 285
  insertion of proteoglycans, **144**, 399
  insulin gene targeting to hepatic cells, **149**, 163
  interaction with
    cells
      electron microscopy studies, **149**, 138
      fluorescence studies, **149**, 137
      measurement, **149**, 141
      radioactive marker studies, **149**, 136
    plant protoplasts, **148**, 37
    protoplasts, **149**, 183
  ligand-bearing
    in detection of
      anti-erythrocyte subgroup antibodies, **149**, 212
      antinuclear antibodies, **149**, 210
      antistreptococcal antibodies, **149**, 211
      rheumatoid factors, **149**, 209
    preparation, **149**, 203
  microemulsified
    capture of cytosine arabinoside, efficiency, **149**, 68
    diameter, **149**, 72
    permeability to cytosine arabinoside, **149**, 70
    preparation, **149**, 66
  Na^+,K^+-ATPase-containing
    ion transport, assay, **156**, 144
    preparation, **156**, 143
  nucleic acid delivery to protoplasts
    assessment, **149**, 181
    procedure, **149**, 180
  nucleic acid transfer into tobacco protoplasts, **148**, 42
  phosphatidylinositol-containing
    modulation of PI turnover in macrophages, **141**, 252
    opsonization, **141**, 248
    preparation, **141**, 247
  phospholipid-containing
    preparation, **141**, 462
    tumoricidal effects, **141**, 464
  photolabile, *see* Photosomes
  photoreactive, preparation for analysis of membrane-penetrating toxins, **165**, 351
  plant protoplast treatment with, **148**, 36
  preparation, **148**, 35
    for analysis of channel-mediated nonelectrolyte fluxes, **148**, 469

with biotinylated phospholipids, **149**, 121
for DNA encapsulation, **149**, 158
for hemoglobin encapsulation, **149**, 187
for phagocytosis studies, **149**, 125
size, role in biological uses, **149**, 65
temperature- and pH-sensitive
lipid selection for syntheses, **149**, 81
preparation, **149**, 83
purity, **149**, 82
in vitro testing, **149**, 84
in vivo testing, **149**, 87
5-Lipoxygenase
assay, **163**, 345
properties, **163**, 347
purification from human neutrophils, **163**, 346
15-Lipoxygenase
assay, **163**, 345
properties, **163**, 347
purification from human neutrophils, **163**, 346
Liquid chromatography
carbohydrate monomers and oligomers
column packing and operation, **160**, 153
column preparation, **160**, 151
instrumentation, **160**, 148
overview, **160**, 145
packing materials, **160**, 149
resin preparation, **160**, 149
system troubleshooting, **160**, 154
Lissamine Rhodamine B sulfonyl chloride
activation of agarose, **135**, 81
Lithium
in assay of pituitary inositol phospholipid turnover, **141**, 149
atomic absorption spectrometry, **158**, 137
Lithium chloride
–amide system, as solvent for chitin, **161**, 404
differential RNA precipitation, **152**, 48
Lithium β-sulfopyruvate
synthesis, **143**, 221
Liver, *see also* Hepatocytes
acute-phase response, assay of mediating factors, **163**, 587
acyl-CoA levels, assay, **166**, 67
bovine, purification of
adenosylhomocysteinase, **143**, 380
branched-chain α-keto acid dehydrogenase, **166**, 299
calcimedins, **139**, 24
calregulin, **139**, 37
dihydropteridine reductase, **142**, 107
monoamine oxidase B, **142**, 633
NADPH-specific dihydropteridine reductase, **142**, 112
nonspecific phospholipid transfer protein, **157**, 370
phenylalanine 4-monooxygenase, **142**, 39
cell lines, description, **151**, 19
chicken, purification of
4-hydroxyphenylpyruvate dioxygenase, **142**, 150
lactose-binding lectins, **138**, 513
failure, treatment with immobilized enzymes, **137**, 450
granuloma as model of inflammation and repair, overview, **163**, 605
human, purification of
betaine–homocysteine methyltransferase, **143**, 385
cystathionine β-synthase, **143**, 391
dihydropteridine reductase, **142**, 120
4-hydroxyphenylpyruvate dioxygenase, **142**, 139
monoamine oxidase B, **142**, 645
phenylalanine 4-monooxygenase, **142**, 29
sulfatide/G_{M1} activator, **138**, 811
intracellular protein degradation
measurement in perfused organ, **166**, 464
regulation by amino acids, **166**, 468
murine, isolation of genomic DNA, **152**, 181
ovine, purification of dihydropteridine reductase, **142**, 99
perfused, branched-chain α-keto acid dehydrogenase flux rates, measurement, **166**, 484
porcine, purification of
cysteamine dioxygenase, **143**, 411

4-hydroxyphenylpyruvate dioxygenase, **142**, 135
selenocysteine β-lyase, **143**, 416
rabbit
biosensors based on derived tissues, **137**, 147
purification of
amine *N*-methyltransferase, **142**, 662
branched-chain α-keto acid dehydrogenase kinase complex, **166**, 316
monoamine oxidase, **142**, 624
protein phosphatase 2C, **159**, 423
rat
assay of branched-chain α-keto acid dehydrogenase, **166**, 204
cell targeting of liposome-encapsulated insulin gene, **149**, 163
expression of liposome-encapsulated insulin gene, **149**, 161
GLC analysis of isethionic acid, **143**, 173
HPLC assay of thiols, **143**, 111
isolation of
endoplasmic reticulum, **157**, 613
gap junctions, **139**, 447
Golgi membranes, **157**, 612
lysosomes, **157**, 591
40S ribosomes, **164**, 579
subcellular organelles, **141**, 94
purification of
acetyl-CoA:α-glucosaminide *N*-acetyltransferase, **138**, 609
aromatic-amino-acid–glyoxylate aminotransferase, **142**, 275
calcimedins, **139**, 25
cAMP phosphodiesterase, **159**, 740, 753
cathepsin T, **142**, 285
cysteine dioxygenase, **143**, 396
cysteinesulfinate decarboxylase, **143**, 406
isovaleryl-CoA dehydrogenase, **166**, 377
α-ketoisocaproate dioxygenase, **166**, 291
2-methyl branched-chain acyl-CoA dehydrogenase, **166**, 363
D-methylmalonyl-CoA hydrolase, **166**, 396
DL-methylmalonyl-CoA racemase, **166**, 403
Mg^{2+}-dependent protein phosphatase, **159**, 440
monoamine oxidase, **142**, 624
phenylalanine 4-monooxygenase, **142**, 8, 19
protein A, **143**, 401
Liverwort
purification of galacturan 1,4-α-galacturonidase, **161**, 377
Lolium multiflorum, *see* Ryegrass
Lucerne
purification of
α-galactosidase, **160**, 630
β-mannanase B, **160**, 600
Luciferase
bacterial, *see* Alkanal monooxygenase (FMN-linked)
firefly, *see* *Photinus*-luciferin 4-monooxygenase (ATP-hydrolyzing)
reporter gene, application in cyanobacteria, **167**, 746
Lucifer Yellow
as fluid-phase marker, endocytotic uptake, **146**, 413
Lung
elastin synthesis in culture, analysis, **144**, 219
human, purification of
dipeptidyl carboxypeptidase I, **163**, 201
eosinophil chemotactic factors, **162**, 171
rabbit, purification of indoleamine *N*-methyltransferase, **142**, 670
rat, purification of
cGMP-binding phosphodiesterase, **159**, 722
lactose-binding lectins, **138**, 511
Lupine
yellow, purification of adenosylhomocysteinase, **143**, 431
Lupinus luteus, *see* Lupine
Luteinizing hormone
pituitary cell release, effects of protein kinase C activators and A23187, **141**, 429

Lutoids
preparation by
density gradient ultracentrifugation, **148**, 90
differential ultracentrifugation, **148**, 89
low-speed centrifugation, **148**, 89
tonoplasts
biochemical composition, **148**, 91
enzymatic activities, **148**, 91
preparation, **148**, 90, 107
Lycopersicon esculentum, *see* Tomato
Lymphatic system
drug targeting with multiple emulsions, **149**, 54
Lymphoblasts
human, transfection with hypoxanthine phosphoribosyltransferase gene, **149**, 23
Lymphocyte chemotactic factors
characterization, **162**, 160
purification, **162**, 154
Lymphocyte receptors, *see also specific receptors*
analysis, **150**, 394
ESR analysis
applications, **150**, 432
methodology, **150**, 420
isolation, **150**, 392
with lectins, **150**, 463
labeling methods, **150**, 408
nonpermeant covalent labels for, **150**, 399
overview, **150**, 377
radiolabeling, **150**, 390
solubilization, **150**, 392
Lymphocytes
accumulation during inflammation, measurement, **162**, 315
allogeneic and autologous reactions in mixed cultures, analytical methodology, **150**, 83
antigen-activated, negative selection, **147**, 278
B, *see* B cells
chemically defined growth media, **150**, 134
derived fibroblast chemoattractant, **163**, 703
derived fibroblast growth factors
assays, **163**, 716
production, **163**, 723
purification, **163**, 728
human, chromosome suspension preparation from, **151**, 293
IM-9
insulin and growth hormone receptors, analysis, **150**, 701
insulin-like growth factor I receptors, assay, **146**, 238
locomotion, mitogens as activators, **162**, 187
mitogen-treated, biochemical changes
cell isolation and culturing methods for analysis, **150**, 33
cell proliferation and morphology studies, **150**, 69
DNA studies, **150**, 58
enzyme studies, **150**, 47, 75
experimental conditions for analysis, **150**, 30
lipid metabolism studies, **150**, 70
RNA studies, **150**, 65
production of antibodies *in vitro*
human systems, **150**, 309
murine systems, **150**, 304
proliferation, assessment by enzyme immunoassay, **150**, 117
stimulation with
concanavalin A derivatives, **150**, 27
proteolytic enzymes, **150**, 109
zinc ions, **150**, 112
T, *see* T cells
Lymphoid cells
TdT$^+$ precursors, generation *in vitro*, **150**, 363
Lymphokines
production by sensitized T cells, **163**, 368
Lymphotoxin
gene isolation, **150**, 753
Lyophilization
cyanobacteria, **167**, 124
δ-Lysin
staphylococcal
gel electrophoretic analysis, **165**, 20
hemolytic assay, **165**, 21
purification, **165**, 17
L-Lysine
assay with biosensor, **137**, 53
Lysogangliosides
characterization, **138**, 337

radiolabeling, **138**, 333
synthesis, **138**, 324, 329
Lysophospholipase
identity with Charcot–Leyden crystal protein, **163**, 41
immunochemical assay, **163**, 41
localization in eosinophils and basophils, **163**, 43
properties, **163**, 39
purification from human eosinophils, **163**, 36
radiochemical assay, **163**, 34
Lysoretinoylphosphatidylcholine
synthesis, **149**, 104
Lysosomes
acidification
fluorescence assay, **157**, 607
radiochemical assay, **157**, 596
density gradient centrifugation, **157**, 605
fluorescent labeling, **157**, 604
preparation from
phagocytic cells, **157**, 594
rat liver, **157**, 591
Lysozyme
bound to surgical biomaterial, characterization, **137**, 550
phage T4, structure and thermal stability, **154**, 511
structural changes, analysis with immobilized Pronase, **135**, 598
thiol–disulfide interchange in, thiol pK_a values, **143**, 134
Lysyl hydroxylase, *see* Procollagen-lysine,2-oxoglutarate 5-dioxygenase
Lyxose
quantitative TLC, **160**, 172

M

α-Macroglobulin
and complexes with trypsin, as chemotactic agents, preparation, **162**, 185
α_2-Macroglobulin
antibodies, preparation, **163**, 489
assays, **147**, 64; **163**, 490
growth factor-conjugated
preparation, **147**, 388
transmembrane delivery, **147**, 390
metabolic labeling and quantitative immunoprecipitation, **163**, 493
as PDGF-binding protein, **147**, 51
properties, **163**, 495
purification from rat serum, **163**, 486
trypsin-protein esterase activity, assay, **163**, 488
Macromolecules
cartilage-associated
biosynthesis and turnover, **145**, 345
extraction, **145**, 342
interactions, **145**, 356
localization methods, **145**, 350
overview, **145**, 339
purification, **145**, 342
Macrophage chemotactic factors
characterization, **162**, 149
purification, **162**, 144
Macrophages
chemoattractant receptors, assay, **162**, 240
derived fibroblast growth factors
assays, **163**, 716
production, **163**, 723
purification, **163**, 726
J774, calmodulin-binding protein, assay and purification, **139**, 148
liposome-mediated activities, evaluation *in vivo*, **149**, 150
membrane changes during phagocytosis, ESR analysis, **150**, 441
migration, spectrophotometric assay, **147**, 145
peritoneal
cultivation, **141**, 251; **149**, 129; **163**, 596
phosphatidylinositol turnover, modulation, **141**, 252
uptake of liposomes, **149**, 131
rabbit
preparation of cytoplasmic extracts, **162**, 249
purification of
actin, **162**, 250
actin-modulating proteins, **162**, 254
role in inflammation and repair, hepatic granuloma model, **163**, 609
Magnesium
atomic absorption spectrometry, **158**, 132, 365
binding to Na^+,K^+-ATPase, kinetics, **156**, 239

in crystallization of membrane-bound Na^+,K^+-ATPase, **156**, 81
effect on phosphoinositide hydrolysis, **141**, 120
induced changes in Ca^{2+}-ATPase, fluorescence assay, **157**, 217
–NH_4^+–polyamine system, for poly(Phe) synthesis with near *in vivo* features, **164**, 650
Magnetic fields
effect on ferromagnetic erythrocyte ghosts, **149**, 307
Maitotoxin
effects on
fractional $^{45}Ca^{2+}$ efflux, **141**, 71
gross $^{45}Ca^{2+}$ uptake, **141**, 67
net $^{45}Ca^{2+}$ flux, **141**, 70
Maize
amyloplasts, isolation from endosperm, **148**, 226
protoplasts, electroporation-mediated uptake of
DNA, **153**, 359
RNA, **153**, 360
L-Malate
production by dehydrogenase–NAD gel, **136**, 29
Malate dehydrogenase
as marker for chromosome detection in somatic cell hybrids, **151**, 187
in multienzymatic NADH recycling, **136**, 70, 74
NAD-coimmobilized
assay, **136**, 28
L-malate production with, **136**, 29
preparation, **136**, 26
Malate dehydrogenase (oxaloacetate-decarboxylating) ($NADP^+$)
as marker for chromosome detection in somatic cell hybrids, **151**, 187
Malbranchea pulchella
purification of
xylanase, **160**, 672
β-xylosidase, **160**, 684
Maleimides
N-substituted, labeling of Ca^{2+},Mg^{2+}-ATPase, **157**, 251
4-(*p*-Maleimidophenyl)butyrylphosphatidylethanolamine
synthesis, in protein conjugation to liposomes, **149**, 113
L-Malic acid
industrial production by immobilized bacteria, **136**, 455
Malic enzyme, *see* Malate dehydrogenase (oxaloacetate-decarboxylating) ($NADP^+$)
Malignancies
hematologic, treatment with platelet-delivered vinca alkaloids, **149**, 323
Malt
assay of (1→3),(1→4)-β-D-glucan, **160**, 549
Manganese
atomic absorption spectrometry, **158**, 138
effect on
cGMP-binding phosphodiesterase, **159**, 717
inositol phosphate accumulation in hippocampal slices, **141**, 200
as paramagnetic probe for Na^+,K^+-ATPase, **156**, 357
Manganese peroxidase
assays, **161**, 259, 266
fungal production, **161**, 265
properties, **161**, 260
purification from *Phanerochaete chrysosporium*, **161**, 259, 267
Mannan
acetolysis products, HPLC, **138**, 112
α-Mannanase
assay, **160**, 620
properties, **160**, 623
purification from *Rhodococcus erythropolis*, **160**, 621
β-Mannanase
assay, **160**, 76, 599, 611
in preparation of manno- and glucomannooligosaccharides, **160**, 518
properties
Aspergillus enzyme, **160**, 608
Bacillus enzyme, **160**, 604
guar seed enzyme, **160**, 602
Helix enzyme, **160**, 603
Irpex enzyme, **160**, 609
lucerne seed enzyme, **160**, 601
Streptomyces enzyme, **160**, 614
purification from
Aspergillus niger, **160**, 604
Bacillus subtilis, **160**, 603
guar seeds, **160**, 601

Helix pomatia, **160**, 602
Irpex lacteus, **160**, 609
lucerne seeds, **160**, 600
Streptomyces, **160**, 613
D-Mannans
reductive cleavage, **138**, 29
β-1,4-Mannooligosaccharides
preparation from copra mannan, **160**, 518
β-D-Mannopyranosides
synthesis, **160**, 515
Mannose-6-phosphate isomerase
as marker for chromosome detection in somatic cell hybrids, **151**, 188
α-Mannosidase
as marker for chromosome detection in somatic cell hybrids, **151**, 188
types I and II
assay, **138**, 780
properties, **138**, 784
purification from *Aspergillus*, **138**, 782
β-Mannosidase
activity staining in gels, **160**, 616
assay, **160**, 590, 616
β-D-mannopyranoside substrates, synthesis, **160**, 515
properties, **160**, 593, 617
purification from
guar seeds, **160**, 591
Helix pomatia, **160**, 616
Mannosidosis
associated oligosaccharides, HPLC, **138**, 107
Mannosyl-glycoprotein endo-β-*N*-acetylglucosaminidase
type F
assay, **138**, 771
properties, **138**, 775
purification from *Flavobacterium*, **138**, 772
type H
properties, **138**, 769
purification from *Streptomyces plicatus*, **138**, 765
Maple syrup urine disease
fibroblasts, assays for branched-chain α-keto acid dehydrogenase, **166**, 135
Marchantia polymorpha, *see* Liverwort
Marfey's reagent
in HPLC of (3*R*)- and (3*S*)-β-leucine, **166**, 14
Markers
fluid-phase, for analysis of endocytosis, **146**, 402
selectable, for plant transformation, **153**, 253
Mass spectrometry
chemical ionization
carbohydrate analysis, **138**, 62
overview, **138**, 60
in combination with immobilized cells and enzymes, **137**, 421
cyanobacterial hydrogen exchange and nitrogen uptake, **167**, 467
direct chemical ionization
glycosphingolipid analysis, **138**, 79
oligosaccharide analysis, **138**, 71
overview, **138**, 66
fast atom bombardment
glycophosphosphingolipids, **138**, 194
sialic acids, **138**, 155
–GLC, sialic acids, **138**, 150
–HPLC, sialic acids, **138**, 153
inductively coupled plasma
detection limits, **158**, 211
instrumentation, **158**, 206
interferences
nonspectral, **158**, 215
spectral, **158**, 212
methods, **158**, 217
signal considerations, **158**, 207
instrumentation, **138**, 59
oxygen exchange in cyanobacteria, **167**, 686
platelet-activating factor, **163**, 52
vanadium, **158**, 416
Mast cell growth factor
gene isolation, **150**, 751
Mast cells
allergic reaction mediators, overview, **163**, 3
human, purification of proteases, **163**, 313
isolation from
guinea pig lung, **162**, 512
human intestine, **162**, 509
human lung, **162**, 510
rat intestine, **162**, 507
rat peritoneum, **162**, 503
products, measurement, **162**, 514
Mastigocladus laminosus
purification of hydrogenases, **167**, 505

Matrices
immobilization, *see* Immobilization matrices
Maxicell analysis
plasmid-coded rRNA, **164**, 691
Medicago sativa, *see* Lucerne
Melanoma growth stimulatory activity
assays, **146**, 113
purification, **146**, 117
Melilotus, *see* Sweet clover
Melittin
affinity matrix
calmodulin purification, **139**, 10
optimization, **139**, 5
preparation, **139**, 4
quantitative properties, **139**, 5
S-100 protein purification, **139**, 14
Melting
DNA, computer simulation, application to gel electrophoresis, **155**, 482
Membrane potential
neutrophil, fluorescence assay, **162**, 297
Membrane potential–dye technique
measurement of ionic permeability of
endoplasmic reticulum vesicles, **157**, 430
sarcoplasmic reticulum vesicles, **157**, 430, 452
Membrane reactor
for cell immobilization by entrapment
diffusional considerations, **135**, 374
flat plate units, **135**, 379
hollow fiber units, **135**, 375
membrane characteristics, **135**, 373
multimembrane systems, **135**, 383
operation, **135**, 384
spiral wound units, **135**, 381
tubular systems, **135**, 381
retention of pyridoxal 5′-phosphate, **136**, 492
Membranes
basement, *see* Basement membranes
cerebral
endoneuraminidase effects, **138**, 634
sialyltransferase distribution, **138**, 629
chloroplast
galactolipid synthesis, **148**, 346
isolation, **148**, 296
lipids
class analysis, **148**, 307
fatty acid analysis, **148**, 303
isolation, **148**, 301
molecular species, analysis, **148**, 311
quantification, **148**, 315
phosphatidylglycerol synthesis, **148**, 342
chloroplast envelope
immunochemical characterization, **148**, 214
level of cross-contamination, estimation, **148**, 216
polypeptide analysis, **148**, 200, 213
preparation, **148**, 211
chromaffin granule, calelectrin binding, **139**, 34
chromoplast, carotenogenic enzymes, solubilization and reconstitution, **148**, 392
cyanobacterial
lipid analysis, **167**, 251
vesicles, size and surface charge analysis, **167**, 691
glyoxysomal
isolation, **148**, 505
lipid analysis, **148**, 527
polypeptide analysis, **148**, 528
Golgi, preparation from rat liver, **157**, 612
junctional face, of sarcoplasmic reticulum, isolation, **157**, 46
leukocyte, chemoattractant receptors, assay, **162**, 240
lysosomal, with acidification activity, preparation, **157**, 599
mitochondrial
inner system, vesicles
matrix inclusion assay, **148**, 449
polarity assay, **148**, 446
preparation, **148**, 448
right-side-out and inside-out, separation from intermediate polarity vesicles, **148**, 451
isolated channels
conductance, **148**, 470
gating properties, **148**, 474
ion selectivity, **148**, 472
nonelectrolyte selectivity, **148**, 473
lipids
analysis, **148**, 475
temperature effects, **148**, 485

outer system, isolation
characterization of membrane fractions, **148**, 461
establishment of membrane lysis conditions, **148**, 457
preliminary considerations, **148**, 454
procedure, **148**, 460
phosphoglycerides
content, **148**, 479
fatty acyl composition, **148**, 481
metabolism, **148**, 483
transmembrane distribution, **148**, 482
phosphoinositide turnover, hormone-regulated, assay, **141**, 122
phospholipid, planar, preparation for functional analysis of channels, **148**, 465
photosynthetic, isolation from *Spirulina platensis*, **167**, 355
plant cell
electron microscopy
dehydration and embedding, **148**, 615
fixation, **148**, 600, 611
postfixative treatments, **148**, 603
sectioning, **148**, 617
staining of sections, **148**, 618
isolation and characterization, **148**, 635
lipid synthesis, pesticide effects, **148**, 667
sterols
analysis, **148**, 636
cellular sites of synthesis, **148**, 643
composition, modification, **148**, 647
intracellular distribution, **148**, 639
plasma, *see* Plasma membranes
platelet
crude preparations, **157**, 352
purification, **157**, 354
preparation from labeled GH_3 cells, **141**, 124
–protein kinase C complex
hormone-induced stabilization, assay, **141**, 405
phorbol ester-induced stabilization, assay, **141**, 400
renal medullary
crude, preparation, **156**, 36
fractionation, **156**, 35
Na^+,K^+-ATPase assay, **156**, 36
vesicles, separation from open membrane fragments, **156**, 36
salivary gland, phosphoinositide breakdown, hormonal activation, **141**, 255
thylakoid, *see* Thylakoids
transverse tubule, sidedness and integrity, **157**, 64
vacuolar
preparation from
Neurospora crassa, **157**, 563
yeast, **157**, 545
properties, **157**, 546
vesicles
preparation, **146**, 376, 380
transport assay, **146**, 380
dl-Menthol
stereoselective esterification by immobilized lipase, **136**, 299
dl-Menthyl succinate
stereoselective hydrolysis by immobilized yeast cells, **136**, 293
Mercaptoacetate
gas chromatography, **143**, 178
Mercaptoethanol
reduction of disulfides, **143**, 249
thiol–disulfide interchange in, thiol pK_a values, **143**, 132
3-Mercaptolactate
gas chromatography, **143**, 178
3-Mercaptopyruvate
gas chromatography, **143**, 178
Mercury
atomic absorption spectrometry, **158**, 139
atomic fluorescence spectrometry, **158**, 235
Merocyanine 540
in assessment of phospholipid asymmetry of loaded erythrocytes, **149**, 282
Mesothelial cells
pleural, cultivation for elastin synthesis studies, **144**, 236
Metabolism
branched-chain amino acids
analysis by radiochemical HPLC, **166**, 79
studies with rat hindquarter preparations, **166**, 476

branched-chain α-keto acids, analysis in perfused heart and liver, **166**, 484
cAMP
analysis in intact cells, **159**, 50
role of cAMP export, **159**, 85
carbon in cyanobacteria, analysis with radioisotopes, **167**, 539
collagens, analytical methods, **163**, 140, 674
cyclic nucleotides in intact cells, monitoring, **159**, 60
cysteine
and derivatives, in microorganisms, **143**, 454
in mammals, **143**, 372
in plants, **143**, 420
hydrogen and nitrogen in cyanobacteria, **167**, 459, 467
inborn errors, treatment with immobilized enzymes, **137**, 445
inositol phospholipids
in PC12 cells, measurement, **141**, 169
in pituitary cells
assay with Li^+, **141**, 149
TLC analysis, **141**, 156
in platelets, measurement, **141**, 176
intracellular calcium, TRH-induced changes, measurements, **141**, 36
leucine, relative carbon flux measurement, **166**, 127
methionine
and derivatives, in microorganisms, **143**, 456
in mammals, **143**, 369
in plants, **143**, 422
nitrogen
in cyanobacterial mats, **167**, 474
in synchronously growing unicellular cyanobacteria, **167**, 484
organic solutes in cyanobacteria, NMR analysis, **167**, 515
phosphatidylinositol in macrophages, modulation by liposomes, **141**, 244
phosphoglycerides in plant mitochondria, **148**, 483
phosphoinositides
assay of subcellular sites, **141**, 92
hormone-regulated
in membranes, assay, **141**, 122
in permeabilized cells, assay, **141**, 118
TRH-stimulated, analysis, **141**, 100
platelet-activating factor, associated enzymes, assays, **141**, 379
proteins, intracellular degradation, regulation by amino acids, **166**, 461
proteoglycans, analysis, **144**, 412
sialic acid *O*-acetyl groups, **138**, 611
sulfur in cyanobacteria, **167**, 572
Metabolite fluxes
across organelles, analysis by rapid filtration technique, **148**, 675
Metallocollagenase
assay, **163**, 325
properties, **163**, 314
purification from human neutrophils, **163**, 319
Metallogelatinase
assay, **163**, 326
properties, **163**, 314
purification from human neutrophils, **163**, 320
Metallothionein
coding sequences, sodium bisulfite mutagenesis, **154**, 423
Metals
adventitious, elimination from solutions
by dithizone extraction, **158**, 7
by metal-chelating resins, **158**, 10
analysis in biological materials
aqueous standards, **158**, 57
quality control, **158**, 61
reference materials, **158**, 58
exchange-inert, introduction into enzymes, **158**, 95
exchange in multimetal enzymes, selectivity
metal displacement, **158**, 85
metal insertion, **158**, 85
metal removal, **158**, 83
interactions with gel filtration media
effect of eluent ionic strength, **158**, 18
elimination by alkaline reduction, **158**, 19
quantitation, **158**, 16
ionic, buffer systems
associated analytical complications, **158**, 53

capacity, **158**, 39
free metal ion concentration, calculation, **158**, 36
sources of uncertainty, **158**, 40
future developments, **158**, 54
ligands, interaction with metals in enzyme active sites, resulting complications, **158**, 42
removal from enzymes, **158**, 24
substitution in enzymes, **158**, 71
Metal shadowing
in analysis of cyanobacterial fine structure, **167**, 159
Metanephrine
enzymatic radioimmunoassay in urine, **142**, 566
Metastases
treatment with liposome-activated macrophages, **149**, 154
Methanesulfonic acid
as solvent for chitin, **161**, 407
Methanococcus vannielii
ribosomal protein A, sequence analysis, **164**, 561
Methanol oxidase
assay, **161**, 323
fungal production, **161**, 323
properties, **161**, 325
purification from *Phanerochaete chrysosporium*, **161**, 324
Methidiumpropyl–EDTA • iron(II)
in footprinting of protein–DNA contacts, **155**, 543
comparison with hydroxyl radical method, **155**, 553
Methionine
and derivatives, metabolism in microorganisms, **143**, 456
dietary, quantification, **143**, 300
enantiomers, analytical resolution, **143**, 167
L-form, deuteration and tritiation, **143**, 296
metabolism in
mammals, **143**, 369
plants, **143**, 422
Methionine γ-lyase
assay, **143**, 460
properties, **143**, 462
purification from
Aeromonas sp., **143**, 462
Pseudomonas putida, **143**, 460
Methotrexate
markers for plant transformation
application, **153**, 270
construction, **153**, 268
resistant mammalian cells, selection, role of gene amplification, **151**, 88
4-Methoxybenzoate monooxygenase
components
assays, **161**, 287
interactions, **161**, 291
properties, **161**, 288
purification from *Pseudomonas putida*, **161**, 284
substrate specificity and reaction mechanisms, **161**, 292
importance in bacterial metabolism, **161**, 281
3-Methoxy-1-butanol
racemic, enzymatic preparation, **136**, 125
Methoxycarbonyl phosphate
enzymatic synthesis, **136**, 270
3-Methoxy-4-hydroxyphenylglycol
GC–MS assay in brain and body fluids, **142**, 577
HPLC assay in brain, **142**, 547
Methylation
calmodulin, assay, **139**, 674
cDNA, in linker ligation, **152**, 345
DNA *in vivo*, effects on *in vitro* restriction, **152**, 139
and oxidation, in lignin analysis, **161**, 91
proteins, calmodulin-dependent, assay, **139**, 668
Methylation analysis
glycolipids, **138**, 13
2-Methyl branched-chain acyl-CoA dehydrogenase
assay, **166**, 361
properties, **166**, 368
purification from rat liver, **166**, 363
(*E*)-2-Methylbutenoate
reduction by *Clostridium* La 1, **136**, 315
S-Methyl-L-cysteine
deuteration and tritiation, **143**, 296
DL-α-Methylcysteinesulfinic acid
synthesis, **143**, 278

α-4-*O*-Methyl-D-glucuronidase
assays, **160**, 561
3-Methylglutaconyl-CoA hydratase
assay, **166**, 214
properties, **166**, 217
6-Methyl-5-hepten-2-ol
racemic, enzymatic preparation, **136**, 132
1-Methyl-3-isobutylxanthine
effect on cGMP-binding phosphodiesterase, **159**, 720
3-Methyl-2-ketobutyric acid
HPLC, **166**, 20
3-Methyl-2-ketovaleric acid
HPLC, **166**, 20
4-Methyl-2-ketovaleric acid
HPLC, **166**, 20
Methylmalonate-semialdehyde dehydrogenase
assay, **166**, 390
properties, **166**, 393
purification from *Pseudomonas aeruginosa*, **166**, 391
Methylmalonic acid
GC–MS assay, **166**, 47
D-Methylmalonyl-CoA hydrolase
assay, **166**, 394
properties, **166**, 399
purification from rat liver, **166**, 396
L-Methylmalonyl-CoA mutase
assay, **166**, 407
properties, **166**, 413
purification from human placenta, **166**, 410
DL-Methylmalonyl-CoA racemase
assay, **166**, 401
properties, **166**, 406
purification from rat liver, **166**, 403
Methylnitrocatechol sulfate
arylsulfatase assay with, **143**, 214
N-Methyl-*N'*-nitro-*N*-nitrosoguanidine
cyanobacterial mutagenesis, **167**, 716
3-Methyl-1-pentanol
racemic, enzymatic preparation, **136**, 126
N-Methyl-4-phenyl-1,2,3,6-tetrahydropyridine
as monoamine oxidase substrate in brain, **142**, 622
Methylselenol
assay with 1-fluoro-2,4-dinitrobenzene, **143**, 32
Methylsulfinylcarbanion
solution preparation, **138**, 14
4-Methylumbelliferone sulfate
arylsulfatase assay with, **143**, 216
4-Methylumbelliferyl-β-D-lactoside
preparation, **160**, 47
1-Methyluric acid
preparation with immobilized xanthine oxidase, **136**, 260
1-Methylxanthine
oxidation by immobilized xanthine oxidase, **136**, 260
Metronidazole
enrichment of cyanobacterial mutants, **167**, 726, 732
Mevalonic acid
conversion to prenyl pyrophosphates *in vitro*, **167**, 331
Mevastatin
as dolichol synthesis inhibitor, **138**, 699
Micelles
detergent, associated proteoglycans, analysis, **144**, 398
reverse, enzyme-containing
apolar substrate conversion, regulating parameters, **136**, 225
characterization, **136**, 197
enzyme activity, **136**, 221
enzyme conformation and activity, **136**, 199
enzyme intrinsic rate parameters, **136**, 222
multienzymatic reactions in, **136**, 228
preparation, **136**, 189, 219
Microcapsules
multienzyme-containing
preparation, **136**, 68, 78
recycling of
dextran-coupled NADH, **136**, 74
free NADH, **136**, 70, 77
free NADPH, **136**, 70
Microcells
mediated chromosome transfer
general strategy, **151**, 315
overview, **151**, 313
procedures, **151**, 317
required materials and reagents, **151**, 316
Microcloning
mammalian metaphase chromosomes

applications, **151**, 514
dissection procedure, **151**, 509
DNA extraction, **151**, 510
instrumentation, **151**, 504
ligation to vector, **151**, 513
overview, **151**, 503
preparation of chromosome spreads, **151**, 504
prospects, **151**, 515
recovery of ligated extracts and *in vitro* packaging, **151**, 513
restriction enzyme digestion, **151**, 513

Microcultures
antibody-producing cell clones, **150**, 240

Microcystis
carotene oxygenase assays, **167**, 336

Microelectrodes
in chemical analysis of cyanobacterial communities, **167**, 647
oxygen, Clark-type, construction, **167**, 641
sulfide, construction, **167**, 645

Microemulsification
liposome preparation by, **149**, 66

Microencapsulation
enzyme immobilization by, **135**, 61
pancreatic islet cells, **137**, 575

Microfilaments
actin, network in leukocytes, regulation, **162**, 247

Microinjection
calmodulin into cultured cells, **139**, 826
erythrocyte-mediated, macromolecules into eukaryotic cells, **149**, 293
mRNA, and complementation of mutant cells
applications, **151**, 379
overview, **151**, 371
procedure, **151**, 373
nucleic acids into *Xenopus* oocytes
cell preparation, **152**, 284
instrumentation, **152**, 276

Microinjector
for *Xenopus* oocytes
construction, **152**, 277
operation, **152**, 279
storage, **152**, 282

Microorganisms
immobilized, in water purification, **137**, 697
mineralization of radiolabeled lignins, analysis, **161**, 29
transformation of radiolabeled lignocelluloses, analysis, **161**, 30

Microprobes
fiber optic, for analysis of light spectral gradients in cyanobacterial communities, **167**, 650

Microscopy, *see also specific techniques*
with microbeam device, in necrotaxis assay, **162**, 122
neutrophil chemotaxis on plane 2-D substrata, **162**, 35

Microsomes
cerebral, purification of Na^+,K^+-ATPase, **156**, 71
enriched in T tubule/sarcoplasmic reticulum complex, preparation, **157**, 469
hepatic, biosensors based on, **137**, 153
neutrophil, preparation, **163**, 352
plant, tonoplast isolation from, **148**, 107
skeletal muscle, homogenization and preparation, **157**, 54

Microtiter assay
critical drug levels for enhancement of gene amplification, **151**, 94

Microtiter wells
solid-phase assay for glycolipid receptors, **138**, 227

Microtubule-associated proteins
binding to calmodulin, analytical methods, **139**, 836

Microultrafiltration
assay for calmodulin–phosphofructokinase binding, **139**, 750

Milk
bovine
heat-sterilized, treatment with immobilized sulfhydryl oxidase, **136**, 423
membrane vesicle preparation from, **143**, 507
sulfhydryl oxidase purification, **143**, 508
human, growth factor purification, **146**, 43

Millet
mitochondria, isolation from leaves, **148**, 425

Mine drainage
acid, treatment with immobilized *Thiobacillus ferrooxidans*, **136**, 530

Mineral acids
strong, solvolysis of cellulose, **160**, 12
Mineralization
radiolabeled lignins by microorganisms, analysis, **161**, 29
Minerals
removal from crustacean chitin, **161**, 418
Minielectrodes
calcium-selective
applications, **141**, 34
evaluation, **141**, 31
precautions, **141**, 33
preparation, **141**, 27
thermostatted incubation vessel for, **141**, 29
Mitochondria
artichoke
inner membrane vesicles
matrix inclusion assay, **148**, 449
polarity assay, **148**, 446
preparation, **148**, 448
right-side-out and inside-out, separation from intermediate polarity vesicles, **148**, 451
isolation from tubers, **148**, 444
calcium-transporting systems, properties, **157**, 8
cauliflower, two-dimensional protein separation, **148**, 630
DNA, in construction of cloning vectors for eukaryotes, **153**, 366
extracts
branched-chain α-keto acid dehydrogenase complex assay, **166**, 175
branched-chain α-keto acid dehydrogenase kinase assay, **166**, 166
hepatic
P_i–P_i exchange, analytical technique, **148**, 681
preparation for monoamine oxidase B purification, **142**, 630
mung bean, purification from hypocotyls, **148**, 416
Neurospora crassa, preparation, **157**, 563
pea, purification from leaves, **148**, 417
placental, preparation for monoamine oxidase A purification, **142**, 629
plant
chlorophyll-free, preparation, **148**, 412
cytochrome-*c* oxidase
properties, **148**, 501
purification, **148**, 497
integrity assessment criteria, **148**, 407
isolated channels
conductance, **148**, 470
gating properties, **148**, 474
ion selectivity, **148**, 472
nonelectrolyte selectivity, **148**, 473
isolation from
CAM species, **148**, 431
C_3 species, **148**, 422
C_4 species, **148**, 425
membrane lipids
analysis, **148**, 475
temperature effects, **148**, 485
membrane phosphoglycerides
content, **148**, 479
fatty acyl composition, **148**, 481
metabolism, **148**, 483
transmembrane distribution, **148**, 482
outer membrane isolation
characterization of membrane fractions, **148**, 461
establishment of membrane lysis conditions, **148**, 457
preliminary considerations, **148**, 454
procedure, **148**, 460
succinate dehydrogenase (ubiquinone)
properties, **148**, 494
purification, **148**, 493
washed, preparation, **148**, 404
potato
ADP/ATP transport, analytical technique, **148**, 680
purification, **148**, 409
ubiquinone synthesis, **148**, 486
spinach, isolation from leaves, **148**, 434
yeast
entrapment with urethane prepolymers, **135**, 242
prepolymer-entrapped, applications, **135**, 245
Mitogenesis
associated phosphorylation of ribosomal protein S6, analysis, **164**, 575
Mitogens
lymphocyte biochemical responses to, **150**, 29

as lymphocyte locomotion activators, **162**, 187
T-cell
assay, **150**, 4
parameters affecting lymphocyte transformation, **150**, 4
Mixed lymphocyte reactions
allogeneic and autologous
analysis in
humans, **150**, 85
mice, **150**, 86
culture set up, **150**, 87
significance and application, **150**, 90
in basal serum-free medium, **150**, 131
in cloning of alloreactive T cells, **150**, 335
Mixtures
water–organic solvent, enzymatic urea synthesis in, **136**, 234
Models
calcium-induced conformational transition of troponin C, **139**, 627
peptide, for structure–function analysis of proteins, **154**, 473
Molar concentration
mRNA, determination, **152**, 262
Molecular dynamics
in protein design, **154**, 430
Molecular mechanics
in protein design, **154**, 430
Molecular structure
calbindin D, **139**, 585
calmodulin, **139**, 633
cAMP phosphodiesterase-inhibiting acylpeptides, **159**, 499
quaternary, Na^+,K^+-ATPase, determination with glutaraldehyde, **156**, 333
secondary
RNA, computer analysis, **164**, 779
5S RNA, **164**, 732
tertiary
globular proteins, effect of point mutations, **154**, 498
5S RNA, **164**, 732
three-dimensional
cyanobacterial ferredoxins, **167**, 400
phage T4 lysozyme, **154**, 511
troponin C, **139**, 610
Molecular titration assay
transcript prevalence levels in RNA populations, **152**, 633
Molecular weight
aerolysin, **165**, 197
bone glycoproteins, **145**, 277
chitosan, distribution determination by HPLC, **161**, 447
chloroplast cytochromes, **148**, 268
coniferyl-alcohol dehydrogenase, **161**, 306
crustacean chitin, determination, **161**, 422
damselysin, **165**, 188
H^+-ATPase from clathrin-coated vesicles, **157**, 642
isovaleryl-CoA dehydrogenase, **166**, 384
2-methyl branched-chain acyl-CoA dehydrogenase, **166**, 369
Na^+,K^+-ATPase, **156**, 13
proteoglycan link protein, **144**, 409
proteoglycans, **144**, 360
variant isovaleryl-CoA dehydrogenase, heterogeneity, **166**, 162
vulnificolysin, **165**, 182
Moloney murine leukemia virus
in gene isolation by viral tagging, **152**, 474
insertional mutagenesis of mammalian cells
applications, **151**, 499
procedure, **151**, 491
reverse transcriptase, first-strand cDNA synthesis with, **152**, 321
Molybdenum
atomic absorption spectrometry, **158**, 139, 376
colorimetric assay, **158**, 372
neutron activation analysis, **158**, 380
Monensin
as glycoprotein transport inhibitor, **138**, 700
Monilia
cultivation, **160**, 456
purification of cellobiose dehydrogenase, **160**, 458
Monkey feeding test
for staphylococcal enterotoxin
enterotoxin administration, **165**, 326
enterotoxin dosage, **165**, 327
overview, **165**, 324
susceptibility to second enterotoxin, **165**, 328
temporary resistance, **165**, 328

Monoamine oxidase, *see* Amine oxidase (flavin-containing)
Monobromobimane
labeled thiols, HPLC analysis, **143**, 85
Monoclonal antibodies
in analysis of calmodulin-regulatory systems, **139**, 518
5-bromo-2-deoxyuridine, in assay for lymphocyte proliferation, **150**, 117
calmodulin-binding protein, preparation, **139**, 153
to calmodulin and dependent enzymes
characterization, **139**, 509
production, **139**, 506
carbohydrate-specific
human, overview, **138**, 492
murine and rat, overview, **138**, 484
carboxypeptidase A
enzyme immobilization with, **135**, 165
preparation, **135**, 164
CD2, preparation, **150**, 538
collagenase, production, **144**, 167
collagen, characterization and isolation, **145**, 148
elastin, production, **144**, 249
epidermal growth factor
competitive binding, **146**, 18
effect on EGF activities, **146**, 19
production, **146**, 13
specificity and dissociation constant, **146**, 17
epidermal growth factor receptor
characterization, **146**, 68
effects on
cell growth *in vitro*, **146**, 74
cell growth *in vivo*, **146**, 79
receptor internalization, **146**, 77
receptor protein kinase, **146**, 78
production, **146**, 67
purification, **146**, 72
α-hemolysin, preparation, **165**, 143
hemopexin, preparation, **163**, 545
insulin-like growth factor I
affinity and specificity, **146**, 211
assay, **146**, 210
characterization, **146**, 213
production, **146**, 207
lactate dehydrogenase
enzyme immobilization with, **135**, 169
preparation, **135**, 168
in localization of ribosomal proteins, **164**, 516
monoamine oxidase B
complex with enzyme
characterization, **142**, 647
chromatography, **142**, 646
preparation, **142**, 646
Na^+,K^+-ATPase
applications, **156**, 407
isotype identification, **156**, 98
preparation, **156**, 96, 98, 394
screening procedures, **156**, 399
nerve growth factor
production, **147**, 187
properties, **147**, 190
OKT3, as lymphocyte locomotion activators, **162**, 188
parvalbumin, production, **139**, 554
phospholamban, purification, **139**, 520
prolyl 4-hydroxylase, preparation, **144**, 102
transferrin receptor
applications, **147**, 276
biological effects, **147**, 272
derivation and characterization, **147**, 266
listing, **150**, 727
Monocytes
culture for production of hepatocyte-stimulating factor, **163**, 595
human, purification of hepatocyte-stimulating factor, **163**, 598
infiltration during inflammation, quantitation, **162**, 311
Monoglycosylceramides
perbenzoylated, HPLC, **138**, 117
5′-*N*-Monomethoxytrityl-5′-amino-5′-deoxythymidine
synthesis, **155**, 271
Mononuclear cells
insulin-like growth factor I receptors, assay, **146**, 235
Monophenol monooxygenase
assay, **142**, 160, 165
properties, **142**, 164, 168
purification from
murine melanoma, **142**, 156
Neurospora crassa, **142**, 166
Mono S cation exchanger
in chromatography of polygalacturonases, **161**, 396

Monosodium urate
microcrystalline, induced chemotactic factor, isolation and assay, **162**, 193
Mouse
inbred and congenic strains, overview, **150**, 163
strain differences related to lupus research, **162**, 386
MS, *see* Mass spectrometry
MSA, *see* Multiplication stimulating activity
Mucor rouxii
purification of chitin deacetylase, **161**, 512
arom Multifunctional enzyme
assays, **142**, 333
catalyzed reactions, **142**, 327
coupling enzymes, **142**, 328
properties, **142**, 338
purification from *Neurospora crassa*, **142**, 337
substrates, **142**, 328
Multiple sulfatase deficiency
enzymatic diagnosis, **138**, 745
Multiplication stimulating activity
carrier protein
assays, **146**, 271
properties, **146**, 275
purification, **146**, 273
Mung bean
chorismate mutase purification, **142**, 455
mitochondria, purification from hypocotyls, **148**, 416
peroxisomes
isolation, **148**, 517
β-oxidation, assay methods, **148**, 520
thiols, HPLC with fluorometry, **143**, 92
Mung bean nuclease
hairpin loop cleavage in cDNA, **152**, 334
Muscle
aortic and tracheal, cGMP-dependent protein kinase activation, **159**, 150
protein kinase regulatory subunit phosphorylation in, analysis, **159**, 143
rabbit
biosensors based on derived tissues, **137**, 146
purification of
high-molecular-weight protein phosphatase, **159**, 368
parvalbumin, **139**, 764
phosphatases C-I and C-II, **159**, 360
phosphorylase phosphatase, **159**, 377
protein phosphatase-1, **159**, 395, 404
protein phosphatase-2A, **159**, 395, 400
protein phosphatase-2B, **159**, 409
protein phosphatase-2C, **159**, 418
protein phosphatase inhibitor-1 and inhibitor-2, **159**, 427
rat, assay of branched-chain α-keto acid dehydrogenase, **166**, 209
skeletal, *see* Skeletal muscle
Mutagenesis
in *Anacystis nidulans*, **153**, 211
application of
polymerase chain reaction *in vitro*, **155**, 345
segmented paper method, **154**, 253
cAMP-dependent protein kinase regulatory subunit I gene, **159**, 334
CHO cells, **151**, 6
cyanobacteria
chemical methods, **167**, 716
recombinational methods, **167**, 721
by transposition, **167**, 719
by uv irradiation, **167**, 718
defined DNA regions, with degenerate oligonucleotides, **155**, 558
embryonal carcinoma cells, **151**, 78
with ethyl methanesulfonate, **147**, 371
HeLa cells, **151**, 45
insertional, in gene isolation, **151**, 489
by retroviral tagging, **152**, 469
lymphoma S49 cells, **151**, 15
myeloma cells, **151**, 53
oligonucleotide-directed
via gapped duplex DNA, **154**, 350
for modification of human tumor necrosis factor gene, **154**, 403
with M13 vectors, **154**, 382
without phenotypic selection, **154**, 367
5S RNA, **164**, 737
with two primers and ssDNA template, **154**, 329
phage T4 lysozyme gene, **154**, 513
single-stranded strategy, **155**, 204
site-specific, calmodulin gene, **139**, 291
sodium bisulfite-induced, metallothionein coding sequences, **154**, 423

Tn*5*-induced
in directed DNA sequencing, **154**, 187
DNA segments cloned into multicopy plasmids, **154**, 178
in gene replacement experiments, **154**, 188
in identification of nitrogen fixation genes in *Rhizobium*, **154**, 197
in protein mapping and determination of transcription direction, **154**, 188
transplason, of λgt 11 clones, **154**, 125
transposon, gene cloning by, **152**, 491
Mutants
cellular
amino acid transport-defective, isolation, **151**, 136
drug-resistant
genetic analysis, **151**, 119
problems associated with isolation, **151**, 117
suitable drug concentration for isolation, **151**, 114
glucose oxidative pathway-defective, isolation, **151**, 135
glucose transport- and glycolysis-defective, isolation, **151**, 132
resistant to transferrin receptor antibodies, selection, **147**, 276
selection
application of replica plating, **151**, 104
role of gene amplification, **151**, 88
suicide selection, **151**, 121
temperature-sensitive, isolation, **151**, 145
CHO cell
endocytosis-defective
genotypic analysis, **138**, 467
isolation, **138**, 460
quantitative testing, **138**, 466
repetitive cloning of dihydrofolate reductase gene, **151**, 451
cyanobacterial
in energetic pathways, enrichment, **167**, 728
growth-impaired, enrichment, **167**, 725
photosynthetically impaired, enrichment, **167**, 726
glycosylation-defective animal cells, biochemical characterization, **138**, 443
hybridoma and myeloma cells, detection by immunoplate assay, **151**, 54
insertional, retrovirally generated
analysis and recovery, **152**, 478
selection, **152**, 471
kdp, of *E. coli*, isolation, **157**, 656
lymphoma S49 cells, listing, **151**, 17
Neurospora crassa, without cell walls, H^+-ATPase purification, **157**, 574
phage T4 lysozyme
random forms, characterization, **154**, 527
site-directed forms, characterization, **154**, 529
pheochromocytoma PC12 cells
generation, **147**, 213
selection, **147**, 214
5S RNA, interactions with proteins, **164**, 745
sulfometuron methyl-hypersensitive microbes, isolation, **166**, 105
sulfometuron methyl-resistant microbes and plants, isolation, **166**, 103
yeast
altered in cAMP cascade system
applications, **159**, 41
biochemical characterization, **159**, 33
isolation and genetic identification, **159**, 28
ura3$^-$, selection with 5-fluoroorotic acid, **154**, 164
Mutations
antibiotic resistance, in rRNA genes of *E. coli*
isolation and characterization, **164**, 676
mapping and sequencing, **164**, 680
branched-chain α-keto acid dehydrogenase, Western blot analysis, **166**, 109
collagen
assessment by R-loop analysis of procollagen mRNA, **145**, 235
detection by 2-D gel electrophoresis, **145**, 204
conditional, in essential genes, isolation by plasmid shuffling, **154**, 167
cyclic nucleotide phosphodiesterase, analysis in *Drosophila*, **159**, 786
deletion
generation in yeast chromosomes, **153**, 246

nested, in dideoxy DNA sequencing, **152**, 542
ordered, generation for DNA sequencing, **155**, 119, 168
primer-directed, in 5S rRNA, **164**, 710
transposon-promoted, in DNA sequencing, **155**, 177
ura3⁻, introduction into yeast chromosomes, **154**, 172
DNA, obtainment of large number in defined region, **154**, 415
gene, isolation
by insertional mutagenesis, **151**, 489
strategies, **151**, 329
Kdp-ATPase gene
isolation, **157**, 661
mapping and complementation analysis, **157**, 664
movement
onto F-100 episome, **157**, 662
by transduction, **157**, 661
sequencing, **157**, 666
photosystem II reaction center, construction in *Synechocystis*, **167**, 766
point
effect on globular protein folding, **154**, 498
in 5S RNA, effects on
molecular structure, **164**, 743
tRNA and mRNA affinity of 70S ribosomes, **164**, 747
pro-α2(I) collagen gene, characterization, **145**, 213
ribosomal protein L24 gene, analysis, **164**, 706
rRNA, and effects on single messenger translation, analysis, **153**, 432
Myelin basic protein
induction of experimental allergic encephalomyelitis, **162**, 417
purification from guinea pig spinal cord, **162**, 413
Myeloperoxidase
assay as measure of leukocyte infiltration in Arthus reaction, **162**, 480
neutrophil granule-associated, assay, **162**, 547
Myocytes
vascular, elastin synthesis
analysis *in vitro*, **144**, 225
cell cultivation for analysis, **144**, 234, 236
Myofibroblasts
mechanical and adhesive properties, analysis by
elastic rubber substratum method, **163**, 624
haptotactic substratum and related methods, **163**, 631
interference reflection microscopy, **163**, 638
Myogenesis
transferrin effects, assay, **147**, 293
Myoglobin
produced in *E. coli*, properties, **153**, 480
Myosin
light chain produced in *E. coli*, properties, **153**, 480
preparation from macrophages, **162**, 259
Myosin heavy chain kinase
calmodulin-dependent
assays, **139**, 106
isolation from chicken intestine, **139**, 110
properties, **139**, 113
Myosin light chain kinase
autophosphorylation, role in enzyme regulation, **139**, 696
calmodulin-binding domain
assay, **139**, 115
corresponding peptides
effects on calmodulin-dependent enzymes, **139**, 125
interactions with calmodulin, **139**, 124
preparation from enzyme, **139**, 119
structure–function studies, **139**, 123
synthesis, **139**, 120
nucleotide-binding site, affinity labeling, **139**, 188
Myosin light chain kinase inhibitors
effects on transmembrane Ca^{2+} signaling, **139**, 576
Myosin phosphatase
assay, **159**, 447
properties, **159**, 447
purification from turkey gizzard, **159**, 448
Myotrophic assay
transferrin effects on
acetylcholine receptors, **147**, 298

acetylcholinesterase, **147**, 296
DNA synthesis, **147**, 295
myogenesis, **147**, 293
Myxococcus xanthus
purification of protein S, **139**, 380

N

NAD (nicotinamide adenine dinucleotide)
coimmobilization with alcohol dehydrogenase, **136**, 56
dehydrogenase-coimmobilized
applications, **136**, 29
preparation, **136**, 26
immobilized analogs, applications, **136**, 7
modification in N-6 position, **136**, 3
polyethylene glycol-bound
preparation, **136**, 39
recycling in enzyme reactor, **136**, 40
polymerizable derivatives, synthesis, **136**, 22, 35
NAD+ ADP-ribosyltransferase
activation in oxidant-exposed cells, assay, **163**, 337
NAD+ glycohydrolase, *see* NAD+ nucleosidase
NADH (reduced nicotinamide adenine dinucleotide)
assay with
enzyme reactor, **137**, 284
microsomal electrode, **137**, 156
nylon tube-immobilized enzymes, **137**, 167
flow-injection analysis with immobilized enzymes, **137**, 176
multienzymatic recycling in microcapsules, **136**, 70, 74, 77
polyethylene glycol-bound, preparation, **136**, 39
regeneration in enzyme–membrane reactor, **136**, 9
NADH oxidase
in multienzymatic NADH recycling, **136**, 73
NADH-putidamonooxin oxidoreductase
properties, **161**, 288
purification from *Pseudomonas putida*, **161**, 286
NAD+ kinase
immobilization by radiation polymerization, **135**, 148
NAD+ nucleosidase
cholera toxin ADP-ribosylation in presence of, **165**, 241
diphtheria toxin activity as, assay, **165**, 224
NADPH (reduced nicotinamide adenine dinucleotide phosphate)
assay with
microsomal electrode, **137**, 156
nylon tube-immobilized enzymes, **137**, 167
multienzymatic recycling in microcapsules, **136**, 70
NAD(P)H dehydrogenase (FMN)
luciferase-coimmobilized
bioluminescent assays with, **136**, 88
preparation, **136**, 85
NAD(P)H dehydrogenase (quinone)
assay, **161**, 271
properties, **161**, 272
tonoplast-associated
kinetic constants, **148**, 103
purification, **148**, 102
NAD(P)H:FMN oxidoreductase, *see* NAD(P)H dehydrogenase (FMN)
NAD(P)H-quinone reductase, *see* NAD(P)H dehydrogenase (quinone)
1,2-Naphthoquinone
dihydropteridine reductase affinity purification with, **142**, 121
Narcissus pseudonarcissus, *see* Daffodil
Necrotaxis
granulocytes
assay, **162**, 115
chamber construction for analysis, **162**, 122
Neisseria gonorrhoeae
purification of IgA protease, **165**, 118
Neomycin phosphotransferase
assay in plant transformants, **153**, 303
Nephelometry
amyloid P component, **163**, 532
cellulase, **160**, 117
complement components C3, C5, C3a, and C5a, **162**, 584
immuno-, *see* Immunonephelometry
laser, C-reactive protein, **163**, 399
sulfate, **143**, 5
xylanase, **160**, 124
Nephritis
antiglomerular basement membrane

animal models, **162**, 441
autologous phase, **162**, 448
heterologous phase, **162**, 446
induction, **162**, 442
overview, **162**, 440
Heymann, *see* Heymann nephritis
murine lupus-associated, tests, **162**, 395
Nerve growth factor
cDNA isolation, **147**, 194
monoclonal antibodies
production, **147**, 187
properties, **147**, 190
role in tissue repair processes, **163**, 713
treatment of PC12 cell cultures, **147**, 210
two-site enzyme immunoassay, **147**, 167
Nerves
mammalian, neurotoxin effects
at axon membrane sites of action potential propagation, **165**, 255
on impulse conduction, **165**, 258
at intrasynaptic sites, **165**, 266
at postsynaptic sites, **165**, 267
at presynaptic sites, **165**, 263
Neural cell adhesion molecules
polysialylated, deglycosylation, **138**, 636
stimulation of polysialic acid synthesis, **138**, 631
Neuraminidase, *see* Sialidase
Neurites, *see* Axons
Neurons
vagal sensory, selective destruction by Shiga toxin, analysis, **165**, 250
Neurospora crassa
purification of
arogenate, **142**, 426, 515
arom multifunctional enzyme, **142**, 337
chitinase, **161**, 472
chorismate synthase, **142**, 363
H^+-ATPase in high yield, **157**, 574
mitochondria and plasma and vacuolar membranes, **157**, 563
monophenol monooxygenase, **142**, 166
prephenate, **142**, 426
Neurotoxicity
electroencephalographic assay, **165**, 270
Neurotoxins
microbial, effects on
axon membrane sites of action potential propagation, **165**, 255
intrasynaptic sites, **165**, 266
nerve impulse conduction, **165**, 258
postsynaptic sites, **165**, 267
presynaptic sites, **165**, 263
Neutron activation analysis
aluminum, **158**, 291
applications, **158**, 284
detection of emitted radiation, **158**, 274
limitations, **158**, 283
molybdenum, **158**, 380
quality of analytical data, **158**, 281
radioactive decay, **158**, 274
sensitivity, **158**, 278
theory and mathematical formulas, **158**, 269
tungsten, **158**, 380
vanadium, **158**, 406
Neutron scattering
deuterated 50S ribosomal subunits, preparation and assay methods for, **164**, 131
low-angle, topography of 30S ribosomal subunit proteins, **164**, 117
Neutrophil chemotactic factor
characterization, **162**, 164
purification, **162**, 162
Neutrophil elastase, *see* Leukocyte elastase
Neutrophils
accumulation during inflammation, measurement, **162**, 307
activation, relevant ionic events
fluorescence assays, **162**, 292
ionic transport measurements, **162**, 283
aggregation assay, **162**, 72
calcium levels, glycogen phosphorylase as noninvasive probe, **141**, 18
C5a receptor detection, **150**, 624
chemoattractant-induced
activation, role of G protein, **162**, 276
biochemical changes, analysis, **162**, 279
phospholipid metabolism, assay, **162**, 273
chemoattractant receptors, assay, **162**, 240
chemokinesis, agarose microdroplet assay, **162**, 75
chemotaxis
agarose gel assay, **162**, 50

inhibitors, **162**, 223
orientation chamber assay, **162**, 65
semiautomated assay with image
analyzer, **162**, 79
visual assays on plane 2-D substrata,
162, 35
degranulation, overview, **162**, 538
granule constituents, assay, **162**, 546
granule-enriched subcellular fractions,
preparation, **162**, 543
granule exocytosis, **162**, 549
human
activation assays, **141**, 263
aggregation assay, **141**, 374
isolation and subcellular fractionation,
163, 310
leukotriene B$_4$ catabolism, HPLC
analysis, **141**, 355
plasma membrane isolation, **141**, 262
polyphosphoinositide breakdown, N
protein role, **141**, 261
preparation, **141**, 262, 373
purification of
lipoxygenases, **163**, 346
proteases, **163**, 313
infiltration in Arthus reaction, measurement, **162**, 480
isolation from human blood, **162**, 540
leukotriene B$_4$ receptors, assay, **163**, 340
and subcellular fractions, preparation,
162, 272
synergistic responses to site-selective
cAMP analogs, **159**, 123
Nickel
atomic absorption spectrometry, **158**,
139
electrothermal procedure, **158**, 382
Nick translation
methodology, **152**, 91
Nicotiana, *see* Tobacco
Nidogen
basement membrane-associated
properties, **145**, 364
purification, **145**, 384
Niemann–Pick disease
enzymatic diagnosis, **138**, 736
Nitrate
tonoplast transport, measurement, **148**,
112
Nitrilase
acrylamide production with, **136**, 530

assay, **136**, 524
properties, **136**, 527
Nitrobenzene
mediated oxidation
in lignin analysis, **161**, 89
in lignin characterization, **161**, 106, 117
7-Nitrobenz-2-oxa-1,3-diazole
in staining assay for actin behavior in
leukocytes, **162**, 268
Nitrocatechol sulfate
arylsulfatase assay with, **143**, 209
Nitrogen
ammonia, assay with enzyme transistor,
137, 241
effects on alkaloid production by *Claviceps purpurea*, **136**, 325
liquid, preservation of cyanobacteria,
167, 123
^{15}N, labeling of 5S RNA, **164**, 159
Nitrogenase
cyanobacterial
assays, **167**, 460, 475
electron donation to, assay, **167**, 499
induction, **167**, 463
Nitrogen fixation
during anoxygenic photosynthesis in
cyanobacteria, assay, **167**, 382
associated genes
in cyanobacterial DNA hybridizations,
167, 803
identification in *Rhizobium* by Tn5
mutagenesis, **154**, 197
cyanobacterial, assays, **167**, 459, 467
in cyanobacterial mats, assays, **167**, 474
in synchronously growing unicellular
cyanobacteria, assay, **167**, 484
S-*o*-Nitrophenyl-L-cysteine
synthesis, **142**, 419
p-Nitrophenyl-β-D-glucoside
β-glucosidase assay with, **160**, 109
p-Nitrophenyl phosphate
in cytochemical localization of Na$^+$,K$^+$-
ATPase, **156**, 418
4-Nitrophenyl sulfate
arylsulfatase assay with, **143**, 214
(*p*-Nitrophenyltriazenylethylenediamine)-
ouabain
synthesis, **156**, 326
Nitroprusside
in spectrophotometric assay of thiols,
143, 59

Nitroquinol sulfate
arylsulfatase assay with, **143**, 210
1-Nitroso-2-napthol
phenylalanine 4-monooxygenase assay with, **142**, 6
Nitrosourea
calmodulin modification with, **139**, 423
2-Nitro-5-thiocyanobenzoic acid
cyanylation of ribosomal proteins for preparation of fragments, **164**, 261
2-Nitro-5-thiosulfobenzoate
peptide and protein disulfide analysis with, **143**, 115
Nitrous acid
in spectrophotometric assay of thiols, **143**, 63
Nitroxide
spin labels as paramagnetic probes for Na^+,K^+-ATPase, **156**, 357
NMR, *see* Nuclear magnetic resonance
Nocardia
N. otitidis-caviarum, purification of restriction enzyme *Not*I, **155**, 18
N. rhodocrous
entrapment with
photocrosslinkable resin prepolymers, **135**, 238
urethane prepolymers, **135**, 243
prepolymer-entrapped, applications, **135**, 247
Nojirimycin
as glycoprotein processing inhibitor, **138**, 690
Nopaline synthase
assay in monocot transformants, **153**, 311
Noradrenaline, *see* Norepinephrine
Norepinephrine
enzymatic radioimmunoassay
in plasma, **142**, 559, 568
in tissues, **142**, 565
in urine, **142**, 565
HPLC assay
in brain, **142**, 539, 547
in plasma, **142**, 521
Norepinephrine *N*-methyltransferase, *see* Phenylethanolamine *N*-methyltransferase
Normetanephrine
enzymatic radioimmunoassay in urine, **142**, 566
Northern blotting
Aplysia RNA with calmodulin cDNA, **139**, 284
methodology, **152**, 572
RNA from wild-type *Anacystis nidulans* and transformants, **153**, 230
Nostoc
akinete formation in culture, **167**, 225
isolation and characterization of cytochrome b_6/f genes, **167**, 779
N. muscorum, purification of
ferredoxin–thioredoxin reductase, **167**, 426
thioredoxin *f*, **167**, 422
thioredoxin *m*, **167**, 418
Nuclear magnetic resonance
in analysis of cyanobacterial bioenergetics, **167**, 509
application to plant tissues
accumulation of spectra, **148**, 688
compartmental analysis, **148**, 695
data interpretation, **148**, 691
overview, **148**, 683
tissue selection, **148**, 685
biologically relevant isotopes, overview, **135**, 513
^{13}C
cellodextrins, **160**, 36, 37
cyanobacterial carbohydrates, **167**, 550
enriched model lignins, **161**, 54
lignin characterization by, **161**, 158
^{113}Cd, S-100 protein, **139**, 186
cyanobacterial organic osmolytes, **167**, 528
^1H
in analysis of
initiation factor 3 binding to RNA cloacin fragment, **164**, 252
ribosome structure and function, **164**, 148
calmodulin–peptide interaction analysis
exchange experiments, **139**, 203
nuclear Overhauser experiments, **139**, 204
peptide secondary structure determination, **139**, 205
titration experiments, **139**, 199
high-resolution, 16S rRNA colicin fragments, **164**, 199

lignin characterization by, **161**, 152
S-100 protein, **139**, 182
immobilized cells
methodology, **135**, 526
probe arrangements, **135**, 522
Na^+,K^+-ATPase
nuclear relaxation with paramagnetic probes, **156**, 355
transferred nuclear Overhauser enhancement measurements, **156**, 367
^{31}P, in analysis of cyanobacterial energetic status, **167**, 667
pulsed Fourier transform technique, overview, **161**, 139
5S RNA and fragments, sample preparation for, **164**, 158
sialic acids, **138**, 157
two-dimensional, oligosaccharides
primary structure, **138**, 42
secondary structure, **138**, 58
Nuclease
digestion of
intra-RNA crosslinks, **164**, 293
RNA–protein crosslinks, **164**, 303
Nuclease protection assay
in mapping gene transcripts, **152**, 619
protein–DNA interactions *in vitro*, **152**, 728
Nuclease S_1
in characterization of pro-α2(I) collagen gene mutation, **145**, 213
cutting of hairpin loop in second-strand cDNA synthesis, **152**, 328
digestion of cDNA for analysis of residual hairpin loops, **152**, 333
limited digestion of 5S RNA, **164**, 726
in mapping gene transcripts, **152**, 623
and synthetic oligonucleotides, in RNA mapping, **154**, 87
Nuclease V_1
limited digestion of 5S RNA, **164**, 727
in probing RNA structure, monitoring by primer extension, **164**, 482
Nuclei
GH_3 cell, detection of
Ca^{2+}/calmodulin-dependent protein kinase, **139**, 665
calmodulin-binding proteins, **139**, 664
hepatic, fluorescent dextran influx assay, **141**, 451
isolated, transcription rate analysis, **152**, 238
preparation for RNA isolation and analysis, **152**, 235
RNA extraction, **152**, 236
soybean, isolation
with milk homogenizer, **148**, 539
from protoplasts, **148**, 537
Nucleic acids
autoradiography
^{32}P, **152**, 57
^{35}S, **152**, 55
biotin-labeled, in detection of transcription in transfected cells, **151**, 530
delivery to synchronized cell protoplasts, **149**, 182
encapsulation in liposomes, **149**, 177
fluorography
^{32}P, **152**, 57
^{35}S, **152**, 55
immobilized, hybridization, theory and practice, **152**, 399
liposome-mediated delivery to protoplasts
assessment, **149**, 181
procedure, **149**, 180
microinjection into *Xenopus* oocytes, instrumentation, **152**, 276
probes
generation and use, application of subclones, **152**, 513
radioactive, in screening colonies or plaques, **152**, 415
for screening DNA libraries, overview, **152**, 393
quantification
absorbance methods, **152**, 49
fluorescence methods, **152**, 50
^{32}P-based methods, **152**, 51
Nucleocytoplasmic transport
fluorescence photobleaching assay, **141**, 447
Nucleophiles
for oligosaccharide synthesis, **138**, 375
Nucleoside-diphosphatase
assay, **135**, 155
preparation, **135**, 155
Sepharose-coupled
continuous reaction with, **135**, 160
preparation, **135**, 157
properties, **135**, 157

Nucleoside monophosphates
degradation to glycerol-3-phosphate, **159**, 71
Nucleoside phosphorylase, *see* Purine-nucleoside phosphorylase
Nucleosides
modified, in tRNA, analysis, **155**, 392
5′-Nucleotidase
in cyclic nucleotide phosphodiesterase histo- and cytochemistry, **159**, 478
laminin binding properties, **144**, 506
Nucleotides
affinity-labeled, in 23S rRNA, identification, **164**, 365
HPLC, in analysis of oxidant injury, **163**, 338
Nucleotide sequence
calmodulin gene 5′ end, **139**, 226
calmodulin gene exons, **139**, 232, 239
C5a gene, **162**, 654
elastin gene, **144**, 263
α-isopropylmalate synthase gene, **166**, 422
Kdp-ATPase mutant genes, **157**, 666
Na^+,K^+-ATPase α subunit cDNA, **156**, 389
pAA3.7X *gal* region, **155**, 202
partial, *Aplysia* calmodulin gene, **139**, 289
Phanerochaete chrysosporium DNA, variation between strains, **161**, 219
pKK175-6 cloning region, **153**, 57
pKK232-8 cloning region, **153**, 59
protein S gene, **139**, 388
RNA, computer analysis, **164**, 765
rRNA, in phylogenetic analysis, **164**, 793
16S rRNA, for phylogenetic studies of cyanobacteria, **167**, 132, 138
slow/cardiac troponin C cDNA, **139**, 331, 335
Synechococcus ATP synthase genes, **167**, 798
tRNA, analysis, **155**, 381
Nutrient dosing
intermittent, control of immobilized bacterial cell stability, **137**, 626
Nutrients
cellular, interactions with growth factors, kinetic analysis, **147**, 392
Nutrition
regulation of cellular phosphatidylinositol, analysis, **141**, 234
Nylon
agar gel-supported netting, plant cell entrapment, **135**, 414
alginate gel-supported netting, plant cell entrapment, **135**, 415
isonitrile derivatives, preparation, **135**, 92
microcapsules containing enzymes, preparation, **136**, 69
Nylon tubes
coimmobilization of alcohol dehydrogenase and NAD, **136**, 59
immobilized bioluminescent enzymes, assays with, **137**, 161
immobilized enzymes, routine analysis with, **137**, 288

O

Oat
(1→3),(1→4)-β-D-glucan assay, **160**, 548
H^+-ATPase purification from
root plasma membranes, **157**, 538
tonoplasts, **148**, 123
high-purity plasma membranes, isolation from leaves, **148**, 558
Octaethylene glycol dodecyl monoether
solubilization of
Ca^{2+}-ATPase in monomeric form, **157**, 268
Na^+,K^+-ATPase, **156**, 72
solubilized Na^+,K^+-ATPase, incorporation into liposomes, **156**, 156
2-Octanol
racemic, enzymatic preparation, **136**, 132
Octopine synthase
assay in monocot transformants, **153**, 311
n-Octyl-Sepharose 4B
xanthine oxidase immobilization with, **136**, 258
Odontoblasts
organ cultures for analysis of dentin proteins, **145**, 302
Oligomycin
partial inhibition of Na^+,K^+-ATPase, **156**, 10
Oligonucleotides
adaptors, ligation to cDNA, **152**, 348

AT-rich probes, hybridization to genomic DNA, **152**, 447
binding to rRNA, quantitative estimation, **164**, 449
chemical synthesis, **146**, 24
complex formation with antibodies and ribosomal subunits, **164**, 502
degenerate, in mutagenesis of defined DNA regions, **155**, 558
directed mutagenesis
via gapped duplex DNA, **154**, 350
with M13 vectors, **154**, 382
for modification of human tumor necrosis factor gene, **154**, 403
without phenotypic selection, **154**, 367
5S RNA, **164**, 737
with two primers and ssDNA template, **154**, 329
fluorescent
in DNA sequencing, **155**, 286
synthesis, **155**, 263
linkers
in directional cDNA cloning, **152**, 349
ligation to cDNA, **152**, 344, 351
long probes, gene cloning with, **152**, 443
protective group removal and purification, **146**, 32
purified, analysis, **146**, 34
radiolabeling by homopolymeric tailing, **152**, 342
random-sequence, in determination of consensus sequences, **155**, 568
rRNA-complementary, synthesis and modification, **164**, 499
–rRNA complexes, site-specific hydrolysis, **164**, 440
for screening recombinant DNA libraries
design, **152**, 434
hybridization conditions, **152**, 437
hybridization procedures, **152**, 438
short, in probing ribosome structure and function, **164**, 401
solid-phase sequencing methods, **155**, 301
synthesis
on cellulose disks, **154**, 234
by phosphoramidite method, **154**, 287
by segmented paper method, **154**, 253
synthetic
in cloning of yeast calmodulin gene, **139**, 248
in construction of DNA double strands, **154**, 245
deprotection, **154**, 241, 277, 308
5′-end phosphorylation, **154**, 282
hybridization techniques, **154**, 94
in identification of ligninase cDNA clones, **161**, 228
in isolation of cyanobacterial photosynthetic apparatus genes, **167**, 762
as probes for calmodulin-encoding DNA, **139**, 299
purification, **154**, 243, 279, 309
in screening sarcoplasmic reticulum protein cDNA libraries, **157**, 297
in site-specific mutagenesis of calmodulin gene, **139**, 291
in S_1 nuclease mapping of RNA, **154**, 87
Oligopeptide receptors
on macrophages, neutrophils, and membranes, assay, **162**, 240
Oligosaccharides
asparagine derivatives
attachment to proteins, **138**, 409
incorporation into glutamine residues, **138**, 413
noncovalent coupling to proteins, **138**, 418
chemical synthesis
glycosylating agents for, **138**, 363
nucleophiles for, **138**, 375
overview, **138**, 359
reaction mechanisms, **138**, 377
direct chemical ionization MS, **138**, 71
effective size data, **138**, 84
fibronectin, preparation, **144**, 429
glycophosphosphingolipid
carbohydrate analysis, **138**, 193
purification, **138**, 192
release, **138**, 191
sequence analysis, **138**, 194
immunoaffinity purification, **138**, 307
immunostaining on TLC plates, **138**, 208
lectin affinity chromatography, **138**, 244
perbenzoylated, HPLC, **138**, 94
phytoalexin-eliciting, preparation, **138**, 721
proteoglycan, analysis, **144**, 370
sialyl, *see* Sialyl oligosaccharides

two-dimensional NMR analysis
primary structure, **138**, 42
secondary structure, **138**, 58
Oligosaccharyl sulfatase
assay, **138**, 818
properties, **138**, 824
purification from keyhole limpet, **138**, 819
Oligouronides
with PIIF activity, preparation, **138**, 719
Oncomodulin
antiserum preparation, **139**, 158
immunohistochemical localization, **139**, 162
isolation from Morris hepatoma, **139**, 157
overview, **139**, 209
Oocytes
Xenopus laevis
microinjection of nucleic acids
cell preparation, **152**, 284
instrumentation, **152**, 276
mRNA translation after microinjection, **152**, 288
purification of lectins, **138**, 516
Opsonization
phosphatidylinositol-containing liposomes, **141**, 248
Optical emission spectrometry
inductively coupled plasma
calibration, **158**, 204
instrumentation, **158**, 192, 196
interferences, **158**, 197
optimization, **158**, 198
sample introduction, **158**, 193
system performance, **158**, 200
Orchitis
postvasectomy autoimmune
animal model evaluation, **162**, 467
induction, **162**, 464
Organ cultures
cartilage, for analysis of macromolecular metabolism, **145**, 348
elastin biosynthesis in, **144**, 216; **163**, 668
odontoblasts, for analysis of dentin proteins, **145**, 302
Organelles
immobilization with crosslinked proteins, **152**, 454
metabolite fluxes, analysis by rapid filtration technique, **148**, 675
Organic compounds
volatile, in cyanobacteria, quantitative trace analysis, **167**, 609
Organic synthesis
applications of immobilized cells and enzymes, **136**, 117
Organosolv
pretreatment of biomass materials for enhancing cellulose availability, **160**, 6
Orientation chamber
in assay of leukocyte chemotaxis
chamber description, **162**, 65
procedures, **162**, 67
Oryza sativa, *see* Rice
Oscillatoria limnetica
anoxygenic photosynthetic reactions, assays, **167**, 380
purification of hydrogenases, **167**, 503
Osmolality
effect on light scattering of endoplasmic and sarcoplasmic reticulum vesicles, **157**, 428
Osmosis
reverse, in preparation of metal-free water, **158**, 5
sensitivity of skeletal muscle triads, **157**, 59
Osmotic lysis
pinosomes, **149**, 42
in toxin–antitoxin method for isolating heterokaryons and cell hybrids
description, **151**, 239
optimization, **151**, 244
Osteoblasts
related cells, proliferation, hormonal effects, **145**, 334
Osteogenesis imperfecta
collagen I mRNAs in fibroblasts, analysis, **145**, 223
Ouabain
in assay for Na^+,K^+-ATPase sidedness in liposomes, **156**, 160
binding sites in Na^+,K^+-ATPase, autoradiography, **156**, 428
binding to transverse tubules, **157**, 66
derivatives for Na^+,K^+-ATPase photoaffinity labeling, **156**, 325

tritiated, binding to Na^+,K^+-ATPase
assay in intact cells, **156**, 219, 223
calcium effects, **156**, 223
corroboration of data, **156**, 228
equilibrium binding conditions, **156**, 219
equilibrium binding constant, **156**, 219, 221
potassium effects, **156**, 222
specificity, **156**, 215
Ouchterlony gel diffusion
Evonymus lectin assay by, **138**, 552
Overlay techniques
in analysis of cAMP-dependent protein kinase RII subunit-binding proteins, **159**, 185
Oxalate
assay with enzyme thermistor, **137**, 190
Oxaloacetate decarboxylase
immobilization and assay, **136**, 489
Oxidants
induced cell injury in inflammation, *in vitro* analysis, **163**, 328
Oxidation
bacterial, in treatment of acid mine drainage, **136**, 530
carboxypeptidase Y, **135**, 144
liberation of protein-bound thiols by, **143**, 127
after methylation, in lignin analysis, **161**, 91
1-methylxanthine by immobilized xanthine oxidase, **136**, 260
nitrobenzene-mediated
in lignin analysis, **161**, 89
in lignin characterization, **161**, 106, 117
permanganate-mediated, in lignin characterization, **161**, 107
permanganate–periodate-mediated, in lignin characterization, **161**, 124
xanthine by immobilized xanthine oxidase, **136**, 259
β-Oxidation
metabolites
gradient system for resolution, **166**, 84
preparation for HPLC analysis, **166**, 88
peroxisomal, assay methods, **148**, 520
Oxidoreductases
conformation and activity in reverse micelles, **136**, 208
3-Oxoacyl-CoA thiolase
peroxisomal, assay, **148**, 524
2-Oxo-4-methylpentanoate
enzymatic reduction, **136**, 316
L-2-Oxothiazolidine 4-carboxylate
as intracellular cysteine delivery system, **143**, 319
Oxygen
dissolved, ESR measurements in cyanobacteria
intracellular values, **167**, 675
total values, **167**, 673
effects on
cGMP-binding phosphodiesterase, **159**, 717
cyanobacterial pH gradients, **167**, 682
exchange in cyanobacteria, MS analysis, **167**, 686
transfer rates in celite-immobilized fungal cultures, **135**, 329
uptake rates in
celite-immobilized fungal cultures, **135**, 327
immobilized pepper cells, **135**, 417
Oxygenases
mixed-function, analysis in mitogen-treated lymphocytes, **150**, 50
Oxygen electrodes
in assay for
cysteamine dioxygenase, **143**, 411
cysteine dioxygenase, **143**, 399
methanol oxidase, **161**, 323
protein A, **143**, 402
sulfhydryl oxidase, **143**, 120
taurine dehydrogenase, **143**, 496
vanillate *O*-demethylase, **161**, 295
vanillate hydroxylase, **161**, 275
N-(1-Oxyl-2,2,6,6-tetramethyl-4-piperidinyl)iodoacetamide
preparation, **135**, 503
spin labeling of immobilized chymotrypsin, **135**, 504
1-Oxyl-2,2,6,6-tetramethyl-4-piperidinyl methylphosphonofluoridate
preparation, **135**, 503
spin labeling of immobilized chymotrypsin, **135**, 504

Ozone
  pretreatment of biomass materials for enhancing cellulose availability, **160**, 4

P

P 2420
  in assay of factor XII, **163**, 177
Palmitoylacetylglycerol
  enzymatic hydrolysis, assay, **141**, 311
  preparation, **141**, 306
1-Palmitoyl-2-retinoyl-*sn*-glycero-3-phosphocholine
  synthesis, **149**, 105
Pancreas
  artificial, preparation by islet microencapsulation, **137**, 575
  purification of
    branched-chain-amino-acid aminotransferase
      canine enzyme, **166**, 280
      human enzyme, **166**, 278
      rat enzyme, **166**, 276
    prokallikrein A, **163**, 125
    prokallikrein B, **163**, 119
Pancreatic elastase
  affinity chromatography at subzero temperatures, **135**, 531
Pancreatic ribonuclease
  denatured, sulfhydryl oxidase assay with, **143**, 511
  immobilized, thermal analysis, **137**, 592
  in molecular titration assay of transcript prevalence levels, **152**, 643
  in probing RNA structure, monitoring by primer extension, **164**, 482
  produced in *E. coli*, properties, **153**, 481
  thermal transitions, analysis with immobilized carboxypeptidase, **135**, 597
Panicum miliaceum, *see* Millet
Papain
  immobilized, peptide synthesis with, **136**, 183
  immobilized system, oscillatory phenomena, **135**, 557
  thiol–disulfide interchange in, thiol pK_a values, **143**, 134
Papalucandin
  as glycosylation inhibitor, **138**, 705
Paper chromatography
  inositol phosphates, **141**, 183
  pyrimidine dimers, in analysis of calmodulin role in DNA repair, **139**, 719
  two-dimensional, cyanobacterial carbon metabolites, **167**, 546
Paper electrophoresis
  alanine sulfodisulfane, **143**, 225
  in analysis of glycosphingolipid biosynthesis, **138**, 580
  in assay for α-4-*O*-methyl-D-glucuronidase, **160**, 567
Papillomavirus
  bovine, calmodulin expression vectors based on, analytical methods, **139**, 642
Partitioning
  aqueous two-phase
    inside-out thylakoid vesicles, **148**, 255
    plant mitochondria
      phase system design, **148**, 435
      procedure, **148**, 438
      special techniques, **148**, 440
    plant plasma membranes, **148**, 554, 558
Parvalbumin
  electrophoretic assay, **139**, 765
  enzyme-linked immunosorbent assays, **139**, 766
  fiber type-specific distribution, **139**, 774
  gene cloning from rat, **139**, 317
  immunolocalization in animal tissues, **139**, 552
  mRNA, *in vitro* assay, **139**, 772
  neural control of expression, **139**, 775
  purification from rabbit muscle, **139**, 764
Passive transfer
  adjuvant arthritis in rats, **162**, 353
  contact sensitivity in mice, **163**, 365
  type II collagen arthritis in mice, **162**, 370
Pathogenesis
  chronic arthritis in rabbits, **162**, 383
Pathology
  chronic arthritis in rabbits, **162**, 381
Patulin
  production by immobilized *Penicillium* spp., **136**, 329
PDGF, *see* Platelet-derived growth factor

Pea
amyloplasts, isolation from epicotyls, **148**, 172
chloroplasts
envelope polypeptide analysis, **148**, 200
integrity criteria, **148**, 154
isolation, **148**, 149
mitochondria
chlorophyll-free, isolation from leaves, **148**, 412
purification from leaves, **148**, 417
purification of
β-glucanases, **160**, 384
3-phosphoshikimate 1-carboxyvinyl-transferase, **142**, 350, 353
Pectate lyase
analysis with activity stain, **161**, 330
assay, **161**, 334
Pectinesterase
assay, **161**, 355, 366
chromatography with Spheron ion exchangers, **161**, 388
properties, **161**, 359, 369
purification from
Botrytis cinerea, **161**, 367
Phytophthora infestans, **161**, 356
Pectin lyase
applications, **161**, 354
assay, **161**, 351
chromatography with Spheron ion exchangers, **161**, 388
nomenclature and terminology, **161**, 350
properties, **161**, 354
purification from *Phoma medicaginis*, **161**, 352
Penicillamine
colorimetric assay, **143**, 190
HPLC assay, **143**, 186
Penicillin
G form
assay with optoelectronic sensor, **137**, 346
production by celite-immobilized *Penicillium chrysogenum*, **135**, 328
production by immobilized *Penicillium* spp., **136**, 329
V form, assay with enzyme thermistor, **137**, 208
Penicillin acylase, *see* Penicillin amidase
Penicillin amidase
benzylpenicillin deacylation in aqueous two-phase system, **137**, 665
coupled to cellulose acetate beads
applications, **135**, 291
assay, **135**, 289
immobilized, penicillin synthesis with, **136**, 284
Penicillinase, *see* β-Lactamase
Penicillium
immobilized, patulin and penicillin production, **136**, 329
P. chrysogenum
celite-immobilized spores
bubble column fermentation, **135**, 326
fermentation in 200-liter tower fermentor, **135**, 330
oxygen transfer rates, **135**, 329
oxygen uptake rates, **135**, 327
penicillin production, **135**, 328
preparation, **135**, 322
shake flask cultures, **135**, 324
single-bead cultures, **135**, 323
mitochondrial cloning vectors for, **153**, 370
purification of ATP-sulfurylase and APS kinase, **143**, 344
P. funiculosum
immobilization, **137**, 689
production of cellulolytic enzymes, **137**, 690
P. pinophilum
cultivation, **160**, 398
purification of cellobiohydrolases, **160**, 188, 400
Pentafluorobenzenesulfonyl chloride
activation of agarose, **135**, 82
Pepsin
coupling to 1,1′-carbonyldiimidazole-activated support, **135**, 115
immobilized, in analysis of protein digestibility changes, **135**, 601
Peptidases
immobilized, in analysis of protein digestibility changes, **135**, 601
as markers for chromosome detection in somatic cell hybrids, **151**, 189
Peptide-N^4-(*N*-acetyl-β-glucosaminyl)asparagine amidase
assay, **138**, 771

properties, **138**, 775
purification from *Flavobacterium*, **138**, 772
Peptide:*N*-glycosidase F
deglycosylation of polysialylated neural cell adhesion molecule, **138**, 636
Peptide mapping
collagens
with cyanogen bromide, **145**, 175, 188
by limited proteolysis, **145**, 177
two-dimensional, proteoglycan link protein, **144**, 409
Peptides
calmodulin-binding, myosin light chain kinase
effects on calmodulin-dependent enzymes, **139**, 125
interactions with calmodulin, **139**, 124
preparation, **139**, 119
structure–function studies, **139**, 123
synthesis, **139**, 120
calmodulin-binding sequences
characterization, **139**, 463
identification, **139**, 456
–calmodulin interactions, NMR analysis, **139**, 197
chemotactic
agonist CHO-Met-Leu-Phe-OH, synthesis, **162**, 135
antagonist Boc-Phe-Leu-Phe-Leu-Phe-OH, synthesis, **162**, 138
preparation from fibronectin, **162**, 214
connective tissue-activating, *see* Connective tissue-activating peptide
disulfides, assay with 2-nitro-5-thiosulfobenzoate, **143**, 115
elastin-derived, as fibroblast chemoattractants, **163**, 702
hydroxyproline-containing, as fibroblast chemoattractants, **163**, 701
immunogenic, calmodulin antibody preparation with, **139**, 394
model, for structure–function analysis of proteins, **154**, 473
Na^+,K^+-ATPase-derived, phosphorylation and separation, **156**, 295
Na^+,K^+-ATPase α subunit, preparation, **156**, 91
proline-containing, sequential hydrolysis with immobilized aminopeptidases, **136**, 170
signal, sequences from gram-positive bacteria, **153**, 510
synthetic
in assay of growth factor-stimulated kinases, **146**, 353
inhibiting cAMP-dependent protein kinase
assay, **159**, 181
description, **159**, 173
preparation, **159**, 177
purity criteria, **159**, 179
specificity, **159**, 180
structural requirements, **159**, 175
as probes for cell surface receptor function, **144**, 479
T-cell receptor D_β/J_β, synthetic, antibody production, **150**, 647
tetanus toxin-derived, preparation, **165**, 89
Peptide synthesis
deblocking with immobilized carboxypeptidase Y, **136**, 157
with immobilized proteases, **136**, 284
kinetic approach, **136**, 179
thermodynamic approach, **136**, 184
Peptidoglycans
cyanobacterial cell wall, isolation, **167**, 178
Peptidyltransferase
in deuterated ribosomal 50S subunit, assay, **164**, 146
functional domain in ribosomes, analysis, **164**, 352, 358
photoaffinity labeling in ribosomes, **164**, 361
Perbenzoylation
oligosaccharides, **138**, 97, 99
Perchloric acid
in purification of hemopexin, **163**, 543
Perfusion
rat hindlimb, for branched-chain amino acid metabolism studies, **166**, 476
Peritoneum
rat, long-term glucose sensor implantation, **137**, 68
Periwinkle
immobilized cells, NMR studies, **135**, 526
Permanganate
in spectrophotometric assay of thiols, **143**, 60

Permeability
gas, cyanobacterial heterocysts, measurement, **167**, 665
ionic
cyanobacteria, ESR analysis, **167**, 512
endoplasmic reticulum vesicles, measurements, **157**, 417
sarcoplasmic reticulum vesicles, measurements, **157**, 417, 437
solute, cyanobacteria, measurement with gas vesicles, **167**, 664
Permeabilization
adipocytes with digitonin, **159**, 194
Anacystis nidulans to electrolytes, **167**, 259
bacteria, **137**, 638
cellular, for analysis of hormone-regulated phospholipid turnover, **141**, 112
fungal cells, **137**, 638
plant cells, **137**, 639
Permethylation
glycolipids, **138**, 16
Peroxidase
coupling to antikininogen IgG, **163**, 254
ethylenediamine adducts, coupling to carboxyalkyl agarose, **135**, 145
Peroxisomes
mung bean, isolation from hypocotyls, **148**, 517
nonglyoxysomal, isolation on Percoll gradients, **148**, 519
β-oxidation, assay methods, **148**, 520
yeast
entrapment with
photocrosslinkable resin prepolymers, **135**, 234
urethane prepolymers, **135**, 242
immobilization in crosslinked proteins, **135**, 471
prepolymer-entrapped, applications, **135**, 245
Pertussis toxin
ADP-ribosyltransferase activity, comparison with cholera toxin activity, **165**, 242
bioassay, **165**, 125
gel electrophoretic analysis, **165**, 125
as probe of
G protein role in phagocyte chemotactic activation, **162**, 276
N protein-mediated polyphosphoinositide breakdown, **141**, 266
purification, **165**, 122
Pesticides
effects on lipid synthesis in plant cell membranes
general analytical methods, **148**, 667
specific analytical methods, **148**, 671
Petunia
genetically altered plants, regeneration, **153**, 329
protoplast preparation, transformation, and culture, **153**, 327
pH
effect on bacterial L-aspartic acid production in reactor, **136**, 469
intracellular
fluorescence assay in neutrophils, **162**, 296
growth factor-induced changes, measurement, **147**, 415
measurement by acid uptake, **146**, 391
oscillations in immobilized papain system, measurement, **135**, 558
Phagocytes
isolation of lysosomes, **157**, 594
Phagocytosis
liposomes by macrophages, **149**, 131
Phagokinetic assay
capillary endothelial cell migration, **147**, 136
Phallacidin
in staining assay for actin behavior in leukocytes, **162**, 268
Phanerochaete chrysosporium
DNA preparation and characterization, **161**, 211
growth and harvest, **161**, 241
immobilization, **137**, 689
lignin biodegradation, assays
based on ethylene production, **161**, 79
with polymeric dyes, **161**, 74
lignin demethoxylation, assays, **161**, 83
maintenance and spore production in culture, **161**, 239
mRNA preparation and characterization, **161**, 221
production of lignin peroxidase, **137**, 690
purification of
glucose oxidase, **161**, 309
ligninase, **161**, 243

manganese peroxidase, **161**, 259, 267
methanol oxidase, **161**, 324
pyranose 2-oxidase, **161**, 318
Phase diagram
pseudo-ternary, determination, **136**, 217
Phaseolus aureus, *see* Mung bean
Phaseolus vulgaris, *see* Kidney bean
Phase systems
for purification of mitochondria, design, **148**, 435
o-Phenanthroline
–copper, in detection of Na^+,K^+-ATPase subunit interactions, **156**, 347
sec-Phenethyl alcohol
racemic, enzymatic preparation, **136**, 132
Phenol
degradation with immobilized *Candida tropicalis*, **137**, 706
in large- and small-scale nucleic acid extraction, **152**, 33
Phenolphthalein bissulfate
arylsulfatase assay with, **143**, 215
Phenol sulfotransferase, *see* Aryl sulfotransferase
Phenothiazine
–calmodulin adduct
preparation, **139**, 409
properties, **139**, 413
specificity for calmodulin-regulated proteins, **139**, 417
Phenotyping
haptoglobin, **163**, 453
Phenylalanine
L-form, postprephenate biochemical pathways to, **142**, 472
production by immobilized *E. coli*, **136**, 497
Phenylalanine ammonia-lyase
assay, **142**, 244, 249
overview, **142**, 242
properties, **142**, 252
purification from
Rhodotorula glutinis, **142**, 246
soybean, **142**, 249
reactor, preparation from dialyzer, **137**, 486
Phenylalanine 4-monooxygenase
amino acid composition
bovine hepatic enzyme, **142**, 42
chromobacterial enzyme, **142**, 49
human hepatic enzyme, **142**, 33
assay, **142**, 5, 18, 28, 35, 50
properties
bovine hepatic enzyme, **142**, 40
chromobacterial enzyme, **142**, 48, 55
human hepatic enzyme, **142**, 32
rat hepatic enzyme, **142**, 13, 16, 26
purification from
bovine liver, **142**, 39
Chromobacterium violaceum, **142**, 46, 52
human liver, **142**, 29
rat liver, **142**, 8, 19
tetrahydropterin as cofactor, **142**, 15
Phenylethanolamine *N*-methyltransferase
assay, **142**, 656
catecholamine radioimmunoassay with, **142**, 559
preparation from
bovine adrenal gland, **142**, 554
rabbit adrenal gland, **142**, 657
properties, **142**, 658
Phenylobacterium immobile
purification of arogenate dehydrogenase, **142**, 516
Te-Phenyltellurocysteine
preparation, **143**, 245
Te-Phenyltellurohomocysteine
preparation, **143**, 245
Pheophytins
formation from chlorophylls, **148**, 374
quantitation, **148**, 375
specific absorption coefficients, **148**, 376
spectral characteristics, **148**, 378
Pheresis
low-density lipoproteins
procedure, **137**, 474
system components, **137**, 468
system description, **137**, 467
pH gradients
in cyanobacteria, ESR measurements, **167**, 517, 677
Phoma
P. hibernica, purification of cellulase–hemicellulase complex, **160**, 333
P. medicaginis var. *pinodella*, purification of pectin lyase, **161**, 352
Phorbol 12,13-dibutyrate
binding assay
in intact cells, **141**, 278
by mixed micellar method, **141**, 287
in subcellular fractions, **141**, 282

Phorbol esters
in analysis of protein kinase C-mediated phosphorylation in intact cells, **141**, 414
effects on inositol phosphate accumulation in hippocampal slices, **141**, 199
Phorbol 12-myristate 13-acetate
and A23187, synergistic effects in pituitary cells, **141**, 429
Phormidium
P. laminosum, isolation of oxygen-evolving photosystem II particles, **167**, 273
P. luridum, purification of C-phycocyanin, **167**, 300
Phosphatases
analysis in mitogen-treated lymphocytes, **150**, 75
Na^+,K^+-ATPase-associated
assay, **156**, 109
description, **156**, 7
Phosphate
exchange reactions in sarcoplasmic reticulum vesicles, **157**, 183, 197
Na^+,K^+-ATPase-liberated, assay, **156**, 107
^{32}P-labeled
in analysis of cellular phosphoinositide pools, **141**, 85
exchange with ATP in clathrin-coated vesicles, assay, **157**, 636
incorporation into inositol phospholipids, **141**, 170
purification, **157**, 194
P_i–P_i exchange in mitochondria, analytical technique, **148**, 681
production after Ca^{2+}-ATPase phosphorylation, **157**, 168
Phosphatidate cytidylyltransferase
in plant endoplasmic reticulum
assay, **148**, 591
properties, **148**, 592
Phosphatidic acid
quantitation in *sn*-1,2-diacylglycerol assay, **141**, 296
Phosphatidylcholine
adenylate cyclase reconstitution with, **139**, 780
assay with enzyme thermistor, **137**, 223
liposomes with, tumoricidal effects, **141**, 459
preparation for Na,K-pump reconstitution by cholate dialysis, **156**, 128
Phosphatidylethanolamine
biotinylated
liposome preparation with, **149**, 122
synthesis, **149**, 121
Phosphatidylethanolamine methyltransferase
in plant endoplasmic reticulum
assay, **148**, 589
properties, **148**, 590
Phosphatidylethanolamine:L-serine phosphatidyltransferase
in plant endoplasmic reticulum
assay, **148**, 590
properties, **148**, 591
Phosphatidylglycerol
chloroplast-synthesized, degradation, **148**, 341
synthesis by
chloroplast envelopes, **148**, 342
intact chloroplasts, **148**, 339
type II pneumonocytes, alteration by manipulation of
inositol availability, **141**, 229
intracellular CMP, **141**, 224
Phosphatidylglycerophosphatase
in plant endoplasmic reticulum
assay, **148**, 593
properties, **148**, 593
Phosphatidylinositol
cellular, nutritional regulation, analysis, **141**, 234
liposomes with, tumoricidal effects, **141**, 459
synthesis by type II pneumonocytes, alteration by manipulation of
inositol availability, **141**, 229
intracellular CMP, **141**, 224
turnover in macrophages, modulation by liposomes, **141**, 244
Phosphatidylinositol 4,5-bisphosphate
breakdown
chemoattractant receptor-induced, G protein role, **162**, 276
in neutrophils, N protein role, **141**, 261
^{32}P-labeled
hydrolysis, **141**, 95
preparation, **141**, 95

Phosphatidylinositol:*myo*-inositol phosphatidyltransferase
in plant endoplasmic reticulum
assay, **148**, 595
properties, **148**, 596
Phosphatidylinositol kinase
assay, **141**, 212
properties, **141**, 215
purification from rabbit reticulocytes, **141**, 213
Phosphatidylinositol 4-phosphate
^{32}P-labeled
hydrolysis, **141**, 95
preparation, **141**, 95
Phosphatidylserine
biotinylated
liposome preparation with, **149**, 122
synthesis, **149**, 121
Phosphodiesterase
calmodulin-dependent, antibodies
characterization, **139**, 509
production, **139**, 506
cGMP-binding
detection and basic properties, **159**, 712
modulation, **159**, 716
purification from
lung, **159**, 722
platelets, **159**, 715
cGMP-inhibited and -stimulated forms from bovine heart
assay, **159**, 688
comparison, **159**, 698
identification, **159**, 689
purification, **159**, 692
hormone-regulated, assay in GH_3 cell membranes, **141**, 124
isozymes, analysis with monoclonal antibodies, **139**, 524
5′-Phosphodiesterase
Eupergit-immobilized
preparation, **136**, 518
properties, **136**, 518
5′-ribonucleotide production with, **136**, 521
Phosphodiesterase inhibitors
effects on transmembrane Ca^{2+} signaling, **139**, 574
Phosphoenolpyruvate
enzymatic synthesis, **136**, 268
Phosphofructokinase
immobilized system
global behavior, numerical prediction, **135**, 564
oscillatory phenomena, **135**, 560
interaction with calmodulin
binding assays, **139**, 749
enzyme activity assays, **139**, 754
enzyme polymerization studies, **139**, 760
Phosphoglucomutase
as marker for chromosome detection in somatic cell hybrids, **151**, 190
6-Phosphogluconate dehydrogenase
as marker for chromosome detection in somatic cell hybrids, **151**, 191
Phosphoglucose mutase
immobilized, UDP-glucose synthesis with, **136**, 279
Phosphoglycerides
in plant mitochondrial membranes
content, **148**, 479
fatty acyl composition, **148**, 481
metabolism, **148**, 483
transmembrane distribution, **148**, 482
synthesis in plant endoplasmic reticulum, **148**, 585
Phosphoglycolate phosphatase
as marker for chromosome detection in somatic cell hybrids, **151**, 191
Phosphoinositide kinase
membrane-associated, assay, **141**, 108
Phosphoinositides
breakdown in salivary gland cell-free system, hormonal activation, **141**, 255
cellular metabolism
assay of subcellular sites, **141**, 92
TRH-stimulated, analysis, **141**, 100
cellular pools, analysis with
myo-[^{3}H]inositol, **141**, 89
$^{32}P_i$, **141**, 85
hormone-regulated turnover
in membranes, **141**, 122
in permeable cells, **141**, 118
Phospholamban
amino acid composition, **157**, 366
assay, **157**, 365
immunochemical studies, **157**, 144
monoclonal antibodies
in assessment of phospholamban role in membrane vesicles, **139**, 530
purification, **139**, 520

phosphorylation
calmodulin-dependent, effect on cardiac sarcoplasmic reticulum Ca^{2+}-ATPase, **157**, 134
cAMP-dependent, effect on cardiac sarcoplasmic reticulum Ca^{2+}-ATPase, **157**, 128
properties, **157**, 127
purification from cardiac sarcoplasmic reticulum, **157**, 138, 360
structural characteristics, **157**, 145
Phospholipase A_2
bacterial, assay, **165**, 299
inhibition by calcium-binding protein CAB-27, **139**, 78
inhibitory lipocortins, analytical methods, **163**, 23
Phospholipase C
clostridial
hemolytic assay, **165**, 294
titrimetric assay, **165**, 297
turbidimetric assay, **165**, 296
phosphoinositide-specific, membrane-associated, assay, **141**, 108
purification from *Clostridium perfringens*, **165**, 91
Phospholipase D
bacterial, assay, **165**, 301
Phospholipids
asymmetry in loaded erythrocytes
bioassay, **149**, 288
dye-removal assay, **149**, 286
maintenance, **149**, 292
staining assay, **149**, 282
in hepatic extracts, TLC, **141**, 237
in neutrophil membranes, chemoattractant-induced metabolism, assay, **162**, 273
planar membranes, preparation for functional analysis of channels, **148**, 465
plant-derived, molecular species, HPLC, **148**, 650
in plant membranes, pesticide effects, **148**, 671
in sarcoplasmic reticulum, content modification, **157**, 369
synthesis in plant endoplasmic reticulum, **148**, 578
as synthetic membrane materials for hemoglobin encapsulation, **149**, 187
transfer protein, purification from spinach, **148**, 661
vesicles
incorporation of H^+-ATPase, **157**, 525
Na^+,K^+-ATPase reconstitution into, **156**, 157
Phosphomannosyl receptor 2
assay, **138**, 504
properties, **138**, 508
purification from bovine testis, **138**, 506
Phosphopeptides
S6, electrophoretic analysis, **146**, 373
Phosphophoryns
dentin-associated
phosphoserine analysis, **145**, 299
purification and characterization, **145**, 296
Phosphoprotein phosphatase
analysis with calmodulin derivatives, **159**, 605
calmodulin-dependent
antibodies
characterization, **139**, 509
production, **139**, 506
holoenzyme reconstitution, **139**, 86
subunit isolation, **139**, 80
high-molecular-weight
properties, **159**, 375
purification from rabbit muscle, **159**, 368
inhibitor-1 and inhibitor-2
assay, **159**, 427
properties, **159**, 436
purification from rabbit muscle, **159**, 430
inhibitor-1 phosphorylated and dephosphorylated forms, copurification, **159**, 435
Mg^{2+}-dependent
assays, **159**, 438
properties, **159**, 444
purification from rat liver, **159**, 440
related enzymes, **159**, 445
resistant substrates, preparation, **159**, 346
serine/threonine-specific, assay, **159**, 341
type 1
assay, **159**, 392
properties, **159**, 406
purification from rabbit muscle, **159**, 395, 404

type 2A
assay, **159**, 392
properties, **159**, 406
purification from rabbit muscle, **159**, 395, 400
type 2B, calcium-dependent
assay, **159**, 409
properties, **159**, 414
purification from rabbit muscle, **159**, 411
type 2C, Mg^{2+}-dependent
assay, **159**, 417
purification from
rabbit liver, **159**, 423
rabbit muscle, **159**, 418
types C-I and C-II
assays, **159**, 359
properties, **159**, 358
purification from rabbit muscle, **159**, 360
tyrosine-specific, assay, **159**, 340
in yeast cAMP cascade system mutants, assay, **159**, 41
Phosphoproteins
M_r 80,000, as marker for protein kinase C activation, **141**, 417
^{32}P-labeled, preparation for protein phosphatase assays, **159**, 344
Phosphoramidites
derivatives, synthesis for DNA sequencing, **155**, 272
in synthesis of oligonucleotides, **154**, 287, 314
N-(5′-Phosphoribosyl-1-)anthranilate
preparation, **142**, 388
Phosphoribosylanthranilate isomerase–indoleglycerol-phosphate synthase
assays, **142**, 389
properties, **142**, 394
purification from *E. coli*, **142**, 392
substrate preparation, **142**, 387
5-Phospho-α-D-ribosyl pyrophosphate
enzymatic synthesis with ATP regeneration, **136**, 278
Phosphorothioate
reduction of disulfides, **143**, 256
Phosphorus
atomic absorption spectrometry, **158**, 140
^{32}P
in assay for nucleic acids, **152**, 51
in autoradiography and fluorography of nucleic acids, **152**, 57
labeled compounds, safety considerations, **152**, 25
labeling of ribosomal subunits, **164**, 289
Phosphorylase *a*
^{35}S-labeled
preparation from phosphorylase *b*, **159**, 348
as substrate for protein phosphatases, **159**, 352
Phosphorylase kinase
autophosphorylation, role in enzyme regulation, **139**, 690
in thiophosphorylation of phosphorylase *b*, **159**, 348
Phosphorylase phosphatase
assay, **159**, 379
characterization, **159**, 386
purification from rabbit muscle, **159**, 381
Phosphorylation
auto-, *see* Autophosphorylation
back-door, Na^+,K^+-ATPase, **156**, 121
assay, **156**, 121
data analysis, **156**, 122
branched-chain 2-keto acid dehydrogenase, analysis, **166**, 10
branched-chain α-keto acid dehydrogenase, inhibition by α-chloroisocaproate, **166**, 121
Ca^{2+}-ATPase
assay, **157**, 230
fluorescence assay, **157**, 215
in sarcoplasmic reticulum vesicles, **157**, 168, 192
cAMP-dependent protein kinase regulatory subunit in intact cells, **159**, 139
cAMP receptor *in vivo*, **159**, 277
front-door, Na^+,K^+-ATPase, **156**, 121
insulin receptors in cell-free system, **150**, 720
Kdp-ATPase, **157**, 677
oligonucleotide 5′-end, **154**, 282
peptides derived from Na^+,K^+-ATPase, **156**, 295
phospholamban
calmodulin-dependent, effect on cardiac sarcoplasmic reticulum Ca^{2+}-ATPase, **157**, 134

cAMP-dependent, effect on cardiac sarcoplasmic reticulum Ca^{2+}-ATPase, **157**, 128
photo-, *see* Photophosphorylation
protein kinase C-mediated, analysis in intact cells, **141**, 412
ribosomal protein S6
during mitogenic response, analysis, **164**, 575
multiple phosphorylated forms, electrophoretic analysis, **146**, 362
sperm proteins, ^{32}P incorporation assay, **139**, 818
stimulation by PDGF-like proteins, **147**, 37
thio-, *see* Thiophosphorylation
tyrosine 3-monooxygenase by protein kinases, **142**, 81
Phosphoserine
in dentin phosphophoryns, assay, **145**, 299
3-Phosphoshikimate 1-carboxyvinyltransferase
assay, **142**, 343, 349
properties, **142**, 346, 352
purification from
E. coli, **142**, 344
pea, **142**, 350, 353
Photinus-luciferin 4-monooxygenase (ATP-hydrolyzing)
coimmobilized with other enzymes, bioluminescent assays with, **136**, 82
Photoaffinity labeling
in analysis of lymphoid cell β-adrenergic receptors, **150**, 492
cAMP-binding protein from *Trypanosoma*, **159**, 292
cAMP receptor, with 8-N_3-[^{32}P]cAMP, **159**, 269
cGMP phosphodiesterase noncatalytic binding sites, **159**, 730
Na^+,K^+-ATPase, with ATP analogs, **156**, 312
peptidyltransferase in ribosomes, **164**, 361
Photolabeling
hydrophobic, membrane-penetrating toxins with photoreactive lipids, **165**, 347
Photolysis
bromodeoxyuridine, in analysis of calmodulin role in DNA repair, **139**, 723
ultraviolet, crosslinked collagen polypeptides, **144**, 138
Photometric assay
inositol 1-phosphate, **141**, 129
plasminogen activator, **163**, 305
plasminogen activator inhibitor, **163**, 307
Photophosphorylation
reconstitution by F_1-ATPase, assay, **167**, 359
Photosomes
chromophore selection for, **149**, 100
light-induced release of contents, **149**, 108
preparation, **149**, 106
Photosynthesis
anoxygenic
in cyanobacteria, **167**, 581
in *Oscillatoria limnetica*
carbon fixation, assay, **167**, 381
effects of inhibitors, **167**, 383
hydrogen evolution, assay, **167**, 382
induction, **167**, 380
nitrogen fixation, assay, **167**, 382
associated oxygen exchange, MS analysis in cyanobacteria, **167**, 686
Photosystem I
associated iron–sulfur proteins, EPR analysis in cyanobacteria, **167**, 432
reaction center complexes, preparation from *Synechococcus*, **167**, 264
Photosystem II
calcium effects, analysis in cyanobacteria, **167**, 280
oxygen-evolving preparations
isolation from
Phormidium laminosum, **167**, 273
Synechococcus, **167**, 267
properties, **167**, 276
–phycobilisome complexes, preparation from
Porphyridium, **167**, 286
Synechococcus, **167**, 289
reaction center complexes, preparation from *Synechococcus*, **167**, 266
reaction center mutations, construction in *Synechocystis* 6803, **167**, 766

pH-stat assay
aromatic-L-amino acid decarboxylase, **142**, 181
sulfatases, **143**, 361
o-Phthalaldehyde
derivatized cysteine and cystine, HPLC, **143**, 141
Phycobiliproteins
cyanobacterial
assay, **167**, 321
bilin prosthetic groups
spectroscopic properties, **167**, 302
structure, **167**, 302
electrophoretic analysis, **167**, 325
overview, **167**, 291
partial purification, **167**, 324
properties, **167**, 303
purification from different species, **167**, 295
Phycobilisomes
characterization, **167**, 309
constituents
absorption spectra, **167**, 319
assays, **167**, 322
isolated
buoyant density, **167**, 317
fluorescence spectra, **167**, 315
light scattering, **167**, 314
–photosystem II complexes, preparation from
Porphyridium, **167**, 286
Synechococcus, **167**, 289
polypeptide composition and topology, **167**, 311
preparation from
Cyanidium caldarium, **167**, 308
cyanobacteria, **167**, 305, 323
Fremyella diplosiphon, **167**, 307
Porphyridium aerugineum, **167**, 308
Pseudoanabaena spp., **167**, 307
red algae, **167**, 305
Synechococcus, **167**, 306, 308
Phycocyanin
absorption spectra, **167**, 319
assay in cyanobacterial cells, **167**, 322
purification from
Agmenellum quadruplicatum, **167**, 301
Anabaena spp., **167**, 295
Fremyella diplosiphon, **167**, 298
Phormidium luridum, **167**, 300
Phycoerythrin
absorption spectra, **167**, 319
assay in cyanobacterial cells, **167**, 322
purification from *Fremyella diplosiphon*, **167**, 298
Phycoerythrocyanin
purification from *Anabaena*, **167**, 295
Phycomyces
thylakoid membrane preparation from, **167**, 333
Phytoalexins
elicitor, preparation from soybean, **138**, 721
Phytoene
conversion to carotenoids *in vitro*, **167**, 332
Phytohemagglutinin
E_4, glycopeptide chromatography with, **138**, 249
in erythrocyte-mediated microinjection, **149**, 295
L_4, glycopeptide chromatography with, **138**, 251
as lymphocyte locomotion activator, **162**, 188
Phytophthora infestans
purification of pectinesterase, **161**, 356
Phytostat
sycamore cell suspension cultures, **148**, 14
Pineal gland
ovine, purification of arylalkylamine *N*-acetyltransferase, **142**, 585
rat, assay of hydroxyindole *O*-methyltransferase, **142**, 591
Pinosomes
osmotic lysis, **149**, 42
Pisum sativum, see Pea
Pituitary
anterior, dispersed cells
calcium channels, maitotoxin effects, **141**, 63
preparation from rat, **141**, 65
bovine, purification of glial growth factor, **147**, 220
cultured cells
preparation, **141**, 429
synergistic effects of protein kinase C activators and A23187, **141**, 429

rat
cellular inositol phospholipid turnover, assay with Li^+, **141**, 149
extraction of inositol phospholipids, **141**, 165
pK_a
determination for thiols in interchange reactions with disulfides, **143**, 132
Placenta
human, purification of
fibronectin, **144**, 421
fibronectin and vitronectin receptors, **144**, 485
glucocerebrosidase, **149**, 29
L-methylmalonyl-CoA mutase, **166**, 410
monoamine oxidase A, **142**, 631
transferrin receptor, **147**, 243
Plague murine toxin
preparation by
gel electrophoresis, **165**, 165
molecular sieve chromatography, **165**, 164
subunit components, preparation, **165**, 166
Plants, *see also specific plants*
cell immobilization after permeabilization, **137**, 639
cell membranes
electron microscopy, **148**, 599
lipid synthesis, pesticide effects, **148**, 667
sterols, isolation, identification, and biosynthesis, **148**, 632
cell permeabilization, **137**, 639
glycolipid and phospholipid molecular species, HPLC, **148**, 650
green, cell culture initiation from
nonphotosynthetic cells, **148**, 18
photosynthetic cells, **148**, 24
immunolocalization of calmodulin, **139**, 478
mitochondria, isolation from
CAM species, **148**, 431
C_3 species, **148**, 422
C_4 species, **148**, 425
monocot, transformation by *Agrobacterium*, detection, **153**, 305
plasma membranes, isolation, **148**, 542
sulfometuron methyl-resistant mutants, isolation, **166**, 104
tissues, application of NMR methods, **148**, 683
transformation
with binary Ti vectors, **153**, 292
vectors for, **153**, 253, 277
woody
extract and moisture contents, assay, **161**, 112
total lignin content, assay, **161**, 113
Plaque assay
hemolytic, *see* Hemolytic plaque assay
Plasma, *see also* Blood, Serum
amino acids, enzyme reactors for removal, **137**, 479
bovine, purification of factor XII, **163**, 55
branched-chain α-keto acids, GLC assay, **166**, 39
catecholamine metabolites, GC–MS assay, **142**, 577
catecholamines
enzymatic radioimmunoassay, **142**, 559, 568
HPLC assay, **142**, 521
contact activation patterns, fluorometric assay, **163**, 80
cystamine and cysteamine, enzymatic assay, **143**, 150
cysteinesulfinic acid, hypotaurine, and taurine, HPLC, **143**, 155
enzyme–albumin conjugates, clearance, **137**, 572
exudation during inflammation, measurement, **162**, 304
human, purification of
α_1-acid glycoprotein, **163**, 420
α_1-antitrypsin, **163**, 407
arginine carboxypeptidase, **163**, 188
C1, **162**, 632
C2, **162**, 634
C3, **162**, 596, 648
C4, **162**, 596, 633
C5, **162**, 596
C5b–6, **162**, 557
C7, **162**, 561
C8, **162**, 564
C9, **162**, 573

ceruloplasmin, **163**, 446
C$\bar{1}$ inhibitor, **163**, 182
dipeptidyl carboxypeptidase I, **163**, 202
dopamine β-monooxygenase, **142**, 604
factor IX, **137**, 463
factor XII, **163**, 72
factor D, **162**, 650
factor P, **162**, 652
gangliosides, **138**, 300
haptoglobin, **163**, 457
α_2-HS glycoprotein, **163**, 433
kallikrein, **163**, 93
kininogens, **163**, 241
prekallikrein, **163**, 89
leakage in Arthus reaction, measurement, **162**, 481
rat, purification of
fibrinogen, **163**, 476
T-kininogen, **163**, 277
thiols, HPLC/fluorometric assay, **143**, 70
Plasma membranes
associated proteoglycans, analysis, **144**, 394
calcium-transporting systems, properties, **157**, 4
cyanobacterial
cytochrome-*c* oxidase
identification, **167**, 443
oxidation rates, **167**, 441
energization, measurement, **167**, 371
isolation, **167**, 245, 437
EHS tumor, laminin receptor isolation, **144**, 497
erythrocyte
electric perforation, **149**, 249
preparation, **157**, 341
resealing after electric perforation, **149**, 253
eukaryotic, interaction with hemolysin, **165**, 203
GH_3 cell
phosphoinositide kinases, assay, **141**, 111
phosphoinositide phospholipase C, assay, **141**, 109
preparation, **141**, 108
Neurospora crassa, preparation, **157**, 563
neutrophil
chemoattractant-induced phospholipid metabolism, assay, **162**, 273
isolation, **141**, 262; **162**, 272
leukotriene B_4 receptors, assay, **163**, 340
N protein-mediated polyphosphoinositide breakdown, **141**, 269
phospholipid breakdown, G protein role, **162**, 278
plant
high-purity, preparation by phase partition, **148**, 558
isolation procedures, **148**, 548
membrane markers for isolation, **148**, 543
surface labeling by
concanavalin A, **148**, 572
diazotized sulfanilic acid, **148**, 571
enzymatic iodination, **148**, 570
lanthanum, **148**, 573
miscellaneous approaches, **148**, 575
positively charged beads, **148**, 574
pyridoxal phosphate-tritiated borohydride, **148**, 572
UDPglucose, **148**, 574
renal, Na^+,K^+-ATPase latency in, **156**, 30
vesicles, sodium–calcium exchange, measurement, **157**, 505
yeast, isolation, **157**, 518, 530
Plasma perfusion system
with immobilized protein A
in cancer treatment, **137**, 513
cleaning and sterilization, **137**, 513
construction, **137**, 502
Plasmapheresis
–enzyme reactor system, construction, **137**, 489
Plasmids
blotting on nitrocellulose, **151**, 424
CAT gene-containing
preparation, **151**, 383
transfer into eukaryotic cells, **151**, 387
chimeric, life cycle, **155**, 140
coded rRNA, maxicell analysis, **164**, 691
conjugal transfer to cyanobacteria, **153**, 232; **167**, 747
cyanobacterial, listing, **167**, 816

for expression of
bovine growth hormone, **153**, 393
calmodulin
applications, **139**, 314
construction, **139**, 307
overview, **139**, 303
human C5a, **162**, 657
human interferon-γ, **153**, 423
human interleukin-2, **153**, 427
human proinsulin, **153**, 386
human tumor necrosis factor, **153**, 424
5S rRNA, description, **164**, 716
full-length cDNA cloning in, **154**, 28
large- and small-scale isolation for cloning applications, **152**, 164
linearized, homopolymeric tailing with dC, **152**, 340
liposome-mediated transfer and expression in tobacco protoplasts, **148**, 42
multicopy, DNA segments cloned into, Tn*5* mutagenesis, **154**, 178
pAA-PZ, bifunctional, construction for DNA sequencing, **155**, 217
pAA3.7X, in DNA sequencing, **155**, 180
pBR322, in DNA sequencing, **155**, 139
pDS5, in transcription–translation system for protein analysis, **155**, 418
pEMBL, single-stranded, in DNA sequencing, **155**, 111
pGA482, restriction map, **153**, 295
pGA492, restriction map, **153**, 296
pGA580, restriction map, **153**, 298
pGSH160, restriction map, **153**, 286
pGSJ280, restriction map, **153**, 286
pGV451
biological properties, **155**, 234
and derived vectors, in DNA sequencing, **155**, 237
pGV941, construction, **153**, 284
pHE6, with λ P_L promoter and cI857 repressor, construction, **153**, 482
pJO184, bacterial–yeast shuttle, construction, **153**, 47
pJRD158, *E. coli*, construction, **153**, 40
pJRD184, *E. coli*, construction, **153**, 45
pJRD215, wide-host-range, construction, **153**, 48
pKC420, description, **153**, 171
pKC513, description, **153**, 171
pKC531, description, **153**, 172
pKC561, description, **153**, 170
pKK5-1, in 5S RNA production, **164**, 159
pKK175-6 and pKK232-8, in selection and analysis of prokaryotic promoters, **153**, 54
pKUN, in production of ssDNA, **153**, 12
pLcII(nic$^-$), for eukaryotic protein expression in *E. coli*, **153**, 463
pMH158, bacterial–yeast shuttle, construction, **153**, 47
pMON200
major regions and restriction sites, **153**, 274
restriction map, **153**, 256
pMON237
construction, **153**, 258
with methotrexate resistance coding sequence, construction, **153**, 268
pMON316, construction, **153**, 261
pMON505
major regions and restriction sites, **153**, 276
restriction map, **153**, 264
pMON530
construction, **153**, 261
with hygromycin resistance coding sequence, construction, **153**, 266
pMR, in open reading frame cloning strategy, **154**, 129
pSV2-CAT, collagen gene expression analysis with, **144**, 67
RNA synthesized from, in DNA sequencing, **152**, 563
shuffling, in isolation of conditional mutations in essential genes, **154**, 167
single-stranded, production with M13 phages, **153**, 3
Streptomyces, as cloning vectors
listing, **153**, 131
overview, **153**, 118
restriction maps, **153**, 142
shotgun cloning with, **153**, 119
Ti, derived plant transformation vectors, **153**, 254, 292
two-cistron expression, construction, **153**, 409

yeast
  library construction for gene cloning, **152**, 492
  overview, **152**, 484
Plasmin
  immobilization on
    aldehyde–Sephadex, **137**, 556
    soluble dextran, **137**, 557
  Sephadex-immobilized, thrombolytic activity, **137**, 562
Plasminogen activator
  assays, **163**, 294
  in bone cells, effects of hormones, **145**, 333
  tissue-type
    assay, **163**, 305
    purification, **163**, 300
  urokinase-type
    bound to surgical biomaterial, characterization, **137**, 549
    immobilization on acrylamide–acrylic acid copolymer, **137**, 558
    immobilized
      antithrombogenicity of tubes treated with, **137**, 539
      assay, **137**, 532
      clinical applications of tubes treated with, **137**, 540
      fibrinolytic activity *in vivo*, **137**, 543
      preparation, **137**, 530
      stability, **137**, 532
    modification with
      fibrinogen, **137**, 560
      heparin, **137**, 559
    purification, **163**, 301
Plasminogen activator inhibitors
  assay, **163**, 298, 307
  purification, **163**, 301
Plastids
  isolated
    intactness assessment, **148**, 239
    purity assessment, **148**, 238
  isolation
    from cauliflower buds, **148**, 235
    in density gradients of Percoll and other silica sols, **148**, 157
Plastocyanin
  in cyanobacteria and chloroplasts, sequence comparisons, **167**, 148

Plastoquinone
  in chloroplast membranes, analysis, **148**, 391
  synthesis in chloroplasts, **148**, 390
Platelet-activating factor
  bioassays, **163**, 49
  extraction and identification, **163**, 46
  mass spectrometric analysis, **163**, 52
  metabolism, assays of associated enzymes, **141**, 379
  overview, **163**, 44
Platelet-derived growth factor
  binding proteins
    analytical methods, **147**, 56
    characteristics, **147**, 51
  bioassay, **147**, 15
  cDNA cloning by differential colony hybridization, **147**, 80
  as fibroblast chemoattractant, **163**, 702
  modulated proteins, identification, **147**, 85
  properties, **147**, 8
  purification, **147**, 5, 9, 15
  radioassay, **147**, 4
  radioiodination, **147**, 15
  radioreceptor assay, **147**, 19
  related proteins
    binding to receptors, **147**, 35
    electrophoretic transfer, **147**, 26
    functional characterization, **147**, 32
    immunoprecipitation, **147**, 24
    stimulation of protein phosphorylation, **147**, 37
  role in tissue repair processes, **163**, 709
Platelets
  agglutination by thrombospondin, assay, **144**, 445
  arachidonate uptake, assay, **141**, 353
  cGMP-binding phosphodiesterase purification, **159**, 715
  human
    extraction of inositol trisphosphate, **141**, 144
    preparation of crude membranes, **157**, 352
    purification of
      actin, **139**, 847
      Ca^{2+}-ATPase, **157**, 355
      cAMP phosphodiesterase, **159**, 772

connective tissue-activating peptide III, **163**, 743
membranes, **157**, 354
transforming growth factor β, **146**, 156, 160
inositol phospholipid metabolism, measurement, **141**, 176
microvascular deposition during inflammation, measurement, **162**, 312
multifunctional Arg–Gly–Asp receptor
isolation, **144**, 486
properties, **144**, 477
porcine
preparation of crude membranes, **157**, 352
purification of
Ca^{2+}-ATPase, **157**, 357
membranes, **157**, 354
rabbit, in bioassays for platelet-activating factor, **163**, 49
thrombospondin purification, **144**, 439
as vinca alkaloid carriers
clinical applications, **149**, 316
overview, **149**, 312
prospects for utilization, **149**, 324
Pleurisy
carrageenan-induced
analytical methods, **162**, 335
comparison with other exudative models, **162**, 338
Ploidy
yeast chromosomes, visual assays, **155**, 351
Pneumonocytes
type II, glycerophospholipid synthesis, alteration methods, **141**, 222
Polarization assay
leukocyte chemotaxis
advantages, **162**, 21
procedures, **162**, 18
mitogen-activated lymphocyte locomotion, **162**, 190
Polarographic assay
α-ketoisocaproate dioxygenase, **166**, 290
sulfur compounds, **143**, 38
Poly(*N*-acetylgalactosamine) deacetylase
assay, **161**, 515
properties, **161**, 517
purification from *Aspergillus parasiticus*, **161**, 516
Polyacrylamide
immobilization of *Brevibacterium ammoniagenes*, **136**, 457
isonitrile derivatives
ligand coupling, **135**, 97
preparation, **135**, 93, 96
microparticles with entrapped asparaginase
hemofilter with, **137**, 493
preparation, **137**, 492
Poly(acrylamide-co-*N*-acryloxysuccinimide)
enzyme immobilization with, **136**, 275
preparation, **136**, 274
Polyacrylamide gels
bead-forming procedure, **135**, 226
buffer gradient, in DNA sequencing, **155**, 93
cell immobilization in, **135**, 227
detection of
collagen peptides, **145**, 199
β-D-glucanase and β-D-xylanase, **160**, 69
in DNA dideoxy sequencing, **155**, 76
entrapment of living microbial cells, **135**, 198
immobilized chymotrypsin, reactivation, **135**, 594, 595
for nucleic acid electrophoresis, preparation, **152**, 66
Polyacrylamide–hydrazide gels
cell immobilization in, **135**, 218
preparation, **135**, 217
Polyalcohols
thin-layer chromatography, **160**, 159
Polyamide, *see* Nylon
Polyamines
metabolism in mitogen-treated lymphocytes, analysis of associated enzymes, **150**, 55
–Mg^{2+}–NH_4^+ system, for poly(Phe) synthesis with near *in vivo* features, **164**, 650
Polyazetidine
immobilization of *E. coli*, **136**, 500
Polyclonal antibodies
collagenase, production, **144**, 166
collagen, characterization and isolation, **145**, 148
desmosine, production, **144**, 250

fibrinogen, production, **163**, 479
hemopexin, preparation, **163**, 545
Na^+,K^+-ATPase, preparation, **156**, 95
parvalbumin, production, **139**, 553
prolyl 4-hydroxylase, preparation, **144**, 101

Polyethylene
–collagen composite material, with bound enzymes
characterization and applications, **137**, 549
preparation, **137**, 546

Polyethylene glycol
bound NAD(H), preparation, **136**, 39
1,1′-carbonyldiimidazole-activated, protein coupling, **135**, 116
precipitation of
bacteriophages in lysates, **152**, 159
E. coli α-hemolysin, **165**, 399

Polyethylene glycol sulfate
preparation, **136**, 49

Polyethyleneimine
in κ-carrageenan immobilization of *Brevibacterium flavum*, **136**, 460

Polygalacturonase
analysis with activity stain, **161**, 330
assay, **161**, 333, 337, 361, 371
isozymes, chromatography with mono S cation exchanger, **161**, 396
properties, **161**, 364, 372
purification from
Botrytis cinerea, **161**, 372
Corticium rolfsii, **161**, 363

Polyglucose granules
cyanobacterial, research methodology, **167**, 196

Poly(L-glutamic acid)
immobilization of *Achromobacter* protease I, **136**, 165

Polyhydrazides
preparation, **138**, 439

Poly-β-hydroxybutyrate granules
cyanobacterial, research methodology, **167**, 202

Polylysine
D-form, in synthesis of nondegradable carbohydrate ligands, **138**, 407
in erythrocyte-mediated microinjection, **149**, 299
L-form
and alginate, islet microencapsulation with, **137**, 577
in plaque assay for antigen-activated B cells, **150**, 223
template-free ribosomal synthesis from lysyl-tRNA, **164**, 634

Polymerase chain reaction
in specific DNA synthesis *in vitro*, **155**, 335

Polymeric dyes
lignin biodegradation assays with, **161**, 74

Polymerization
actin, calmodulin effects, **139**, 852
coniferyl alcohol, **161**, 71
phosphofructokinase, analysis, **139**, 760
radiation-induced, in enzyme and microorganism immobilization, **135**, 146

Polymers
effect on immobilized enzyme stability, **137**, 595
organic, as semiconductors in bioelectrocatalysis, **137**, 437
polyhydroxylic, isonitrile derivatives
coupling reactions, **135**, 91
ligand coupling, **135**, 97
preparation, **135**, 98

Polymix
in measurement of translational kinetic parameters, **164**, 612

Polymorphism
genetic
α_1-acid glycoprotein, **163**, 427
α_1-antitrypsin, **163**, 409
ceruloplasmin, **163**, 450
haptoglobin, **163**, 466
hemopexin, **163**, 554
α_2-HS glycoprotein, **163**, 436

Polymyxin B
bound to surgical biomaterial, characterization, **137**, 551

Polynucleotide kinase
phosphorylation of
adaptors, **152**, 348
linkers, **152**, 344
T4, treatment of nucleic acids
5′-end radiolabeling, **152**, 100
3′-phosphate removal, **152**, 102

Polyoxins
as glycosylation inhibitors, **138**, 705
Polypeptides
bone morphogenetic
amino acid composition, **146**, 301
bioassay, **146**, 308
preparation, **146**, 302
chloroplast envelope
antibody preparation, **148**, 207
electrophoretic analysis, **148**, 200
immunochemical characterization in
envelope membrane fractions, **148**, 211
intact plastids, **148**, 208
collagen, crosslinked
chromatographic assay, **144**, 135
fluorescent detection in gels, **144**, 137
composition in vacuolar membrane H^+-ATPase, **157**, 558
design, application of structural profiles, **154**, 450
glyoxysomal membrane-associated, analysis, **148**, 528
inactive, plague murine toxin, isolation, **165**, 168
template-free ribosomal synthesis from
aminoacyl-tRNA, **164**, 631
misacylated tRNA, **164**, 645
thylakoid, 2-D gel electrophoresis, **148**, 623
Polyphenylalanine
poly(U)-dependent synthesis
in assay for ribosomal 50S subunit, **164**, 145
with near *in vivo* features, **164**, 650
Polyphosphate granules
cyanobacterial, research methodology, **167**, 199
Polypropylene
–collagen composite material, with bound enzymes
characterization and applications, **137**, 549
preparation, **137**, 546
Polyribonucleotide synthase (ATP)
T4, RNA 3′-end labeling with, **152**, 103
Polyribosomes
associated collagen mRNAs, *in vitro* translation, **144**, 93
Polysaccharides
1,1′-carbonyldiimidazole-activated, preparation, **135**, 108
complex formation with Congo Red, in assay of β-D-glucanases, **160**, 59
cyanobacterial cell wall, isolation, **167**, 179
dye-labeled, for endohydrolase assays, **160**, 74
isonitrile derivatives
four-component coupling to ligands, **135**, 97
preparation, **135**, 92, 95
pectic, inducing plant defense responses, preparation, **138**, 715
permethylated, reductive cleavage, **138**, 26
Polysialic acid
gel electrophoresis, endoneuraminidase effects, **138**, 631
synthesis, stimulation by neural cell adhesion molecule, **138**, 631
Polysialic acid depolymerase
bacteriophage
colominic acid depolymerization, **138**, 789
purification, **138**, 786
Poly-α-2,8-sialosyl sialyltransferase
assay, **138**, 628
distribution in cerebral membranes, **138**, 629
Polysomes
analysis, cytoplasm preparation for, **152**, 232
endoplasmic reticulum-bound
characterization, **148**, 584
isolation
overview, **148**, 582
procedure, **148**, 584
membrane-bound, mRNA isolation, **152**, 241
mRNA isolation, **152**, 253
specific, isolation with antibodies, **152**, 248
Polysorbate 80
solubilization of Ca^{2+}-ATPase in oligomeric form, **157**, 265
Polyurethane
entrapment of

chloroplasts, **135**, 447
cyanobacteria, **135**, 448
foam, entrapment of cyanobacteria, **167**, 631
immobilization of *E. coli*, **136**, 464
reticulate, entrapment of plant cells, **135**, 415
Polyuridine
dependent poly(Phe) synthesis, in assay for ribosomal 50S subunit, **164**, 145
Polyvinyl chloride
immobilized urokinase
clinical applications, **137**, 540
preparation, **137**, 530
Polyvinyl foam
entrapment of cyanobacteria, **167**, 631
Polyvinylpyrrolidone, *see* Povidone
Pores
cyanobacterial, ultrastructural analysis, **167**, 169
Porphyridium
P. aerugineum, purification of phycobilisomes, **167**, 308
P. cruentum, isolation of photosystem II–phycobilisome complex, **167**, 286
Porphyrin
binding to hemopexin, assay, **163**, 562
Potassium
assay with dry reagent chemistries, **137**, 417
atomic absorption spectrometry, **158**, 141
binding to Na^+,K^+-ATPase
during ATP hydrolysis, assay, **157**, 233
centrifugation assay, **156**, 230
characteristics, **156**, 235
kinetics, **156**, 244
concentration, effect on Na,K-pump reconstitution by sonication, **156**, 149
effect on
[^{3}H]ouabain binding to Na^+,K^+-ATPase, **156**, 222
Na^+,K^+-ATPase, **156**, 2
efflux from erythrocytes, vanadate effects, **156**, 261
intracellular levels, measurement, **146**, 387
–sodium exchange in vesicles, assay, **156**, 181
–sodium pump, *see* ATPase
transport
in cyanobacteria, assay, **167**, 566
in erythrocyte inside-out vesicles, assay, **156**, 175
after Na,K-pump reconstitution
assay, **156**, 144
coupling ratio, **156**, 153
electrogenicity, **156**, 154
time course at different temperatures, **156**, 151
Potassium *tert*-butoxide
dimethyl sulfoxide-dissolved, preparation, **135**, 95
Potassium permanganate
mediated oxidation, in lignin characterization, **161**, 107
–sodium periodate oxidation, in lignin characterization, **161**, 124
Potato
mitochondria
purification from tubers, **148**, 409
ubiquinone synthesis, **148**, 486
Povidone
radiolabeled, as fluid-phase marker
endocytotic uptake, **146**, 412
overview, **146**, 404
preparation, **146**, 406
Precipitation
acid, in analysis of asialoorosomucoid intracellular degradation, **149**, 9
bacteriophages in lysates, **152**, 159
elastin, **144**, 210
endoglucanase D, **160**, 359
Escherichia α-hemolysin, with polyethylene glycol, **165**, 399
immuno-, *see* Immunoprecipitation
nucleic acids, **152**, 41
salt-mediated, collagens, **163**, 686
transferrin–receptor complex, **147**, 267
trimethylselenonium ion, with ammonium reineckate, **143**, 197
Precipitation assay
C1q receptor, **150**, 561
protein-bound cyclic nucleotides, **159**, 147

Precipitation inhibition assay
Amphicarpaea bracteata lectin, **138**, 561
Precipitin ring test
Pseudomonas exotoxin A, **165**, 151
Precipitin technique
Evonymus lectin assay, **138**, 552
Prekallikrein, *see* Prokallikrein
Prenyl alcohols
TLC and radioassay, **167**, 332
Prenyl pyrophosphates
formation from mevalonic acid *in vitro*, **167**, 331
Prenylquinones
chemical synthesis, **148**, 383
cyanobacterial
distribution and function, **167**, 271
HPLC assay, **167**, 270
synthesis in chloroplasts, localization, **148**, 390
Prephenate
assay, **142**, 431
HPLC, **142**, 427
preparation, **142**, 426, 441
Prephenate aminotransferase
assay, **142**, 480
gel electrophoresis, **142**, 483
properties, **142**, 487
purification from tobacco, **142**, 484
Prephenate dehydratase
–chorismate mutase
assays, **142**, 432
properties, **142**, 435
purification from *E. coli*, **142**, 433
monofunctional
assay, **142**, 507
properties, **142**, 511
purification from *Bacillus subtilis*, **142**, 509
Prephenate dehydrogenase
–chorismate mutase
assays, **142**, 442
properties, **142**, 448
purification from *E. coli*, **142**, 444
substrate preparation, **142**, 440
monofunctional
assay, **142**, 503
properties, **142**, 507
purification from *Alcaligenes eutrophus*, **142**, 504
Primer extension
in analysis of calmodulin gene 5′ end, **139**, 227
cDNA, in mapping 5′ termini of gene transcripts, **152**, 629
in monitoring chemical and enzymatic probing of RNA structure, **164**, 481
Primers
for dideoxy DNA sequencing, design rules, **152**, 539
mRNA fragments, second-strand cDNA synthesis with, **152**, 330
oligo(dT), first-strand cDNA synthesis with, **152**, 316
–vector molecules, in full-length cDNA cloning, **152**, 371
Print immunofixation
after isoelectric focusing, α_1-antitrypsin, **163**, 412
Procollagen
genes, restriction fragment length polymorphisms, analysis, **145**, 205
mRNA, R-loop analysis for assessment of collagen mutations, **145**, 235
type I, C-terminal propeptide, purification, **145**, 406
type III, N-terminal propeptide, purification, **145**, 407
various types, isolation and characterization, **144**, 42
Procollagen-lysine,2-oxoglutarate 5-dioxygenase
general description, **144**, 109
Procollagen-proline,2-oxoglutarate 4-dioxygenase
antibody preparation, **144**, 101
cDNA clones, preparation, **144**, 103
properties, **144**, 103
purification, **144**, 97
subunits, isolation, **144**, 100
Profilin
preparation from macrophages, **162**, 254
Progesterone
enzymatic reduction in two-phase system, **136**, 154
Proinsulin
production by
E. coli, enhancement technique, **153**, 385

nonproliferating entrapped bacteria, **137**, 632
Prokallikrein
plasma-type
activation by factor XII or XIIa, **163**, 67
assay, **163**, 87
with synthetic substrates, **163**, 170
properties, **163**, 91
purification, **163**, 89
structure, **163**, 91
tissue-type
assays, **163**, 116
properties, **163**, 127
purification from rat pancreas, **163**, 119
purity criteria, **163**, 127
Prokaryotes
screening for cellulose- and hemicellulose-degrading enzymes, **160**, 180
Prolactin
gene, regulation by Ca^{2+}/calmodulin, **139**, 655
Prolamellar bodies
etioplast
characterization, **148**, 251
isolation, **148**, 250
separation from prothylakoids, **148**, 249
Proliferating cell nuclear antigen
amino acid composition, **150**, 153
detection in cells, **150**, 153
purification, **150**, 148
Proline
^3H-labeled
in analysis of fibroblast collagen production, **163**, 644
in selection of amino acid transport-defective mutants, **151**, 136
Prolyl 4-hydroxylase, *see* Procollagen-proline,2-oxoglutarate 4-dioxygenase
Promoters
analysis with
binary Ti vectors, **153**, 292
expression vectors, **153**, 452
E. coli, consensus sequence determination, **155**, 576
λ phage, in construction of pHE6 expression plasmid, **153**, 482
prokaryotic, plasmids for selection and analysis, **153**, 54
T5-derived, in transcription–translation system for protein analysis, **155**, 416
Pronase
coupling to 1,1′-carbonyldiimidazole-activated support, **135**, 115
immobilized, in analysis of protein structural changes, **135**, 598
(*R*)-Propanediol
preparation with *Candida* and *Alcaligenes*, **136**, 316
Propiothetin bromide
synthesis, **143**, 227
Proplastids
isolation from *Euglena gracilis*, **148**, 169
sec-Propylene chlorohydrin
racemic, enzymatic preparation, **136**, 133
Propylene glycol
qualitative TLC, **160**, 173
Prostaglandin A_1
regulation of cAMP export, **159**, 83
Prostaglandin E_2
production by hypothalamic cells *in vitro*, in assay for endogenous pyrogens, **163**, 509
Prostaglandins
antisera
characterization, **141**, 326
preparation, **141**, 323
biosynthesis inhibition by lipocortins, assay, **163**, 30
derivatization, **141**, 348
GC analysis, **141**, 348
GC–MS analysis, **141**, 349
HPLC analysis, **141**, 347
purification, **141**, 332
radioimmunoassay, **141**, 323
Protein A
assay, **143**, 402
collodion–charcoal-immobilized
in cancer treatment, **137**, 513
preparation, **137**, 502
commercial, enterotoxin radioimmunoassay in, **137**, 514
immobilized, interaction with IgG, ellipsometric study, **137**, 385
isolation from *Staphylococcus aureus*, **137**, 459

matrix-bound, adsorption of factor VIII antibodies, **137**, 458
from *Methanococcus* ribosomes, sequence analysis, **164**, 561
in plaque assay for Ig-secreting cells, **150**, 217
properties, **143**, 401
purification from rat liver, **143**, 401
–Sepharose, in antibody-mediated carboxypeptidase A immobilization, **135**, 165
staphylococcal, as lymphocyte locomotion activator, **162**, 188
Proteinase
collagenase-rich, in bone cell isolation, **145**, 311
conformation and activity in reverse micelles, **136**, 201
cyanophycin-hydrolyzing, assay, **167**, 212
immobilized, applications, **135**, 75
stimulation of lymphocytes, **150**, 109
V8, in peptide mapping of collagens, **145**, 177
Proteinase inhibitor inducing factor
galacturonic acid-rich, preparation, **138**, 717
preparation, **138**, 716
Proteinase inhibitors
endothelial cell growth-stimulating, purification, **147**, 153
in purification of protein phosphatases C-I and C-II, **159**, 365
Protein C
human, two-cistron expression vector, construction, **153**, 409
Protein determination
Bradford method, modification for cAMP phosphodiesterase assay, **159**, 747
cells immobilized in crosslinked polyacrylamide–hydrazide gels, **135**, 219
Protein disulfide-isomerase
assays, **144**, 112
properties, **144**, 114
purification from chick embryo, **144**, 113
Protein engineering
application of
molecular mechanics and dynamics, **154**, 430
structural profiles, **154**, 450
Protein-glutamine γ-glutamyltransferase
glycosyl-asparagine incorporation into glutamine residues, **138**, 413
Protein kinase
activation in assays of cyclic nucleotides and analogs, **159**, 78
analysis in mitogen-treated lymphocytes, **150**, 75
Ca^{2+}/calmodulin-dependent, detection in GH_3 cell nuclear matrix, **139**, 665
calmodulin-dependent, autophosphorylation, role in enzyme regulation, **139**, 690
cAMP-dependent
cAMP-free and cAMP-bound holoenzyme, purification, **159**, 202
catalytic subunit
cDNA cloning, **159**, 311
fluoresceinated, in regulatory subunit cytochemistry, **159**, 255
free, direct cytochemical localization, **159**, 236
cellular compartmentalization, immunogold electron microscopy, **159**, 225
inhibition by cAMP antagonist (*Rp*)-cAMPS, **159**, 163
isozymes, hormonal activation, analysis, **159**, 97, 105
purification from *Drosophila*, **159**, 215
regulatory subunit I, expression and mutagenesis in *E. coli*, **159**, 325
regulatory subunit II
binding proteins, identification, **159**, 183
cDNA cloning, **159**, 318
comparison with cAMP-binding protein from *Trypanosoma*, **159**, 294
phosphorylation in intact cells, **159**, 139
regulatory subunits
comparison with catabolite gene activator protein, **159**, 284
direct cytochemical localization, **159**, 255
preparation by nondenaturing method, **159**, 208
role in cellular regulation, molecular genetic analysis, **159**, 299

synergistic responses to site-selective cAMP analogs in intact cells, **159**, 118
synthetic peptide inhibitors, **159**, 173
in thiophosphorylation of histone, **159**, 350
in yeast cAMP cascade system mutants, assay, **159**, 39
cGMP-dependent, activation in intact tissues, **159**, 150
different types, phosphorylation of tyrosine 3-monooxygenase, **142**, 81
Protein kinase C
assay, **141**, 403, 426
–membrane complex
hormone-induced stabilization, assay, **141**, 405
phorbol ester-induced stabilization, assay, **141**, 400
phosphorylation activity in intact cells, analysis by
cell treatment with enzyme modulators, **141**, 414
enzyme assay in crude tissue extracts, **141**, 421
production of enzyme-deficient cells, **141**, 419
use of phosphoprotein marker for enzyme activation, **141**, 417
purification from rat brain, **141**, 426
redistribution in leukocytes in response to chemoattractants, assay, **162**, 282
structure–function analysis with diacylglycerol analogs, **141**, 313
Protein kinase C inhibitors
effects on transmembrane Ca^{2+} signaling, **139**, 577
Protein kinase inhibitors
fluoresceinated
in localization of cAMP-dependent protein kinase free catalytic subunit, **159**, 247
preparation, **159**, 238
Protein phosphatase, *see* Phosphoprotein phosphatase
Protein S
gene
cloning, **139**, 383
expression, **139**, 387
sequencing, **139**, 385
purification from *Myxococcus*, **139**, 380
Proteins
actin-binding, purification from macrophages, **162**, 257
actin-modulating, purification from macrophages, **162**, 254
activator
for branched-chain α-keto acid dehydrogenase, assay, **166**, 183
for glycolipid hydrolysis
assays, **138**, 797
purification, **138**, 810
acute-phase
description, **163**, 374
gene expression, regulation, **163**, 381
induction, mediating system, **163**, 378
inflammation-induced changes in serum, analysis, **163**, 567
physiological role, **163**, 382
regulation in cultured hepatocytes, analysis, **163**, 584
structural and functional properties, identification, **163**, 576
ADP-ribosylation by cholera toxin, analysis, **165**, 235
biotin-binding, complex formation with oligosaccharide, **138**, 422
bone-associated, immunoprecipitation, **145**, 321
bone morphogenetic, *see* Bone morphogenetic protein
bound thiols, liberation, **143**, 127
calcium-binding, *see* Calcium-binding proteins
calmodulin-binding, *see* Calmodulin-binding proteins
calmodulin-binding sequences
characterization, **139**, 463
identification, **139**, 456
–calmodulin complex
properties, **139**, 145
purification from chicken intestine, **139**, 137
calmodulin-dependent methylation, assay, **139**, 668
cAMP-binding, from *Trypanosoma cruzi*
assay, **159**, 287

comparison with protein kinase regulatory subunit II, **159**, 294
partial purification, **159**, 289
photoaffinity labeling, **159**, 292
separation from protein kinase, **159**, 290
cAMP-dependent protein kinase RII subunit-binding, identification, **159**, 183
carrier, as supportive receptors on lymphoid cells, overview, **150**, 385
catabolite gene activator, crystallization with cAMP for structural analysis, **159**, 278
cationic, production of serum sickness in rodents, **162**, 499
cell-free translation, **148**, 275
Charcot–Leyden crystal, identity with lysophospholipase, **163**, 41
chlorophyll–carotenoid, isolation and pigment content, overview, **148**, 353
coding regions on DNA, identification by open reading frame cloning strategy, **154**, 129
complement, *see* Complement
complement C1 inhibitor, purification, **162**, 636
complement C4b-binding, purification, **162**, 637
complement C4b/C3b inactivator, purification, **162**, 638
contractile, in leukocytes, analysis, **162**, 246
coupling to
carbonate-activated carriers, **135**, 84
1,1′-carbonyldiimidazole-activated support, **135**, 116
glycosyl-asparagine derivatives, **138**, 409
covalent attachment to liposomes, **149**, 111
C-reactive, *see* C-reactive protein
crosslinking with RNA in *E. coli* ribosomes, **164**, 298
crustacean chitin-associated
assay, **161**, 421
removal, **161**, 418
cyanobacterial
extracellular
preparation, **167**, 600
qualitative analysis, **167**, 601
quantitative analysis, **167**, 603
outer membrane, isolation, **167**, 181
rigid layer, isolation, **167**, 180
cyclic nucleotide-binding, precipitation assay, **159**, 147
cystine-binding
cystine assay with, **143**, 146
purification from *E. coli*, **143**, 144
cytosolic, required for cholera toxin activity
assays, **165**, 246
partial purification, **165**, 247
delipidated, radioimmunoassay, **138**, 198
denatured, as chemotactic agents, **162**, 180
dentin-associated noncollagenous, preparation and extraction, **145**, 290
disulfides, assay with 2-nitro-5-thiosulfobenzoate, **143**, 115
DNA-binding, gel electrophoretic assay, **151**, 551
–DNA complexes, hydroxyl radical footprinting, **155**, 537
entrapment in erythrocytes, **149**, 226, 235
extracellular matrix, analysis in biological fluids, **145**, 391
foreign
bacterial production, application of open reading frame vectors, **154**, 156
expressed by vaccinia virus recombinants, analysis, **153**, 557
expression in *E. coli*, **153**, 492
preparation from λgt 11 clones, **154**, 121
produced in *E. coli* as fusion proteins, digestion, **153**, 475
secretion in *E. coli*, **153**, 499
synthesis in *E. coli* via pLcII(nic$^-$) vector, **153**, 464
fusion, Na^+,K^+-ATPase α subunit, characterization, **156**, 383
globular, folding, effect of point mutations, **154**, 498
growth regulator-induced, assay, **147**, 427
guanine nucleotide-binding, *see* G proteins

heterologous
expression in
higher eukaryotes, assays, **152**, 698
yeast, **152**, 673
unfused, expression in *E. coli*, **153**, 416
IL-2 receptor precursor, analysis and post-translational modification, **150**, 687
immobilized
inactivation mechanisms, **135**, 590
reactivation mechanisms, **135**, 591
interactions with
DNA
in vitro assays, **152**, 721
in vivo footprinting, **152**, 735
in yeast extracts, gel electrophoretic assay, **155**, 528
mutant 5S RNA, analysis, **164**, 745
5S RNA, analysis, **164**, 733
intracellular degradation, regulation by amino acids, **166**, 461
iron–sulfur
associated with cyanobacterial photosystem I, EPR spectra, **167**, 432
in cyanobacterial cells, EPR spectra, **167**, 431
kallikrein-binding, Western blot analysis, **163**, 141
leucine-binding, purification from *E. coli*, **166**, 249
link
and proteoglycans, newly synthesized, isolation, **144**, 376
proteoglycan-specific
biological function, **144**, 410
characterization, **144**, 406
isolation, **144**, 404
peptide mapping on silica, **144**, 409
and lipids, ratio in sarcoplasmic reticulum membranes, control during reconstitution, **157**, 308
matrix, extraction from bovine bone, **146**, 286
membrane-associated
or secreted, recognition of encoding mRNAs, **152**, 303
sedimentation and binding studies with ultracentrifuge, **150**, 447
thrombin-sensitive, assay, **147**, 157
membrane translocation, analysis with transcription–translation system, **155**, 427
microtubule-associated, *see* Microtubule-associated proteins
mitochondrial, two-dimensional separation, **148**, 630
MSA carrier
assays, **146**, 271
properties, **146**, 275
purification, **146**, 273
N, *see* G proteins
noncollagenous
in bone, overview, **145**, 256
in dentin, overview, **145**, 260
noncovalent coupling to glycosyl-asparagine derivatives, **138**, 418
oligomeric and self-associating, purification with immobilized subunits, **135**, 484
oxidant-induced damage in inflammation, *in vitro* analysis, **163**, 329
PDGF-binding
analytical methods, **147**, 56
characteristics, **147**, 51
PDGF-like
binding to receptors, **147**, 35
electrophoretic transfer, **147**, 26
functional characterization, **147**, 32
immunoprecipitation, **147**, 24
stimulation of protein phosphorylation, **147**, 37
PDGF-modulated, identification, **147**, 85
phospholipid transfer
assay, **148**, 662
nonspecific, in sarcoplasmic reticulum phospholipid content modification, **157**, 369
properties, **148**, 665
purification from spinach, **148**, 664
phosphorylation in mitogen-treated lymphocytes, analysis, **150**, 75
produced by cell-free translation
autoradiographic analysis in chloroplasts, **148**, 291
uptake by isolated chloroplasts, **148**, 284
–protein crosslinks, identification in *E. coli* ribosomes, **164**, 64

proteoglycan core, characterization, **144**, 354
recombinant, expression and analysis in *E. coli*, **152**, 661
ribosomal
affinity-labeled, identification, **164**, 347
archaebacterial, HPLC for microsequence analysis, **164**, 542
binding to RNA
filter assay, **164**, 203
gel mobility shift assay, **164**, 215
sucrose gradient assay, **164**, 215
binding to rRNA
analysis with α-sarcin, **164**, 468
enzymatic and chemical probing, **164**, 456
exposed on ribosome surface, identification, **164**, 434
fluorescent labeling, **164**, 176
genes in chloroplast DNA, cloning and identification, **164**, 748
interactions with rRNAs, electron microscopic analysis, **164**, 61
ion-exchange HPLC, **164**, 532
isolated, concentration, **164**, 139
L7/L12, structure, effect of elongation factor, **164**, 154
L24, gene mutations, analysis, **164**, 706
localization on *E. coli* ribosomal subunits, **164**, 503
purification, **164**, 179
radiolabeled, 2-D electrophoresis, **164**, 431
reversed-phase HPLC, **164**, 523
and rRNA, reconstitution, **164**, 141
rRNA-recognizing fragments, isolation, **164**, 258
5S-binding, purification, **164**, 171
single, preparation from large subunit, **164**, 134
S6, phosphorylation during mitogenic response, analysis, **164**, 575
30S subunit, neutron scattering topography, **164**, 117
surface exposure, dependence on structural state of ribosome, **164**, 436
total
deuterated, isolation, **164**, 138
preparation from large subunit, **164**, 133
–RNA complexes
analysis, **164**, 315
formation via soluble carbodiimide, **164**, 311
purification, **164**, 311
S6
multiple phosphorylated forms, electrophoretic analysis, **146**, 365
phosphopeptide analysis, **146**, 373
S-100
affinity purification, **139**, 14
antibody to subunits, production, **139**, 497
immunoperoxidase staining in cells and tissues, **139**, 502
preparation of forms a and b, **139**, 496
purification from bovine brain, **139**, 170
spectroscopic analyses
circular dichroism, **139**, 177
fluorescence, **139**, 180
nuclear magnetic resonance, **139**, 182
UV absorbance and difference, **139**, 174
sarcoplasmic reticulum, cDNA cloning, **157**, 289
secretion in gram-positive bacteria, genetic engineering, **153**, 507
serum amyloid A, *see* Serum amyloid A protein
serum, foreign, production of serum sickness in mice, **162**, 499
sialyloligosaccharide-binding, specificity analysis, **138**, 162
in solution
inactivation mechanisms, **135**, 586
reactivation mechanisms, **135**, 587
spanning, isolation from skeletal muscle triads, **157**, 83
spermatozoal, phosphorylation and ^{32}P incorporation assay, **139**, 818
structural changes, analysis with immobilized enzymes, **135**, 596
structure–function analysis with peptide models, **154**, 473
sulfatide-binding, characterization, **138**, 473

truncated, analysis with transcription–translation system, **155**, 430
urinary excretion in serum sickness, measurement, **162**, 498
Protein sequencing
computer programs for, **155**, 582
ribosomal proteins from archaebacteria, **164**, 542
Protein synthesis
analysis with transcription–translation system, **155**, 424
in bone cell cultures, analysis, **145**, 316
eukaryotic proteins in *E. coli*, **153**, 464
inhibition by
Pseudomonas exotoxin A
in vitro assay, **165**, 228
in vivo assay, **165**, 230
Shiga toxin *in vitro*, **165**, 233
in *Synechococcus*, heat shock effects, **167**, 625
Protein-tyrosine kinase
growth factor-stimulated, assay, **146**, 355
Proteoglycans
aggregates, stabilization by link proteins, **144**, 411
aggregation with hyaluronate, assays, **144**, 361
associated oligosaccharides, analysis, **144**, 370
basement membrane-associated
properties, **145**, 364
purification, **145**, 380
binding to link proteins, analysis, **144**, 411
bone-associated
characterization, **145**, 267
extraction, **145**, 263
purification, **145**, 265
radiolabeling *in vivo*, **145**, 268
in cartilage
large aggregating molecules, **144**, 307
small molecules, **144**, 312
catabolism and turnover, analysis, **144**, 412
cell-associated forms, **144**, 314
composition, **144**, 338
conformational studies, **145**, 77
core protein characterization, **144**, 354
detergent micelle-associated, analysis, **144**, 398
early studies, **144**, 305
electron immunohistochemistry, **145**, 92, 99
electron microscopy, **145**, 72
extraction, **144**, 320
glycosaminoglycan chains, characterization, **144**, 342
gradient centrifugation, **144**, 307
heparan sulfate-containing forms, **144**, 315
intracellular forms, **144**, 317
isolation and purification from rat granulosa cells, **138**, 279
membrane-associated
functions, **144**, 396
hydrophobic chromatography, **144**, 398
insertion into liposomes, **144**, 399
intercalated molecules, **144**, 394
isolation, **144**, 397
peripheral molecules, **144**, 395
solubilization, **144**, 396
newly synthesized
aggregating forms, gel electrophoresis, **144**, 391
core protein precursor and link protein
immunoprecipitation, **144**, 382
isolation based on aggregate-forming ability, **144**, 376
hyaluronate-binding ability, **144**, 381
in non-cartilage tissues
large molecules, **144**, 312
small molecules, **144**, 313
overview, **145**, 71
purification, **144**, 324
separation, **144**, 331
size determination, **144**, 358
solution studies, **145**, 76
substructures, identification, **144**, 366
Proteolysis
bone morphogenetic protein, polypeptide generation, **146**, 302
in characterization of radiolabeled lignins, **161**, 25
controlled, Ca^{2+}-ATPase, **139**, 801
cyanobacterial gas vesicle proteins, **167**, 219
fibronectin, in preparation of chemotactic fragments, **162**, 218
fusion proteins produced in *E. coli*, **153**, 475

limited
aromatic-amino-acid aminotransferase, **142**, 264
kininogens, **163**, 247
membrane-bound Na^+,K^+-ATPase, for structural and conformational analyses, **156**, 291
in preparation of anthranilate synthase partial complex, **142**, 380
ribosomal proteins, for generation of rRNA-recognizing fragments, **164**, 259
thrombospondin, **144**, 441
Prothylakoids
etioplast
characterization, **148**, 251
separation from prolamellar bodies, **148**, 249
Protocatechuic acid
identification as product of vanillate *O*-demethylase activity, **161**, 299
Proton extrusion
respiratory, measurement in cyanobacteria, **167**, 363
Proton fluxes
in sarcoplasmic reticulum vesicles, measurement, **157**, 468
Protopectinase
assay, **161**, 336
properties, **161**, 345
purification from
Galactomyces reessii, **161**, 340
Kluyveromyces fragilis, **161**, 338
Trichosporon penicillatum, **161**, 343
Protoplast fusion
interspecific transfer of partial nuclear genomic information, **148**, 46
verification, **148**, 50
Protoplasts
interaction with liposomes, **149**, 183
nucleic acid delivery by liposomes
assessment, **149**, 181
procedure, **149**, 180
plant
amyloplast isolation from, **148**, 230
anchorage to microcarriers, **135**, 424
applications, **135**, 433
characterization, **135**, 426
monitoring, **135**, 425
association with liposomes, assessment, **148**, 37
coupling to hydrous titanium(IV) oxide and derivatives, **135**, 363
direct gene transfer to, **153**, 313
disruption, **148**, 33
electroporation-mediated introduction of DNA and RNA, **153**, 336, 351
fractionation for amyloplast isolation, **148**, 222
immobilized
applications, **135**, 439
preparation, **135**, 435
viability assay, **135**, 436
intactness after preparation, **148**, 31
interaction with liposomes, **148**, 37
isolation, **135**, 423, 434; **148**, 188
for vacuole preparation, **148**, 57
liposome-mediated nucleic acid transfection, **148**, 42
lysis for vacuole preparation
isotonic shear, **148**, 63
osmotic, **148**, 61
polybase-induced, **148**, 62
nuclei, isolation, **148**, 537
plasmid DNA expression, assay, **148**, 42
preparation from
castor bean endosperm, **148**, 30
grape leaves, **148**, 28
maize, **148**, 229
soybean, **148**, 220
sycamore cell cultures, **148**, 31
tobacco cell cultures, **148**, 35; **149**, 177
tobacco leaves, **148**, 41
rapid fractionation for recovery of chloroplasts and cytosol, **148**, 192
release of organelles, **148**, 191
treatment with liposomes, **148**, 36
vacuole preparation from, **148**, 76, 83
viral RNA expression, assay, **148**, 42
from synchronized cells, nucleic acid delivery, **149**, 182
Pseudoanabaena spp.
purification of phycobilisomes, **167**, 307
Pseudomonas
cosmid gene bank, construction and screening, **153**, 51

P. aeruginosa
purification of
branched-chain keto acid dehydrogenase, **166**, 346
exotoxin A, **165**, 147
lipoamide dehydrogenase-valine, **166**, 349
methylmalonate-semialdehyde dehydrogenase, **166**, 391
P. dacunhae
κ-carrageenan-immobilized
L-alanine production in reactor, **136**, 477
L-aspartate β-decarboxylase activity, stabilization, **136**, 476
preparation, **136**, 474
P. diminuta, purification of arogenate dehydratase, **142**, 501
P. fluorescens, purification of cellulases, **160**, 202
P. putida
cloning of branched-chain keto acid dehydrogenase genes, **166**, 350
purification of
anthranilate synthase, **142**, 302
methionine γ-lyase, **143**, 460
4-methoxybenzoate monooxygenase components, **161**, 284
purification of
ω-amino acid–pyruvate aminotransferase, **143**, 501
4-hydroxyphenylpyruvate dioxygenase, **142**, 145
tryptophan side chain oxidases I and II, **142**, 203
Pseudomonas exotoxin A
ADP-ribosyltransferase activity, assay, **165**, 226
cytotoxicity, assay, **165**, 227
inhibition of protein synthesis
in vitro assay, **165**, 228
in vivo assay, **165**, 230
precipitin ring test, **165**, 151
properties, **165**, 151
purification, **165**, 147
Psoralen
crosslinking of rRNA, **164**, 330
Pulse labeling
fibroblasts, for analysis of collagen production, **163**, 646
Purified protein derivative
as lymphocyte locomotion activator, **162**, 188
Purine-nucleoside phosphorylase
as marker for chromosome detection in somatic cell hybrids, **151**, 188
Purines
metabolism in mitogen-treated lymphocytes, analysis of associated enzymes, **150**, 51
Purpura
idiopathic thrombocytopenic, treatment with platelet-delivered vinca alkaloids, **149**, 316
Putidamonooxin
properties, **161**, 289
purification from *Pseudomonas putida*, **161**, 284
Pycnoporus cinnabarinus
purification of
β-*N*-acetylhexosaminidase, **161**, 464
chitinase, **161**, 468
Pyranose oxidase
amidination, **137**, 602
amino groups, assay, **137**, 603
assay, **137**, 601; **161**, 317
crosslinked
characterization, **137**, 609
gel electrophoresis, **137**, 604
preparation, **137**, 602
glucosone effects, **137**, 608, 611
properties, **161**, 319
purification from *Phanerochaete chrysosporium*, **161**, 318
thermodenaturation, **137**, 605, 615
Pyridoxal 5′-phosphate
bound Sepharose
coupling of nucleoside diphosphatase, **135**, 157
preparation, **135**, 156
retention in membrane reactor, **136**, 492
3-(2-Pyridyldithio)propionyl-calmodulin
preparation, **159**, 612
properties and applications, **159**, 614
Pyridyldithiopropionylphosphatidylethanolamine
synthesis, in protein conjugation to liposomes, **149**, 112

Pyrimidine dimers
paper chromatography, in analysis of calmodulin role in DNA repair, **139**, 719
Pyrimidines
metabolism in mitogen-treated lymphocytes, analysis of associated enzymes, **150**, 51
Pyrocarbonic acid diethyl ester
modification of 5S RNA, **164**, 727
in probing of
large RNA molecules, **164**, 476
rRNA–protein interactions, **164**, 464
Pyrogens
endogenous
assays, **163**, 502
chemical characterization, **163**, 497
measurement during fever, **163**, 501
overview, **163**, 495
Pyrophosphatase
tonoplast-associated, assay and properties, **148**, 114
Pyroxylin
microcapsules containing enzymes, preparation, **136**, 68
Pyruvate dehydrogenase
assay in mitogen-treated lymphocytes, **150**, 49
–branched-chain 2-keto acid dehydrogenase complex
assay, **166**, 331
properties, **166**, 337
purification from *Bacillus subtilis*, **166**, 332
Pyruvate kinase
in assay of Na^+,K^+-ATPase, **156**, 116
immobilization by radiation polymerization, **135**, 148
immobilized, in synthesis of 5-phospho-α-D-ribosyl pyrophosphate, **136**, 278
as marker for chromosome detection in somatic cell hybrids, **151**, 191
Pyruvate synthase
as electron donor to nitrogenase, assay in cyanobacteria, **167**, 500

Q

Quin 2
in analysis of TRH-induced changes in cellular calcium metabolism, **141**, 47
cell loading for calcium mobilization analysis, **141**, 58

R

Radioactivity
^{32}P-labeled nucleic acids, measurement, **152**, 51
Radiochemical assay
acetolactate synthase I, **166**, 436
acetyl-CoA:α-glucosaminide *N*-acetyltransferase, **138**, 608
adenosylhomocysteinase, **143**, 430
adenylylsulfate–ammonia adenylyltransferase, **143**, 354
amine *N*-methyltransferase, **142**, 661
arginine carboxypeptidase, **163**, 187
arogenate dehydratase, **142**, 497
arylalkylamine *N*-acetyltransferase, **142**, 584
ATP-sulfurylase, **143**, 338, 341
bacterial phospholipases, **165**, 299
betaine–homocysteine methyltransferase, **143**, 384
branched-chain α-keto acid dehydrogenase, **166**, 140, 193, 204, 298
component E1, **166**, 143, 147
component E2, **166**, 143, 152
potential interfering substances, **166**, 196
branched-chain α-keto acid dehydrogenase kinase, **166**, 171
potential sources of error, **166**, 169
branched-chain α-keto acid dehydrogenase phosphatase, **166**, 321
inhibitor protein, **166**, 322
calmodulin-stimulated phosphodiesterase, **159**, 548
cAMP-binding protein from *Trypanosoma*, **159**, 287
cAMP-dependent protein kinase, **159**, 112
cAMP phosphodiesterase, **159**, 737, 745, 752
chitinase, **161**, 424, 460, 471, 480
chitin deacetylase, **161**, 511, 521
chitin synthase, **138**, 643
chloramphenicol acetyltransferase in eukaryotic cells, **151**, 389
cholera toxin, **165**, 171

cyclic nucleotide phosphodiesterase, **159**, 457
cystathionine β-lyase, **143**, 444
cystathionine β-synthase, **143**, 390
cysteinesulfinate decarboxylase, **143**, 405
dipeptidyl carboxypeptidase I, **163**, 197
DNA-directed RNA polymerase in cyanobacteria, **167**, 592
DNA methylases, **155**, 35
dopamine β-monooxygenase, **142**, 608
elastase, **144**, 289
endo-β-*N*-acetylglucosaminidase, **138**, 773
endochitinase, **161**, 498
1,3-β-glucan synthase, **138**, 637
H^+-ATPase, **157**, 515, 636
histamine from mast cells, **162**, 518
histone phosphatase, **159**, 360
hydroxyindole *O*-methyltransferase, **142**, 591
3-hydroxy-3-methylglutaryl-CoA lyase, **166**, 219
4-hydroxyphenylpyruvate dioxygenase, **142**, 133, 134, 149
hypotaurine aminotransferase, **143**, 183
indoleamine 2,3-dioxygenase, **142**, 192
indoleamine *N*-methyltransferase, **142**, 669
β-isopropylmalate dehydrogenase, **166**, 227
isovaleryl-CoA dehydrogenase, **166**, 156, 375
α-ketoisocaproate dioxygenase, **166**, 289
leucine–tRNA ligase, **166**, 262
lipoxygenases, **163**, 345
lysophospholipase, **163**, 34
lysosomal acidification, **157**, 596
α-mannosidases I and II, **138**, 780
3-methylglutaconyl-CoA hydratase, **166**, 214
D-methylmalonyl-CoA hydrolase, **166**, 394
L-methylmalonyl-CoA mutase, **166**, 407
DL-methylmalonyl-CoA racemase, **166**, 401
monoamine oxidase, **142**, 647
monophenol monooxygenase, **142**, 160
myosin light chain kinase, **139**, 115
myosin phosphatase, **159**, 447
norepinephrine *N*-methyltransferase, **142**, 656
oligosaccharyl sulfatase, **138**, 818
phenylalanine ammonia-lyase, **142**, 249
phenylalanine 4-monooxygenase, **142**, 35
phosphatidylinositol kinases, **141**, 212
phospholamban, **157**, 365
phospholipid transfer protein, **148**, 662
phosphorylase phosphatase, **159**, 359, 379
platelet-derived growth factor, **147**, 4
PNGase F, **138**, 773
poly(*N*-acetylgalactosamine) deacetylase, **161**, 515
prephenate aminotransferase, **142**, 480
protein kinase C, **141**, 426
protein phosphatase
  Mg^{2+}-dependent form, **159**, 438
  type 1, **159**, 392
  type 2A, **159**, 392
  type 2B, **159**, 409
  type 2C, **159**, 417
protein phosphatase inhibitor-1 and inhibitor-2, **159**, 427
shikimate kinase, **142**, 355
sulfotransferases, **143**, 205
tryptophan *C*-methyltransferase, **142**, 236
tyrosine 3-monooxygenase, **142**, 65, 73
vanillate hydroxylase, **161**, 274
Radiography
  assessment of
    adjuvant arthritis, **162**, 345
    streptococcal cell wall-induced arthritis in rats, **162**, 376
    type II collagen-induced arthritis
      in mice, **162**, 365
      in rats, **162**, 360
Radioimmunoassay
  amyloid P component, **163**, 535
  automated, cAMP and cGMP, **159**, 45
  bacterial chemotactic factor, **162**, 207
  calbindin D, **139**, 539
  calmodulin antibodies, **139**, 402
  calmodulin-binding protein, **139**, 152
  calmodulin and dependent enzymes, **139**, 511
  calregulin, **139**, 43
  with chromium-51, leukocyte chemotaxis, **162**, 59
  collagen antibodies, **145**, 163
  complement components C3a and C5a, **162**, 587

connective tissue-activating peptide III, **163**, 739
C-reactive protein, **163**, 394
CR2 receptor, **150**, 597
cyclic nucleotides in cyanobacteria, **167**, 587
delipidated proteins, **138**, 198
desmosine, **144**, 254
diphtheria toxin fragment A–hormone hybrid, **165**, 209
elastin, **144**, 251
enterotoxins in commercial protein A, **137**, 514
enzymatic, *see* Enzymatic radioimmunoassay
extracellular matrix proteins in biological fluids, **145**, 396
fibroblast growth factor, **147**, 110
glycolipids, **138**, 200
hemopexin, **163**, 548
IgE production in human lymphocyte cultures, **150**, 315
insulin-like growth factor I, **146**, 220, 229
kinins, **163**, 275
α_2-macroglobulin, **163**, 491
PDGF-binding proteins, **147**, 60
phosphomannosyl receptor 2, **138**, 504
prostaglandins, **141**, 323
serotonin from mast cells, **162**, 516
serum amyloid A protein, comparison with ELISA, **163**, 521
staphylococcal epidermolytic toxin, **165**, 337
tissue kallikrein, **163**, 130
tissue prokallikrein, **163**, 117
T-kinin, **163**, 274
T-kininogen, **163**, 276
transferrin, **147**, 242
transferrin receptor, **147**, 261
Radioimmunoblotting
in detection of antigen-specific B-cell colonies, **150**, 297
Radioimmunoprecipitation
collagen, **163**, 684
elastin in explant cultures, **163**, 671
Radioiodination
aprotinin, for kallikrein assay, **163**, 164
calmodulin, **139**, 838; **159**, 669
complement C5a, **150**, 619
crystal-induced chemotactic factor, **162**, 196
diferric transferrin, **147**, 248
epidermal growth factor, **146**, 13
growth factor–α_2-macroglobulin conjugates, **147**, 389
insulin-like growth factor II, **146**, 249
leukocytes, in purification of IgG Fc receptors, **150**, 533
platelet-derived growth factor, **147**, 15
proteins in RNA–protein complexes, **164**, 315
T-cell receptors, **150**, 677
transferrin receptor, **147**, 258
transforming growth factor
type α, **146**, 145
type β, **146**, 175
Radioisotope flux–Millipore filtration technique
for measurement of ionic permeability of
endoplasmic reticulum vesicles, **157**, 419
sarcoplasmic reticulum vesicles, **157**, 419, 443
Radioisotopes
in analysis of cyanobacterial carbon metabolism, **167**, 539
Radioisotopic assay
arachidonate release by cultured cells, **141**, 354
pore-forming toxins in cultured cells, **165**, 278
Radiolabeling
bone proteoglycans *in vivo*, **145**, 268
C3, in assay for complement alternative pathway activity, **162**, 643
CD2 antigens, **150**, 539
cells in culture, **147**, 431
collagen peptides generated by cyanogen bromide digestion, **145**, 188
DNA probes for *in situ* hybridization to chromosomes, **151**, 283
GH_3 cells, for lipid turnover assay, **141**, 101
inositol phospholipids, **141**, 158
insulin-like growth factor I, **146**, 217
leukocytes, for *in vivo* chemotaxis assays, **162**, 104
lignins, **161**, 65
lipids, in liposome–cell interaction studies, **149**, 136
lymphocyte surface receptors, **150**, 390
lysogangliosides, **138**, 333

L-methionine with deuterium and tritium, **143**, 296
S-methyl-L-cysteine with deuterium and tritium, **143**, 296
with nonpermeant reagents, in analysis of lymphocyte receptors, **150**, 399
phosphoproteins, for protein phosphatase assays, **159**, 344
ribosomal subunits with ^{32}P, **164**, 289
5S RNA 3′ and 5′ ends, **164**, 726
T-cell receptors before isolation, **150**, 671
thyroglobulin, in analysis of spontaneous autoimmune thyroiditis, **162**, 472
transferrin receptors, **150**, 729
Radioligand assay
insulin-like growth factor II, **146**, 248
Radioreceptor assay
growth hormone, **150**, 701
insulin, **150**, 701
insulin-like growth factor II, **146**, 261
kinin B_1 receptors, **163**, 224
kinin B_2 receptors, **163**, 225
leukotriene B_4 receptors on neutrophils, **163**, 340
platelet-derived growth factor, **147**, 13
platelet-derived growth factor-binding proteins, **147**, 60
transferrin receptor, **147**, 239, 247
transforming growth factor
type α, **146**, 96, 104, 116
type β, **146**, 99, 181
transforming growth factor β receptor
monolayer assay, **146**, 169
suspension assay, **146**, 172
Reactors, *see specific reactors*
Recombination
chromosomal, in transformation of cyanobacteria, **153**, 216
in cyanobacterial mutagenesis, **167**, 721
Reconstitution
adenylate cyclase with
β-adrenergic receptors, **139**, 788
calmodulin, **139**, 787
multilamellar liposomes, **139**, 780
branched-chain 2-keto acid dehydrogenase, **166**, 308
Ca^{2+}-ATPase
by cholate dialysis, **139**, 796
from plasma membranes, **157**, 347
from platelets, **157**, 355
by Triton X-100 removal, **139**, 796
calmodulin-dependent protein phosphatase, **139**, 86
cardiac sarcoplasmic reticulum, **157**, 314
carotenogenic enzymes solubilized from chromoplast membranes, **148**, 392
H^+-ATPase, **157**, 525, 540
from chromaffin granules, **157**, 622
from clathrin-coated vesicles, **157**, 642
from endoplasmic reticulum and Golgi membranes, **157**, 617
liposomes with toxin pores, **165**, 285
Na^+,K^+-ATPase
by cholate dialysis, **156**, 127
by freeze–thaw sonication, **156**, 141
into phospholipid vesicles, **156**, 157
ribosomal 30S subunits containing fragmented 16S rRNA, **164**, 448
ribosomes from labeled components, **164**, 185
5S RNA fragment 1, **164**, 166
rRNA and ribosomal total protein fractions, **164**, 141
sarcoplasmic reticulum, control of lipid/protein ratio, **157**, 302
thylakoid lipids, **148**, 324
total, 70S ribosomes from *E. coli*, **164**, 278
vesicles by freeze–thaw sonication, **156**, 179
Rectal gland
shark, purification of Na^+,K^+-ATPase, **156**, 43
Redox potentials
chloroplast cytochromes, **148**, 266
Red pepper
chromoplasts, isolation from fruits, **148**, 176
Reduction
disulfides
direct method, **143**, 251
in electrochemical reduction cell, **143**, 257
with substitution, **143**, 253
by thiols, **143**, 246
electroenzymatic and electromicrobial, in preparation of chiral compounds, **136**, 302
electrolytic, thiols, **143**, 99

liberation of protein-bound thiols by, **143**, 128
Reductive cleavage
permethylated polysaccharides, **138**, 26
Reflectometry
in biomedical analysis
applications, **137**, 376
instrumentation, **137**, 374
principles, **137**, 373
Regeneration
genetically altered plants, **153**, 319
Reinecke salt
precipitation of trimethylselenonium ion, **143**, 197
Remazol Brilliant Blue
labeled polysaccharides, for endohydrolase assays, **160**, 74
–xylan, as soluble chromogenic xylanase substrate
applications, **160**, 538
preparation, **160**, 536
Reovirus type 3 receptors
isolation and characterization, **150**, 554
Replica plating
CHO cell colonies
application to detection of antigen variants, **151**, 108
miscellaneous applications, **151**, 111
procedure, **151**, 106
in isolation of temperature-sensitive mutants, **151**, 148
Repressors
λ phage, complexes with DNA, hydroxyl radical footprinting, **155**, 541
Resin prepolymers
emulsion-type photocrosslinkable, entrapment of biocatalysts, **135**, 239
hydrophilic photocrosslinkable
entrapment of biocatalysts, **135**, 234
properties, **135**, 233
structures, **135**, 232
hydrophobic photocrosslinkable
entrapment of biocatalysts, **135**, 237
properties, **135**, 233
structures, **135**, 232
physicochemical properties, effect on immobilized biocatalysts, **135**, 250
Resins
ion-exchange, in preparation of metal-free water, **158**, 5
photocrosslinkable, yeast immobilization with, **136**, 381
as support in
insulin-like growth factor I synthesis, **146**, 199
transforming growth factor α synthesis, **146**, 131
Resistance
antibiotic, related mutations in *E. coli* rRNA genes, **164**, 673
drug
cell selection for, role of gene amplification, **151**, 88
in CHO cells, selection and dominance analysis, **151**, 113
Respiration
cyanobacterial
associated oxygen exchange, MS analysis, **167**, 686
associated proton extrusion, measurement, **167**, 363
Restriction enzyme analysis
miniaturized, sorted chromosomes, **151**, 311
Restriction enzymes
apparent recognition specificities, alteration with DNA methylases, **155**, 41
commercially available, listing, **152**, 121
definition and nomenclature, **152**, 113
*Dpn*I, site-specific DNA cleavage at 8-, 9-, and 10-bp sequences, **155**, 22
in *E. coli* K12, overview, **152**, 130
in generation of proper ends for subsequent steps, **152**, 120
*Hgi*CI
cleavage sites, determination, **155**, 4
in mismatch cloning, **155**, 9
purification from *Herpetosiphon giganteus*, **155**, 4
mapping with, **152**, 118
*Mbo*I
digestion of chromosomal DNA for cloning into cosmid vector, **152**, 205
partial DNA digestion, **152**, 184
*Not*I
assay, **155**, 16
properties, **155**, 21
purification from *Nocardia otitidiscaviarum*, **155**, 18

*Rsr*II
assay, **155**, 11
properties, **155**, 14
purification from *Rhodopseudomonas sphaeroides*, **155**, 11
*Sau*3A, partial DNA digestion, **152**, 184
selection and use, **152**, 115
*Sfi*I
assay, **155**, 16
properties, **155**, 21
purification from *Streptomyces fimbriatus*, **155**, 20
specific fragmentation of genomic DNA, **155**, 463
Restriction fragment length polymorphisms
Phanerochaete chrysosporium DNA, use for genetic mapping, **161**, 219
procollagen gene, analysis, **145**, 205
Restriction mapping
cosmid vectors for, **152**, 604
macroregions of mammalian genomes, strategies, **151**, 461
protected probe fragments in nuclease protection assays, **152**, 626
Restriction maps
calmodulin locus in trypanosome, **139**, 275
construction strategies, overview, **152**, 118
bacterial–yeast shuttle vectors
pJO184, **153**, 49
pMH158, **153**, 47
E. coli vectors
pJRD158, **153**, 41
pJRD184, **153**, 46
pAA-PZ, **155**, 220
pAA3,7X, **155**, 187
pEMBL, **155**, 113
pGA482, **153**, 295
pGA492, **153**, 296
pGA580, **153**, 298
pGSH160, **153**, 286
pGSJ280, **153**, 286
pIJ486, **153**, 120
pIJ487, **153**, 120
pKK175-6, **153**, 56
pKK232-8, **153**, 58
pPUC29, **153**, 205
pUC303, **153**, 205
pUH24, **153**, 203
Streptomyces phage vectors, **153**, 154
Streptomyces plasmid vectors, **153**, 142
wide-host-range vectors
pJRD215, **153**, 51
RSF1010, **153**, 49
Restriction site banks
application to construction of plasmid cloning vectors, **153**, 36
Reticulin
electron immunohistochemistry, **145**, 87
Reticulocytes
cell-free system for mRNA translation, **152**, 269
protein synthesis in lysates, inhibition by Shiga toxin, **165**, 233
rabbit, purification of phosphatidylinositol kinases, **141**, 213
Retinoids
lipoidal
photosome preparation with, **149**, 106
stock solutions, storage and assay, **149**, 105
synthesis, **149**, 101
Retroviruses
hypoxanthine phosphoribosyltransferase gene-containing vectors
arrangement in transfected cells, **149**, 19
construction, **149**, 13
transfection protocol, **149**, 15
transfer into
bone marrow cells, **149**, 24
lymphoblasts, **149**, 23
infection of embryonal carcinoma cells, **151**, 81
insertional mutagenesis, in gene isolation, **152**, 469
in murine lupus, analytical procedures, **162**, 390
Reverse transcriptase, *see* RNA-directed DNA polymerase
RFLP, *see* Restriction fragment length polymorphisms
Rheumatoid factors
assay with ligand-bearing liposomes, **149**, 209
Rhizobium
nitrogen fixation genes, identification by Tn*5* mutagenesis, **154**, 197

Rhodamine
elastase assay with, **144**, 293
Rhodococcus
R. erythropolis, purification of
coniferyl-alcohol dehydrogenase, **161**, 303
α-mannanase, **160**, 621
S-6, gel-entrapped
acrylamide production, **136**, 530
nitrilase properties, **136**, 527
preparation, **136**, 525
Rhodopseudomonas sphaeroides
purification of restriction enzyme *Rsr*II, **155**, 11
Rhodospirillum rubrum
cultivation and immobilization, **137**, 669
hydrogen production in fuel cell, **137**, 671
Rhodotorula
R. glutinis, purification of phenylalanine ammonia-lyase, **142**, 246
R. minuta var. *texensis*, immobilized, stereoselective hydrolysis of *dl*-menthyl succinate, **136**, 293
Ribonuclease
contamination
prevention in RNA preparation, **152**, 20
reduction in mRNA isolation from membrane-bound polysomes, **152**, 242
in probing rRNA–protein interactions, **164**, 464
Ribonuclease A, *see* Pancreatic ribonuclease
Ribonuclease H
site-specific hydrolysis of rRNA–oligonucleotide complexes, **164**, 440
Ribonuclease T_1
limited digestion of 5S RNA, **164**, 726
in molecular titration assay of transcript prevalence levels, **152**, 643
in probing of RNA structure, monitoring by primer extension, **164**, 482
Ribonucleosides
–vanadyl complexes, in isolation of cytoplasmic RNA, **152**, 227
5′-Ribonucleotides
production with immobilized 5′-phosphodiesterase, **136**, 517
Ribose
quantitative TLC, **160**, 172
Ribose-phosphate pyrophosphokinase
immobilized, in synthesis of 5-phospho-α-D-ribosyl pyrophosphate, **136**, 278
Ribosomes
affinity labeling, **164**, 341
bacterial, crystallographic and image reconstruction studies, **164**, 95
Escherichia coli
intra-RNA crosslinking in, **164**, 290
kinetic intermediates in *in vitro* assembly, isolation, **164**, 270
^{32}P-labeled subunits, preparation, **164**, 289
preparation, **164**, 240
protein–protein crosslinks, identification, **164**, 64
RNA–protein crosslinking in, **164**, 298
70S, total reconstitution, **164**, 278
and subunits active in tRNA binding, preparation, **164**, 658
subunit surfaces, localization of ribosomal proteins, **164**, 503
functional domains, definition, **164**, 352
function, ^{1}H NMR analysis, **164**, 148
inactivation by Shiga toxin *in vitro*, **165**, 235
P and A sites, binding of tRNA, assay, **164**, 597
peptidyltransferase, photoaffinity labeling, **164**, 361
proteins, *see* Proteins, ribosomal
reconstitution from labeled components, **164**, 185
30S
DNP–hapten-labeled at 5′-end of 16S RNA, preparation, **164**, 85
with fragmented 16S rRNA, reconstitution, purification, and biological activity, **164**, 448
protein separation, **164**, 536
protein topography, neutron scattering study, **164**, 117
16S rRNA 5′ terminal region in, accessibility, **164**, 451
40S, preparation from rat liver, **164**, 579
70S
affinity for tRNA and mRNA, effect of 5S RNA point mutations, **164**, 747

16S rRNA 5′ terminal region in, accessibility, **164**, 451
S6
multiple phosphorylated forms, electrophoretic analysis, **146**, 362
phosphopeptide analysis, **146**, 369
specialized system
applications, **153**, 450
construction, **153**, 434
properties, **153**, 445
structural analysis
electron microscopy and computer image processing, **164**, 3
^{1}H NMR, **164**, 148
scanning transmission electron microscopy, **164**, 49
structure and function, analysis with cDNA probes, **164**, 401
subunits
complex formation with antibodies and oligonucleotides, **164**, 502
electron micrographs, statistical image analysis, **164**, 35
hapten-labeled at RNA 3′-end, preparation and immunocomplex formation, **164**, 82
50S
deuterated, preparation and assays for neutron scattering analysis, **164**, 131
protein separation, **164**, 541
single proteins, preparation, **164**, 134
total proteins, preparation, **164**, 133
surface topography, analysis by hot tritium bombardment, **164**, 426
template-free polypeptide synthesis from aminoacyl-tRNA, **164**, 631
translation activity, kinetics, **164**, 611
translation initiation complexes, extension inhibition analysis, **164**, 419
translocation, pre-steady-state kinetics, **164**, 581
tRNA binding sites, affinity labeling, **164**, 372
Rice
amyloplasts, isolation from endosperm, **148**, 172
Ricin
A subunit
growth factor-conjugated
cell variant isolation with, **147**, 386
preparation, **147**, 384
purification, **147**, 383
Ricinus communis, *see* Castor bean
Rivanol, *see* Ethacridine
R-loop analysis
procollagen mRNA, for assessment of collagen mutations
principle, **145**, 236
procedures, **145**, 244
required materials and reagents, **145**, 242
RNA
antisense, inhibition of endogenous genes, **151**, 519
Aplysia
analysis with calmodulin cDNA, **139**, 284
isolation, **139**, 282
binding to ribosomal proteins
filter assay, **164**, 203
gel mobility shift assay, **164**, 215
sucrose gradient assay, **164**, 215
changes in mitogen-treated lymphocytes, **150**, 65
crosslinking with proteins in *E. coli* ribosomes, **164**, 298
crosslinks, isolation and identification, **164**, 319
cyanobacterial, isolation from transformants, **153**, 228
cytoplasmic, isolation in presence of vanadyl complexes, **152**, 227
detection in culture by *in situ* hybridization, **151**, 539
–DNA hybridization, in chemical probing of large RNA molecules, **164**, 475
dot blotting, **152**, 585
electroporation-mediated transfer into plant protoplasts, **153**, 336, 351
3′-end labeling with T4 RNA ligase, **152**, 103
5′-end labeling with ^{32}P, **152**, 100
in fibroblasts, hybridization for analysis of collagen production, **163**, 648
gel electrophoresis
in agarose, **152**, 75
in polyacrylamide, **152**, 64
hybridization to oligonucleotides, **154**, 106

intramolecular crosslinking in *E. coli* ribosomes, **164**, 290
isolation with guanidinium salts, **152**, 219
large-scale phenol extraction, **152**, 39
mapping with S_1 nuclease and synthetic oligonucleotides, **154**, 87
messenger
  affinity of 70S ribosomes for, effect of 5S RNA point mutations, **164**, 747
  calcium protease, distribution and complexity, **139**, 373
  collagen
    assay in cell cultures, **147**, 413
    assay by *in vitro* translation, **144**, 84, 91
  collagenase, isolation and *in vitro* translation, **144**, 170
  collagen type I
    in osteogenesis imperfecta fibroblasts, analysis, **145**, 223
    steady-state levels, assay, **144**, 79
  elastase, cloning of sequences, **144**, 304
  elastin
    fractionation, **144**, 275
    isolation, **144**, 271
    tissue levels, assay, **144**, 223
  epidermal growth factor
    preparation, **147**, 197
    sources, **147**, 195
  fibroblast-activating factor, isolation and translation, **163**, 729
  fragments as primers in second-strand cDNA synthesis, **152**, 330
  haptoglobin, tissue localization, **163**, 468
  hybrid selection and translation arrest, **152**, 567
  insulin-like growth factor
    biological sources, **146**, 240
    expression analysis, **146**, 247
    isolation, **146**, 242
  interaction zone on ribosomal 30S subunits, analysis, **164**, 354
  interleukin-2 receptor, purification, **150**, 693
  isolation from polysomes, **152**, 241, 253
  laminin
    abundance in cells and tissues, **144**, 468
    assay, **144**, 469
    isolation, **144**, 465
  from lignin-degrading fungus
    characterization, **161**, 225
    extraction from mycelium, **161**, 222
    separation from other nucleic acids, **161**, 224
  microinjection and complementation of mutant cells
    applications, **151**, 379
    overview, **151**, 371
    procedure, **151**, 373
  molar concentration determination, **152**, 262
  nerve growth factor
    preparation, **147**, 197
    sources, **147**, 195
  9-kDa calcium-binding protein
    characterization, **139**, 348, 352
    purification, **139**, 341, 350
  parvalbumin, *in vitro* assay, **139**, 772
  preparation for full-length cDNA cloning, **154**, 5
  procollagen, R-loop analysis for assessment of collagen mutations, **145**, 235
  prolactin, assay for analysis of Ca^{2+}/calmodulin gene regulation, **139**, 659
  rare species, cloning by subtraction technique, **152**, 425
  ribosomal binding sites, **164**, 418
  serum amyloid A protein, measurement in tissues, **163**, 511
  single species, ribosome direction to, in analysis of rRNA mutations, **153**, 432
  specific types, prevalence levels in RNA populations, assay, **152**, 633
  translation
    in injected *Xenopus* oocytes, **152**, 288
    in reticulocyte cell-free system, **152**, 267
    in vitro, characterization of products, **152**, 296
  translation initiation complexes, extension inhibition analysis, **164**, 419
microinjection into *Xenopus* oocytes, cell preparation for, **152**, 284
nuclear
  extraction, **152**, 236

transcription rate analysis *in vitro*, **152**, 238
nucleotide sequences, computer analysis, **164**, 765
poly(A), isolation from *Phanerochaete chrysosporium*, **161**, 229
poly(A)$^+$
fractionation, **152**, 259
purification, **152**, 254
precipitation with
ethanol, **152**, 44
high salt concentrations, **152**, 48
preparation
associated RNase contamination, prevention, **152**, 20
and characterization, overview, **152**, 215
–protein complexes
analysis, **164**, 315
formation via soluble carbodiimide, **164**, 311
purification, **164**, 311
from recombinant vaccinia virus-infected cells, analysis, **153**, 557
removal of
3′-phosphates, **152**, 102
5′-phosphates, **152**, 99
ribosomal
affinity-labeled, identification, **164**, 348
antibody probes for, **164**, 493
complementary oligonucleotides, synthesis and modification, **164**, 499
crosslinking by psoralen, **164**, 330
electron microscopic analysis
conformational changes, **164**, 59
interactions with ribosomal proteins, **164**, 61
mass measurements, **164**, 53
radius of gyration, **164**, 58
3′-end modification, **164**, 499
with hapten, **164**, 82
fluorescent labeling, **164**, 177
genes in *E. coli*, antibiotic resistance mutations, **164**, 673
interaction with initiation factor 3, physical studies, **164**, 238
mutations and effects on single messenger translation, analysis, **153**, 432
naked, electron microscopy, **164**, 77
oligonucleotide binding, quantitative estimation, **164**, 449
–oligonucleotide complexes, site-specific hydrolysis, **164**, 440
in phylogenetic analysis, **164**, 793
plasmid-coded, maxicell analysis, **164**, 691
purification, **164**, 180
ribosomal protein binding
analysis with α-sarcin, **164**, 468
enzymatic and chemical probing, **164**, 456
and ribosomal total proteins, reconstitution, **164**, 141
5S
and fragments, preparation for NMR and crystallography studies, **164**, 158
internally labeled by hapten, preparation, **164**, 87
primer-directed deletions, **164**, 710
structural and functional studies, **164**, 721
16S
5′-end modification with DNP–hapten, **164**, 85
sequencing, in phylogenetic studies of cyanobacteria, **167**, 132, 138
3′ terminal cloacin fragment, preparation, **164**, 244
16S colicin fragments
gel electrophoretic analysis, **164**, 196
^{32}P-labeled, isolation, **164**, 197
properties, **164**, 198
purification, **164**, 191
23S, photoaffinity labeling, **164**, 363
23S and 5S
deuterated, isolation, **164**, 138
separation, **164**, 132
specific cleavage at 7-methylguanine residue, **164**, 454
specific fragments, preparation, **164**, 221
topography, immunoelectron microscopy, **164**, 80
structural analysis via probing monitored by primer extension, **164**, 481
synthesized from plasmids containing phage promoters, in DNA sequencing, **152**, 563

termini, gel electrophoretic analysis, **155**, 433
tobacco mosaic virus
encapsulation into liposomes, **148**, 41
expression in tobacco protoplasts, assay, **148**, 42
transfection of tobacco protoplasts, **148**, 42
transfer
affinity of 70S ribosomes for, effect of 5S RNA point mutations, **164**, 747
amino acid-specific, HPLC, **155**, 373
aminoacyl-, in template-free ribosomal polypeptide synthesis, **164**, 631
binding to ribosomal P and A sites, assay, **164**, 597
binding to ribosome preparations, assay, **164**, 665
binding to ribosomes
analysis, **164**, 355
assays, **164**, 408
displacement and release during ribosomal translocation, kinetics, **164**, 591
dissociation from ribosomes, kinetics, **164**, 593
doubly modified, preparation and use, **164**, 95
fluorescent labeling, **164**, 177
misacylated, in template-free ribosomal polypeptide synthesis, **164**, 645
modified, functional activity, **164**, 389
modified nucleosides, analysis, **155**, 392
purification, **164**, 181
ribosomal binding sites, **164**, 416
affinity labeling, **164**, 372
sequence analysis, **155**, 381
RNA-directed DNA polymerase
in analysis of
affinity-labeled 23S rRNA, **164**, 368
rRNA protein binding sites, **164**, 463
in direct sequencing of denatured plasmid DNA, **152**, 558
viral
first-strand cDNA synthesis with, **152**, 321
second-strand cDNA synthesis with, **152**, 328
RNA ligase, *see* Polyribonucleotide synthase (ATP)
RNA mapping
by nuclease protection assays and cDNA primer extension, **152**, 611
RNA polymerase
assay in mitogen-stimulated lymphocytes, **150**, 67
DNA-directed, *see* DNA-directed RNA polymerase
SP6, *in vitro* RNA synthesis with, **155**, 397
RNA probes
for genomic walking, synthesis, **152**, 607
preparation with phage SP6 transcription system, **155**, 519
single-stranded, preparation for measuring transcript prevalence levels, **152**, 637
for Southern and Northern blotting, **152**, 577
RNA synthesis
with SP6 RNA polymerase *in vitro*, **155**, 397
in *Synechococcus*, heat shock effects, **167**, 625
Rod outer segments
purification of
cGMP phosphodiesterase, **159**, 707
G protein subunits, **159**, 703
Rosette assay
C1q receptor, **150**, 565
CR2 receptor, **150**, 589
factor H receptor, **150**, 599
IgE Fc receptors on lymphoid cells, **150**, 514
IgG Fc receptors on leukocytes, **150**, 529
Rubber tree
lutoids, preparation, **148**, 89
lutoid tonoplasts
biochemical composition, **148**, 91
enzymatic activities, **148**, 91
preparation, **148**, 90, 107
Rubidium
binding to Na^+,K^+-ATPase, centrifugation assay, **156**, 230
–congener exchange in vesicles, assay, **156**, 188

electrogenic transport by Na^+,K^+-ATPase, **157**, 244
influx into erythrocytes, vanadate effects, **156**, 255
isotope spaces and flux rates in sarcoplasmic reticulum vesicles, **157**, 422
net flux in vesicles, assay, **156**, 188
occluded ions, release from Na^+,K^+-ATPase, measurement, **156**, 281
^{86}Rb, efflux from vesicles, assay, **156**, 184
–rubidium exchange in vesicles, assay, **156**, 185
Ruminococcus albus
cultivation, **160**, 216, 393, 409
purification of
cellobiosidase, **160**, 393
cellulase, **160**, 219
β-glucosidase, **160**, 410
Ryanodine
binding to cardiac junctional sarcoplasmic reticulum, **157**, 104
Ryanodine receptors
sarcoplasmic reticulum-associated
binding assay, **157**, 491
characteristics, **157**, 502
purification
cardiac preparations, **157**, 500
skeletal muscle preparations, **157**, 493
Ryegrass
protoplast preparation, transformation, and culture, **153**, 329

S

Saccharification
industrial
immobilized enzyme system, **136**, 377
liquid enzyme system, **136**, 374
Saccharomyces cerevisiae
calcium alginate-immobilized, industrial ethanol production, **136**, 394
cAMP cascade system mutants
applications, **159**, 41
biochemical characterization, **159**, 33
isolation and genetic identification, **159**, 28
cell walls, coupling to modified Separon H 1000, **135**, 351
chromosome ploidy, visual assays, **155**, 351
coentrapment with β-glucosidase, **137**, 642
coupling to
hydrous transition metal oxides, **135**, 358
modified Separon H 1000, **135**, 348
transition metal-activated supports, **135**, 367
DNA-binding activities in extracts, gel electrophoretic assay, **155**, 528
–*E. coli* shuttle restriction site bank vectors, construction, **153**, 47
encapsulation with lactase, **137**, 646
expression and secretion vectors for, **153**, 516
5-fluoroorotic acid as selective agent, **154**, 164
gene isolation by complementation, **152**, 481
heterologous gene expression, **152**, 673
immobilization in κ-carrageenan, **135**, 196
immobilized on wood chips
application to ethanol fermentation, **135**, 340
characteristics, **135**, 336
flow velocity effects, **135**, 338
preparation, **135**, 334
microsomal fraction, immobilization and properties, **137**, 680
mitochondrial cloning vectors for, **153**, 370
permeabilized cells, immobilization and properties, **137**, 680
polyacrylamide gel-entrapped
enzymatic activity, **135**, 208, 209
microcultural analysis, **135**, 202
preparation, **135**, 199
viability, **135**, 202, 209
purification of
O-acetylserine–*O*-acetylhomoserine sulfhydrylase, **143**, 480
chitin synthase, **138**, 645
cytochrome P450, **137**, 679
1,3-β-glucan synthase, **138**, 638
H^+-ATPase from
plasma membranes, **157**, 528, 537
vacuolar membranes, **157**, 552

thiosulfate reductase, **143**, 351
Zn-containing cyclic nucleotide phosphodiesterases, **159**, 777
site-directed chromosomal rearrangements, **153**, 243
sulfometuron methyl-resistant mutants, isolation, **166**, 104
Saccharomyces lactis
entrapment by cellulose acetate beads, **135**, 291
applications, **135**, 292
Saccharomyces paradoxus
coupling to modified Separon H 1000, efficiency, **135**, 346
Saccharomyces spp.
resin-immobilized
industrial ethanol production, **136**, 385
preparation, **136**, 381
Saccharomyces uvarum
adsorption to solid supports
applications, **135**, 317
effects on
growth kinetics, **135**, 316
metabolic activity, **135**, 314
kinetics, **135**, 310
methodology, **135**, 308
Salivary glands
cell-free system, phosphoinositide breakdown, hormonal activation, **141**, 255
submandibular, purification of kallikrein, **163**, 150
Salmonella
diarrheal toxin, ileal loop test, **165**, 318
S. typhimurium, purification of
acetolactate synthase II, **166**, 448
anthranilate synthase–phosphoribosyltransferase and subunits, **142**, 372
tryptophan synthase, **142**, 402
Salt gland
duck, purification of Na^+,K^+-ATPase, **156**, 46
Salts
chaotropic, effects on cGMP-binding phosphodiesterase, **159**, 717
Sand bed reactor
biofilm-fluidized
description, **135**, 301
examples, **135**, 305
operation, **135**, 302
α-Sarcin
in analysis of rRNA–protein interactions, **164**, 468
Sarcolemma
vesicles
characterization, **157**, 31
preparation from
canine heart, **157**, 85
rabbit skeletal muscle, **157**, 27
Sarcoplasmic reticulum
associated proteins, cDNA cloning, **157**, 289
ATP synthesis
analytical methods, **157**, 190
from Ca^{2+} gradient by Ca^{2+}-ATPase, **157**, 220
Ca^{2+}-ATPase
crystallization, **157**, 272
kinetic and equilibrium characterization, **157**, 154
calcium-binding protein, purification and crystallization, **157**, 321
calcium channels
observation in planar lipid bilayers, **157**, 480
purification, **157**, 493
calcium release
calcium-induced
kinetic analysis, **148**, 681
measurement, **141**, 10
inositol trisphosphate-induced, measurement, **141**, 15
measurement in skinned muscle fibers, **157**, 12
measurement techniques, **141**, 5
models, **141**, 4
calcium-transporting systems, properties, **157**, 7
cardiac
Ca^{2+}-ATPase assays, **157**, 117, 128, 134
calcium transport assays, **157**, 112, 128, 134
cAMP-dependent phospholamban phosphorylation, **157**, 123
characteristics after reconstitution, **157**, 317
junctional and longitudinal fractions
characteristics, **157**, 102
preparation, **157**, 101

preparation and structural characteristics, **157**, 109
purification of
calcium channel, **157**, 500
phospholamban, **157**, 138, 360
solubilization and reconstitution, **157**, 315
vesicles
characteristics, **157**, 96
preparation, **157**, 92
characteristics after reconstitution, **157**, 312
coupling to transverse tubules, **157**, 469
junctional face membrane isolation, **157**, 46
light and heavy fractions, isolation from rabbit skeletal muscle, **157**, 36
phospholipid content, modification, **157**, 369
reconstitution, control of lipid/protein ratio, **157**, 302
transverse tubule-mediated Ca^{2+} release, induction and measurement, **157**, 472
vesicles
calcium fluxes, measurement, **157**, 464
characterization, **157**, 42
ionic permeability
ANS fluorescence measurements, **157**, 455
effects of drugs, **157**, 464
fluorescence-quenching measurements, **157**, 455
light-scattering measurements, **157**, 427, 446
membrane potential–dye measurements, **157**, 430, 452
radioisotope flux–Millipore filtration measurements, **157**, 419, 443
isolation, **157**, 418
proton fluxes, measurement, **157**, 468
spherical, permeation characteristics, **157**, 439
voltage-induced Ca^{2+} uptake, **157**, 248
Schistosoma japonicum
eggs, purification of eosinophil chemotactic factor, **162**, 176
Schizophyllum commune
cultivation, **160**, 435
purification of
acetylxylan esterase, **160**, 705
β-glucosidase, **160**, 435
xylanase A, **160**, 661
Schizosaccharomyces pombe
mitochondrial cloning vectors for, **153**, 370
purification of
O-acetylhomoserine sulfhydrylase, **143**, 466
H^+-ATPase from plasma membranes, **157**, 513
Schwann cells
rat, cultures
maintenance and purification, **147**, 218
preparation, **147**, 218
proliferation assay, **147**, 219
Sclerotium rolfsii
cultivation, **160**, 378, 450
purification of
carboxymethylcellulase, **160**, 378
cellobiose dehydrogenase, **160**, 450
1,4-β-D-glucan cellobiohydrolase, **160**, 310
β-glucosidase, **160**, 426
β-xylosidase, **160**, 680
Scopolia japonica
purification of α-L-arabinofuranosidase, **160**, 713
Screening
cDNA expression libraries in mammalian cells, **154**, 25
colonies and plaques, with radioactive nucleic acid probes, **152**, 415
DNA libraries
by antibody recognition, **152**, 396
as clonal populations, **152**, 398
with nucleic acid probes, **152**, 393
by phenotypic expression, **152**, 396
immuno-, *see* Immunoscreening
λgt libraries for specific sequences, considerations, **152**, 367
Na^+,K^+-ATPase antibodies by various procedures, **156**, 399
phage libraries, **152**, 414
plasmid libraries, **152**, 408
+/−, in isolation of differentially expressed genes, **152**, 424
recombinant DNA libraries, with oligonucleotide probes, **152**, 432
Southern and Northern blots, **152**, 576

Sea urchin
  purification of Spec proteins, **139**, 360
Sedimentation assay
  actin properties in presence of modulating proteins, **162**, 260
  calmodulin–phosphofructokinase binding, **139**, 752
  ribosomal protein–RNA complexes, **164**, 215
Sedimentation equilibrium analysis
  membrane proteins, **150**, 447
  21-kDa calcium-binding protein, **139**, 95
Sedum praealtum, *see* Stonecrop
Seeds
  Erythrina, lectin purification, **138**, 544
  Evonymus, lectin purification, **138**, 553
  Vicia, lectin purification, **138**, 538
Selected ion monitoring
  in GC–MS assay of catecholamine metabolites, **142**, 571
Selection
  5-fluoroorotic acid-based, in yeast molecular genetics, **154**, 164
  mammalian cell mutants
    application of replica plating, **151**, 104
    role of gene amplification, **151**, 88
    suicide method, **151**, 121
    with tritium-labeled precursors, **151**, 131
  role in site-specific mutagenesis, **154**, 367
  sib, *see* Sib selection
  temperature-sensitive cell mutants, **151**, 145
Selenium
  assay by
    atomic absorption spectrometry, **158**, 140, 399
    fluorometry, **158**, 397
    gas chromatography–mass spectrometry, **158**, 393
    gas–liquid chromatography, **158**, 401
  deficiency in rats, production, **143**, 307
  volatilization, **143**, 34
Selenocysteine
  assay with Gaitonde's reagent, **143**, 148
  properties, **143**, 242
  Se-substituted forms, preparation, **143**, 292
  synthesis, **143**, 240
Selenocysteine β-lyase
  assay, **143**, 415
  properties, **143**, 418, 494
  purification from
    Citrobacter freundii, **143**, 493
    porcine liver, **143**, 416
L-Selenocystine
  preparation, **143**, 295
Selenodjenkolic acid
  properties, **143**, 244
  synthesis, **143**, 243
L-Selenohomocystine
  preparation, **143**, 295
Seminal fluid
  extracellular matrix proteins, analysis, **145**, 396
S150 enzymes
  tRNA-free, preparation, **164**, 664
Separon H 1000
  covalent coupling of yeast cells, **135**, 341
Sepharose
  –blue dextran, preparation and regeneration, **159**, 773
  –protein A, in antibody-mediated carboxypeptidase A immobilization, **135**, 165
Sepharose 4B
  copper-binding capacity, quantitation, **158**, 17
  immobilization of
    chymotrypsin, **135**, 504
    20β-hydroxysteroid dehydrogenase subunits, **135**, 477
    protein subunits, **135**, 487
  immobilized trypsin, reactivation, **135**, 594
  pyridoxal phosphate-bound
    coupling of nucleoside diphosphatase, **135**, 157
    preparation, **135**, 156
Sepharose CL-4B
  chlorocarbonate-activated
    active groups, determination, **135**, 88
    affinity chromatography with, **135**, 89
    bound trypsin, coupling yield and activity, **135**, 89
    coupling of proteins and ligands, **135**, 88
    preparation, **135**, 87

immobilization of
alkaline phosphatase subunits, **135**, 493
β-galactosidase, **136**, 231
Sepharose CL-6B
1,1′-carbonyldiimidazole-activated
enzyme coupling, **135**, 115
preparation, **135**, 108
protein purification with, **135**, 113
Sepharose C_7-NH_2
synthesis, **142**, 417
Septum
bovine nasal, type II collagen preparation from, **162**, 358
Serotonin
brain, HPLC assay, **142**, 539, 547
mast cell-derived, assay, **162**, 515
Serratia marcescens
coupling to hydrous transition metal oxides, **135**, 358
purification of chitinase, **161**, 460
Serum, *see also* Blood, Plasma
acute-phase proteins, inflammation-induced changes, analysis, **163**, 567
aluminum levels, measurement, **158**, 295
chicken, purification of transferrin, **147**, 291
chromium levels, measurement, **158**, 339
cobalt levels, measurement, **158**, 349
fetal bovine, effect on elastin synthesis, **144**, 237
glucose
assay with enzyme reactor, **137**, 263
flow enthalpimetric assay, **137**, 230
human, isolation of
C3a, C4a, and C5a, **162**, 617
C5a, **150**, 617
serum amyloid P component, **163**, 531
insulin-like growth factor I levels, regulation, **146**, 225
lipids, assay with enzyme thermistor, **137**, 221
rat, purification of α_2-macroglobulin, **163**, 486
uric acid, assay with enzyme reactor, **137**, 263
Xenopus, purification of lectins, **138**, 518
Serum albumin
assay with optoelectronic sensor, **137**, 343
bovine
cationized, induction of glomerulonephritis, **162**, 449
in production of
acute serum sickness, **162**, 484
chronic serum sickness, **162**, 488
as chemotactic agent, preparation, **162**, 181
interaction with antibody, ellipsometric studies, **137**, 371
Serum amyloid A protein
assay, **163**, 519
mRNA, measurement in tissues, **163**, 511
Serum amyloid P component
isolation from human serum, **163**, 531
quantitation, **163**, 532
Serum sickness
acute, production in rabbits with bovine serum albumin, **162**, 484
chronic, production in
mice, with foreign serum proteins, **162**, 499
rabbits, with bovine serum albumin, **162**, 488
rats, with bovine serum albumin, **162**, 492
passive form, **162**, 501
production in mice, rabbits, and rats with cationic proteins, **162**, 499
Sheaths
cyanobacterial, isolation, **167**, 183
Shiga toxin
assays, **165**, 158
cytotoxicity, **165**, 232
inhibition of protein synthesis *in vitro*, **165**, 233
production, **165**, 153, 399
purification, **165**, 155, 401
selective destruction of vagal sensory neurons, analysis, **165**, 250
Shigella
purification of Shiga toxin, **165**, 155
S. dysenteriae
diarrheal toxin, ileal loop test, **165**, 319
purification of toxin, **165**, 250
Shikimate dehydrogenase
assay, **142**, 316
properties, **142**, 318
purification from *E. coli*, **142**, 317

Shikimate kinase 1
assay, **142**, 355
properties, **142**, 359
purification from *E. coli*, **142**, 357
Shikimate kinase 2
assay, **142**, 355
partial purification from *E. coli*, **142**, 359
Shikimate pathway
associated reactions, **142**, 326
Shikimate 3-phosphate
preparation for *arom* multifunctional enzyme assay, **142**, 329
Showdomycin
as inhibitor of lipid-linked saccharide pathway, **138**, 675
Sialate *O*-acetylesterase
isolation, **138**, 622
properties, **138**, 623
Sialic acids
O-acetyl groups
enzymatic hydrolysis, **138**, 620
enzymatic incorporation, **138**, 617
migration, **138**, 615
cell-surface, labeling with hydrazides, **138**, 441
chromatographic analyses
GLC, **138**, 149
GLC–MS, **138**, 150
HPLC, **138**, 144
HPLC–MS, **138**, 153
TLC, **138**, 143
colorimetric assay, **138**, 138
content in CD2 antigens, **150**, 547
fast atom bombardment MS, **138**, 155
histochemical analysis, **138**, 158
isolation and purification, **138**, 135
NMR spectroscopy, **138**, 157
oligomers, isolation, **138**, 790
removal from glucocerebrosidase before cell targeting, **149**, 31
specific lectins
from *Cancer antennarius*, **138**, 520
from *Limax flavus*, **138**, 527
Sialidase
diagnostic assay for galactosialidosis, **138**, 760
with galactose oxidase, stimulation of T cells, **150**, 105
glucocerebrosidase digestion before cell targeting, **149**, 31
Sialyl oligosaccharides
binding proteins, specificity analysis, **138**, 162
chemical synthesis, **138**, 402
and phosphates, isolation from bovine colostrum and human urine, **138**, 289
Sialyltransferase
erythrocyte resialylation with, **138**, 163
glycosphingolipid-synthesizing, assays, **138**, 601
polysialosyl glycoprotein detection with, **138**, 185
preparation from *E. coli*, **138**, 181
Sib selection
application to gene isolation
fractionation of cosmid library
serial procedure, **151**, 447
simple procedure, **151**, 446
grid analysis, **151**, 448
overview, **151**, 445
in detection of hybridoma and myeloma cell variants, **151**, 61
Silanization
inorganic surface
in biomolecule immobilization, **137**, 383
procedure, **137**, 382
Silica, *see* Silicon dioxide
Silicon
atomic absorption spectrometry, **158**, 141
Silicon dioxide
porous
immobilization of
dextransucrase, **136**, 245
yeast cells, **135**, 308
transition metal-activated, coupling of yeast cells, **135**, 367
silanized surfaces
in biomolecule immobilization, **137**, 383
preparation, **137**, 382
sols, in purification of chloroplasts, **148**, 179
supports, attached deoxynucleosides, synthesis, **154**, 300
vitreous, in preparation of metal-free water, **158**, 5

Silicon wafers
immobilization of biomolecules
application to ellipsometry, **137**, 392
procedures, **137**, 389
Silver
atomic absorption spectrometry, **158**, 141
Silver staining
lipopolysaccharide monomers, **138**, 271
Silybum marianum, *see* Thistle
Simian sarcoma virus
platelet-derived growth factor-like gene product, analysis, **147**, 22
Sinapyl alcohol
preparation, **161**, 73
Size-exclusion chromatography
high-resolution, cellooligomers, **160**, 41
lignin–carbohydrate complexes
choice of solvents and columns, **161**, 193
conventional method, **161**, 195
high-performance method, **161**, 197
Skeletal growth factor
bioassay, **146**, 283
purification from bovine bone, **146**, 287
Skeletal muscle
sarcolemmal vesicles
characterization, **157**, 31
preparation, **157**, 27
sarcoplasmic reticulum
isolation of
junctional face membrane, **157**, 46
light and heavy fractions, **157**, 36
purification of
calcium-binding protein, **157**, 321
calcium channel, **157**, 493
skinned fibers
calcium release, measurement, **157**, 16
preparation, **157**, 13
transverse tubules, isolation, **157**, 68
triads
characterization, **157**, 58
isolation, **157**, 51
Skin
rat
elicitation of reverse passive Arthus reaction, **162**, 479
purification of sulfhydryl oxidase, **143**, 512
Xenopus, purification of lectins, **138**, 518
Skinning
skeletal muscle fibers, **157**, 13
Skin window technique
and modifications, for chemotaxis assays *in vivo*, **162**, 105
Snail
purification of
β-mannanase, **160**, 602
β-mannosidase, **160**, 616
sulfatases, **143**, 363
Sodium
atomic absorption spectrometry, **158**, 141
binding to Na^+,K^+-ATPase
during ATP hydrolysis, assay, **157**, 233
centrifugation assay, **156**, 230
characteristics, **156**, 235
kinetics, **156**, 244
–calcium exchange in plasma membrane vesicles, measurement, **157**, 505
concentration, effect on Na,K-pump reconstitution by sonication, **156**, 147
effect on Na^+,K^+-ATPase, **156**, 2
efflux from
dialyzed giant axons, vanadate effects, **156**, 252
erythrocytes, vanadate effects, **156**, 259
resealed erythrocyte ghosts, vanadate effects, **156**, 263
intracellular levels, measurement, **146**, 387
isotope spaces and flux rates in sarcoplasmic reticulum vesicles, **157**, 422
^{22}Na, in assay for uptake of liposome-encapsulated agents, **149**, 143
occluded ions, release from Na^+,K^+-ATPase, measurement, **156**, 281
–potassium exchange in vesicles, assay, **156**, 181
–potassium pump, *see* ATPase
–sodium exchange in vesicles, assay, **156**, 185
transport
in cyanobacteria, assay, **167**, 570
in erythrocyte inside-out vesicles, assays, **156**, 175
in isolated transverse tubules, **157**, 79

after reconstitution of Na,K pump
assay, **156**, 144
coupling ratio, **156**, 153
electrogenicity, **156**, 154
temperature effects, **156**, 151
time course at different temperatures, **156**, 151
in sarcolemmal vesicles, assay, **157**, 34
across tonoplasts, measurement, **148**, 112
uncoupled flux in vesicles, assay, **156**, 185
Sodium alginate, *see* Algin
Sodium amalgam
reduction of disulfides, **143**, 252
Sodium bisulfite
mutagenesis of metallothionein coding sequences, **154**, 423
Sodium borohydride
tritiated, and galactose oxidase, in labeling of lymphocyte receptors, **150**, 412
Sodium *tert*-butoxide
dimethyl sulfoxide-dissolved
preparation, **135**, 95
preparation of isonitrile-derivatized polymers with, **135**, 95
Sodium dodecyl sulfate
effect on elastase activity, **144**, 295
in isolation of plague murine toxin subunits, **165**, 166
Sodium *p*-mercurihydroxybenzoate
enrichment of cyanobacterial mutants, **167**, 733
Sodium metaperiodate
–potassium permanganate oxidation, in lignin characterization, **161**, 124
stimulation of T cells, **150**, 105
Sodium orthovanadate
effect on phosphoinositide hydrolysis, **141**, 120
Sodium periodate, *see* Sodium metaperiodate
Soft agar growth assay
transforming growth factors
culture medium selection, **146**, 347
general methodology, **146**, 344
indicator cells used, **146**, 342
rapid assay procedure, **146**, 351
Solanum aviculare, *see* Kangaroo apple
Solanum tuberosum, *see* Potato
Solid-phase immunoisolation
transferrin receptors, **150**, 732
Solid-phase synthesis
insulin-like growth factor I, **146**, 199
kinins, **163**, 263
transforming growth factor α, **146**, 128
Solubilization
Ca^{2+}-ATPase
in monomeric and oligomeric forms, **157**, 261
from platelets, **157**, 355
cardiac sarcoplasmic reticulum, **157**, 315
carotenogenic enzymes from chromoplast membranes, **148**, 392
collagen type IV
by limited digestion, **145**, 371
by reduction, **145**, 369
H^+-ATPase from yeast plasma membranes, **157**, 520, 530
Kdp-ATPase, **157**, 671
leukotriene B_4 receptor, **141**, 377
lymphocyte membrane proteins, **150**, 392
membrane proteoglycans, **144**, 396
monoamine oxidase from mammalian liver, **142**, 623
Na^+,K^+-ATPase, **156**, 72, 133
particulate cAMP phosphodiesterase, **159**, 740
sarcoplasmic reticulum, role in lipid/protein ratio control during reconstitution, **157**, 308
transferrin receptors, **150**, 730
Solubilization–precipitation assay
intracellular asialoorosomucoid–receptor dissociation, **149**, 5
intracellular asialoorosomucoid segregation from receptor, **149**, 7
Solutions
aqueous, of metals and ligands, computer calculations of ionic concentrations, **157**, 378
Solvents
metal complex, solvolysis of cellulose, **160**, 12
Solvent transfer
in purification of staphylococcal δ-lysin, **165**, 19

Solvolysis
cellulose, with cadoxen, **160**, 11
sulfatides, **138**, 479
Somatomedin C, *see* Insulin-like growth factor I
Sonication
DNA, generation of subfragments for sequencing, **155**, 58
freeze–thaw, reconstitution of
Na^+,K^+-ATPase, **156**, 141
vesicles, **156**, 179
Sorfix
immobilization of
fungal mycelium, **135**, 355
plant cells, **135**, 355
yeast cells, **135**, 354
Sorghum
purification of chorismate mutase, **142**, 459
Southern blotting
affinity-labeled 23S rRNA, **164**, 366
Aplysia DNA, with calmodulin cDNA, **139**, 287
methodology, **152**, 572
miniprep DNA from cyanobacterial transformants, **153**, 227
Soybean
amyloplasts, isolation from suspension-cultured cells, **148**, 218
cell walls
isolation of phytoalexin elicitor, **138**, 722
preparation, **138**, 721
nuclei, isolation from suspension-cultured cells, **148**, 535
purification of
chitinase and chitobiase, **161**, 492
phenylalanine ammonia-lyase, **142**, 249
Spectrofluorometric assay
4-methylumbelliferone release from glycosides, **160**, 48
Spectrography
dc arc emission, vanadium assay, **158**, 414
Spectrometer
ESR, in analysis of lymphoid cell receptors, **150**, 425
Spectrometry, *see specific techniques*
Spectrophotometric assay
acetolactate synthase, **166**, 230
isozyme II, **166**, 446
isozyme III, **166**, 455
β-*N*-acetylhexosaminidase, **161**, 463
O-acetylserine–*O*-acetylhomoserine sulfhydrylase, **143**, 479
acid proteases, **160**, 501
anthranilate synthase–phosphoribosyltransferase-associated glutaminase activity, **142**, 371
arogenate dehydratase, **142**, 498
arogenate dehydrogenase, **142**, 514
aromatic amine dehydrogenase, **142**, 650
aromatic-amino-acid aminotransferase, **142**, 267
aromatic-L-amino acid decarboxylase, **142**, 180
aromatic-amino-acid–glyoxylate aminotransferase, **142**, 274
arom multifunctional enzyme, **142**, 333
arylformamidase, **142**, 226
asparagusate reductase, **143**, 517
ATP-sulfurylase, **143**, 336, 340, 342
branched-chain-amino-acid aminotransferase, **166**, 270
branched-chain keto acid dehydrogenase, **166**, 345
branched-chain α-keto acid dehydrogenase, **166**, 181, 193, 298, 309
component E3, **166**, 143
branched-chain 2-keto acid dehydrogenase, **166**, 303
branched-chain α-keto acid dehydrogenase kinase, **166**, 170
branched-chain 2-keto acid dehydrogenase–pyruvate dehydrogenase complex, **166**, 331
cell migration, **147**, 144
cellobiose dehydrogenase, **160**, 444, 448, 457
cellobiose dehydrogenase (quinone), **160**, 464
cellobiose phosphorylase, **160**, 468
cellulosomes, **160**, 474
ceruloplasmin, **163**, 443
chitinase, **161**, 467, 493, 496
chitobiase, **161**, 494
chitosanase, **161**, 502, 506
chorismate mutase, **142**, 442, 451, 463
$C\bar{1}$ inhibitor, **163**, 181

coniferyl-alcohol dehydrogenase, **161**, 302
cyanobacterial hydrogenases, **167**, 508
cystathionine β-lyase, **143**, 445, 484
D-cysteine desulfhydrase, **143**, 450
cytochrome b_6/f complex, **167**, 348
cytochrome P450, **137**, 678
3-dehydroquinate dehydratase, **142**, 321
dehydroquinate synthase, **142**, 308
dextransucrase, **138**, 651
N,N'-diacetylchitobiase, **161**, 524
dihydropteridine reductase, **142**, 104, 118, 127
dopamine β-monooxygenase, **142**, 597, 603
endopectate lyase, **161**, 381
exo-1,4-β-glucanases, **160**, 128
factor XII, **163**, 57
factor XIIa, **163**, 72
F_1-ATPase, **167**, 351
ferredoxin–thioredoxin reductase, **167**, 424
β-D-glucanases, **160**, 60
glucose oxidase, **161**, 308
H^+-ATPase, **157**, 515, 622
horseradish peroxidase, in pinosome osmotic lysis evaluation, **149**, 47
indoleamine 2,3-dioxygenase, **142**, 193
indoleglycerol-phosphate synthase, **142**, 391
isopropylmalate dehydratase, **166**, 424
β-isopropylmalate dehydrogenase, **166**, 226, 430
α-isopropylmalate synthase, **166**, 415
isovaleryl-CoA dehydrogenase, **166**, 376
kallikrein in plasma, **163**, 87
kynurenine-oxoglutarate aminotransferase, **142**, 218
leucine dehydrogenase, **166**, 283
ligninase, **161**, 242
lignin demethoxylation by *Phanerochaete chrysosporium*, **161**, 83
lignin depolymerization by *Streptomyces*, **161**, 252
lipoamide dehydrogenase-valine, **166**, 345
manganese peroxidase, **161**, 259, 266
methionine γ-lyase, **143**, 460
2-methyl branched-chain acyl-CoA dehydrogenase, **166**, 361
methylmalonate-semialdehyde dehydrogenase, **166**, 390
4-methylumbelliferone release from glycosides, **160**, 48
monoamine oxidase, **142**, 628
monophenol monooxygenase, **142**, 165
NAD(P)H dehydrogenase (quinone), **161**, 271
p-nitrophenyl phosphate phosphatase, **159**, 360
pectin lyase, **161**, 351
phenylalanine ammonia-lyase, **142**, 244, 249
phenylalanine 4-monooxygenase, **142**, 50
phosphoribosylanthranilate isomerase, **142**, 391
3-phosphoshikimate 1-carboxyvinyltransferase, **142**, 343, 349
polygalacturonase, **161**, 371
prephenate aminotransferase, **142**, 481
prephenate dehydratase, **142**, 432, 507
prephenate dehydrogenase, **142**, 443
pyranose 2-oxidase, **161**, 317
stopped-flow, Ca^{2+} release from sarcoplasmic reticulum, **157**, 470
Streptococcus ATPase, **157**, 683
sulfatases, **143**, 362
sulfhydryl oxidase, **143**, 509
sulfite, **143**, 12
β-sulfopyruvate, **143**, 222
thiols
  oxidation-based methods, **143**, 45
  substitution-based methods, **143**, 60
thioredoxin *f*, **167**, 422
thioredoxin *m*, **167**, 411, 417
thiosulfate reductase, **143**, 350
tryptophanase, **142**, 414
tryptophan *C*-methyltransferase, **142**, 236
tryptophan side chain oxidases I and II, **142**, 201
tryptophan synthase, **142**, 401
vanillate *O*-demethylase, **161**, 295
vanillate hydroxylase, **161**, 276
Spectrophotometry
  absorption, immobilized alcohol dehydrogenase, **135**, 547
  carotenoids, **148**, 366
  chlorophylls, **148**, 359
  pheophytins, **148**, 378
  vanadium, **158**, 418

Spectroscopy, *see also specific techniques*
dc plasma atomic emission, vanadium assay, **158**, 415
extracellular matrix proteins, principles, **145**, 22
inductively coupled plasma atomic emission, vanadium assay, **158**, 412
organelles immobilized in crosslinked proteins, **135**, 467
21-kDa calcium-binding protein, **139**, 97
UV absorbance and difference, S-100 protein, **139**, 174
Spermatozoa
antibodies, in postvasectomy autoimmune orchitis, detection, **162**, 466
motility analysis
digital image methods, **139**, 813
manual methods, **139**, 816
preparation for reactivation, **139**, 809
proteins, phosphorylation and ^{32}P incorporation assay, **139**, 818
reactivation, **139**, 812
Spermine
precipitation of DNA, **152**, 46
Spheron ion exchangers
in chromatography of pectic enzymes, **161**, 388
Sphingolipidoses
enzymatic diagnosis, **138**, 727
Sphingomyelin phosphodiesterase
diagnostic assay for Niemann–Pick disease, **138**, 736
Spinach
chloroplasts
envelope polypeptides, analysis, **148**, 200
integrity criteria, **148**, 154
isolation, **148**, 149, 165
tocopherol and plastoquinone synthesis, **148**, 383
mitochondria, isolation from leaves, **148**, 422, 434
purification of
cystathionine β-lyase, **143**, 446
D-cysteine desulfhydrase, **143**, 450
phospholipid transfer protein, **148**, 661
thylakoids, isolation from leaves, **148**, 320
Spinacia oleracea, *see* Spinach
Spinal cord
guinea pig, purification of myelin basic protein, **162**, 413
Spin labeling
immobilized chymotrypsin, **135**, 504
Na^+,K^+-ATPase
class II groups, **156**, 374
sulfhydryl groups, **156**, 373
Spin probes
in ESR studies of lymphoid cell receptors, **150**, 420
Spirulina
S. maxima, purification of hydrogenase, **167**, 503
S. platensis
isolation of photosynthetic membrane vesicles, **167**, 355
purification of
cytochrome *c*-554, **167**, 394
F_1-ATPase, **167**, 357
Spleen
human, purification of connective tissue-activating peptide Ib, **163**, 744
murine, isolation of
erythroblasts, **147**, 342
genomic DNA, **152**, 181
rat, expression of liposome-encapsulated insulin gene, **149**, 161
Spleen cells
azobenzene arsonate-coupled, preparation, **163**, 361
trinitrophenyl-coupled, preparation, **163**, 360
Splenic focus assay
monoclonality and donor origin of antibodies produced in splenic fragment cultures, **150**, 203
overview, **150**, 196
splenic fragment culture methodology and procedures for, **150**, 199
variations, **150**, 207
Sponges
implants as models of inflammation
N-acetylglucosaminidase levels, **162**, 330
associated fluid, assay of mediators, **162**, 332
leukocyte infiltration estimation, **162**, 329
preparation and implantation, **162**, 327

Spores
Curvularia lunata, entrapment with photocrosslinkable resin prepolymers, **135**, 236
Penicillium chrysogenum
immobilization on celite, **135**, 322
preparation, **135**, 322
Sporocytophaga myxococcoides
culture and harvest of cells, **160**, 339
purification of cellulases, **160**, 341
Sporotrichum
S. pulverulentum
cultivation, **160**, 370
growth and preparation of mycelial extract, **161**, 272
purification of
acid proteases, **160**, 501
cellobiose dehydrogenase (quinone), **160**, 464
endo-1,4-β-glucanases, **160**, 370
β-glucosidases, **160**, 419
vanillate hydroxylase, **161**, 277
S. thermophile
cultivation, **160**, 444
purification of cellobiose dehydrogenase, **160**, 445
Spreading factor
role in tissue repair processes, **163**, 710
Squash
yellow, biosensors based on derived tissues, **137**, 149
Staining
cellulolytic enzyme activity in gels, **160**, 135
chromosomes after *in situ* hybridization, **151**, 286
collagen peptides in gels, **145**, 199
immuno-, *see* Immunostaining
immunogold, in electron microscopy of cAMP-dependent protein kinase, **159**, 230
immunohistochemical, tissue kallikrein, **163**, 150
immunoperoxidase, *see* Immunoperoxidase staining
leukocytes, in assay for actin behavior, **162**, 268
β-mannosidase activity in gels, **160**, 616
NADPH-specific dihydropteridine reductase, **142**, 113
negative
in analysis of cyanobacterial fine structure, **167**, 160
in electron microscopic assay for microbial toxins, **165**, 357
Staphylococcus
diarrheal toxin, ileal loop test, **165**, 322
S. aureus, purification of
alpha toxin, **165**, 3
enterotoxins, **165**, 43
epidermolytic toxin, **165**, 33
γ-hemolysin, **165**, 8
leukocidin, **165**, 23
δ-lysin, **165**, 17
protein A, **137**, 459
toxic shock syndrome toxin-1, **165**, 37
Starch
enzymatic degradation in aqueous two-phase system, **137**, 661
hydrolyzates, TLC, **160**, 176
Steam
pretreatment of biomass materials for enhancing cellulose availability, **160**, 9
Steroid hormones
as elastin synthesis modulators, **144**, 243
Steroid sulfotransferase
binding assay, **143**, 206
Sterols
plant membrane
analysis, **148**, 636
cellular sites of synthesis, **148**, 643
composition modification, **148**, 647
intracellular distribution, **148**, 639
Steryl-sulfatase
assay, **143**, 362
properties, **143**, 365
purification from *Helix pomatia*, **143**, 363
Stomach
mammalian, isolation of H^+,K^+-ATPase, **157**, 649
Stonecrop
mitochondria, isolation from leaves, **148**, 431
Storage
DNA libraries, **152**, 407
T cells in liquid nitrogen, **150**, 332
Streptavidin
–biotin, in visualization of ribosomal crosslink sites, **164**, 396

Streptococci
  cell walls, induction of arthritis in rats, **162**, 373
Streptococcus
  S. faecalis, ATPase purification, **157**, 684
  S. pneumoniae, β-galactosidase purification, **149**, 34
  S. pyogenes, purification of
    laminin receptor, **144**, 506
    streptolysin S, **165**, 62
  S. sanguis, dextransucrase purification, **138**, 652
Streptokinase
  dextran-immobilized, clinical studies, **137**, 563
Streptolysin O
  hemolytic assay, **165**, 58
  production, **165**, 53
  properties, **165**, 58
  purification, **165**, 55
Streptolysin S
  hemolytic assay, **165**, 63
  production, **165**, 61
  properties, **165**, 63
  purification, **165**, 62
Streptomyces
  assay of acetylxylan esterase, **160**, 551
  cultivation on lignocellulose, **161**, 37, 250
  DNA, cloning and analysis with cosmid shuttle vectors, **153**, 166
  lignin depolymerization, assays, **161**, 252
  β-mannanase-producing forms, isolation, **160**, 612
  plasmid and phage vectors for gene cloning and analysis, **153**, 116
  production of acid-precipitable polymeric lignin, **161**, 39
  purification of
    β-mannanase, **160**, 613
    xylanase, **160**, 649
  S. aureofaciens, chorismate mutase purification, **142**, 466
  S. cattleya, celite-immobilized, continuous fermentation, **135**, 332
  S. fimbriatus, restriction enzyme *Sfi*I purification, **155**, 20
  S. flocculus, tryptophan *C*-methyltransferase purification, **142**, 239
  S. griseus, chitosanase purification, **161**, 507
  S. parvulus, arylformamidase purification, **142**, 227
  S. phaeochromogenes, cystathionine γ-lyase purification, **143**, 488
  S. plicatus, Endo H purification, **138**, 765
  S. rimosus
    entrapment with urethane prepolymers, **135**, 244
    prepolymer-entrapped, applications, **135**, 247
Streptovirudin
  as inhibitor of lipid-linked saccharide pathway, **138**, 664
Strictosidine
  synthesis by immobilized strictosidine synthase, **136**, 349
Strictosidine synthase
  immobilization, **136**, 345
  isolation from periwinkle, **136**, 344
Strongylocentrotus purpuratus, *see* Sea urchin
Strontium
  and lead, in cytochemical localization of Na^+,K^+-ATPase, **156**, 418
Structural profiles
  in rational design of polypeptides, **154**, 450
Structure–activity relationships
  botulinum neurotoxins, **165**, 76
  proteins, analysis with peptide models, **154**, 473
Subcloning
  molecular
    applications, **152**, 512
    general strategies, **152**, 517
    techniques, **152**, 519
Substance P
  octapeptide, sequential hydrolysis with immobilized aminopeptidases, **136**, 175
Succinate dehydrogenase (ubiquinone)
  mitochondrial
    properties, **148**, 494
    purification from sweet potato, **148**, 493
N-Succinimidyl-3-(2-pyridyldithio)propionate

modification of human chorionic gonadotropin, **165**, 206
reversible protein immobilization with, **135**, 138
Succinyl-L-trialanine-*p*-nitroanilide
in affinity chromatography of
leukocyte elastase, **135**, 535
pancreatic elastase, **135**, 531
Sucrose
assay with
biosensor, **137**, 52
enzyme thermistor, **137**, 190
hydrolysis, enzyme thermistor monitoring and control, **137**, 308
radiolabeled, as fluid-phase marker
endocytotic uptake, **146**, 412
overview, **146**, 404
preparation, **146**, 408
tonoplast transport, measurement, **148**, 112
Sucrose gradients
alkaline, for DNA repair analysis
calibration, **139**, 728
computer analysis, **139**, 729
Sugar acids
thin-layer chromatography, **160**, 159
Sugar nucleotides
transport into Golgi- and rough endoplasmic reticulum-derived vesicles, assays, **138**, 709
Sugars
analysis in cellulase assays, **160**, 90
chromatography, in characterization of radiolabeled lignins, **161**, 24
thin-layer chromatography, **160**, 159
Sulfane
sulfur analysis by cyanolysis
cold method, **143**, 26
with cupric ion, **143**, 28
hot method, **143**, 27
Sulfanilic acid
diazotized, labeling of plant plasma membranes, **148**, 571
Sulfate
cellular
activation, **143**, 329
transfer and reduction, **143**, 332
ion-exchange HPLC assay, **143**, 7
nephelometric assay, **143**, 5
turbidimetric assay, **143**, 5
Sulfate adenylyltransferase
assays, **143**, 335
catalyzed reaction, **143**, 334
purification from *Penicillium chrysogenum*, **143**, 344
Sulfatide/G_{M1} activator
assays, **138**, 799
purification, **138**, 810
Sulfatides
binding characteristics, **138**, 483
lipid extracts for detection, **138**, 476
solid-phase binding assay, **138**, 477
solvolysis, **138**, 479
specificity criteria, **138**, 480
thin-layer chromatography, **138**, 478
Sulfhydryl groups
Ca^{2+},Mg^{2+}-ATPase, labeling with maleimides and iodoacetamide derivative, **157**, 251
Na^{+},K^{+}-ATPase
differential spin labeling, **156**, 373
modifications, **156**, 302
prelabeling for ESR, **156**, 372
Sulfhydryl oxidase
assay, **143**, 119, 509, 511
flavor modification of heat-sterilized milk, **136**, 430
immobilization, **136**, 425
preparation from whey, **136**, 423
properties, **143**, 514
purification from
bovine milk, **143**, 508
rat skin, **143**, 512
reactors, characteristics, **136**, 427
Sulfide
assay with sulfide-specific electrode, **143**, 29
Sulfinoalanine decarboxylase
assay, **143**, 405
properties, **143**, 408
purification from rat liver, **143**, 406
Sulfite
assay with
S-(2-amino-2-carboxyethylsulfonyl)-L-cysteine, **143**, 17
fuchsin, **143**, 15
sulfite oxidase, **143**, 11
Sulfite oxidase
sulfite assay with, **143**, 11
Sulfitolysis
disulfides, **143**, 254

Sulfometuron methyl
hypersensitive microbial mutants, isolation, **166**, 105
microbial susceptibility to, **166**, 99
properties, **166**, 98
resistant microbial and plant mutants, isolation, **166**, 103
β-Sulfopyruvate
assay, **143**, 222
lithium salt, synthesis, **143**, 221
Sulfotransferases
Ecteola-cellulose chromatography, **143**, 204
ion-pair extraction, **143**, 202
radiochemical assay, **143**, 205
thin-layer chromatography, **143**, 203
Sulfoxides
chemistry and preparation, **143**, 281
Sulfur
compounds, polarography, **143**, 38
inorganic and organic compounds, additional articles dealing with, **143**, 523
metabolism in cyanobacteria, **167**, 572
reactions in biosphere, **143**, 330
^{35}S
in autoradiography and fluorography of nucleic acids, **152**, 55
labeled compounds, safety considerations, **152**, 25
Sulfurtransferases
chromogenic substrates, **143**, 236
fluorogenic substrates, **143**, 239
Superoxide dismutase
as marker for chromosome detection in somatic cell hybrids, **151**, 192
metal-hybrid, preparation, **158**, 88
Surface plasmon resonance
in biomedical analysis, **137**, 376
Surgery
biomaterials for, application of immobilized enzymes, **137**, 545
Swainsonine
as glycoprotein processing inhibitor, **138**, 685
Sweet clover
vacuoles, preparation from protoplasts, **148**, 76
Sycamore
cell suspension culture techniques, **148**, 3
protoplasts, preparation from suspension-cultured cells, **148**, 31
tonoplasts, isolation from vacuoles, **148**, 79
Symbiotic associations
cyanobacterial
cultivation, **167**, 114
isolation of bacteria, **167**, 119
reconstitution, **167**, 117
Synechococcus
ATP synthase genes
cloning, **167**, 795
sequencing, **167**, 798
axenic culture and maintenance, **167**, 107
batch culture, **167**, 107
cytochrome-*c* oxidase, Cu^{2+} detection by EPR, **167**, 450
cytochromes *c*-550 and *c*-553, purification from *S. vulcanus*, **167**, 395
energetic status, NMR analysis, **167**, 667
enrichment culture, **167**, 106
heat shock response, characterization, **167**, 624
isolation, **167**, 106
photosystem I and II, isolation, **167**, 263
photosystem II–phycobilisome complex, isolation, **167**, 289
phycobilisomes, purification, **167**, 306, 308
sampling, **167**, 106
synchronous culture, **167**, 109
Synechocystis
cultivation and harvesting, **167**, 192
fimbriae
electron microscopic analysis, **167**, 192
isolation, **167**, 194
photosystem II reaction center, construction of mutations, **167**, 766
plasma membranes, isolation, **167**, 248
transformation, **167**, 712
Synephrine
radioiodinated
catecholamine radioimmunoassay with, **142**, 559
preparation, **142**, 557
Synovial fluid
extracellular matrix proteins, analysis, **145**, 396

Syringaldehyde
  carbon isotope-labeled, preparation, **161**, 52
Systemic lupus erythematosus
  murine
    anti-DNA autoantibodies, analytical procedures, **162**, 399
    nonspecific immunologic defects, analytical procedures, **162**, 396
    overview, **162**, 385
    retroviruses and autoimmunity, analytical procedures, **162**, 390
    strains used in research, **162**, 386

T

Tailing
  homopolymeric
    double-stranded cDNA with dG, **152**, 339
    linearized plasmid DNA with dC, **152**, 340
    in oligonucleotide labeling, **152**, 342
Talaromyces
  T. byssochlamydoides, purification of xylanase, **160**, 676
  T. emersonii
    cultivation, **160**, 363, 438
    purification of
      endo-1,4-β-glucanases, **160**, 365
      β-glucosidase, **160**, 439
Tamoxifen
  effect on cellular DNA synthesis
    de novo pathway analysis, **146**, 337
    scavenger pathway analysis, **146**, 334
Tank reactor
  continuously stirred, for immobilized organelles, **135**, 463
Tannic acid
  in κ-carrageenan immobilization of *Brevibacterium flavum*, **136**, 461
Tannin, *see* Tannic acid
Tartryl diazide
  crosslinking of aspartate transcarbamoylase, **135**, 571
Taurine
  HPLC assay, **143**, 155
Taurine dehydrogenase
  assay, **143**, 496
  properties, **143**, 498
  purification from bacteria, **143**, 497
Taxonomy
  cyanobacteria, **167**, 33
    application of DNA/DNA hybridization, **167**, 125
T-cell receptors
  –antigen complex, antibodies, activation of T cells, **150**, 103
  antigen-specific
    gene cloning, **150**, 655
    isolation, **150**, 650
    reactive antisera, **150**, 629
  MHC-restricted putative, gene isolation, **150**, 662
  non-MHC-restricted
    assay, **150**, 679
    heteroantisera, production, **150**, 679
    isolation, **150**, 673
    radioiodination, **150**, 677
    SDS–PAGE, **150**, 678
    source, **150**, 668
    storage, **150**, 677
T cells
  activation with antibodies to T-cell antigen–receptor complex, **150**, 103
  alloreactive, cloning, **150**, 333
  antigen-specific, cloning, **150**, 325
  in basal serum-free medium
    antigen-specific responses, **150**, 131
    mitogen responses, **150**, 130
  clones and hybrids, karyotypic analysis, **150**, 352
  cyclosporine–calmodulin interaction
    flow cytometric analysis, **139**, 683
    fluorometric analysis, **139**, 680
  cytotoxic, *see* Cytotoxic T cells
  with distinct functions, cloning, **150**, 342
  interleukin-2 receptors, detection, **150**, 682
  membrane Thy-1 glycoprotein, isolation with lectins, **150**, 473
  role in inflammation and repair, hepatic granuloma model, **163**, 609
  sensitized, proliferation assay, **163**, 367
  stimulation with periodate or neuraminidase/galactose oxidase, **150**, 105
Tellurium
  atomic absorption spectrometry, **158**, 142

Temperature
effects on
bacterial L-aspartic acid production in reactor, **136**, 468
cellulose bead formation, **135**, 271
cGMP-binding phosphodiesterase, **159**, 717
mitochondrial membrane lipids, **148**, 485
subzero, enzyme studies with immobilized substrates, **135**, 528
Terminal deoxynucleotidyltransferase
positive lymphoid precursor cells, generation *in vitro*, **150**, 363
Terminal transferase
DNA 3′-end labeling with, **152**, 104
oligonucleotide radiolabeling with, **152**, 342
Testis
bovine, purification of
cytosolic factor required for cholera toxin activity, **165**, 247
phosphomannosyl receptor 2, **138**, 506
porcine, purification of calspermin, **139**, 131
rat
distribution of cyclic nucleotide phosphodiesterase isoforms, **159**, 679
purification of
myo-inositol-1-phosphatase, **141**, 140
L-*myo*-inositol-1-phosphate synthase, **141**, 131
Testosterone
enzymatic oxidation in two-phase system, **136**, 155
Tetanus toxin
peptide C, preparation, **165**, 89
purification from *Clostridium tetani*, **165**, 86
two-dimensional crystallization, **165**, 389
Tetracyanoquinodimethane
complex salts, in bioelectrocatalysis, **137**, 438
Tetracycline
enrichment of cyanobacterial mutants, **167**, 734
12-*O*-Tetradecanoylphorbol-13-acetate
enzymatic hydrolysis, assays, **141**, 308
induced stabilization of protein kinase C–membrane complex, assay, **141**, 400
inhibition of cell–cell communication, fluorescence photobleaching assay, **141**, 439
12-*O*-Tetradecanoylphorbol-13-acetate hydrolase
preparation, **141**, 303
TPA hydrolysis, assays, **141**, 308
Tetraglycosylceramides
perbenzoylated, HPLC, **138**, 117
Tetrahydropterin
as cofactor of phenylalanine 4-monooxygenase, **142**, 15
Tetramethylammonium chloride
in hybridization of genomic DNA to oligonucleotide probes, **152**, 447
4-(2,2,6,6-Tetramethylpiperidine-1-oxyl)-*m*-fluorosulfonylbenzamide
preparation, **135**, 503
spin labeling of immobilized chymotrypsin, **135**, 504
Thallium
atomic absorption spectrometry, **158**, 142
Theophylline
assay with dry reagent chemistries, **137**, 418
Thermal analysis
chitin and chitosan acetylation degree, **161**, 446
Thermoascus aurantiacus
purification of cellulases, **160**, 303
Thermoase
immobilized
in aspartame production
batchwise system, **136**, 514
continuous column system, **136**, 513
preparation, **136**, 507
quantitation, **136**, 511
Thermodenaturation
catalase, **137**, 605
pyranose oxidase, **137**, 605, 615
Thermolysin
chloroplast treatment for analysis of envelope polypeptides, **148**, 195
immobilized, peptide synthesis with, **136**, 185
metal-hybrid, preparation, **158**, 94

Thermomonospora fusca
cultivation, **160**, 316
purification of cellulases, **160**, 316
Thermostability
phage T4 lysozyme, **154**, 511
Thin-layer chromatography
adenosylhomocysteinase, **143**, 379
in analysis of glycosphingolipid biosynthesis, **138**, 568
in assay for
cyclic nucleotides in cyanobacteria, **167**, 586
leukotriene B_4 20-hydroxylase, **163**, 350
α-4-*O*-methyl-D-glucuronidase, **160**, 566
carotene oxygenase, **167**, 339
carotenoids, **167**, 334
cellulose hydrolyzates, **160**, 176
chitin hydrolyzates, **160**, 176
chloroplast membrane lipids
class separation, **148**, 307
molecular species separation, **148**, 311
continuous-development, glycosphingolipids, **138**, 125
cyanobacterial membrane lipids
class separation, **167**, 253
molecular species separation, **167**, 257
gangliosides, **138**, 303
glycolipid receptors, overlay assay, **138**, 221
with immunostaining
glycolipids, **138**, 129, 197
oligosaccharides, **138**, 208
inositol phospholipids, **141**, 166, 172, 178, 180
kinin analogs, **163**, 271
lignin degradation intermediates
analytical procedure, **161**, 200, 207
preparative procedure, **161**, 204, 208
phospholipids in hepatic extracts, **141**, 237
platelet-activating factor, **163**, 46
polyalcohols, **160**, 159
prenyl alcohols, **167**, 332
prostaglandin A_1 metabolite, in analysis of cAMP export, **159**, 90
prostaglandins, **141**, 334
sialic acids, **138**, 143
starch hydrolyzates, **160**, 176
sugar acids, **160**, 159
sugars, **160**, 159
sulfatides, **138**, 478
sulfotransferases, **143**, 203
thylakoid lipids, **148**, 322
two-dimensional
cyanobacterial carbon metabolites, **167**, 547
tRNA-derived nucleotides, **155**, 385
xylan hydrolyzates, **160**, 176
xylobiose and xylooligomers, **160**, 534
Thioacidolysis
in lignin analysis, **161**, 95
in lignin characterization, **161**, 104
Thiobacillus ferrooxidans
immobilized, in treatment of acid mine drainage, **136**, 530
Thiobarbituric acid
in assay for dehydroquinate synthase, **142**, 308
Thiocyanate
colorimetric assay, **143**, 23
Thiocyanide
^{14}S-labeled, uptake by clathrin-coated vesicles, assay, **157**, 638
α-Thiodeoxynucleotides
in DNA sequencing, **155**, 166
Thiodisaccharides
chemical synthesis, **138**, 403
Thioglycolic acid
in lignin quantitation, **161**, 99
Thioinosine triphosphate
derivatives
as affinity labels for Na^+,K^+-ATPase, **156**, 310
preparation, **156**, 303
Thiols
dinitrophenol derivatives, HPLC, **143**, 101
electrolytic reduction, **143**, 99
fluorometric assay with fluorobenzoxadiazoles, **143**, 67
hepatic, HPLC with electrochemical detection, **143**, 110
interchange with disulfides
reaction rates, **143**, 130
Brønsted plots, **143**, 132
thiol equilibrium constants, **143**, 138
thiol pK_a values, **143**, 132
labeling with bromobimanes
analytical applications, **143**, 82
procedures, **143**, 80

monobromobimane-labeled, HPLC analysis, **143**, 85
protein-bound, liberation, **143**, 127
purification from *Halobacterium halobium*, **143**, 97
reduction of disulfides, **143**, 246
spectrophotometric assay
oxidation-based methods, **143**, 45
substitution-based methods, **143**, 60
Thiol status
cellular, diamide treatment effects, **143**, 265
Thionyl chloride
activation of succinamidopropyl-glass, **135**, 132
Thiophosphorylation
histone, **159**, 350
phosphorylase *b*, **159**, 348
Thioredoxin *f*
assays, **167**, 420
properties, **167**, 422
purification from *Nostoc muscorum*, **167**, 422
Thioredoxin *m*
assays, **167**, 411, 417
properties, **167**, 414, 419
purification from
cyanobacteria, **167**, 413
Nostoc muscorum, **167**, 418
Thiosulfate
assay with *S*-(2-amino-2-carboxyethylsulfonyl)-L-cysteine, **143**, 17
colorimetric assay, **143**, 24
Thiosulfate reductase
assay, **143**, 350
properties, **143**, 353
purification from yeast, **143**, 351
Thistle
cell coupling to hydrous titanium(IV) oxide and derivatives, **135**, 363
Thrombin
role in tissue repair processes, **163**, 711
sensitive membrane proteins, assay, **147**, 157
Thrombogenesis
inhibition in urokinase-treated tubes
Chandler loop testing method, **137**, 539
mechanism, **137**, 540
Thrombolytic therapy
immobilized enzymes for, **137**, 552
Thrombosis
microvascular, during inflammation, quantitation, **162**, 312
Thrombospondin
hemagglutination assay, **144**, 443
heparin-binding domain, isolation, **144**, 442
isolation, **144**, 439
platelet agglutination assay, **144**, 445
proteolysis, **144**, 441
Thylakoids
Aphanocapsa and *Phycomyces*, preparation, **167**, 333
associated lipids, isolation and reconstitution, **148**, 319
cyanobacterial
anoxygenic photosynthetic reactions, assays, **167**, 384
cytochrome-*c* oxidase
identification, **167**, 443
oxidation rates, **167**, 441
isolation, **167**, 437
inside-out vesicles
generation by fragmentation, **148**, 254
properties, **148**, 258
separation by aqueous two-phase partition, **148**, 255
lettuce
immobilized in crosslinked proteins
activity assays, **135**, 457
activity yields, **135**, 460
application of reactors, **135**, 463
electron microscopy, **135**, 468
preparation, **135**, 457
spectroscopy, **135**, 467
immobilized, preparation and stability, **137**, 587
isolation, **135**, 455
Phormidium laminosum, preparation, **167**, 273
polypeptides, 2-D gel electrophoresis, **148**, 623
spinach, isolation from leaves, **148**, 320
Synechococcus, preparation, **167**, 263
Thymidine kinase
gene transferred from cosmid library, rescue, **151**, 427
mRNA, detection by microinjection and complementation of mutant cells, **151**, 377

Thyroglobulin
circulating antibodies, detection, **162**, 473
oligosaccharides, species differences, HPLC analysis, **138**, 104
purification from rat thyroid, **162**, 472
radiolabeling, **162**, 472
Thyroid
histopathology, in analysis of spontaneous autoimmune thyroiditis, **162**, 473
rat, purification of thyroglobulin, **162**, 472
Thyroiditis
spontaneous autoimmune
analysis, **162**, 472
animal model evaluation, **162**, 475
Thyrotropin-releasing hormone
induced changes in cellular calcium metabolism, measurements, **141**, 36
stimulated lipid turnover, measurement, **141**, 100
Thyroxine
assay with dry reagent chemistries, **137**, 418
Tieghemella orchidis
polyacrylamide gel-entrapped mycelium
enzymatic activity, **135**, 211
viability, **135**, 203
Tin
atomic absorption spectrometry, **158**, 142
Tissue cultures
articular cartilage explants, **144**, 414
elastin synthesis
ELISA, **163**, 672
radioimmunoprecipitation assay, **163**, 671
embryonic, in analysis of basement membrane biosynthesis, **145**, 389
mammalian
RNA detection by *in situ* hybridization, **151**, 539
transcription detection *in situ* with biotin-labeled probes, **151**, 530
spleen fragment, in splenic focus assay, **150**, 200
Tissues
aluminum levels, measurement, **158**, 296
amyloid-rich, isolation of amyloid P component, **163**, 528
animal
acyl-CoA levels, assay, **166**, 67
biosensors based on, **137**, 141
branched-chain α-keto acid dehydrogenase assays, **166**, 189
branched-chain α-keto acid levels, GLC assay, **166**, 39
calbindin D, immunolocalization, **139**, 541
calelectrins, purification, **139**, 32
calmodulin
immunoperoxidase staining, **139**, 495
purification, **139**, 10
collagens, isolation, **145**, 171
cysteinesulfinic acid, hypotaurine, and taurine, HPLC, **143**, 155
mineralized, overview, **145**, 255
parvalbumin, immunolocalization, **139**, 552
S-100 protein
immunoperoxidase staining, **139**, 495
purification, **139**, 14
catecholamines, enzymatic radioimmunoassay, **142**, 565
cobalt levels, measurement, **158**, 350
insulin-like growth factor I levels, assay, **146**, 227
insulin-like growth factor mRNAs, isolation, **146**, 242
nerve growth factor assay, **147**, 183
plant
application of NMR methods, **148**, 683
biosensors based on, **137**, 149
preparation for branched-chain amino and keto acid analysis, **166**, 9
repair, in inflammation
hepatic granuloma as model, **163**, 605
role of growth factors, **163**, 707
specific gene expression, *in situ* hybridization assay, **152**, 649
Titanium
activated supports
coupling of yeast cells, **135**, 367
derivatized, enzyme coupling, **135**, 127
activation of proteinaceous matrices, **135**, 122

chelated enzymes, glutaraldehyde crosslinking, **135**, 126
in enzyme immobilization
enzyme coupling chemistry, **135**, 117
metal-link method, **135**, 119
hydrous oxides
coupling of microbial cells, **135**, 358
enzyme chelation, **135**, 123
Titration
clostridial alpha toxin, **165**, 297
in determination of chitin and chitosan acetylation degree, **161**, 443
staphylococcal γ-hemolysin, **165**, 10
TLC, *see* Thin-layer chromatography
Tobacco
genetically altered plants, regeneration, **153**, 326, 332
leaf fragments, transformation, **153**, 287
protoplasts
association with liposomes, assessment, **148**, 37
liposome-mediated nucleic acid transfection, **148**, 42
plasmid DNA expression, assay, **148**, 42
preparation
from leaves, **148**, 41
from suspension-cultured cells, **148**, 35
and transformation and culture, **153**, 321, 331
treatment with liposomes, **148**, 36
viral RNA expression, assay, **148**, 42
purification of
arogenate dehydrogenase, **142**, 490
prephenate aminotransferase, **142**, 484
Tocopherol
analysis in chloroplast membranes, **148**, 391
and intermediates, synthesis in chloroplasts
associated enzymes, purification, **148**, 391
enzymatic assay, **148**, 387
identification and purification of labeled products, **148**, 388
Tolerance
to adjuvant arthritis, induction, **162**, 354
5′-*O*-*p*-Toluenesulfonylthymidine
synthesis, **155**, 269
Tomato
purification of
1-aminocyclopropane-1-carboxylate synthase, **143**, 428
chitinase, **161**, 487
proteinase inhibitor inducing factor, **138**, 716
Tonoplasts
characteristics, **148**, 108
H^+-ATPase
assay, **148**, 126; **157**, 584
properties, **148**, 128; **157**, 584
purification, **148**, 124; **157**, 580
subunit composition, **148**, 127
isolation from
beet microsomes, **148**, 107
beet vacuoles, **148**, 106
broken plant vacuoles, **148**, 85
rubber tree lutoids, **148**, 107
sycamore cell vacuoles, **148**, 79
lutoid
biochemical composition, **148**, 91
enzymatic activities, **148**, 91
preparation, **148**, 90
Mg^{2+}-ATPase, **148**, 110
porters, **148**, 112
pyrophosphatase, **148**, 114
vesicles
applications to H^+-coupled transport, **148**, 118
characteristics, **148**, 117
preparation from
beet, **148**, 115
oat, **148**, 123
1-Tosyloxy-3-isocyanopropane
in preparation of isonitrile-derivatized polymers, **135**, 95
synthesis, **135**, 94
Toxicity assay
cholera toxin, **165**, 245
diphtheria toxin, **165**, 219
Toxin–antitoxin technique
in isolation of
cybrids, **151**, 233
heterokaryons and cell hybrids
applications, **151**, 249
description, **151**, 239
osmotic lysis conditions, optimization, **151**, 244
procedure, **151**, 247

Toxins, *see also specific toxins*
associated pores, incorporation into liposomes, **165**, 285
bacterial, phospholipase activity, assay, **165**, 298
cytolytic, *Vibrio*-produced
bioassay, **165**, 177
properties, **165**, 181, 187
purification, **165**, 179, 186
diarrheal, ileal loop fluid accumulation test, **165**, 306
epidermolytic, staphylococcal
antiserum preparation, **165**, 338
bioassay, **165**, 334
dot immunoblotting assay, **165**, 336
ELISA, **165**, 337
Ouchterlony immunodiffusion assay, **165**, 334
properties, **165**, 35
purification, **165**, 33
purity criteria, **165**, 34
radial immunodiffusion assay, **165**, 337
radioimmunoassay, **165**, 337
storage, **165**, 34
erythrogenic, streptococcal
bioassay, **165**, 66
production, **165**, 65
properties, **165**, 66
purification, **165**, 65
hemolytic, assay, **165**, 213
membrane-penetrating, analysis with photoreactive lipids, **165**, 347
microbial, electron microscopic assay
freeze-fracture and freeze-etching technique, **165**, 374
negative staining technique, **165**, 357
pore-forming, in cultured cells, radioisotopic assay, **165**, 278
toxic shock syndrome, type 1
antiserum preparation, **165**, 340
chick embryo assay, **165**, 344
in detection of endotoxin
in vitro method, **165**, 304
in vivo method, **165**, 303
immunochemical assays, **165**, 339
production, **165**, 339
properties, **165**, 42
purification by
ion-exchange chromatography and gel filtration, **165**, 40
isoelectric focusing, **165**, 38
Toyopearl
thermoase immobilization with, **136**, 507
Tracer methods
in calculation of phospholipid transfer, **157**, 693
Transcription
collagen type I genes, assay, **144**, 75
direction, determination based on Tn*5* mutagenesis, **154**, 188
DNA templates, in preparation of specific rRNA fragments, **164**, 226
initiation and termination signal structures, analysis with expression vectors, **153**, 452
rate analysis in isolated nuclei, **152**, 238
reverse, crosslinked RNA, **164**, 325
in transfected cells, *in situ* detection with biotin-labeled probes, **151**, 530
Transcription factors
TFIIIA, complexes with DNA, hydroxyl radical footprinting, **155**, 544
Transcription–translation system
T5 promoter-based, for protein analysis *in vitro* and *in vivo*, **155**, 416
Transduction
phage-mediated, cDNA expression libraries into mammalian cells, **151**, 434
Transfection
eukaryotic cells
for functional analysis of cloned genes
calcium phosphate method, **152**, 693
DEAE-dextran method, **152**, 692
for identification of gene regulatory elements, **152**, 713
murine C127 cells with virus-based calmodulin expression vectors, **139**, 645
transferrin receptor gene cloning by, **147**, 281
Transferrin
cell binding, effects of receptor antibodies, **147**, 275
cell culture assay, **147**, 310
growth-inhibitory activity assay, **147**, 309
growth-stimulatory activity assay, **147**, 306
immunocytochemical localization, **147**, 299

iron-saturated, cell binding assay, **147**, 311
mediated iron uptake, effects of receptor antibodies, **147**, 274
myotrophic assays, **147**, 293
purification from chicken serum, **147**, 291
radioimmunoassay, **147**, 242
radioiodinated diferric
preparation, **147**, 248
receptor assay with, **147**, 250
saturation with iron, **147**, 248
Transferrin receptors
antiserum
evaluation, **147**, 257
preparation, **147**, 255
gene cloning by transfection, **147**, 281
lymphoid cell
biochemical characterization, **150**, 735
detection, **150**, 726
expression, correlation with cell proliferation, **150**, 734
molecular biological studies, **150**, 739
overview, **150**, 723
purification, **150**, 732
monoclonal antibodies
applications, **147**, 276
biological effects, **147**, 272
derivation and characterization, **147**, 266
polyclonal antisera, preparation, **147**, 246
properties, **147**, 245
purification, **147**, 243, 254
radioassay, **147**, 239, 247
radioimmunoassay, **147**, 261
radioiodination, **147**, 258
Transformation
Anacystis nidulans, **153**, 209
competent *E. coli* cells with M13 DNA, **152**, 162
cyanobacteria, **167**, 703
by chromosomal recombination, **153**, 216
E. coli with double-cut vectors, **152**, 354
monocots by *Agrobacterium*, detection, **153**, 305
petunia protoplasts, **153**, 328
plants
with binary Ti vectors, **153**, 292
vectors for, **153**, 253, 277
ryegrass protoplasts, **153**, 330
Synechocystis, **167**, 712
tobacco protoplasts, **153**, 323, 331
yeast, in gene isolation by complementation
overview, **152**, 482
procedure, **152**, 497
Transforming growth factor α
purification from transformed cells, **146**, 106
radioiodination, **146**, 145
radioreceptor assays, **146**, 96, 104, 116
synthetic
bioassays, **146**, 142
characterization, **146**, 141
purification, **146**, 140
solid-phase preparation method, **146**, 128
Transforming growth factor β
purification, **146**, 156
large-scale, **146**, 160
quantitation, **146**, 162
radioiodinated
binding to
intact cells, **146**, 177
solubilized receptors, **146**, 182
preparation, **146**, 175
radioreceptor assays, **146**, 99, 181
role in tissue repair processes, **163**, 710
storage, **146**, 163
Transforming growth factor receptors
type α, affinity labeling
in intact cells, **146**, 147
in isolated membranes, **146**, 149
type β
affinity labeling, **146**, 184
radioreceptor assay, **146**, 169
Transforming growth factors
soft agar growth assays, **146**, 341
Transglutaminase, *see* Protein-glutamine γ-glutamyltransferase
Translation
arrest by mRNA hybridization, **152**, 571
in cell-free systems
characterization of products, **152**, 296
protein precursors for chloroplast transport
reticulocyte lysate system, **148**, 279; **152**, 267
wheat germ system, **148**, 276

collagenase mRNA *in vitro*, **144**, 170
collagen mRNAs *in vitro*, **144**, 84
initiation complexes, extension inhibition analysis, **164**, 419
in injected *Xenopus* oocytes, **152**, 288
kinetic parameters, measurement, **164**, 611
parvalbumin mRNA *in vitro*, **139**, 772
single messenger in *E. coli*, effects of rRNA mutations, **153**, 432

Transplacement
site-directed, in identification of nitrogen fixation genes in *Rhizobium*, **154**, 214

Transport assay
ammonium ion in cyanobacteria
direct method, **167**, 558
indirect method, **167**, 559
branched-chain amino acids in *E. coli*, **166**, 245
branched-chain amino and 2-keto acids in mammalian cells, **166**, 252
carbohydrates in cyanobacteria, **167**, 551
cations in cyanobacteria, **167**, 561
in intact cell monolayers, **146**, 382
isotope fluxes, design considerations, **156**, 180
in membrane vesicles, **146**, 380
Na^+,K^+-ATPase, **156**, 120, 136, 175

Transposition
in cyanobacterial mutagenesis, **167**, 719

Transposons
mutagenesis, gene cloning by, **152**, 491
promoted deletions, in DNA sequencing, **155**, 177
Tn*5*, mutagenesis
in directed DNA sequencing, **154**, 187
DNA segments cloned into multicopy plasmids, **154**, 178
in gene replacement experiments, **154**, 188
in identification of nitrogen fixation genes in *Rhizobium*, **154**, 197
in protein mapping and determination of transcription direction, **154**, 188

Tresyl chloride
activated supports
enzyme coupling, **135**, 68
preparation, **135**, 67

TRH, *see* Thyrotropin-releasing hormone

Triacylglycerol lipase
celite-adsorbed, triglyceride interesterification with, **136**, 405
entrapment with
photocrosslinkable resin prepolymers, **135**, 237
urethane prepolymers, **135**, 241
immobilized, stereoselective esterification of *dl*-menthol, **136**, 299
in preparation of
racemic esters, **136**, 134
racemic secondary alcohols, **136**, 127
prepolymer-entrapped, applications, **135**, 245

Triads
skeletal muscle
characterization, **157**, 58
isolation, **157**, 51
junction reformation, **157**, 81
spanning protein, isolation, **157**, 83
transverse tubules, preparation, **157**, 70

Tributylphosphine
reduction of disulfides, **143**, 253

Trichoderma
T. koningii, purification of cellulase, **160**, 224
T. lignorum
cultivation, **160**, 664
purification of xylanase and β-xylosidase, **160**, 665
T. reesei
mycelium, coupling to modified Sorfix, **135**, 355
purification of
cellobiohydrolases, **160**, 188
cellulase, **160**, 237
T. viride, purification of
cellulases, from mutant QM 9414 strain, **160**, 245
β-xylosidase, **160**, 689

Trichosporon
T. cutaneum, as BOD sensor component, **137**, 125
T. penicillatum, purification of protopectinase, **161**, 343

Trifluoperazine
^3H-labeled, binding to calmodulin, assay, **159**, 656

Trifluoromethanesulfonic acid
glycoprotein deglycosylation with, **138**, 343

Triglycerides
assay with enzyme thermistor, **137**, 221
interesterification with celite-adsorbed lipase, **136**, 405
Triglycosylceramides
perbenzoylated, HPLC, **138**, 117
Trimethylselenonium ion
HPLC assay, **143**, 198
precipitation-based assay, **143**, 197
Trinitrobenzenesulfonic acid
in assay of catalase and pyranose oxidase amino groups, **137**, 603
in spectrophotometric assay of thiols, **143**, 58
Triose-phosphate isomerase
as marker for chromosome detection in somatic cell hybrids, **151**, 192
Trisacryl GF-2000
1,1′-carbonyldiimidazole-activated
enzyme coupling, **135**, 115
IgG purification with, **135**, 114
Tris(hydroxymethyl)aminomethane, *see* Tromethamine
Trislactoside
preparation, **138**, 425
Tromethamine
cluster glycoside preparation with, **138**, 425
Trophic factors
T-cell, as ancillary signals in B-cell activation by anti-Ig, **150**, 101
Tropoelastin
as fibroblast chemoattractant, **163**, 702
immunoprecipitation, **144**, 256
isolation, **144**, 179; **163**, 665
Troponin C
produced in *E. coli*, properties, **153**, 481
slow/cardiac, from quail
cDNA cloning, **139**, 328, 334
codon sequence, **139**, 331, 335
structure, **139**, 333
X-ray crystallography, **139**, 610
Troponin I
binding by *Acanthamoeba* calmodulin, assay, **139**, 62
Trypanosoma
T. brucei, immunolocalization of calmodulin, **139**, 262
T. cruzi, partial purification of cAMP-binding protein, **159**, 289
Trypsin
bound to surgical biomaterial, characterization, **137**, 549
coupling to
carbonate-activated Sepharose CL-4B, **135**, 89
1,1′-carbonyldiimidazole-activated support, **135**, 115
isonitrile-derivatized polymers, **135**, 100
immobilization by radiation polymerization, **135**, 151
immobilized
in analysis of protein digestibility changes, **135**, 601
in peptide synthesis, **136**, 183, 185, 284
–α-macroglobulin complexes as chemotactic agents, preparation, **162**, 185
purification on 1,1′-carbonyldiimidazole-activated support, **135**, 113
Sepharose-coupled
catalytic activity, **135**, 593
preparation, **135**, 592
reactivation, **135**, 594
thermoinactivation, **135**, 593
Trypsin inhibitor
coupling to
tresyl-agarose, **135**, 72
tresyl-silica, **135**, 70, 72
Trypsin-protein esterase
activity of α_2-macroglobulin, assay, **163**, 488
Tryptase
assay, **163**, 327
properties, **163**, 314
purification from human mast cells, **163**, 321
L-Tryptophan
synthesis from chorismate
in bacteria, **142**, 293
in fungi and higher eukaryotes, **142**, 299
Tryptophanase
activation, **142**, 417
assay, **142**, 414
properties, **142**, 420
purification from *E. coli*, **142**, 415
Tryptophan *C*-methyltransferase
assay, **142**, 236
catalyzed reaction, **142**, 235

properties, **142**, 240
purification from *Streptomyces flocculus*, **142**, 239
Tryptophan 5-monooxygenase
assay, **142**, 83, 89, 93
extraction from murine mastocytoma, **142**, 90
properties, **142**, 86, 92, 95
purification from
murine mastocytoma, **142**, 94
rat brain stem, **142**, 84
Tryptophan side chain oxidase
types I and II
assay, **142**, 201
catalyzed reactions, **142**, 195
properties, **142**, 208
purification from *Pseudomonas* sp., **142**, 203
Tryptophan synthase
$\alpha_2\beta_2$ complex
crystal growth, **142**, 407
purification from bacteria, **142**, 403
apo-β_2 subunit preparation, **142**, 406
assay, **142**, 400
catalyzed reactions, **142**, 399
holo-β_2 subunit isolation, **142**, 405
properties, **142**, 407
α subunit
isolation, **142**, 406
separation from β_2 subunit, **142**, 405
Tubules
longitudinal, isolation from skeletal muscle, **157**, 36
transverse
composition, **157**, 73
coupling to sarcoplasmic reticulum, **157**, 469
enzymatic characteristics, **157**, 64
ion transport, **157**, 79
isolation as free vesicles, **157**, 72
mediated Ca^{2+} release from sarcoplasmic reticulum, induction and measurement, **157**, 472
membrane sidedness and integrity, **157**, 64
morphology, **157**, 77
orientation, **157**, 77
ouabain binding, **157**, 66
preparation from skeletal muscle triads, **157**, 70
Tumor cells
basophilic leukemia
in mediator release studies, **162**, 537
RBL-1, purification of leukotriene C_4 synthase, **163**, 355
breast cancer MCF-7, DNA synthesis, mitogen effects, **146**, 332
breast carcinoma ZR-75-1, derived fibroblast chemoattractant, **163**, 706
calcium-binding proteins, analysis, **139**, 207
chromosome banding analysis, **151**, 267
cytotoxicity of phospholipid-containing liposomes, assays, **141**, 464
epidermoid carcinoma A431
EGF receptor
biosynthesis, **146**, 49
internalization, antibody effects, **146**, 77
purification, **146**, 85
growth
effects of EGF receptor antibodies, **146**, 74, 79
in serum-containing medium, **146**, 89
in serum-free medium, **146**, 90
mitogen-induced pH_i changes, assay, **147**, 424
variants with altered growth responses, isolation, **146**, 91
HeLa, *see* HeLa cells
hepatoma
description of various lines
BW, **151**, 25
H4II EC3, **151**, 26
HA22T/VGF, **151**, 38
Hep 3B, **151**, 34
Hep G2, **151**, 34
Hepa, **151**, 20
HTC, **151**, 29
MH_1 C_1, **151**, 30
PLC/PRF/5, **151**, 32
fibrinogen biosynthesis studies, **163**, 485
purification of endothelial cell growth factors, **147**, 153
SK HEP-1, derived growth factor, purification, **147**, 102
hybridoma
in culture, overview, **151**, 50

growth conditions, **151**, 51
mutants, detection by immunoplate assay, **151**, 54
variants, detection by sib selection technique, **151**, 61
induction of delayed hypersensitivity in mice, **163**, 363
lymphoma
lactoferrin and transferrin assays with, **147**, 306
S49, β-adrenergic receptors, photoaffinity labeling study, **150**, 492
mammary C127, virus-based calmodulin expression in, **139**, 645
mastocytoma, tryptophan 5-monooxygenase
extraction, **142**, 90
purification, **142**, 94
melanoma, purification of
monophenol monooxygenase, **142**, 156
tissue-type plasminogen activator, **163**, 300
myeloma
in culture, overview, **151**, 50
growth conditions, **151**, 51
mutagenesis, **151**, 53
mutants, detection by immunoplate assay, **151**, 54
RNA distribution, **152**, 247
variants, detection by sib selection technique, **151**, 61
osteogenic sarcoma
MG-63, isolation of
fibronectin receptor, **144**, 483
vitronectin receptor, **144**, 485
UMR lines, properties, **145**, 327
pheochromocytoma PC12
culture, **147**, 208
inositol phospholipid metabolism, measurement, **141**, 169
mutants
generation, **147**, 213
selection, **147**, 214
nerve growth factor treatment, **147**, 210
neurite-enriched cultures, **147**, 215
pituitary GH_3
lipid turnover, TRH-stimulated, measurement, **141**, 100
permeability assessment, **141**, 112
permeabilization, **141**, 112
phosphoinositide turnover, hormone-regulated, assay, **141**, 118
prolactin gene expression, Ca^{2+}/calmodulin regulation, **139**, 655
radiolabeling, **141**, 101
and TtT, TRH-induced changes in calcium metabolism, measurements, **141**, 36
Raji, isolation of Epstein–Barr virus/C3d receptor, **150**, 549
T-cell leukemic HUT 102B2, purification of
interleukin-2 receptor, **150**, 691
interleukin-2 receptor mRNA, **150**, 693
teratocarcinoma
in analysis of basement membrane biosynthesis, **145**, 391
EC line
characteristics, **151**, 68
culture conditions, **151**, 69
differentiated derivatives, **151**, 77
differentiation markers, **151**, 78
differentiation *in vitro*, **151**, 74
DNA transfer into, **151**, 81
mutagenesis, **151**, 78
retroviral infection, **151**, 81
EK line, establishment, **151**, 73
intercellular communication, inhibition by phorbol ester, assay, **141**, 442
T lymphoma S49
in analysis of cAMP turnover, **159**, 53
counterselection, **151**, 13
frozen storage, **151**, 13
growth conditions, **151**, 10
origin, **151**, 9
mutagenesis, **151**, 15
mutant and variant forms, **151**, 17
Tumor necrosis factor
characterization as pyrogen, **163**, 498
gene
isolation, **150**, 753
modification by site-specific mutagenesis, **154**, 403
human, expression plasmid construction, **153**, 424
Tumors
chondrosarcoma, derived growth factor, purification, **147**, 100

metastases, treatment with liposome-activated macrophages, **149**, 154
Morris hepatoma, oncomodulin isolation, **139**, 157
pheochromocytoma, purification of tyrosine 3-monooxygenase, **142**, 76
Tungsten
colorimetric assay, **158**, 373
neutron activation analysis, **158**, 380
Tunicamycin
as inhibitor of lipid-linked saccharide pathway, **138**, 664
Turbidimetric assay
cellulase, **160**, 117
clostridial alpha toxin, **165**, 296
selenocysteine β-lyase, **143**, 415
sulfate, **143**, 5
Turbidostat
sycamore cell suspension cultures, **148**, 12
Turgor pressure
in cyanobacteria
ESR analysis, **167**, 513
measurement with gas vesicles, **167**, 661
Tween 80, *see* Polysorbate 80
Two-phase systems
aqueous, cell and enzyme immobilization in, **137**, 657
Tyrosinase, *see* Monophenol monooxygenase
L-Tyrosine
postprephenate biochemical pathways to, overview, **142**, 472
Tyrosine aminotransferase
cathepsin T assay with, **142**, 280
Tyrosine kinase, *see* Protein-tyrosine kinase
Tyrosine 3-monooxygenase
assay, **142**, 57, 63, 72
properties, **142**, 62, 68, 79
purification from
bovine adrenal gland, **142**, 60
rat adrenal gland, **142**, 66
rat pheochromocytoma, **142**, 76

U

Ubiquinones
synthesis in plant mitochondria, **148**, 486
UDPglucose
labeling of plant plasma membranes, **148**, 574
UDPglucose pyrophosphorylase, *see* UTP–glucose-1-phosphate uridylyltransferase
Ultracentrifugation
in analysis of
extracellular matrix proteins, principles, **145**, 10
laminin, **145**, 37
in purification of
branched-chain α-keto acid dehydrogenase, **166**, 311
branched-chain 2-keto acid dehydrogenase multienzyme complex, **166**, 334
Ultracentrifuge
air-driven, for sedimentation and binding studies of membrane proteins, **150**, 447
Ultrafiltration
enzyme immobilization by, **135**, 62
β-glucanases, **160**, 386
micro-, *see* Microultrafiltration
in preparation of metallocarboxypeptidases, **158**, 76
Ultrastructure
cyanobacterial
analysis by
freeze-drying, **167**, 159
freeze-etching, **167**, 158
metal shadowing, **167**, 159
negative staining, **167**, 160
observation in thin sections, **167**, 160
Ultraviolet detection
with HPLC, disulfides and thiols as dinitrophenol derivatives, **143**, 101
Ultraviolet spectroscopy
first derivative, determination of chitin and chitosan acetylation degree, **161**, 444
Urate oxidase
reactors
preparation, **137**, 261, 289
uric acid assay with, **137**, 263, 295
Urea
assay with
enzyme thermistor, **137**, 189
enzyme transistor, **137**, 242, 258
hybrid biosensor, **137**, 135

solutions, protein structural changes in, analysis with immobilized Pronase, **135**, 598

synthesis with urease in water–organic solvent mixtures, **136**, 234

Urease

coupling to isonitrile-cellulose, **135**, 101

gelatin-immobilized, properties, **135**, 297

in multienzymatic NAD(P)H recycling, **136**, 71, 77

transistor

preparation, **137**, 256

urea assay with, **137**, 258

as urea sensor component, **137**, 135

urea synthesis with, in water–organic solvent mixtures, **136**, 234

Urethan

prepolymers

entrapment of

biocatalysts, **135**, 241

cyanobacteria, **167**, 633

general structure, **135**, 241

properties, **135**, 242

effects on immobilized biocatalysts, **135**, 250

Urethane, *see* Urethan

Uric acid

assay with enzyme reactor, **137**, 295

continuous-flow system, **137**, 263

flow–injection system, **137**, 268

preparation with immobilized xanthine oxidase, **136**, 259

Uricase, *see* Urate oxidase

Uridine

radiolabeled, in assay of pore-forming toxins in cultured cells, **165**, 280

Uridine diphosphate glucose

enzymatic synthesis with ATP regeneration, **136**, 279

Uridine-5'-diphosphoglucose, *see* Uridine diphosphate glucose

Urine

aluminum levels, measurement, **158**, 296

calcium

ionized, measurement by ion-selective electrodes, **158**, 332

total, flame atomic absorption spectrometry, **158**, 314

catecholamine metabolites, GC–MS assay, **142**, 577

catecholamines, enzymatic radioimmunoassay, **142**, 565

chromium levels, measurement, **158**, 337

extracellular matrix proteins, analysis, **145**, 396

human

isolation of sialyl oligosaccharides and phosphates, **138**, 289

purification of

connective tissue-activating peptide V, **163**, 743

epidermal growth factor, **146**, 6

erythropoietin, **147**, 333

kallikrein, **163**, 112, 162

latent kallikrein, **163**, 99

sulfatide/G_{M1} activator, **138**, 812

rat

protein excretion in serum sickness, measurement, **162**, 498

purification of kallikrein, **163**, 135

trimethylselenonium ion, assay methods, **143**, 197

Urokinase, *see* Plasminogen activator

Uterus

bovine, purification of estradiol receptor-kinase, **139**, 737

UTP–glucose-1-phosphate uridylyltransferase

immobilized, UDP-glucose synthesis with, **136**, 279

V

Vaccinia virus

as vector for foreign gene expression, **153**, 545

Vacuoles

Neurospora crassa, membrane preparation, **157**, 563

plant

adenine nucleotides, HPLC, **148**, 132

isolation from protoplasts, **148**, 76, 83

preparation by

direct isolation from tissues, **148**, 65

isotonic shear lysis of protoplasts, **148**, 63

osmotic lysis of protoplasts, **148**, 61

polybase-induced lysis of protoplasts, **148**, 62

polybase procedure, suitability for different plants, **148**, 77

purity determination after isolation, **148**, 84
purity, integrity, and stability criteria
in compartmentation analyses, **148**, 67
in transport studies, **148**, 72
tonoplast isolation from, **148**, 79, 85, 106
yeast, membrane vesicles
preparation, **157**, 545
properties, **157**, 546
Vanadate
in assay for Na^+,K^+-ATPase sidedness in liposomes, **156**, 162
in crystallization of membrane-bound Na^+,K^+-ATPase, **156**, 81
effects on Na,K-pump activities in
dialyzed giant axons, **156**, 252
erythrocytes, **156**, 255, 259, 261
resealed erythrocyte ghosts, **156**, 263
inhibition of Na^+,K^+-ATPase, **156**, 9
Vanadium
assay by
atomic absorption spectrometry, **158**, 142, 402
dc arc emission method, **158**, 414
dc plasma atomic emission spectroscopy, **158**, 415
energy-dispersive X-ray fluorescence, **158**, 420
EPR spectroscopy, **158**, 420
inductively coupled plasma method, **158**, 412
mass spectrometry, **158**, 416
neutron activation analysis, **158**, 406
spectrophotometry, **158**, 418
Vanadyl complexes
in isolation of cytoplasmic RNA, **152**, 227
Vanillate *O*-demethylase
assays, **161**, 295
preparation from *Pseudomonas*, **161**, 297
properties, **161**, 300
Vanillate hydroxylase
assays, **161**, 274
properties, **161**, 278
purification from *Sporotrichum pulverulentum*, **161**, 277
Vanillin
carbon isotope-labeled, preparation, **161**, 52
Vanillylmandelic acid, *see* Vanilmandelic acid
Vanilmandelic acid
GC–MS assay in brain and body fluids, **142**, 577
Vasectomy
in rabbits, for induction of autoimmune orchitis, **162**, 464
Vectors
for analysis of transcription initiation and termination signal structures, **153**, 452
bovine papillomavirus-based, for calmodulin expression, analytical methods, **139**, 642
for cAMP-dependent protein kinase expression
construction, **159**, 300
induction, **159**, 306
for cAMP-dependent protein kinase regulatory subunit expression, **159**, 326
for cloning in cyanobacteria, listing, **167**, 824
cosmid
for cloning *dhfr* gene from mutant CHO cells, construction, **151**, 454
for genomic walking and restriction mapping, **152**, 604
shuttle, for cloning and analysis of *Streptomyces* DNA, **153**, 166
direct expression, for yeast, **153**, 527
double *cos* site, preparation, **153**, 82
expression cassette, for plant transformation, **153**, 253
extrachromosomal replication, for yeast, **153**, 518
for foreign protein expression in *E. coli*, **152**, 671; **153**, 492
λ phage, EMBL series, cloning in, **153**, 103
λgt11 phage, elastin cDNA cloning into, **144**, 282
M13
in DNA dideoxy sequencing, **155**, 63, 119
in oligonucleotide-directed mutagenesis, **154**, 382
mitochondrial, for cloning in eukaryotes, **153**, 366

open reading frame, application, **154**, 156
phage, in oligonucleotide-directed mutagenesis, **154**, 329
plasmid
cDNA expression libraries in, immunoscreening, **152**, 451
directional cDNA cloning in, **152**, 349
–primer molecules, in full-length cDNA cloning, **152**, 371
restriction site bank, for cloning in bacteria and yeast, **153**, 34
retroviral, containing hypoxanthine phosphoribosyltransferase gene
arrangement in transfected cells, **149**, 19
construction, **149**, 13
transfection protocol, **149**, 15
transfer into
bone marrow cells, **149**, 24
lymphoblasts, **149**, 23
secretion
for *E. coli*, **153**, 499
for yeast, **153**, 533
single-strand, for dideoxy sequencing of large DNA fragments, **152**, 553
Streptomyces phage- and plasmid-type, for gene cloning and analysis, **153**, 116
Ti plasmid-derived
for cloning in plant cells, **153**, 277
for plant transformation and promoter analysis, **153**, 292
vaccinia virus, for foreign gene expression, **153**, 545
viral, in identification of gene regulatory elements, **152**, 713
Verticillium albo-atrum
purification of chitinase, **161**, 475
Vesicles
cardiac sarcoplasmic reticulum
characteristics after reconstitution, **157**, 317
phospholamban purification, **157**, 360
solubilization and reconstitution, **157**, 315
clathrin-coated, preparation from bovine brain, **157**, 638
cyanobacterial membrane, size and surface charge analysis, **167**, 691
endoplasmic reticulum, ionic permeability measurements, **157**, 417
erythrocyte, inside-out
Na,K-transport, assays, **156**, 175
preparation and adjustment of ionic composition, **156**, 173
everted, preparation from *E. coli*, **157**, 670
gas, *see* Gas vesicles
Golgi- and rough endoplasmic reticulum-derived, sugar nucleotide transport assays, **138**, 709
membrane
preparation, **143**, 507; **146**, 376, 380
transport assay, **146**, 380
mitochondrial inner membrane
matrix inclusion assay, **148**, 449
polarity assay, **148**, 446
preparation, **148**, 448
right-side-out and inside-out, separation from intermediate polarity vesicles, **148**, 451
phospholipid
H^+-ATPase incorporation, **157**, 525
Na^+,K^+-ATPase reconstitution into, **156**, 157
plasma membrane, sodium–calcium exchange, measurement, **157**, 505
reconstituted
ATPase assays, **156**, 135
ATPase transport assays, **156**, 136
Na^+,K^+-ATPase, electron microscopy, **156**, 438
Na,K-pump activities
Na^+–K^+ exchange, **156**, 181
Na^+–Na^+ exchange, **156**, 185
net Rb^+ flux, **156**, 188
Rb^+–congener exchange, **156**, 188
^{86}Rb efflux, **156**, 184
Rb^+–Rb^+ exchange, **156**, 185
uncoupled Na^+ flux, **156**, 185
preparation by freeze–thaw sonication, **156**, 179
sarcolemmal
characterization, **157**, 31
preparation from
canine heart, **157**, 85
rabbit skeletal muscle, **157**, 27
sarcoplasmic reticulum
characteristics, **157**, 96
functional characterization, **157**, 42
ionic permeability measurements, **157**, 417, 437

phosphorylation, **157**, 128
preparation, **157**, 92
reconstitution with control of lipid/protein ratio, **157**, 302
voltage-induced Ca^{2+} uptake, **157**, 248
sealed, incorporated Na,K-pump, density gradient purification, **156**, 139
thylakoid, inside-out
generation by fragmentation, **148**, 254
properties, **148**, 258
separation by aqueous two-phase partition, **148**, 255
tonoplast
applications to H^+-coupled transport, **148**, 118
characteristics, **148**, 117
preparation from
beet, **148**, 115
oat, **148**, 123
transverse tubule, isolation, **157**, 72
yeast vacuolar membrane
preparation, **157**, 545
properties, **157**, 546
Vibrio
V. cholerae
diarrheal toxin, ileal loop test, **165**, 313
purification of cholera toxin, **165**, 171
V. damsela, purification of cytolytic toxin, **165**, 186
V. harveyi, purification of *N,N′*-diacetylchitobiase, **161**, 527
V. parahaemolyticus
diarrheal toxin, ileal loop test, **165**, 319
purification of thermostable direct hemolysin, **165**, 190
V. vulnificus, purification of cytolytic toxin, **165**, 179
Video microscopy
calmodulin and calcium localization in cultured cells, **139**, 830
Viruses, *see also specific viruses*
glycolipid receptors
overlay analysis, **138**, 212
overlay and solid-phase analyses, **138**, 220
Viscosimetric assay
carboxymethylcellulase, **160**, 130
chitinase, **161**, 426
polygalacturonase, **161**, 362
Viscosity
actin solutions in presence of modulating proteins, **162**, 260
Visual assay
chromosome ploidy in yeast, **155**, 351
Vitis vinifera, *see* Grape
Vitronectin
properties, **144**, 432
purification, **144**, 431, 436
role in cell attachment, **144**, 435
Vitronectin receptors
functional analysis with synthetic peptides, **144**, 479
incorporation into liposomes, **144**, 489
isolation from
cultured cells, **144**, 485
placenta, **144**, 485
properties, **144**, 476, 478
Volatilization
selenium, **143**, 34
Voltammetry
analytical procedures, **158**, 258
apparatus, **158**, 252
arsenic(III) assay, **158**, 264
cadmium(II) assay, **158**, 263
linear scan technique, **158**, 250
overview, **158**, 243
pulse techniques, **158**, 248
rotating disk technique, **158**, 245
stripping techniques, **158**, 251
Vulnificolysin
bioassay, **165**, 177
properties, **165**, 181
purification, **165**, 179

W

W-7
^3H-labeled, binding to calmodulin, assay, **159**, 656
Water
metal-free, preparation, **158**, 3
purification with immobilized microorganisms, **137**, 697
Waxes
plant cuticular, pesticide effects, **148**, 670, 673

Western blotting
branched-chain α-keto acid dehydrogenase proteins, **166**, 109
collagen peptides in gels, **145**, 201
collagens, **163**, 685
complement components C3, C5, C3a, and C5a, **162**, 585
erythropoietin, **147**, 338
laminin receptors, **144**, 495
native and recombinant human C5a, **162**, 661
tissue kallikrein, **163**, 140
tropoelastin, **163**, 671
Wet oxidation
pretreatment of biomass materials for enhancing cellulose availability, **160**, 9
Wheat germ
purification of endochitinase, **161**, 498
Wheat germ agglutinin
plant protoplast immobilization with, **135**, 424, 426
Wood chips
cell immobilization by adsorption, **135**, 334
Wood degradation
macroscopic and electron microscopic analysis, **160**, 193
Wort
(1→3),(1→4)-β-D-glucan assay, **160**, 549
Wound healing
acceleration by GHL-Cu growth factor, **147**, 326

X

Xanthine
oxidation by immobilized xanthine oxidase, **136**, 259
substituted, synthesis, **159**, 495
Xanthine oxidase
milk-derived
immobilization, **136**, 258, 259
overview, **136**, 256
oxidation of
1-methylxanthine, **136**, 260
xanthine, **136**, 259
X-ray crystallography
bacterial ribosomes, **164**, 95
calbindin D, **139**, 585
calmodulin, **139**, 632
catabolite gene activator protein–cAMP complex, **159**, 278
chitin, **161**, 435
phage T4 lysozyme, **154**, 520
troponin C, **139**, 610
X-ray fluorescence
aluminum assay by, **158**, 290
energy-dispersive, vanadium assay by, **158**, 420
Xylan
–Remazol Brilliant Blue, as soluble chromogenic xylanase substrate
applications, **160**, 538
preparation, **160**, 536
Xylanase
assays, **160**, 538, 633, 638, 649, 655, 663, 671, 675
detection in
gels, **160**, 539
prokaryotes, **160**, 185
nephelometric assay, **160**, 124
properties
Bacillus enzymes, **160**, 636, 657
Cryptococcus enzyme, **160**, 642
Malbranchea enzyme, **160**, 673
Streptomyces enzymes, **160**, 651
Talaromyces enzyme, **160**, 678
Trichoderma enzyme, **160**, 668
purification from
Bacillus, **160**, 634, 656
Cryptococcus albidus, **160**, 639
Malbranchea pulchella, **160**, 672
Streptomyces, **160**, 649
Talaromyces byssochlamydoides, **160**, 676
Trichoderma lignorum, **160**, 665
type A
assay, **160**, 659
properties, **160**, 661
purification from *Schizophyllum commune*, **160**, 661
β-Xylanase
assays, **160**, 69
Xylans
acetylation, **160**, 552
evaluation, **160**, 554
hydrolysates, TLC, **160**, 176
isolation, **160**, 529
structure, **160**, 528

xylobiose and xylooligomers, preparation, **160**, 530
1,4-β-D-Xylan xylohydrolase, *see* Xylan 1,4-β-xylosidase
Xylan 1,4-β-xylosidase
assays, **160**, 664, 679, 685, 696
properties
Chaetomium enzyme, **160**, 698
Emericella enzyme, **160**, 694
Malbranchea enzyme, **160**, 686
Sclerotium enzyme, **160**, 681
Trichoderma enzymes, **160**, 668, 690
purification from
Chaetomium trilaterale, **160**, 697
Emericella nidulans, **160**, 692
Malbranchea pulchella, **160**, 684
Sclerotium rolfsii, **160**, 680
Trichoderma lignorum, **160**, 665
Trichoderma viride, **160**, 689
Xylobiose
characteristics, **160**, 536
chromatographic analysis, **160**, 532
preparation from xylans, **160**, 530
Xylooligosaccharides
characteristics, **160**, 536
chromatographic analysis, **160**, 532
preparation from xylans, **160**, 530
Xylose
quantitative TLC, **160**, 172
β-Xylosidase, *see* Xylan 1,4-β-xylosidase

Y

Yeasts, *see specific yeasts*
Yellow affinity substance
isolation from *Clostridium thermocellum*, **160**, 495
properties, **160**, 499
Yersinia
Y. enterocolitica, diarrheal toxin, ileal loop test, **165**, 321
Y. pestis, isolation of plague murine toxin, **165**, 162

Z

Zea mays, *see* Maize
Zinc
atomic absorption spectrometry, **158**, 143, 422
in cyclic nucleotide phosphodiesterases, assay, **159**, 784
stimulation of lymphocytes, **150**, 112
Zirconium
hydrous oxides, coupling of microbial cells, **135**, 358
Zymography
in detection of cellobiose dehydrogenase, **160**, 458
with type I collagen fibrils as substrate, **144**, 158

Contributor Index

Boldface numerals indicate volume number.

A

Abell, Creed W., **142**, 242, 638
Aberhart, D. John, **166**, 14, 17
Abita, Jean-Pierre, **142**, 27
Abo, Steven, **164**, 158
Abril, Obsidiana, **136**, 263
Adachi, Anne-Marie, **139**, 518
Adams, Deborah A., **152**, 61
Adams, Douglas O., **143**, 426
Adams, Sherrill L., **144**, 84
Adler, Howard, **143**, 38
Affolter, Hubert, **141**, 25
Aftring, R. Paul, **166**, 201
Agelli, Maria, **163**, 642
Ahmed, Asad, **155**, 177
Ahmed, Syed Ashrafuddin, **142**, 398
Ahn, Yeon S., **149**, 312
Aitken, Alastair, **167**, 145
Aitken, Susan, **146**, 329
Aizawa, Masuo, **137**, 111
Akazawa, Takashi, **148**, 27
Åkerlund, Hans-Erik, **148**, 252
Akiba, Teruhiko, **160**, 655
Akiyama, Shin-ichi, **151**, 38
Al-Awqati, Qais, **157**, 611
Alban, Claude, **148**, 206
Alber, Tom, **154**, 511
Albright, Kevin, **151**, 252
Alemany, Susana, **159**, 390
Aletta, John M., **147**, 207
Alexander, Danny C., **154**, 41
Algarin, Lourdes, **149**, 281
Alibert, Gilbert, **148**, 74
Alic, Margaret, **161**, 74
Allen, Mary Mennes, **167**, 207
Allen, Robert H., **147**, 239; **166**, 393, 400, 407
Allende, Catherine, **159**, 660
Allende, Jorge E., **159**, 660
Allende, Oscar, **159**, 286
Almon, Helmar, **167**, 459
Alonso, Teresa, **143**, 191
Alouf, Joseph E., **165**, 52, 59, 64, 91, 293
Altendorf, Karlheinz, **157**, 668
Alving, Carl R., **141**, 244, 459; **149**, 124
Amano, Junko, **138**, 779
Ambard-Bretteville, F., **148**, 623
Ambrose, Barbara J. B., **151**, 522
Amenta, Peter S., **145**, 133
Amit, Adolfo G., **160**, 355
Amory, Antoine, **157**, 513
An, Gynheung, **153**, 292
Anasetti, Claudio, **150**, 536
Andersen, Jens P., **157**, 261
Anderson, Clark L., **150**, 524
Anderson, Conwell H., **141**, 201
Anderson, Mary E., **143**, 313
Anderson, Wayne B., **141**, 399
Andersson, Bertil, **148**, 252
Andersson, L., **135**, 475; **136**, 117
Anding, C., **148**, 667
Andrews, Glenn C., **162**, 653
Andrews, Jaen E., **148**, 338
Andrews, William V., **141**, 156
Angerer, Lynne M., **152**, 649
Angerer, Robert C., **152**, 649
Anraku, Yasuhiro, **157**, 544
Ansher, Sherry S., **142**, 660
Anton, Ian A., **142**, 315
Antoniades, Harry N., **147**, 22
Antoniu, Bozena, **157**, 469
Antonucci, Tammy K., **166**, 245
Anumula, Kalyan R., **138**, 816
ap Rees, Tom, **148**, 218
Arai, K., **154**, 3
Arai, Motoo, **160**, 274
Arai, Yuji, **139**, 127

Araki, Toshiyoshi, **160**, 583
Araki, Yoshio, **161**, 510, 514
Arbuthnott, John P., **165**, 32, 333, 334
Argentieri, Dennis C., **162**, 339
Argraves, Scott, **144**, 475
Armenta, Richard, **136**, 93
Arnaout, M. Amin, **150**, 602
Arnaud, Philippe, **163**, 400, 418, 431, 441
Arrio, Bernard, **148**, 132; **167**, 691
Arrow, Amy, **155**, 204
Arroyo-Begovich, Angel, **161**, 471
Arwin, Hans, **137**, 366
Asahi, Tadashi, **148**, 491
Askari, Amir, **156**, 345
Assoian, Richard K., **146**, 153
Aubert, Jean-Paul, **160**, 355
Auld, David S., **158**, 13, 71, 110
Auricchio, Ferdinando, **139**, 731
Auriol, D., **136**, 239
Austin, Paul R., **161**, 403
Au-Young, Janice, **138**, 643
Avenard, G., **149**, 242
Avrameas, Stratis, **150**, 117
Awad, William M., Jr., **143**, 384
Azari, Mahmood R., **137**, 675
Azuma, Jun-Ichi, **161**, 12
Azuma, Masaki, **136**, 394

B

Baba, Akiko, **157**, 251
Babu, Y. Sudhakar, **139**, 632
Bader, Johann, **136**, 302
Bahl, Om P., **138**, 341, 350, 816
Bailey, Christopher J., **165**, 32
Bailey, G. S., **163**, 115
Bailey, James E., **135**, 502
Bailey, Philip J., **162**, 327, 478
Baker, David H., **143**, 297
Baker, John O., **158**, 33
Baker, John R., **144**, 401
Baldwin, Julia H., **146**, 163
Ball, William J., Jr., **156**, 87
Ballou, Lisa M., **159**, 377
Balny, Claude, **135**, 528
Bame, Karen J., **138**, 607
Bancroft, Carter, **139**, 655
Banda, Michael J., **144**, 288
Bankier, A. T., **155**, 51
Barabé, J., **163**, 210, 282
Barak, Ze'ev, **166**, 234, 455
Barber, James, **148**, 294
Barber, R., **159**, 50
Barbieri, Joseph T., **165**, 68
Barbotin, Jean-Noël, **135**, 454
Baret, J. L., **136**, 411
Baribault, Thomas J., **147**, 186
Barlow, D. P., **144**, 464
Barnes, David, **146**, 88; **163**, 707
Barnes, Wayne M., **152**, 538
Barnett, Don R., **163**, 452
Barondes, Samuel H., **138**, 510, 516
Barone, A. D., **154**, 287
Bar-Or, Yeshaya, **167**, 616
Barrell, B. G., **155**, 51
Barrett, Alan J., **163**, 240
Barta, Andrea, **164**, 361
Bartetzko, Andreas, **164**, 650
Bartholdi, Marty, **151**, 252
Bartlett, Kim, **166**, 79
Bassham, James A., **167**, 539
Basu, Manju, **138**, 575
Basu, Subhash, **138**, 575
Bateman, John F., **145**, 183
Bates, Paul, **153**, 82
Battisto, Jack R., **150**, 83
Bauch, Bärbel, **161**, 518
Bauerle, Ronald, **142**, 366, 398
Baumann, Heinz, **163**, 566
Baumeister, Kirk, **155**, 139
Bayer, Edward A., **138**, 429; **149**, 119; **160**, 472
Baylink, David J., **146**, 281
Beasley-Leach, Alfreda, **159**, 531
Beaucage, S. L., **154**, 287
Beaugé, Luis A., **156**, 251, 281
Beavo, Joseph A., **159**, 543, 685
Beck, Christoph F., **153**, 452
Beck, Stephan, **155**, 250
Becker, David W., **167**, 280
Beckingham, Kathy, **139**, 230
Beebe, Stephen J., **159**, 74, 118, 531
Beevers, Harry, **148**, 526
Beggs, Mark, **166**, 166, 175
Béguin, Pierre, **160**, 355
Behbahani-Nejad, Iraj, **142**, 414
Behie, L. A., **136**, 329
Beier, N., **159**, 685
Belagaje, Rama M., **153**, 401

Beldman, G., **160**, 243
Belitsina, N. V., **164**, 631
Belkin, Shimshon, **167**, 380, 670, 677
Bell, Graeme I., **146**, 239; **147**, 194
Bell, Robert M., **141**, 287, 294, 313
Bendall, Derek S., **148**, 259; **167**, 272
Bender, Myron L., **135**, 537
Benditt, Earl P., **163**, 510
Benkovic, Stephen J., **142**, 50
Bennett, Alan B., **157**, 579
Benveniste, Pierre, **148**, 632
Benz, Edward, J., Jr., **156**, 379
Beppu, Masatoshi, **150**, 17
Berchtold, Martin W., **139**, 317
Berezin, I. V., **135**, 586
Berg, Torill, **163**, 143
Bergdoll, Merlin S., **165**, 306, 324
Berger, Shelby L., **152**, 49, 215, 227, 262, 307, 316, 335, 337, 399, 415
Berkman, F. A., **145**, 303
Berman, Judith, **155**, 528
Bernhardt, Frithjof-Hans, **161**, 281
Bernheimer, Alan W., **165**, 213
Bernicker, E., **139**, 19
Bernier, Roger, Jr., **160**, 432
Bernstein, Howard, **137**, 515
Bertholf, Roger L., **158**, 289
Beutler, Hans-Otto, **143**, 11
Beyer, P., **148**, 392
Bhakdi, Sucharit, **165**, 285
Bhat, K. Mahalingeshwara, **160**, 87
Bhatt, Ashwin K., **162**, 193
Bibb, Mervyn J., **153**, 116
Bieber, L. L., **166**, 288
Biely, Peter, **160**, 536, 638, 700
Biemesderfer, Daniel, **156**, 392
Bigazzi, Pierluigi E., **162**, 461
Bill, Eckhard, **161**, 281
Billingsley, Melvin L., **159**, 605
Birkbeck, T. Harry, **165**, 16
Birkedal-Hansen, Henning, **144**, 140
Birnbaum, Staffan, **137**, 334, 632
Bishop, Paul D., **138**, 715
Bitensky, Mark W., **159**, 702, 730
Bitter, Grant A., **152**, 673; **153**, 516
Blackmore, Peter F., **159**, 74, 118
Blackshear, Perry J., **141**, 412
Blackwell, John, **161**, 435
Blanchette, Robert A., **160**, 193
Blandin-Savoja, Félicienne, **142**, 27
Blank, Merle, **141**, 379
Bleasdale, John E., **141**, 222
Bligny, Richard, **148**, 3; **167**, 667
Bliven, Marcia L., **162**, 334
Block, Kevin P., **166**, 201
Block, Maryse A., **148**, 206
Blöcker, Helmut, **154**, 221
Blomster-Hautamaa, Debra A., **165**, 37
Blostein, Rhoda, **156**, 171
Blumberg, Daphne D., **152**, 3, 20
Blumel, M., **160**, 560
Blumenthal, Donald K., **139**, 115
Blumwald, Eduardo, **148**, 115
Boeke, Jef D., **154**, 164
Boekema, Egbert, **164**, 35
Boeynaems, Jean-Marie, **159**, 19
Bogdanov, Alexey A., **164**, 440
Böger, Peter, **167**, 459
Boghossian, Jocelyne, **150**, 134
Bohach, G. A., **165**, 137, 302, 399
Bohrer, Alan N., **162**, 79
Boller, Thomas, **161**, 430, 479
Bondurant, Maurice C., **147**, 340
Bonner, C., **142**, 479, 488
Bonner, William M., **152**, 55
Boocock, Martin R., **142**, 325
Bookman, Michael, **150**, 342
Borbely, George, **167**, 592, 622
Borchmann, Annegret, **160**, 528
Borden, Angela, **164**, 673
Borgman, Cheryl A., **138**, 125
Bornman, Chris H., **135**, 421
Botelho, Lynne H. Parker, **159**, 159
Bothe, Hermann, **167**, 496
Boublik, Miloslav, **164**, 49
Boudet, Alain M., **148**, 74
Bourguignon, Jacques, **148**, 403
Bowers, George N., Jr., **158**, 302, 320
Bowers, William E., **150**, 105
Bowes, Jane M., **167**, 272
Bowman, Barbara H., **163**, 452
Bowman, Barry J., **157**, 562
Bowman, Emma Jean, **157**, 562
Boyer, Charles, **148**, 226
Boyle, Michael D. P., **162**, 101
Bradford, Andrew P., **166**, 10
Brand, Jerry J., **167**, 280
Brandhorst, Bruce P., **139**, 354
Brandl, Christopher J., **157**, 289
Brandt, Neil R., **157**, 68

Braun, Marcus, **150**, 536
Brautigan, David L., **159**, 339
Bremer, Jon, **166**, 476
Brentjens, Jan R., **162**, 484
Brewer, Linda M., **139**, 156
Brimacombe, Richard, **164**, 287
Briskin, Donald P., **148**, 542
Britten, Roy J., **152**, 611
Brockes, Jeremy P., **147**, 217
Broda, Paul, **161**, 191, 211, 221
Brodelius, P., **135**, 173, 222, 434, 512
Brooker, Gary, **159**, 45
Brooks, Joan E., **152**, 113
Brosius, Jürgen, **153**, 54
Brouers, M., **167**, 629
Brouquisse, Renaud, **148**, 403
Brown, Dale, **142**, 225
Brown, David B., **151**, 371
Brown, Peter C., **151**, 85
Brown, Sue, **158**, 289
Brownell, A. G., **146**, 294
Brownstein, M., **154**, 3
Bruce, Barry D., **151**, 292
Brunel, Françoise, **153**, 34
Brunow, Gösta, **161**, 65
Brunschwig, J-P., **157**, 68
Brunton, Laurence L., **159**, 83
Brusslan, Judy, **153**, 215
Bryant, Donald A., **167**, 755
Bryers, J. D., **137**, 697
Buchanan, Bob B., **167**, 415
Buchholz, K., **135**, 3
Buck, Melissa A., **164**, 523
Bucke, C., **135**, 175
Buckley, Anne, **163**, 674
Buckley, J. Thomas, **165**, 193
Buckthal, David J., **142**, 132
Bugg, Charles E., **139**, 632
Buhl, Susan, **151**, 50
Bujard, Hermann, **155**, 416
Bülow, Leif, **137**, 334, 632
Bunzel, Robert J., **163**, 474
Burk, Raymond, F., **143**, 307
Burke, T. R., Jr., **139**, 405
Burrous, Mary R., **143**, 235
Buse, Maria Gordon, **166**, 201
Bushuev, V. N., **164**, 148
Buswell, John A., **161**, 271, 274, 294
Butcher, R. W., **159**, 50
Butler, Larry G., **136**, 234
Butler, William T., **145**, 255, 290
Buxton, Denis B., **166**, 484
Byus, Craig V., **159**, 236

C

Cabib, Enrico, **138**, 637, 643; **161**, 424, 457, 460, 498
Caboche, Michel, **148**, 39
Cabot, Myles C., **141**, 301
Cabral, Antonio R., **163**, 731
Cabral, J. M. S., **135**, 117, 357
Cadd, Gary G., **159**, 299
Callis, Judy, **153**, 351
Calton, G. J., **136**, 497
Calvo, Joseph M., **166**, 225, 455
Calzone, Frank J., **152**, 611
Cambier, John, **141**, 53
Cambou, Bernard, **136**, 117
Cammack, R., **167**, 427
Camp, David G., **164**, 401
Campbell, Ernest B., **143**, 166, 279
Campbell, Evelyn, **151**, 252
Campbell, M. L., **136**, 497
Campbell, Margaret A., **144**, 412
Campbell, William B., **141**, 323
Canevascini, Giorgio, **160**, 112, 443
Cann, Pauline A., **164**, 493
Cantor, Charles R., **151**, 461; **155**, 449
Cao, Shuli, **167**, 105
Capel, Malcolm S., **164**, 117, 532
Carafoli, Ernesto, **139**, 791; **157**, 3, 340
Caras, Steve, **137**, 247
Carde, Jean-Pierre, **148**, 599
Carle, Georges F., **155**, 468
Carlson, Donald E., **167**, 415
Carnuccio, Rosa, **163**, 23
Carothers, Adelaide M., **151**, 449
Carpenter, Graham, **146**, 49
Carr, N. G., **167**, 599
Carrea, Giacomo, **135**, 475; **136**, 150; **137**, 161
Carroll, Stephen F., **165**, 68, 218
Carson, M., **141**, 18
Carter, Paul, **154**, 382
Caruthers, M. H., **154**, 287
Caskey, C. Thomas, **151**, 519
Casnellie, John E., **159**, 730
Cassel, Dan, **147**, 415

Cassidy, Paul, **165**, 3
Castenholz, Richard W., **167**, 68, 96
Castle, John E., **161**, 490
Castor, C. William, **163**, 731
Castoria, Gabriella, **139**, 731
Caswell, Anthony H., **157**, 68
Catignani, George L., **135**, 596
Causey, Anthony G., **166**, 79
Cavalieri, S. J., **165**, 137, 399
Cavener, Douglas R., **150**, 746
Cayama, Edmundo, **159**, 286
Celio, Marco R., **139**, 552
Celis, Julio E., **150**, 147
Ceña, Valentin, **142**, 590
Cerovic, Zoran G., **148**, 145
Chabbott, Holly, **141**, 301
Chadwick, Christopher, **157**, 46
Chamberlain, Brian K., **157**, 91
Chambon, Pierre, **154**, 250
Chan, Danny, **145**, 183
Chan, William W.-C., **135**, 569
Chandler, Wayne L., **137**, 479
Chandra, Subhash, **158**, 157
Chang, J. J., **146**, 294
Chang, Shing, **153**, 507
Chang, T. M. S., **136**, 67; **137**, 444
Changchien, Li-Ming, **164**, 258, 270
Chao, Julie, **163**, 128
Chao, Lee, **163**, 128
Chapman, David J., **148**, 294
Chapuis-Cellier, Colette, **163**, 400
Charbonneau, Harry, **159**, 543
Charleson, Stella, **163**, 340
Charp, Paul A., **139**, 715
Chasin, Lawrence A., **151**, 449
Chassaigne, M., **149**, 242
Chasteen, N. Dennis, **158**, 402
Chater, Keith F., **153**, 116
Chaudhuri, Subhendu, **142**, 315, 320, 325
Chauncey, Thomas R., **143**, 350
Cheema-Dhadli, S., **142**, 127
Cheetham, Peter S. J., **136**, 432
Chen, Chen-Loung, **161**, 110, 137
Chen, Victor J., **138**, 418
Cheng, Heung-Chin, **159**, 173
Chereskin, B. M., **167**, 286
Cheung, David T., **138**, 409
Cheung, Wai Yiu, **139**, 79, 505
Chiancone, Emilia, **135**, 484
Chiang, Peter K., **143**, 191, 377
Chiaruttini, C., **164**, 310
Chiba, Hideo, **147**, 328
Chibata, Ichiro, **135**, 189; **136**, 455, 472
Chichkova, Nina V., **164**, 440
Childers, Lisa C., **161**, 524
Chin, Gilbert, **156**, 127
Chin, Jean, **157**, 591
Chiorazzi, Nicholas, **150**, 92
Chipman, David M., **166**, 234
Chirgwin, John M., **152**, 219
Chock, P. Boon, **159**, 3
Chon, Hung-Che, **138**, 575
Chrestin, H., **148**, 87
Chrisman, Ted D., **159**, 118
Christakos, Sylvia, **139**, 534
Christiansen, Jan, **164**, 456, 710
Chritton, Douglas, **151**, 252
Chu, Alice, **157**, 36
Chuang, David T., **166**, 135, 146
Churchill, Mair E. A., **155**, 537
Cidon, Shulamit, **157**, 619
Claeyssens, Marc, **160**, 45, 187
Clark, Douglas S., **135**, 502
Clark, Leland C., Jr., **137**, 68
Clark, Richard A. F., **162**, 214
Clark, Richard B., **150**, 492
Clayton, Linda K., **150**, 602
Clegg, Christopher, H., **159**, 299
Cleland, Neil, **137**, 298
Clement-Metral, Jenny D., **135**, 454; **167**, 286
Cobb, Charles E., **159**, 202
Cobianchi, Fabio, **152**, 94
Coburn, Jenifer, **165**, 246
Cochran, Brent H., **147**, 64
Cochrane, C. G., **163**, 328
Cocquempot, Marie-Françoise, **135**, 454
Coffino, Philip, **151**, 9
Coggins, John R., **142**, 315, 320, 325, 342, 348
Cohen, Alan S., **163**, 523
Cohen, Philip, **159**, 390, 409, 416, 427
Cohn, Melvin, **150**, 129
Colburn, Joel C., **139**, 188
Cole, Anthony L., **160**, 300
Cole, William G., **145**, 183
Coleclough, Christopher, **154**, 64
Collier, R. John, **165**, 68, 218
Collins, Francis S., **151**, 461
Collins, Jimmy H., **139**, 105, 137

Collmer, Alan, **161**, 329
Colman, Robert W., **159**, 772; **163**, 179
Colombani, Paul M., **139**, 677
Colombini, Marco, **148**, 465
Colowick, Sidney P., **146**, 399
Combes, Didier, **137**, 584
Cone, Robert E., **150**, 388, 666
Conn, P. Michael, **141**, 149, 156, 429
Connelly, James A., **142**, 422, 450
Connor, Jerome, **149**, 88
Conrad, Daniel H., **150**, 503
Conrad, Richard, **164**, 258
Conroy, S., **149**, 64
Conzelmann, Ernst, **138**, 792
Cook, Kenneth G., **166**, 10, 303
Cook, William J., **139**, 632
Coombes, Robert V., **159**, 159
Cooney, Charles L., **137**, 515
Cooper, John D. H., **143**, 141
Cooper, Michael E., **159**, 751
Cooper, Neil R., **150**, 548
Cooperman, Barry S., **164**, 341, 523
Corbin, Jackie D., **159**, 74, 118, 202, 531, 722
Corkey, Barbara E., **166**, 55
Cornelius, Flemming, **156**, 156
Cornell, Neal W., **166**, 70
Coronado, Roberto, **157**, 480
Correll, Leslay A., **159**, 299
Cortese, Riccardo, **155**, 111
Costanzi, Carl, **152**, 582
Costello, Brian, **157**, 46
Costlow, Nancy A., **152**, 633
Cotting, Jacques-Alain, **166**, 14
Cotton, R. G. H., **142**, 116
Coughlan, Michael P., **160**, 135, 363, 437, 483
Covès, Jacques, **148**, 206
Cox, Kathleen H., **152**, 649
Cox, Raymond P., **167**, 467
Cox, Rody P., **166**, 135
Cozens, Alison L., **167**, 794
Craig, William S., **156**, 333
Crain, Pamela F., **155**, 373
Cram, L. Scott, **151**, 252
Crans, Debbie C., **136**, 263
Craven, Gary R., **164**, 258, 270, 532
Crawford, Don L., **161**, 18, 35, 175, 183, 249
Crawford, Irving P., **142**, 293, 300
Crawford, Nancy A., **167**, 415
Crawford, Ronald L., **161**, 18, 83, 264
Cremonesi, Piero, **136**, 150
Crisostomo, M. Cristina, **158**, 382
Crowell, Pamela L., **166**, 39
Cullen, Bryan R., **152**, 684
Cummings, Richard D., **138**, 232
Cunningham, Dennis D., **147**, 157
Curtain, Cyril C., **150**, 418
Cushman, John C., **148**, 157
Cybulsky, Andrey V., **162**, 421
Czerny, T., **167**, 361

D

Dahlberg, Albert E., **164**, 691
Dahms, A. Stephen, **161**, 525
Dakour, J., **138**, 307
Dale, George L., **149**, 229
Dale, Roderic M. K., **155**, 204
Dalton, Colin, **148**, 16
Damerval, Thierry, **167**, 803
Damuni, Zahi, **166**, 321
Dana, Nava, **150**, 602
Daniel, Larry W., **165**, 298
Daniel, Peter F., **138**, 94
Daniels, M. J., **136**, 371
Danielsson, Bengt, **137**, 3, 181, 232, 307, 334
Danks, D. M., **142**, 116
Danner, Dean J., **166**, 108, 298
Darling, Allan J., **167**, 410
Darlington, G. J., **151**, 19
Das, Kamal K., **138**, 575
Datta, Dipak B., **164**, 270, 532
Datta, Syamal K., **162**, 385
Daughaday, Carlos C., **162**, 79
Daughaday, William H., **146**, 248
d'Auzac, J., **148**, 87
Davidson, Barrie E., **142**, 432, 440
Davidson, Eric H., **152**, 611
Davidson, Jeffrey M., **144**, 214; **163**, 656, 674
Davidson, Susan, **146**, 42
Davies, Donald H., **161**, 442
Davis, E. Jack, **166**, 476
Davis, Ronald L., **159**, 786
Davis, Ronald W., **154**, 107
Davis, Stanley S., **149**, 51
Davis, Trisha N., **139**, 248
Davison, John, **153**, 34

De, Tripti, **138**, 575
de Agostini, Ariane, **163**, 179
Dean, Gary E., **157**, 601
Dean, William L., **157**, 351
Deaven, Larry L., **151**, 252
de Azavedo, Joyce C. S., **165**, 32, 333, 344
Deblaere, R., **153**, 277
de Boer, Herman A., **153**, 432
de Bruijn, Frans J., **154**, 175
De Bruyne, Clement K., **160**, 45
De Camilli, Pietro, **159**, 183
de Crombrugghe, Benoit, **144**, 61
Deckman, Ingrid C., **164**, 203
Decristoforo, Georg, **137**, 197
Dedman, J. R., **139**, 19
De Felice, Maurilio, **166**, 241
De Feyter, Robert, **142**, 355
DeFoor, Paul, **157**, 369, 693
Degerman, Eva, **159**, 504
DeGrado, William F., **139**, 455
DeGuzman, Anthony, **159**, 356
Deininger, Prescott L., **152**, 371
Dekker, Matthijs, **135**, 79
Dekker, R. F. H., **160**, 454
DeLange, R. J., **146**, 294
De Larco, Joseph E., **146**, 95
De La Rosa, James, **143**, 155
DeLoach, John R., **149**, 235
DeLuca, M., **136**, 82
DeMaster, Eugene G., **143**, 110
de Mazancourt, Philippe, **159**, 766
de Meis, Leopoldo, **157**, 190
Deng, H.-Y., **164**, 174
Denman, Robert, **164**, 372
Dennis, W. H., **149**, 77
Dente, Luciana, **155**, 111
D'Ercole, A. Joseph, **146**, 227
Deshpande, M. V., **160**, 126
Deshpande, V., **160**, 415
Desplan, Claude, **139**, 337
Desrochers, Michel, **160**, 432
Devreotes, Peter, **159**, 267
de Wet, Wouter J., **145**, 235
Dickson, Robert B., **146**, 329
Di Domenico, Marina, **139**, 731
Dietrich, Helmut, **135**, 546
DiGiusto, David L., **141**, 53
DiLella, Anthony G., **152**, 199, 447
Dimon, Bernard, **167**, 686
Dinarello, Charles A., **163**, 495
DiNello, Robert K., **136**, 93
DiNubile, Mark J., **162**, 246
Di Rosa, Massimo, **163**, 23
Di Sabato, Giovanni, **150**, 3
Distler, Jack J., **138**, 504
Di Virgilio, Francesco, **141**, 3
Dixon, John F., **156**, 141
Dixon, Mark C., **157**, 36
Dobberstein, Bernhard, **155**, 416
Dodds, D. R., **154**, 287
Dombroski, Beth A., **155**, 537
Domingo, Derrick, **147**, 265
Donaldson, Robert P., **148**, 505
Doner, Landis W., **160**, 176
Donnelly, Patricia V., **139**, 303
Donohue-Rolfe, Arthur, **165**, 152, 231, 250, 399
Dooijewaard, Gerard, **163**, 80
Dorne, Albert-Jean, **148**, 195
Dosch, Hans-Michael, **150**, 223
Døskeland, Anne P., **142**, 35
Døskeland, Stein Ove, **142**, 35; **159**, 97, 147
Douce, Roland, **148**, 195, 206, 403
Douzou, Pierre, **135**, 528
Doyle, Kevin E., **139**, 230
Draetta, Giulio, **159**, 573
Drapeau, G., **163**, 263
Draper, David E., **164**, 203, 221
Droux, Michel, **167**, 415
Dubbels, R., **142**, 583
Duffaud, Guy D., **153**, 492
Duffel, Michael W., **143**, 149
Dufour, Jean-Pierre, **157**, 513, 528
Dumont, Jacques E., **159**, 19
Duncan, Kenneth, **142**, 320
Dunn, Irving J., **135**, 300
Dupaix, A., **148**, 132
Dupont, Yves, **148**, 675; **157**, 206
Durand, Gilbert, **135**, 307
Durbin, Mary L., **160**, 342
Dustoor, Meher M., **150**, 83
Dutchik, James E., **153**, 69
Dux, L., **157**, 271
Dye, J. L., **142**, 414
Dziadek, Marie, **145**, 363

E

Echeverria, Edgardo, **148**, 226
Edman, Peter, **137**, 491

Edström, Jan-Erik, **151**, 503
Edwards, Gerald E., **148**, 421
Edwards, William T. E., **166**, 20
Egan, Kevin M., **153**, 516
Egebjerg, Jan, **164**, 710
Ehrenberg, Måns, **164**, 611
Eisenberg, Frank, Jr., **141**, 127
Eisenberg, Shlomo, **155**, 528
Ek, Bo, **147**, 3
Ekanger, Ronald, **159**, 97
Elbein, Alan D., **138**, 661
Elhai, Jeff, **167**, 747
Elia, Marinos, **166**, 20
Elks, Martha, **159**, 504
Elledge, Stephen, **154**, 107
Elliott, Steven G., **153**, 516
Elmorjani, Khalil, **167**, 584
Elsas, Louis J., **166**, 108
Emerson, Charles P., Jr., **139**, 326
Emerson, D. L., **163**, 431
Emi, Shigenori, **160**, 719
Emori, Yasufumi, **139**, 363
Emptage, Mark, **166**, 92
Enari, T.-M., **160**, 117
Endo, M., **157**, 12
Enfors, Sven-Olof, **137**, 298
Engel, James Douglas, **152**, 359
Engel, Jürgen, **145**, 3
Enns, Caroline A., **135**, 569
Eoyang, Lillian, **166**, 435
Epstein, Paul, **139**, 217
Epstein, Wolfgang, **157**, 655
Erber, Walter, **167**, 437
Erfle, James D., **160**, 59
Erickson-Viitanen, Susan, **139**, 455
Ericson, Ingemar, **148**, 434, 442
Eriksen, Nils, **163**, 510
Eriksson, Karl-Erik, **160**, 126, 368, 415, 463, 500; **161**, 271, 274, 316, 322
Erneux, Christophe, **159**, 520
Ernst, Ludger, **161**, 47
Ernst, Stephen A., **156**, 213
Esaki, Nobuyoshi, **143**, 148, 245, 291, 415, 459, 493
Eschenfeldt, William H., **152**, 335, 337
Esmann, Mikael, **156**, 43, 72, 105, 278, 371
Espinal, Joseph, **166**, 166, 175
Esser, Alfred F., **162**, 551
Ettayebi, Mohamed, **164**, 673
Evans, Glen A., **152**, 604
Evans, Jillian F., **162**, 72; **163**, 340
Evenson, Merle A., **158**, 351
Expert-Bezançon, A., **164**, 310
Eyre, David, **144**, 115

F

Fabiato, Alexandre, **157**, 378
Fahey, Robert C., **143**, 85, 96
Fain, John N., **141**, 255
Falchuk, K. H., **158**, 422
Faloona, Fred A., **155**, 335
Fang, Tung K., **148**, 505
Fanger, Michael W., **147**, 186
Fankhauser, Heinz, **143**, 354
Fariss, Marc W., **143**, 101
Farley, Robert A., **156**, 291
Farmer, Martha C., **149**, 184
Fassold, E., **157**, 220
Fathman, C. Garrison, **150**, 304, 325, 333
Faulders, Brian, **147**, 3
Fearon, Douglas T., **150**, 579
Feder, Joseph, **135**, 393
Feldman, Ross D., **150**, 492
Feldman, Susan C., **139**, 534
Felix, Hansruedi, **137**, 637
Fellman, Jack H., **142**, 148; **143**, 172, 183, 223
Felsenfeld, Gary, **152**, 735
Fiddes, John C., **152**, 349
Fiedler, Franz, **163**, 257
Field, Katharine G., **167**, 138
Fiers, Walter, **153**, 416
Filoteo, Adelaida G. **157**, 340
Filutowics, Marcin, **159**, 325
Fine, Jo-David, **145**, 148
Fink, Gerald R., **154**, 164
Finne, Jukka, **138**, 289
Firca, Joseph R., **154**, 314
Firgaira, Frank A., **142**, 116
Fischer, Edmond H., **159**, 377
Fischer, R., **142**, 472, 495, 503, 507
Fischer, W., **160**, 159
Fiscus, Ronald R., **159**, 150
Fish, Wayne W., **158**, 357
Fisher, E. F., **154**, 287
Fisher, Larry W., **145**, 269
Fisher, Susan J., **144**, 420
Fishman, Alfred P., **138**, 260

FitzGerald, David J. P., **151**, 139
Flatmark, Torgeir, **142**, 35
Fleischer, Sidney, **157**, 26, 36, 46, 51, 91, 100, 302, 314, 321, 369, 490, 693
Fleischmann, Robert, **151**, 405
Fleminger, G., **136**, 170
Fletcher, Daniel, S., **162**, 478
Fletcher, William H., **159**, 236, 255
Fliegel, Larry, **157**, 289
Flowers, Harold M., **138**, 359
Focella, Antonino, **164**, 372
Fontana, J. D., **160**, 551, 560
Förberg, Cecilia, **137**, 626
Forbes, Brian J. R., **142**, 132
Ford-Hutchinson, Anthony W., **162**, 72; **163**, 340
Fordis, C. Michael, **151**, 382
Forgac, Michael, **156**, 127
Förster, Helga, **161**, 355
Foster, John L., **159**, 215
Foster, S. J., **143**, 195
Foulkes, J. Gordon, **159**, 427
Fraley, R. T., **153**, 253
Franche, Claudine, **167**, 803
Franchi, Arlette, **151**, 131
Francis, Sharron H., **159**, 74, 722
Frank, Joachim, **164**, 3
Frank, Ronald, **154**, 221
Franke, Arthur E., **162**, 653
Frazier, William A., **144**, 438
Freeborn, Betty, **164**, 158
Freeman, Amihay, **135**, 216
Freer, John H., **165**, 16
Freer, Richard J., **162**, 132
Freiburghaus, Christian, **137**, 458
French, Deborah, **151**, 50
French, Sarah, **142**, 366
Friedberg, Devorah, **167**, 736
Friedmann, Theodore, **149**, 10
Frigon, Ronald P., **142**, 603
Frischauf, Anna-Maria, **152**, 180, 183, 190; **153**, 103
Fritz, Hans-Joachim, **154**, 350
Frolik, Charles A., **146**, 95
Fromm, Michael, **153**, 351
Froscher, Barbara G., **150**, 196
Frost, John W., **142**, 306
Fry, Ian V., **167**, 450
Fuchs, Renata, **157**, 601
Fujii, Jun-ichi, **157**, 107
Fujikawa, Kazuo, **163**, 54
Fujisawa, Hitoshi, **142**, 44, 63, 83, 93
Fujiwara, Sakuhei, **145**, 363
Fukui, Saburo, **135**, 230; **136**, 293
Fuller, Gerald M., **163**, 474, 594
Fuller, Ray W., **142**, 655
Furlanetto, Richard W., **146**, 216
Furlong, Clement E., **143**, 144
Fürst, Peter, **157**, 680
Furthmayr, Heinz, **145**, 3
Furusawa, Motohisa, **137**, 225
Furuto, Donald K., **144**, 41
Furuto-Kato, Sumiko, **163**, 230
Fusee, Murray C., **136**, 463
Füssle, Roswitha, **165**, 285

G

Gaber, Bruce P., **149**, 184
Gaertner, Frank H., **142**, 362
Gaitanaris, George A., **151**, 397
Gal, Susannah, **151**, 104
Galbraith, Robert M., **150**, 377
Gallin, John I., **162**, 59
Galloway, Cynthia J., **157**, 601
Gamet, Didier, **167**, 691
Ganong, Barry R., **141**, 313
Gans, Pierre, **167**, 686
Ganther, Howard E., **143**, 32, 195
Gantt, E., **167**, 286
Garber, Leonard J., **143**, 354
Gardeström, Per, **148**, 421, 434, 442
Garland, Donita L., **158**, 145
Garrett, Roger, **164**, 456, 710
Gasking, Angela L., **166**, 20
Gattoni, Maurizio, **135**, 484
Gaucher, G. M., **136**, 329
Gay, Steffen, **144**, 3; **145**, 148
Gaynes, Lynne A., **146**, 259
Gbewonyo, Kodzo, **135**, 318
Gebara, M., **160**, 560
Gee, Adrian P., **162**, 101
Geha, Raif S., **150**, 309
Gehri, Annette, **161**, 479
Geigenmüller, Ute, **164**, 658
Geiger, Reinhard, **163**, 102, 257
Gelfand, Erwin W., **150**, 223
Gellf, Gérard, **135**, 454
Genovesse, Connie, **145**, 223

Gentz, Reiner, **155**, 416
Geoffroy, Christiane, **165**, 52, 64
Gerard, Craig J., **162**, 653
Geras-Raaka, Elizabeth, **141**, 36
Gerhardt, Bernt, **148**, 516
Gershengorn, Marvin C., **141**, 36, 100
Gersten, Merril J., **150**, 129
Gestrelius, S., **136**, 353
Gettys, Thomas W., **159**, 74
Gewirth, Daniel T., **164**, 158
Ghebrehiwet, Berhane, **150**, 558
Ghini, Severino, **137**, 161
Gianazza, Elisabetta, **163**, 441
Giartosio, Anna, **142**, 179
Gibbons, Ian, **136**, 93
Gibson, Jane, **167**, 557
Gibson, Kenneth M., **166**, 214, 219
Giffin, James C., **153**, 516
Gigli, Irma, **162**, 626
Gilchrist, David G., **142**, 450
Gill, D. Michael, **165**, 235, 246
Gill, Gordon N., **146**, 82
Gillard, Baiba, **138**, 300
Gillespie, David, **152**, 582
Gillespie, G. Yancey, **146**, 207
Gillespie, Gerald A., **151**, 461
Ginns, Edward I., **152**, 276
Ginsburg, Victor, **138**, 195
Giovanelli, John, **143**, 419, 443
Giovenco, Silvía, **135**, 282
Giro, M. Gabriella, **163**, 656
Girotti, Stefano, **137**, 161
Gisby, P. E., **135**, 440
Gisler, Roland H., **150**, 251
Giudicelli, Yves, **159**, 766
Glass, David B., **159**, 74
Glazer, Alexander N., **167**, 291, 304
Glenn, Jeffrey K., **161**, 74, 258
Glennie-Holmes, M., **160**, 545
Glitz, Dohn G., **164**, 493
Glynn, I. M., **156**, 281
Go, Masayoshi, **141**, 424
Godse, Damodar D., **142**, 571
Goff, Stephen P., **151**, 489; **152**, 469
Goffeau, André, **157**, 513, 528
Gohda, Eiichi, **142**, 279
Goksøyr, Jostein, **160**, 338
Gold, Larry, **164**, 419
Gold, Michael H., **161**, 74, 258
Goldberg, Nelson D., **159**, 60
Golden, Susan S., **153**, 215; **167**, 714
Goldfinch, Michael J., **137**, 338
Goldin, Stanley M., **156**, 127
Goldman, R. C., **138**, 267
Goldschneider, Irving, **150**, 363
Goldsmith, Lowell A., **143**, 510
Goldstein, Irwin J., **138**, 552, 561
Goldstein, Leon, **135**, 90
Gollop, Natan, **166**, 234
González, Elma, **148**, 526
Goodwin, Gary W., **166**, 189
Goody, R. S., **155**, 166
Gopalakrishna, Rayudu, **141**, 399
Gorden, Phillip, **150**, 701
Gordon, Larry M., **150**, 418
Gordon, Richard K., **143**, 191
Göringer, H. U., **164**, 721
Görisch, Helmut, **142**, 463
Gorman, R. R., **141**, 372
Goronzy, Jorg, **150**, 333
Gospodarowicz, D., **147**, 106
Gottesman, Max E., **151**, 397
Gottesman, Michael M., **151**, 3, 113, 329
Gottschalk, Dieter, **160**, 528
Graeff, Richard M., **159**, 60
Grant, Paul G., **159**, 772
Gray, Gary R., **138**, 26
Gray, Larry D., **165**, 176
Gray, Mark R., **154**, 129
Graziano, Frank M., **162**, 501, 522
Green, Steven H., **147**, 207
Greenbaum, Lowell M., **163**, 272
Greenberg, Richard N., **165**, 126
Greene, Alice A., **143**, 144
Greene, Jonathan R., **152**, 512
Greene, Lloyd A., **147**, 207
Greene, Mark I., **163**, 357
Greene, Warner C., **150**, 682
Greenfield, Sybil, **146**, 163
Greenstein, Lawrence A., **146**, 259
Greer, Karen L., **148**, 274
Griffith, Owen W., **143**, 166, 221, 270, 274, 279, 286, 366, 404
Griffo, Giovanna, **166**, 241
Grisham, Charles M., **156**, 353
Grosse, Elisabeth, **149**, 157
Grotendorst, Gary R., **147**, 144
Gruler, Hans, **162**, 115
Grunberger, George, **150**, 701
Guarente, Leonard, **152**, 512

Gubler, Ueli, **152**, 325, 330
Gudkov, A. T., **164**, 148
Guilbault, George G., **137**, 14
Guillain, F., **157**, 206
Günther, Helmut, **136**, 302
Guranowski, Andrzej, **143**, 430

H

Haeuptle, Marie-Therese, **155**, 416
Hager, Vreni, **142**, 386
Häggström, Lena, **137**, 626
Hagi, Norio, **136**, 503
Hahn-Hägerdal, Bärbel, **137**, 641
Haider, Konrad, **161**, 47
Hainfeld, James F., **164**, 49
Hakanson, Hakan, **137**, 319
Hakomori, Sen-itiroh, **138**, 3, 13
Hall, D. O., **135**, 440; **167**, 501, 629
Hall, David I., **143**, 439
Hall, J. L., **148**, 568
Hall, Janice M., **150**, 3
Hall, Linda M., **159**, 215
Hallenbeck, Patrick C., **138**, 169
Halsey, W. A., Jr., **163**, 328
Hamer, G., **137**, 697
Hamet, Pavel, **159**, 710
Hamilton, Gordon A., **142**, 132
Hamilton, Richard T., **147**, 427
Hammacher, Annet, **147**, 3
Hampel, Arnold, **166**, 260
Hanahan, Douglas, **152**, 349
Handley, Christopher J., **144**, 412
Handlogten, Mary E., **166**, 252
Hannun, Yusuf A., **141**, 287
Hansen, John A., **150**, 536
Hansen, R. Scott, **159**, 543
Hansson, G., **138**, 307
Hanyu, Naohiro, **155**, 379
Hara-Nishimura, Ikuko, **148**, 27
Harauz, George, **164**, 35
Harbron, S., **136**, 138
Hardesty, Boyd, **164**, 174
Hardin, Susan H., **139**, 354
Harford, Joe, **149**, 3
Harkins, Richard N., **146**, 270
Harnly, James M., **158**, 145
Harper, Alfred E., **166**, 39, 201
Harper, Mary E., **151**, 539
Harper, Robert A., **146**, 3
Harrington, Joan, **160**, 575
Harrington, Maureen A., **147**, 400
Harrington, William J., **149**, 312
Harris, Albert K., **163**, 623
Harris, Robert A., **166**, 114, 189
Harrison, S. A., **159**, 685
Harshman, Sidney, **165**, 3
Hartmann, Marie-Andrée, **148**, 632
Hartz, Dieter, **164**, 419
Harwood, John L., **148**, 475
Hascall, Vincent C., **138**, 279
Hasegawa, Hiroyuki, **142**, 88, 103, 111
Haselkorn, Robert, **153**, 215
Hashizume, Shuichi, **147**, 302
Hasselbach, Wilhelm, **157**, 220
Hasselberg, Marion, **156**, 302
Hastings, Kenneth E. M., **139**, 326
Haston, Wendy S., **162**, 3, 17
Hatter, Kenneth, **166**, 389
Hattori, Tsukaho, **148**, 491
Haufler, Ursula, **136**, 280
Havir, Evelyn A., **142**, 248
Hayaishi, Osamu, **142**, 188, 195
Hayashi, Hideo, **162**, 140
Hayashida, Kazuhiro, **163**, 485
Hayashida, Shinsaku, **160**, 323, 675
Hayden, Patrick, **143**, 228
Hayes, Ernest R., **161**, 442
Hayes, P. K., **167**, 213
Haylock, Richard, **161**, 221
Hayman, Edward G., **144**, 430
Hayward, Michael A., **143**, 279
Hearing, Vincent J., Jr., **142**, 154
Hearn, Milton T. W., **135**, 102
Hearst, John E., **164**, 330
Heasley, Lynn E., **159**, 83
Heath, Timothy D., **149**, 111, 135
Hedbys, Lars, **136**, 230
Hedin, Lars, **159**, 318
Hedjran, Farah, **161**, 525
Hedo, Jose A., **150**, 701
Heffelfinger, Sue C., **166**, 298
Heidecker, Gisela, **154**, 28
Heimer, Ralph, **138**, 260
Heinegård, Dick, **144**, 305, 319, 336
Heinonen, Tiina, **137**, 171
Heinz, E., **148**, 650
Heizmann, Claus W., **139**, 207, 552
Heldin, Carl-Henrik, **147**, 3

Helfman, David M., **152**, 349, 451
Helms, Cynthia, **153**, 69
Hemmings, Brian A., **15**, 390
Henikoff, Steven, **155**, 156
Hennighausen, Lothar, **152**, 721
Heptinstall, John, **160**, 264
Herbert, Hans, **156**, 80
Herdman, Michael, **167**, 222, 232, 584, 851
Hernalsteens, J.-P., **153**, 277
Herrmann, Bernhard G., **152**, 180
Herron, Michael J., **162**, 50
Herschman, Harvey R., **147**, 355
Hertzberg, Elliot L., **139**, 445
Hervagault, J. F., **135**, 554
Herzberg, Osnat, **139**, 610
Hess, Allan D., **139**, 677
Hess, John, **142**, 366
Heus, Hans A., **164**, 188
Heusterspreute, Michel, **153**, 34
Hicks, David B., **167**, 350
Hicks, Kevin B., **161**, 410
Hidaka, Hiroyoshi, **139**, 570; **159**, 652
Hieter, Philip, **155**, 351
Higuchi, Takayoshi, **161**, 101, 200
Hikuma, Motohiko, **137**, 124
Hildebrandt, Ellen, **166**, 484
Hilhorst, Riet, **136**, 216
Hill, David E., **155**, 558
Hill, Kenneth E., **163**, 674
Hill, M., **148**, 132
Hill, Walter E., **164**, 401
Hilt, K. L., **158**, 422
Hino, Tsunetoshi, **137**, 545
Hintz, Raymond L., **146**, 234
Hiraga, Akira, **159**, 437
Hiraki, Yuji, **146**, 313
Hirano, Shigehiro, **161**, 408
Hirashima, Mitsuomi, **162**, 140
Hirschbein, Bernard L., **136**, 263
Hirschberg, Carlos B., **138**, 709
Hirschberg, Joseph, **151**, 145
Hirschberger, Lawrence L., **143**, 155
Hitchcock, Michael J. M., **142**, 225; **152**, 276, 284
Hjemdahl, Paul, **142**, 521
Ho, Mat H., **137**, 271
Hobom, Gerd, **155**, 3
Hobson-Frohock, Anthony, **166**, 20
Hodges, Thomas K., **148**, 542
Höfte, H., **153**, 277
Hogan, B. L. M., **144**, 464
Hokin, Lowell E., **156**, 48, 141
Holbrook, J. John, **159**, 208
Hollander, Zeev, **135**, 160
Holley, Robert W., **146**, 163
Holmes, Charles F. B., **159**, 427
Holmquist, Barton, **158**, 6
Holub, Bruce J., **141**, 234
Homareda, Haruo, **156**, 229
Honda, Mitsuo, **162**, 140
Hong, Guo-Fan, **155**, 93
Hon-nami, Hiromi, **160**, 483
Hon-nami, Koyu, **160**, 483
Hood, Leroy E., **154**, 314; **155**, 260
Höök, Magnus, **144**, 394
Hootman, Seth R., **156**, 213
Hooykaas, Paul J. J., **153**, 305
Hopfer, Sidney M., **158**, 382
Hoppe, C. A., **138**, 404
Hopwood, David A., **153**, 116
Horikoshi, Koki, **160**, 655
Horsch, R. B., **153**, 253
Horton, H. Robert, **135**, 130; **143**, 504
Horvath, Suzanna J., **154**, 314
Horwitz, Joel, **141**, 169, 201
Hosokawa, Yu, **143**, 395
Hosono, Kuniaki, **159**, 497
Houk, Janette, **143**, 129
Houmard, Jean, **167**, 318, 808
Houslay, Miles D., **159**, 751
Howard, Bruce H., **151**, 382, 397, 405
Howard, S. Peter, **165**, 193
Howlett, Geoffrey J., **150**, 447
Hsieh, Thomas C.-Y., **138**, 186
Hsiung, Hansen M., **153**, 390
Huan, Ling-Jun, **159**, 346
Huang, Leaf, **149**, 88
Huang, P. C., **154**, 415
Huang, Wu-hsiung, **156**, 345
Huber, Paul W., **164**, 468
Huckle, William R., **141**, 149
Hudson, Graham S., **142**, 440
Huey, R., **150**, 615
Hughes, Stephen H., **152**, 451
Hugli, T. E., **150**, 615
Hui, Anna, **153**, 432
Huisman, William H., **161**, 524
Hulme, Michael A., **160**, 130
Hunger, Hans-Dieter, **155**, 301
Hunkapiller, Michael W., **154**, 314
Hunkapiller, Tim, **154**, 314
Hunt, C. A., **149**, 99

Huo, Y. K., **146**, 294
Huynh, Van-Ba, **161**, 83, 264
Hymel, Lin, **157**, 302
Hynes, Richard O., **144**, 447
Hyslop, P. A., **163**, 328

I

Iandolo, John J., **165**, 43
Ibrahimi, Ibrahim, **155**, 416
Ichiyama, Arata, **142**, 88
Igloi, Gabor L., **155**, 433
Ihler, Garret M., **149**, 221
Iino, M., **157**, 12
Ikariyama, Yoshihito, **137**, 111
Ikeda, Yasuyuki, **166**, 155, 360, 374
Ikemoto, Noriaki, **157**, 469
Ikuta, Shigeru, **146**, 11
Ill, Charles R., **144**, 430
Im, Wha Bin, **157**, 649
Imai, Atsushi, **141**, 100
Imai, Kazuhiro, **143**, 67
Imaizumi, T., **136**, 530
Imajoh, Shinobu, **139**, 363
Inagaki, Masaki, **159**, 652
Inesi, Guiseppe, **157**, 154
Inman, Wendelyn H., **146**, 399
Inouye, Masayori, **139**, 380; **153**, 492
Inouye, Sumiko, **139**, 380
Inui, Makoto, **157**, 100, 107, 314, 490
Inuzuka, Keiichi, **136**, 394
Irino, Shigeaki, **136**, 503
Ishibashi, Hiromi, **163**, 485
Ishida, Tamako A., **159**, 255
Ishihara, Toshiki, **150**, 304
Ishikawa, Tatsuo, **159**, 27
Ishimoto, Makoto, **143**, 227, 496
Ishinami, Chizuko, **139**, 127
Issekutz, Andrew C., **162**, 301
Issekutz, Thomas B., **162**, 301
Ito, Eiji, **161**, 510, 514
Ito, Fumiaki, **147**, 387
Iwami, Kimikazu, **143**, 434

J

Jacewicz, Mary, **165**, 152, 231, 399
Jackson, Julius H., **166**, 230
Jackson, P. David, **152**, 735
Jacobson, Allan, **152**, 254
Jacobson, B. J., **160**, 145
Jacoby, Benjamin, **148**, 105
Jaeger, E., **161**, 301
Jagadish, Mittur N., **154**, 197
Jagus, Rosemary, **152**, 267, 296, 567
Jahnsen, Tore, **159**, 318
Jaken, Susan, **141**, 275
Jakob, Hedwig, **151**, 66
Jakoby, William B., **142**, 660; **143**, 201
Jakubke, Hans-Dieter, **136**, 178
Jakubowski, Hieronim, **143**, 430
James, Michael N. G., **139**, 610
Janata, Jiří, **137**, 247
Janatova, Jarmila, **162**, 579
Jannatipour, Mehrdad, **161**, 524
Janolino, Violeta G., **136**, 423
Janowsky, Aaron, **141**, 192
Jardetzky, Theodore S., **142**, 386
Jasin, Hugo E., **162**, 379
Jastorff, Bernd, **159**, 159
Jedlicki, Eugenia, **159**, 660
Jemiolo, David K., **164**, 691
Jendrisak, Jerry, **152**, 359
Jennings, I., **142**, 116
Jennings, John C., **146**, 281
Jensen, Bent B., **167**, 467
Jensen, Jørgen, **156**, 191
Jensen, R., **142**, 472, 479, 488, 495, 503, 507
Jensen, Villy J., **136**, 356
Jett, Marti, **141**, 459
Jhurani, Parkash, **153**, 432
Jiang, Lian-wei, **141**, 447
Jirků, Vladimir, **135**, 341
Jocelyn, Peter C., **143**, 44, 246
Johannin, Georges, **167**, 691
John, Michael, **160**, 662
Johnson, Carl R., **143**, 281
Johnson, David A., **163**, 240
Johnson, Jean L., **158**, 371
Johnson, Judith P., **150**, 352
Johnson, K. G., **160**, 551, 560
Johnsson, Ann, **147**, 3
Joliff, Gwennaël, **160**, 355
Jolivet-Reynaud, Colette, **165**, 91, 293
Jolly, Douglas J., **149**, 10
Jones, Christopher S., **149**, 143
Jones, Larry R., **157**, 85, 360
Jones, Matthew O., **153**, 516
Jones, Sheelagh M. A., **166**, 10

Jönsson, Ulf, **137**, 381
Jørgensen, Bo Barker, **167**, 639
Jørgensen, Peter Leth, **156**, 29, 80, 291, 430
Joset, Françoise, **167**, 712, 728
Jourdian, George W., **138**, 504
Journet, Etienne-Pascal, **148**, 234
Joy, K. W., **148**, 179
Joyard, Jacques, **148**, 195, 206
Judd, Allan M., **141**, 63
Jung, Chan Y., **149**, 217
Jung, Rudolf, **155**, 301
Jungmann, Richard A., **159**, 225
Jurasek, L., **160**, 659
Jürgens, Uwe Johannes, **167**, 173
Jurgensen, Stewart R., **159**, 3
Jüttner, Friedrich, **167**, 336, 609
Juy, Michel, **160**, 355

K

Kader, Jean-Claude, **148**, 661
Kadin, Harold, **143**, 257
Kadoma, Masaaki, **157**, 107
Kafatos, Fotis C., **152**, 588, 598
Kaiser, E. T., **154**, 473
Kaiser, Robert J., **155**, 260
Kaiser, Rolf, **151**, 503
Kaji, Akira, **160**, 542, 707; **161**, 361
Kakar, Sham S., **156**, 345
Kallas, Toivo, **167**, 779
Kam, Laurance, **155**, 537
Kamarck, Michael E., **147**, 280; **151**, 150
Kambayashi, Junichi, **137**, 529
Kamikubo, Tadashi, **160**, 251
Kamp, Roza Maria, **164**, 542
Kanda, Takahisa, **160**, 403
Kane, Anne V., **165**, 152, 399
Kang, Andrew H., **163**, 694
Kang, Mohinder S., **138**, 637, 643
Kannagi, Reiji, **138**, 3
Kanzaki, Hiroshi, **143**, 486
Kaplan, Aaron, **167**, 534
Kaplan, Allen P., **163**, 3, 68, 85
Kaplan, Jerry, **147**, 247
Karlish, S. J. D., **156**, 179, 271
Karlsson, Karl-Anders, **138**, 212, 220
Karube, Isao, **137**, 131, 255, 668
Kasai, Michiki, **157**, 437
Kasche, Volker, **136**, 280
Kashgarian, Michael, **156**, 392
Katchalski-Katzir, Ephraim, **135**, 160
Katinger, H. W. D., **135**, 399
Kato, Iwao, **165**, 22
Kato, Kanefusa, **139**, 495
Kato, Yukio, **146**, 313
Katoh, Sakae, **167**, 263, 269
Katoh, Tetzuya, **167**, 313
Katz, Edward, **142**, 225
Kaufman, Seymour, **142**, 3, 97
Kauss, Heinrich, **161**, 518
Kawaguchi, K., **138**, 404
Kawaichi, M., **154**, 3
Kawakita, Masao, **157**, 251
Kawakita, Norio, **135**, 154
Kawamori, Ryuzo, **137**, 326
Kawamoto, Richard M., **157**, 68
Kawamoto, Tomoyuki, **146**, 63
Kawashima, Koji, **135**, 146
Kay, Cyril M., **139**, 88, 168
Kaytes, Paul S., **152**, 407
Kazlauskas, Romas J., **136**, 263
Kean, Joanne M., **164**, 221
Keim, Paul, **148**, 535
Keller, Brigitte, **142**, 513
Keller, Eberhard, **142**, 513
Keller, Reinhold, **136**, 517
Kelley, Robert L., **161**, 79, 307
Kelly, Paul T., **139**, 690
Kelly, Theresa, **151**, 50
Kelsoe, Garnett, **150**, 287
Kemp, Bruce E., **159**, 173
Kennedy, John F., **135**, 117, 357
Kennedy, Martha, **165**, 298
Kennelly, Peter J., **139**, 188
Kerc, Ewa, **159**, 446
Kern, Hartmut, **161**, 47
Kerwar, S. S., **162**, 355, 373
Kesselmeier, J., **148**, 650
Keusch, Gerald T., **165**, 152, 231, 250, 399
Khanna, Navin C., **139**, 36, 68
Khatra, Balwant S., **159**, 368
Kiba, Nobutoshi, **137**, 225
Kidd, Vince J., **159**, 318
Kido, Ryo, **166**, 275
Kidwell, William R., **147**, 407
Kieser, Tobias, **153**, 116
Kikkawa, Ushio, **141**, 424
Kikuchi, Kunimi, **159**, 437

Kilberg, Michael S., **166**, 252
Kim, Do Han, **157**, 469
Kimmel, Alan R., **152**, 307, 393, 399, 507, 572
Kimura, James H., **144**, 372
Kincaid, Randall L., **139**, 3; **159**, 457, 557, 605, 627
King, David J., **137**, 675
King, J., **149**, 64
King, Lynn, **165**, 298
King, M. Todd, **166**, 70
Kirchner, H., **150**, 112
Kirk, T. Kent, **161**, 3, 65, 87, 238
Kirley, Terence L., **156**, 87
Kirschner, Kasper, **142**, 386
Kirsh, Richard, **149**, 147
Kison, Ralf, **156**, 302
Kitajima, Satoko, **139**, 495
Kitamikado, Manabu, **160**, 583
Kitamura, Kumpei, **160**, 211
Kitamura, Naomi, **163**, 230
Kivirikko, Kari I., **144**, 96
Kjellbom, Per, **148**, 558
Kjellén, Lena, **144**, 394
Klagsbrun, Michael, **146**, 42, 320; **147**, 95
Klaus, R., **160**, 159
Klausner, Richard D., **149**, 3
Klee, Claude B., **139**, 405; **159**, 573
Klee, H. J., **153**, 253
Klein, Dagmar, **150**, 163
Klein, David C., **142**, 583, 590
Klein, J., **135**, 3, 252, 259
Klein, Jan, **150**, 163
Klein, Peter, **159**, 267
Klein, William, **139**, 354
Kleinig, Hans, **148**, 486
Klevit, Rachel E., **139**, 197
Klibanov, Alexander M., **136**, 117
Klinman, Norman R., **150**, 196
Kluepfel, Dieter, **160**, 180
Kluft, Cornelis, **163**, 80, 170
Klug, Gary A., **139**, 763
Knauer, Daniel J., **146**, 270
Knowles, Jeremy R., **142**, 306
Kobata, Akira, **138**, 84, 779
Koerner, Theodore A. W., **138**, 38
Kohlhaw, Gunter B., **166**, 414, 423, 429
Koide, Yoshinobu, **139**, 127
Kolhinen, Outi, **137**, 171
Kolhouse, J. Fred, **166**, 407
Kollman, Peter, **154**, 430
Kondo, Hiroyuki, **143**, 227, 496
Konings, Ruud N. H., **153**, 12
Konishi, Yasuo, **143**, 115
Könnecke, Andreas, **136**, 178
Konno, Haruyoshi, **161**, 373, 381
Konno, Y., **136**, 530
Kono, Tetsuro, **159**, 745
Kopp, Bettina, **136**, 317
Koppel, Rela, **135**, 160
Kopylov, Alexey M., **164**, 440
Kornfeld, Rosalind, **138**, 536
Korpela, T. K., **166**, 269
Korsching, Sigrun, **147**, 167
Koshcheenko, K. A., **135**, 198
Koshland, Douglas, **155**, 351
Koski, Raymond A., **153**, 516
Kosower, Edward M., **143**, 76, 264
Kosower, Nechama S., **143**, 76, 264
Kössel, Hans, **155**, 433
Kothary, Mahendra H., **165**, 176
Koumoto, Junko, **141**, 424
Koury, Mark J., **147**, 340
Kovachy, Robin J., **166**, 393
Kozak, Kenneth J., **165**, 147
Kramer, Wilfried, **154**, 350
Kranenburg, Hans, **135**, 79
Kraus, Jan P., **143**, 388
Kraus, R. J., **143**, 32, 195
Krauss, Friedrich, **136**, 302
Krebs, Edwin G., **139**, 115, 188
Kreger, Arnold S., **165**, 176
Kressdorf, B., **135**, 252
Kricka, L. J., **136**, 82
Krieg, Joachim, **164**, 575
Krieg, P. A., **155**, 397
Kröger, Manfred, **155**, 3
Krug, Marc S., **152**, 262, 316
Kuchino, Yoshiyuki, **155**, 379
Kuechler, Ernst, **164**, 361
Kuettel, Michael R., **159**, 225
Kuhlemeier, C. J., **153**, 199
Kuhstoss, S., **153**, 166
Kula, Maria-Regina, **136**, 9
Kumazawa, Shuzo, **167**, 484
Kung, Viola T., **149**, 200
Kunkel, Thomas A., **154**, 367
Kuntz, Martha J., **166**, 114
Kuriyama, K., **143**, 164
Kurkdjian, A., **148**, 132

Kurkijärvi, Kalevi, **137**, 171
Kurland, C. G., **164**, 611
Kuroda, Kazuhiko, **147**, 302
Kurzmack, Mark, **157**, 154
Kusakabe, Isao, **160**, 518, 611, 648
Kushner, Irving, **163**, 373
Kwast-Welfeld, Joanna, **159**, 225
Kwiatkowski, Bartłomej, **138**, 786
Kyle, John W., **138**, 575
Kyriatsoulis, Apostolos, **164**, 287

L

Laane, Colja, **136**, 216
Labarca, Rodrigo, **141**, 192
Labeit, S., **155**, 166
Lacapere, J. J., **157**, 206
Lachke, Anil H., **160**, 679
Ladisch, Christine M., **160**, 11
Ladisch, M. R., **160**, 26, 145
Ladisch, Stephan, **138**, 300
Lago, Carmine T., **166**, 241
Laidler, Keith J., **136**, 56
Laine, Roger A., **138**, 186; **144**, 420
Laing, Lance G., **164**, 238
Lam, Y.-H. Peter, **139**, 518
Lambert, John M., **142**, 325
Lamed, Raphael, **160**, 472
Lamptey, J., **135**, 333
Landiss, Leslie R., **159**, 74
Lane, David J., **167**, 138
Lane, John, **143**, 235
Lane, Lois K., **156**, 87
Langer, Robert, **137**, 515
Lang-Hinrichs, Christine, **153**, 366
Langridge, W. H. R., **153**, 336
Lanz, D.-J., **149**, 301
Lanzer, Michael, **155**, 416
Lapetina, Eduardo G., **141**, 176
Lapointe, David S., **166**, 484
Laposata, Michael, **141**, 350
LaRossa, Robert A., **166**, 97
Larreta-Garde, Véronique, **135**, 454
Larsen, Gerald, **143**, 228
Larsson, Christer, **148**, 558
Larsson, Per-Olof, **136**, 230
Lasater, Linda S., **164**, 493
Lau, Yun-Fai, **151**, 426
Lauffenburger, Douglas A., **162**, 85
Lawman, Michael J. P., **162**, 101
Lawrance, Simon K., **151**, 461
Lazdunski, Michel, **156**, 323
Lazo, R., **149**, 64
Le, Anh D., **146**, 63
Leberer, Ekkehard, **139**, 763
Lebo, Roger V., **151**, 292
Lee, Ernest Y. C., **159**, 356
Lee, F., **154**, 3
Lee, James J., **152**, 633
Lee, Lilly, **146**, 259
Lee, Reiko T., **138**, 424
Lee, Sung-Hee Cho, **166**, 476
Lee, Ten-ching, **141**, 379
Lee, William T., **150**, 503
Lee, Y. C., **138**, 404, 424
Lee, Young C., **139**, 834
Leemans, J., **153**, 277
Leffler, Hakon, **138**, 510
Lefkovits, Ivan, **150**, 209, 240
Legrand, Alain, **149**, 157
Leguay, Jean-Jacques, **148**, 3
Lehrach, H., **153**, 103; **155**, 166
Leigh, Roger A., **157**, 579
Leinweber, Franz-Josef, **143**, 15, 160
Leive, L., **138**, 267
Lem, Nora W., **148**, 346
le Maire, Marc, **157**, 261
Leonard, Robert T., **148**, 82, 542
Leonard, Warren J., **150**, 682
Leontis, Neocles B., **164**, 158
Leppla, Stephen H., **165**, 103
Lerch, Konrad, **142**, 165
Lerman, Leonard S., **155**, 482, 501
Lesley, Jayne F., **147**, 265
Lesniak, Maxine A., **150**, 701
Levenson, Corey, **154**, 403
Levenson, Robert, **156**, 379
Lever, Julia E., **146**, 376
Levery, Steven B., **138**, 13
Lewendon, Ann, **142**, 342
Lewis, David, **157**, 154
Lewis, Lowell N., **160**, 342
Lewis, S. A., **158**, 391
Lewitter, F. I., **155**, 582
Li, B. J., **153**, 336
Li, Choh Hao, **146**, 199
Li, G.-X., **137**, 686
Li, Heng-Chun, **159**, 346
Li, Peter P., **142**, 571

Lichtenthaler, Hartmut K., **148**, 350
Lidén, Annika C., **148**, 442
Liebman, Arnold, **164**, 372
Liedvogel, Bodo, **148**, 241
Lietze, A., **146**, 294
Lietzke, Rolf, **164**, 278
Lilja, H., **135**, 512
Lill, Roland, **164**, 597
Lillie, John H., **145**, 171
Lilly, M. D., **136**, 138
Lim, Ramon, **147**, 225
Limauro, Danila, **166**, 241
Lin, A. H., **141**, 372
Lin, Chin-Tarng, **139**, 478
Lin, J. K., **160**, 145
Lin, P., **138**, 404
Lin, Pin-Fang, **151**, 371
Lindner, William A., **160**, 376
Lindsey, Keith, **135**, 410
Lindstedt, Sven, **142**, 139, 143
Lingens, Franz, **142**, 513
Linko, P., **135**, 268; **137**, 686
Linko, Susan, **137**, 686
Linko, Yu-Yen, **135**, 268; **137**, 686
Linsefors, L., **135**, 434
Lipman, Jonathan J., **165**, 254, 270
Lippman, Marc E., **146**, 329
Lipson, Samuel E., **164**, 330
Lis, Halina, **138**, 544
Litosch, Irene, **141**, 255
Litwer, Stuart, **166**, 108
Liu, Mou Chung, **137**, 529
Livesey, Geoffrey, **166**, 3, 20, 282
Livesey, S. A., **145**, 324; **159**, 105
Livingstone, Alexandra, **150**, 325
Ljones, Torbjørn, **142**, 596
Ljungdahl, Lars G., **160**, 483
Ljungh, Åsa, **165**, 200
Lo, Amy C., **160**, 432
Logan, Daniel J., **143**, 149
Login, Ivan S., **141**, 63
Lohmann, Suzanne M., **159**, 183
Lohmeier-Vogel, E. M., **135**, 512
Londesborough, John, **159**, 777
Loomis, Carson R., **141**, 294
Looney, Frank D., **150**, 418
Looney, R. John, **150**, 524
Loontiens, Frank G., **160**, 45
Lopez, A., **136**, 239
Lord, J. Michael, **148**, 576
Loridan, Catherine, **165**, 59
Loskutoff, David J., **163**, 293
Loten, E. G., **159**, 736
Lou, Marjorie F., **143**, 124
Lovejoy, Steve, **147**, 314
Lövgren, Timo, **137**, 171
Lowe, Anson, **157**, 611
Lowe, Christopher R., **137**, 338
Lowe, Peter N., **166**, 330
Lubon, Henryk, **152**, 721
Lucken, Roger N., **165**, 344
Ludwig, Wolfgang, **167**, 132
Luedemann, Mary, **151**, 252
Luisi, P. L., **136**, 188
Lukas, Thomas J., **139**, 393; **157**, 328; **159**, 471
Lum, Jean B., **163**, 452
Lumsden, James, **142**, 325
Lunardi, Joël, **157**, 369, 693
Lund, Patricia, **166**, 3, 282
Lundblad, A., **138**, 307
Lundin, Arne, **137**, 171
Lundström, Ingemar, **137**, 232, 366, 388
Lupski, James R., **153**, 54
Lurquin, Paul F., **148**, 39
Lütke-Brinkhaus, Friedhelm, **148**, 486
Lynch, Dennis C., **152**, 248
Lyons, Russette M., **146**, 270

M

MacCallum, Donald K., **145**, 171
Macdonald, Fraser D., **148**, 218
MacDonald, Raymond J., **152**, 219
MacDougall, Elizabeth A., **139**, 642
Macintyre, Stephen S., **163**, 383
MacKellar, Warren C., **153**, 390
MacKenzie, C. R., **160**, 551, 560, 700
Mackenzie, Ian, **148**, 16
Maclachlan, Gordon, **160**, 382
MacLennan, David H., **157**, 289
MacLeod, Robert M., **141**, 63
MacManus, John P., **139**, 156
Maderazo, Eufronio G., **162**, 223
Maeda, Hidekatsu, **136**, 21
Maeshima, Masayoshi, **148**, 491
Magnani, John L., **138**, 195, 208, 484
Magolda, Ron, **166**, 92
Maisonpierre, Peter C., **139**, 326

Majerus, Philip W., **141**, 350
Makinose, Madoka, **157**, 228
Maksimenko, A. V., **137**, 552
Malarek, David, **164**, 372
Maley, Frank, **138**, 763
Maliarik, M., **138**, 561
Malkin, Richard, **167**, 341, 779
Malmqvist, Magnus, **137**, 381
Maly, Peter, **164**, 287
Mamer, Orval A., **166**, 27, 47
Mandenius, Carl Fredrik, **137**, 307, 388
Mandiyan, Valsan, **164**, 49
Manganiello, Vincent C., **159**, 457, 504
Mangum, John H., **143**, 384
Mani, Rajam S., **139**, 168
Maniatis, Tom, **155**, 501
Mankin, Alexander S., **164**, 440
Mann, Dennis, **139**, 417
Mannella, Carmen A., **148**, 453
Mansouri, Sohrab, **137**, 349
Månsson, Mats-Olle, **136**, 3, 103, 117
Marasco, Wayne A., **162**, 198
March, Paul E., **153**, 492
Marchalonis, John J., **150**, 377, 399, 463, 628
Marconi, Walter, **135**, 282
Marcus, Lionel, **161**, 366
Marcus, Menashe, **151**, 145
Marcus, Yehouda, **167**, 534
Marcus-Samuels, Bernice, **150**, 701
Marcus-Sekura, Carol J., **152**, 276, 284
Maret, Wolfgang, **158**, 79
Marin, B., **148**, 87
Marino, Jean M., **146**, 216
Mark, David F., **154**, 403
Markelonis, George J., **147**, 291
Marmenout, Anne, **153**, 416
Marselle, Lisa M., **151**, 539
Martin, Donna M., **158**, 402
Martin, Francis J., **149**, 200
Martin, Jean-Baptiste, **167**, 667
Martin, Mark T., **158**, 15, 344, 365
Martin, Paul J., **150**, 536
Martin, T. J., **145**, 324; **159**, 105
Martin, Thomas F. J., **141**, 111
Martinek, Karel, **135**, 586; **137**, 615
Martinez-Hernandez, Antonio, **145**, 78, 103, 133
Martin-Pérez, Jorge, **146**, 369
Martins, T. J., **159**, 685
Martonosi, A., **157**, 271
Marui, Masaki, **160**, 648
Massagué, Joan, **146**, 103, 143, 174
Mathew, J. K., **139**, 19
Matis, Louis A., **150**, 342
Matsubara, Hiroshi, **167**, 387
Matsui, Hideo, **156**, 229
Matsumoto, Kunihiro, **159**, 27
Matsuno, Ryuichi, **160**, 251
Matsuo, Masaru, **160**, 671, 684, 696
Matteucci, M., **154**, 287
Matthes, Hans W. Djurhuus, **154**, 250
Matthews, Brian W., **154**, 511
Matthews, C. Robert, **154**, 498
Matthews, Kathleen Shive, **150**, 134
Mattiasson, Bo, **137**, 647, 657
Mauch, Felix, **161**, 430, 479
Maune, John F., **139**, 230
Maunsbach, Arvid B., **156**, 80, 430
Maurer, Andreas, **157**, 321
Maurer, Richard A., **159**, 311
Mavrides, Charalampos, **142**, 253
Mayahara, Hiroshi, **156**, 417
Mayer, Frank, **160**, 483
Mayer, Robert M., **138**, 649
Mayhew, E., **149**, 64
Mayr, Georg W., **139**, 745
Mazaev, A. V., **137**, 552
Mazid, M. Abdul, **136**, 56
Mazliak, Paul, **148**, 661
Mazzara, Gail P., **154**, 129
McBride, L. J., **154**, 287
McClane, Bruce A., **165**, 94
McCleary, Barry V., **160**, 74, 511, 515, 523, 545, 572, 575, 589, 596, 614, 627
McClelland, Alan, **147**, 280
McClelland, Michael, **155**, 22, 32
McCluer, R. H., **138**, 117
McCormick, Mary, **151**, 397, 405, 445
McCoy, Ronald D., **138**, 169, 627
McCracken, Susan, **135**, 492
McCrae, Sheila, **160**, 187
McCubbin, William D., **139**, 88
McDonald, John R., **139**, 88
McDonel, James L., **165**, 94
McDonough, Carol S., **157**, 340
McEvoy, Leslie, **149**, 281, 293
McGill, John R., **151**, 279
McGowan, Clare H., **159**, 416
McHale, Anthony, **160**, 437

McKeehan, K. A., **147**, 152, 392
McKeehan, W. L., **147**, 152, 392
McKelvey, Steven W., **145**, 171
McKerrow, James H., **144**, 288
McKinley, Dana N., **146**, 402
McKnight, G. Stanley, **159**, 299
McNall, Steven, J., **159**, 377
McPheeters, David S., **164**, 419
McReynolds, Richard A., **162**, 339
McVey, J. H., **144**, 464
Means, Anthony R., **139**, 217, 303, 433, 642, 808
Mecham, Robert P., **144**, 232, 246
Mechler, Bernard M., **152**, 241
Medlock, Eugene S., **150**, 363
Meek, Rick L., **163**, 510
Meeks, J. C., **167**, 113
Mehdi, Shujaath, **142**, 306
Mehlhorn, Rolf J., **167**, 509, 670
Meier, Jurg, **135**, 318
Meighen, Edward, **135**, 492
Meinkoth, Judy L., **152**, 91, 572
Meissner, Gerhard, **157**, 417, 480
Meister, Alton, **143**, 313
Mekalanos, John J., **165**, 169
Melchers, Fritz, **150**, 217
Mellman, Ira, **157**, 601
Melton, Douglas A., **152**, 288; **155**, 397
Mencke, Arlene J., **138**, 409
Mendiola-Morgenthaler, Leticia R., **148**, 157
Mendoza, Nancy, **158**, 289
Mendoza, Stanley A., **146**, 384
Merat, Dennis L., **139**, 79
Mercer, Robert W., **156**, 379
Mercola, Mark, **147**, 64
Merkle, Roberta K., **138**, 232
Messing, Joachim, **153**, 3; **154**, 28
Meyerhans, Andreas, **154**, 221
Meyne, Julie, **151**, 252
Mibu, Yasuhiro, **162**, 140
Michel, Robert G., **158**, 222
Middlebrook, John L., **165**, 76
Midura, Ronald J., **138**, 279
Mierendorf, Robert C., **152**, 458, 556, 563
Migliaccio, Antimo, **139**, 731
Mikeš, Otakar, **161**, 385
Miles, Edith Wilson, **142**, 398
Miller, Edward J., **144**, 3, 41
Miller, Harvey, **152**, 145
Miller, Jerry R., **159**, 489, 594
Miller, Joyce F., **147**, 225
Miller, Marian, L., **137**, 68
Miller, Robert H., **166**, 39
Miller, Ronald L., **138**, 527
Millet, Jacqueline, **160**, 355
Mills, W. R., **148**, 179
Milman, Gregory, **153**, 482
Minocherhomjee, Arda-e-viraf M., **139**, 776
Minor, Ronald R., **145**, 171
Miot, Françoise, **159**, 520
Miribel, Laurent, **163**, 418, 431, 441
Miron, Talia, **135**, 84
Mishkind, Michael L., **148**, 274
Miska, Werner, **163**, 102
Mislovičová, Dana, **160**, 536
Misra, Tapan K., **155**, 119
Mitchell, R. D., **157**, 51
Mitsui, Akira, **167**, 105, 484
Mitsutomi, Masaru, **161**, 453
Miura, George A., **143**, 191
Miyada, C. Garrett, **152**, 432; **154**, 94
Mo, Kaiguo, **160**, 323, 675
Moazed, Danesh, **164**, 481
Mobedshahi, M., **160**, 26
Moffat, K., **139**, 585
Mohan, Subburaman, **146**, 281
Molitor, Véronique, **167**, 437, 691
Møller, Ian M., **148**, 442
Møller, Jesper V., **157**, 261
Moloney, Aidan P., **160**, 363
Monaco, Marie E., **141**, 83
Monsan, Pierre, **135**, 307; **136**, 239; **137**, 584
Montecucco, Cesare, **165**, 347
Montgomery, Jane A., **166**, 27, 47
Montibeller, Judith, **139**, 105, 137
Montie, Thomas C., **165**, 162
Monties, Bernard, **161**, 31
Monty, Kenneth J., **143**, 15, 160
Mooibroek, Marilyn, **139**, 518
Mooney, Robert A., **159**, 193
Moore, Anthony L., **148**, 415
Moore, Peter B., **164**, 158
Moore, Peter F., **162**, 320
Moore, Thomas S., Jr., **148**, 585
Moorehead, Thomas J., **142**, 132
Moo-Young, M., **135**, 333
More, John E., **167**, 551

Moreau, Hervé, **165**, 91, 293
Morgan, Edward A., **164**, 673
Mori, Takesada, **137**, 529
Mori, Yutaka, **160**, 483
Morier-Teissier, Elisabeth, **142**, 535
Morihara, Kazuyuki, **136**, 162
Moriizumi, Toyosaka, **137**, 255
Moriyama, Yoshinori, **157**, 619
Morrison, George H., **158**, 157
Mortimore, Glenn E., **166**, 461
Mosbach, Klaus, **135**, 65, 173, 222, 399, 475; **136**, 3, 103, 117, 230, 353; **137**, 3, 181, 232, 334, 388, 443, 583, 632
Moult, John, **139**, 610
Mount, Mark S., **161**, 329
Mousdale, David M., **142**, 348
Moutin, Marie-José, **148**, 675
Mozhaev, V. V., **135**, 586
Mudd, J. Brian, **148**, 338
Mueller, Michael, **155**, 416
Muller-Eberhard, Ursula, **163**, 536
Müller-Esterl, Werner, **163**, 240
Mullis, Kary B., **155**, 335
Mumby, Marc C., **159**, 139
Muneyuki, Ryonosuke, **136**, 162
Murachi, Takashi, **137**, 260, 545
Murad, Ferid, **159**, 150
Murakami, Hiroki, **147**, 302
Murakami, Satoru, **148**, 246
Murao, Sawao, **160**, 274
Murata, Norio, **167**, 245, 251
Murayama, T., **136**, 530
Murphy, Pat, **151**, 371
Murray, Gary J., **149**, 25
Murray, N., **153**, 103
Muthukumaraswamy, Natesa, **162**, 132
Myers, C. D., **150**, 316
Myers, Jeanne C., **145**, 213
Myers, Richard M., **155**, 501
Myllylä, Raili, **144**, 96
Mylvaganam, Ravindra, **149**, 312
Myones, Barry L., **150**, 586

N

Naccache, Paul H., **162**, 283
Naes, Helga, **167**, 605
Nagai, Kiyoshi, **153**, 461
Nagasawa, Toru, **143**, 474, 486
Nagashima, Minoru, **136**, 394
Nagata, Toshiyuki, **148**, 34; **149**, 176
Nagatsu, Toshiharu, **142**, 56
Nakamura, Katsumi, **160**, 211
Nakanishi, Kotoyoshi, **160**, 648
Nakanishi, Nobuo, **142**, 103, 111
Nakanishi, Shigetada, **163**, 230
Nakata, Hiroyasu, **142**, 44, 83, 93
Nakatsubo, Fumiaki, **161**, 57
Namboodiri, M. A. A., **142**, 583
Narang, Saran A., **153**, 385
Narasimhan, Krishna, **137**, 103
Narendranathan, T. J., **136**, 138
Natsoulis, Georges, **154**, 164
Navarro, Jean Marie, **135**, 307
Nawa, Hiroyuki, **163**, 230
Nawa, Yukifumi, **162**, 174
Naylor, Susan L., **151**, 279
Neame, Peter J., **144**, 401
Neet, Kenneth E., **147**, 186
Nelson, Michael, **155**, 32, 41
Nelson, Nathan, **157**, 619
Nelson, Robert D., **162**, 50
Nemerow, Glen R., **150**, 548
Nesbitt, James E., **163**, 474
Neuburger, Michel, **148**, 403
Neuer, Gabriele, **167**, 496
Neumann, Stefan. **136**, 302
Nevins, Joseph R., **152**, 234
Newman, Roland A., **150**, 723
Newton, D. L., **139**, 405
Newton, Gerald L., **143**, 85, 96
Ng, K. W., **145**, 324
Ng, Thomas K., **160**, 351
Nicolas, Jean-François, **151**, 66
Nicolau, Claude, **149**, 157
Niedel, James E., **141**, 294
Nierhaus, Knud H., **164**, 131, 278, 650, 658
Nierras, Concepcion R., **164**, 532
Niggli, V., **139**, 791
Nikolopoulus, G., **149**, 64
Niku-Paavola, M.-L., **160**, 117
Nilsen-Hamilton, Marit, **147**, 427
Nilsson, Inga Marie, **137**, 458
Nilsson, K., **135**, 222, 399
Nilsson, Kjell, **135**, 387
Nilsson, Kurt, **135**, 65
Nimmo, Gillian A., **142**, 325; **159**, 427
Nishi, Kayoko, **164**, 706

Nishida, Atsumi, **161**, 322
Nishikawa, Katsuzo, **146**, 11
Nishikawa, Masakatsu, **159**, 652
Nishimura, Mikio, **148**, 27
Nishimura, Susumu, **155**, 373, 379
Nishizuka, Yasutomi, **141**, 424
Nisizawa, Kazutosi, **160**, 403
Nissley, S. Peter, **146**, 259
Nitschmann, W. H., **167**, 361
Nobel, Bernice, **162**, 484
Noda, Masatoshi, **165**, 22
Noguchi, Sadao, **136**, 394
Noguchi, Tomoo, **142**, 267, 273
Nojima, Shogo, **136**, 380
Noller, Harry F., **164**, 475, 481
Nørby, Jens G., **156**, 116, 191
Noti, John D., **154**, 197
Nowotny, Petra, **164**, 131
Nowotny, Volker, **164**, 131
Noyes, Linda K., **137**, 68
Nozaki, Mitsuhiro, **142**, 650
Nunogaki, Kazuki, **157**, 437
Nurse, Kelvin, **164**, 372
Nylén, Ulf, **137**, 491

O

Obst, John R., **161**, 3, 87
O'Connor, C. David, **155**, 11
Odelhög, Birgit, **142**, 139, 143
Odom, O. W., **164**, 174
Oeltmann, Thomas N., **165**, 204, 250
Ofengand, James, **164**, 372
O'Flaherty, **163**, 44
Ogata, Kenji, **150**, 147
Ogawa, Kazuo, **156**, 417
Ogden, Richard C., **152**, 61
Ogihara, Yasunari, **164**, 748
Øgreid, Dagfinn, **159**, 147
Oh, Tae H., **147**, 291
Ohlson, Sten, **137**, 458
Ohmiya, Kunio, **160**, 391, 408
Ohno, Shigeo, **139**, 363
Ohshiro, Takeshi, **137**, 529
Ohsumi, Yoshinori, **157**, 544
Ohta, Kazuyoshi, **160**, 323, 675
Ohtakara, Akira, **161**, 426, 453, 462, 501, 505
Ojeda, Sergio R., **141**, 323
Oka, Kazuhiro, **142**, 56
Oka, Tatsushi, **136**, 162
Okada, Gentaro, **160**, 259
Okada, Hirosuke, **136**, 34; **160**, 632
Okamoto, Hiroshi, **163**, 272
Okayama, Hiroto, **151**, 434; **154**, 3
Okita, Richard T., **163**, 349
Okubo, Hideo, **163**, 485
Okuno, Sachiko, **142**, 63
Oliphant, Arnold R., **155**, 558, 568
Olivares, José A., **158**, 205
Oliver, K. G., **150**, 316
Oliver, R. L., **167**, 521
Olivier, Andrée R., **164**, 575
Olofsson, Göran, **137**, 381
Olsen, Anne S., **144**, 74
Olsen, Gary J., **164**, 793; **167**, 138
Olson, Helen McKuskie, **164**, 493
Olson, Maynard V., **153**, 69; **155**, 468
Olson, Merle S. **166**, 484
Omata, Tatsuo, **167**, 245
Ono, Tomio, **139**, 127
Oostergetel, Gerrit T., **164**, 49
Orlow, Seth J., **139**, 148
Oronsky, Arnold L., **162**, 355, 373
Osawa, Toshiaki, **150**, 17
Osteryoung, Janet, **158**, 243
Otterness, Ivan G., **162**, 320, 334
Owhashi, Makoto, **162**, 174
Oxender, Dale L., **166**, 245
Oyama, Kiyotaka, **136**, 503
Ozaki, Hachiro, **143**, 470
Ozawa, Takayuki, **157**, 321

P

Pace, Norman R., **167**, 138
Packer, Lester, **167**, 667, 670, 677
Padan, Etana, **167**, 380, 561
Paice, M. G., **160**, 659
Paige, Christopher J., **150**, 251, 257
Palade, Philip, **157**, 51
Pangburn, Michael K., **162**, 639
Pansolli, Paolo, **135**, 282
Pantazis, Panayotis, **147**, 22
Paoletti, Enzo, **153**, 545
Papageorgiou, George C., **167**, 259
Parker, Charles W., **150**, 29, 512
Parker, T. S. **137**, 466

Parkkinen, Jaakko, **138**, 289
Parthasarathy, Ranganathan, **141**, 127
Partridge, N. C., **145**, 324
Pasta, Piero, **135**, 475
Paszczyński, Andrzej, **161**, 83, 264
Patel, Tarun B., **166**, 484
Paterson, Alistair, **161**, 191
Patil, Rajkumar V., **160**, 307, 424, 448
Pato, Mary D., **159**, 446
Patston, Philip A., **166**, 175
Patterson, David, **151**, 121
Patton, Curtis L., **139**, 262
Paul, F., **136**, 239
Paul, Steven M., **141**, 192
Paulsen, Harald, **164**, 581
Paulson, James C., **138**, 162, 520
Paulsson, Mats, **145**, 336, 363
Pavlova, I. N., **160**, 620
Paxton, Ralph, **166**, 189, 313
Peeters, Ben P. H., **152**, 12
Pegg, G. F., **161**, 474, 484
Peltier, Gilles, **167**, 686
Pember, Stephen O., **142**, 50
Peng, Ze-guo, **155**, 214
Penniston, John T., **157**, 340
Percy, Chris, **152**, 458
Pereira, A. N., **160**, 26, 145
Perera, V. Y., **165**, 120
Perez, Mary, **138**, 709
Perham, Richard N., **166**, 330
Periyasamy, Sankaridrug M., **156**, 345
Perkus, Marion E., **153**, 545
Perlman, Robert L., **141**, 169, 201
Perret, Christine, **139**, 337
Peschek, Günter A., **167**, 361, 437, 450, 691
Peterson, Gary L., **156**, 48
Petryniak, Jerzy, **138**, 552
Pette, Dirk, **139**, 763
Pettersson, B., **160**, 368, 500
Pettersson, L. G., **160**, 126
Pettit, Flora H., **150**, 134; **166**, 309
Pfeffer, Diana, **152**, 556, 563
Pfitzner, Ursula, **136**, 342
Piazza, Gary A., **139**, 846
Piccini, Antonia, **153**, 545
Pickart, Loren, **147**, 314
Pidgeon, Charles, **149**, 99
Pierce, James, **146**, 3
Pierce, Mark W., **150**, 602
Pierschbacher, Michael D., **144**, 430, 475
Pihlajaniemi, Taina, **145**, 213
Pike, Beverley L., **150**, 265
Pike, Linda J., **146**, 353
Pike, Marilyn C. **162**, 236
Pine, Richard, **154**, 415
Pitot, Henry C., **142**, 279
Pitt, D., **161**, 350
Plaut, Andrew G., **165**, 117
Pledger, W. Jackson, **147**, 85, 400
Pless, Reynaldo C., **152**, 522
Plommet, Michel, **165**, 8
Plummer, Thomas H., Jr., **138**, 770
Pohl, Fritz M., **155**, 250
Polarek, James W., **157**, 655
Poljak, Roberto J., **160**, 355
Pometto, Anthony L., III, **161**, 35, 175, 183, 249
Pongor, Sándor, **154**, 450
Poole, Ronald J., **148**, 115
Porta, Raffaele, **142**, 668
Porter, Ronald D., **167**, 703
Portsmann, Thomas, **150**, 117
Pösö, A. Reeta, **166**, 461
Poste, George, **149**, 147
Postlethwaite, Arnold E., **163**, 694
Poston, J. Michael, **166**, 127, 130
Potrykus, Ingo, **153**, 313
Poulsen, Lawrence L., **143**, 124
Pouysségur, Jacques, **151**, 131
Powell, Marianne Broome, **163**, 357
Poznansky, Mark J., **137**, 566
Pozzan, Tullio, **141**, 3
Preiss, Jennifer E., **141**, 294
Prestegard, James H., **138**, 38
Price, Carl A., **148**, 157
Prince, Charles W., **145**, 261
Prockop, Darwin J., **144**, 74
Prombona, Anastasia, **164**, 748
Proudlove, Michael O., **148**, 415
Przybyla, Alan E., **152**, 219
Puls, Jürgen, **160**, 528
Purvis, Kenneth, **159**, 675
Puskas, Robert S., **152**, 337
Putkey, John A., **139**, 303
Pyne, Nigel J., **159**, 751
Pytela, Robert, **144**, 475

Q

Qiang, Bo-Qin, **155**, 15

R

Rabon, Edd C., **157**, 649
Racker, Efraim, **157**, 634
Radermacher, Michael, **164**, 3
Raeder, Ute, **161**, 211
Raia, C. A., **135**, 577
Raines, Elaine W., **147**, 48
Rains, Theodore C., **158**, 302
Raleigh, Elisabeth A., **152**, 130
Rall, Leslie B., **146**, 239
Ramakrishnan, V., **164**, 117
Ramaswamy, Sengoda G., **143**, 201
Rammensee, Hans-Georg, **147**, 265
Randall, Stephen K., **148**, 123
Randle, Philip J., **166**, 166, 175
Rangel-Aldao, Rafael, **159**, 286
Ransom, John T., **141**, 53
Rao, K. K., **135**, 440; **167**, 501
Rao, R. Nagaraja, **153**, 166
Rashidbaigi, Abbas, **150**, 492
Rasmussen, Colin D., **139**, 642
Ratan, Rajiv, **139**, 824
Ratcliffe, R. G., **148**, 683
Raum, William J., **142**, 550
Raunio, Raimo, **137**, 171
Ravindranath, M. H., **138**, 520
Ray, Anuradha, **152**, 343
Rea, Philip A., **148**, 115
Reardon, Ellen M., **148**, 157
Rechler, Matthew M., **146**, 259
Rechnitz, Garry A., **137**, 138
Rechsteiner, Martin, **149**, 42
Reddy, C. Adinarayana, **161**, 228, 307
Reddy, Pranhitha, **154**, 129
Redfern, Beth, **143**, 110
Redl, Bernhard, **164**, 64
Redmon, J. Bruce, **159**, 74
Reed, Donald J., **143**, 101
Reed, Jon A., **149**, 281
Reed, Lester J., **166**, 309, 321
Reed, Robert H., **167**, 528
Reeves, John P., **157**, 505
Regoli, D., **163**, 210, 263, 282
Reidler, J., **165**, 389
Reinhold, Leonora, **167**, 534
Reinhold, Vernon N., **138**, 59
Reiss, Paul D., **166**, 70
Remaut, Erik, **153**, 416
Remy, R., **148**, 623
Renneberg, Reinhard, **137**, 29
Renosto, Franco, **143**, 334
Resh, Marilyn D., **156**, 119
Resink, Therese J., **159**, 390
Rest, Richard F., **163**, 309
Revsbech, Niels Peter, **167**, 639
Rexová-Benková, Lubomíra, **161**, 385
Rey, Françoise, **142**, 27
Reyes, A. A., **154**, 87
Reynaerts, A., **153**, 277
Rheinberger, Hans-Jörg, **164**, 658
Rhoten, William B., **139**, 534
Ribbons, Douglas W., **161**, 294
Ricca, Ezio, **166**, 241
Rice, K. C., **139**, 405
Richards, D. E., **156**, 281
Richards, JoAnne S., **159**, 318
Richardson, M. A., **153**, 166
Richerson, Russell B., **143**, 410
Richmond, Ann, **146**, 112
Ridge, Susan C., **162**, 355, 373
Riechmann, Lutz, **136**, 280
Ried, Jeffrey L., **161**, 329
Rieker, James P., **139**, 105
Rindone, W. P., **155**, 582
Riordan, James F., **158**, 3
Rippka, Rosmarie, **167**, 3, 28, 232, 851
Rips, Richard, **142**, 535
Risby, Terence H., **158**, 180
Risse, Gundula, **144**, 490
Risteli, Juha, **145**, 391
Risteli, Leila, **145**, 391
Rittenhouse, Susan Erika, **141**, 143
Rivière, Marie-Emmanuel, **167**, 691
Rivnay, B., **149**, 119
Rizzino, Angie, **146**, 341
Robbins, April R., **138**, 458
Roberson, Marie M., **138**, 516
Robert, Danielle, **161**, 137
Roberts, Daniel M., **139**, 290
Roberts, David D., **138**, 473, 561
Roberts, John D., **154**, 367
Robertson, James M., **164**, 581
Robey, Pamela Gehron, **145**, 269
Robinson, John P., **165**, 85, 389
Robinson, Joseph D., **156**, 236
Robinson, Mitchell, **141**, 379
Robinson, Simon P., **148**, 145, 188
Roby, Claude, **167**, 667
Roche, Patrick A., **142**, 132
Roda, Aldo, **137**, 161
Roff, Calvin F., **138**, 458

Rogers, Gary N., **138**, 162
Rogers, S. G., **153**, 253
Röhme, Dan, **151**, 503
Rolfe, Stephen A., **148**, 259
Rollins, Barrett, **147**, 64
Romanus, Joyce A., **146**, 259
Rombouts, F. M., **160**, 243
Rome, Leonard H., **138**, 607
Romette, J. L., **137**, 44
Roninson, Igor B., **151**, 332
Rönnberg, Inger, **137**, 381
Rönnstrand, Lars, **147**, 3
Roos, David S., **151**, 85
Ropars, C., **149**, 242
Rosbash, Michael, **154**, 129
Rose, Mark D., **152**, 481
Rosen, Ora M., **139**, 148
Rosenberg, Gary B., **139**, 776
Rosenberg, Martin, **152**, 661
Rosenberg, Steven A., **150**, 342
Rosenberg, Susan M., **153**, 95
Rosenbloom, Joel, **144**, 172, 259
Rosenthal, André, **155**, 301
Rosenthal, Nadia, **152**, 704
Ross, Gordon D., **150**, 586
Ross, Russell, **147**, 48
Rossi, Bernard, **156**, 323
Rossi, Mosè, **135**, 577
Roth, Jesse, **150**, 701
Rothenberg, Paul, **147**, 415
Rothermel, John D., **159**, 159
Rotondi, Andrea, **139**, 731
Roughan, Grattan, **148**, 327
Roux, Anne Francoise, **163**, 418
Rowe, David, **145**, 223
Rowe, Paul M., **139**, 667
Rowell, Peter, **167**, 410
Roy, Alexander B., **143**, 207, 361
Roy, Robert G. B., **146**, 112
Royer, G. P., **135**, 141; **136**, 157
Rozengurt, Enrique, **146**, 384; **147**, 40
Rozzell, J. David, **136**, 479
Ruben, Larry, **139**, 262
Ruddle, Frank H., **147**, 280; **151**, 371
Rugh, Susanne, **136**, 356
Rui, Hallgeir, **159**, 675
Rukenstein, Adriana, **147**, 207
Ruoho, Arnold E., **150**, 492
Ruoslahti, Erkki, **144**, 430, 475
Ryan, Clarence A., **138**, 715
Ryan, James W., **163**, 160, 186, 194
Rybka, Russell, **147**, 95

S

Sabourin, P. J., **166**, 288
Sachs, George, **157**, 649
Sadana, Jai C., **160**, 307, 424, 448
Saddler, J. N., **160**, 3
Saeed, Abdul M. K., **165**, 126
Saelinger, Catharine B., **165**, 147, 226
Saidha, Tekchand, **143**, 329, 354
Saito, Akitsugu, **157**, 36, 51, 100
Saito-Nakatsuka, Kimiko, **157**, 251
Sakai, Masahiro, **141**, 92
Sakai, Takuo, **161**, 335
Sakamoto, Reiichiro, **160**, 274
Sakata, T., **136**, 530
Salach, James I., **142**, 627
Salant, David J., **162**, 421
Salas, Margarita, **150**, 112
Salvesen, Guy, **163**, 240
Salviati, Giovanni, **141**, 3
Samejima, Hirotoshi, **136**, 394
Sampson, Phyllis, M., **138**, 260
Samuelsson, Bo E., **138**, 567
Sandberg, Lawrence B., **144**, 196
Sanders, Jane Z., **155**, 260
Sanders, V. M., **150**, 316
Sandhoff, Konrad, **138**, 319, 792
Sandmann, Gerhard, **167**, 329
Sano, Mamoru, **139**, 495
Santoro, Samuel A., **144**, 438
Saraswat, Lakshmi D., **159**, 325
Sargent, Thomas D., **152**, 423
Sasaki, Ryuzo, **147**, 328
Sasaki, Takashi, **160**, 468
Sasse, Joachim, **146**, 320
Sato, J. Denry, **146**, 63
Sato, Naoki, **167**, 251
Sato, Tadashi, **135**, 189
Satoh, Ikuo, **137**, 217
Satoh, Kazuhiko, **167**, 269
Saunders, Jon R., **155**, 11
Sautebin, Lidia, **163**, 23
Sauter, Helmut, **150**, 257
Sauvage, Carol, **147**, 265
Savage, C. Richard, Jr., **146**, 3
Savory, John, **158**, 289

Sawyer, Stephen T., **147**, 340
Saxon, Paul J., **151**, 313
Sburlati, Adriana, **161**, 457
Scanlan, D. J., **167**, 599
Scarborough, Gene A., **157**, 574
Scardi, Vincenzo, **135**, 293
Schaeffer, Valentine H., **143**, 228
Schaeper, Robert J., **138**, 575
Schapira, Marc, **163**, 179
Scharff, Matthew D., **151**, 50
Schauer, Roland, **138**, 132, 611
Scheiner-Bobis, Georgios, **156**, 312
Scheirer, W., **135**, 399
Schejter, Abel, **161**, 366
Scheller, Frieder W., **137**, 29, 152
Scher, Charles D., **147**, 85
Scheraga, Harold A., **143**, 115
Schieder, Otto, **148**, 45
Schiff, Jerome A., **143**, 329, 354
Schildkraut, Ira, **155**, 15, 41
Schilperoort, Rob A., **153**, 305
Schimke, Robert T., **151**, 85
Schindler, Melvin, **141**, 439, 447
Schleef, Raymond R., **163**, 293
Schlegel, Robert A., **149**, 281, 293
Schlievert, Patrick M., **165**, 37, 302, 339
Schlingmann, Merten, **136**, 517
Schloss, John V., **166**, 92, 445, 455
Schluter, Samuel F., **150**, 628
Schmer, Gottfried, **137**, 479
Schmid, Georg, **160**, 38
Schmidt, Ahlert, **143**, 449; **167**, 572
Schmidt, Azriel, **144**, 61
Schmidt, Gregory W., **148**, 274
Schmidt, James J., **165**, 76
Schmidt, Jürgen, **160**, 662
Schneider, Donald L., **157**, 591
Schneider, George J., **167**, 592
Schneider, H., **160**, 560, 700
Schneider, Jay W., **156**, 379
Schneider, Jerry A., **143**, 144
Schneider, Klaus, **153**, 452
Schneider-Bernlöhr, Helga, **135**, 546
Schnier, Joachim, **164**, 706
Schoner, Brigitte E., **153**, 401
Schoner, Ronald G., **153**, 401
Schoner, Wilhelm, **156**, 302, 312
Schraufstätter, Ingrid U., **163**, 328
Schrier, Stanley L., **149**, 260
Schroer, K., **138**, 307
Schubert, Florian, **137**, 29, 152
Schülein, Martin, **160**, 234
Schulte, Roberta, **147**, 265
Schultz, Jerome S., **137**, 349
Schulz, Tanya, **159**, 318
Schwartz, Arnold, **156**, 201
Schwartz, J. H., **139**, 277
Schwarzbauer, Jean E., **144**, 447
Schwarzmann, Günter, **138**, 319
Schwellnus, Konrad, **154**, 221
Scott, Clay W., **159**, 139
Scott, James, **146**, 239; **147**, 194
Scouten, William H., **135**, 30, 79
Scrimgeour, K. G., **142**, 127
Scully, R. R., **139**, 19
Searles, Lillie L., **166**, 225
Seedorf, Udo, **139**, 763
Segal, David M., **150**, 478
Segel, Irwin H., **143**, 334
Seiler, Steven, **157**, 26, 36
Selby, Mark J., **147**, 194
Seligman, Paul A., **147**, 239
Sena, Salvador F., **158**, 320
Sephel, Gregory C., **144**, 214
Serrano, Ramón, **157**, 533
Seto-Ohshima, Akiko, **139**, 495
Seubert, Peter A., **143**, 334
Seville, Mark, **159**, 208
Seyfred, Mark A., **141**, 92
Sha'afi, Ramadan I., **162**, 283
Shacter, Emily, **159**, 3
Shahak, Yosepha, **167**, 380
Shak, Steven, **141**, 355
Shaked, Ze'ev, **137**, 599
Shameer, I., **160**, 545
Shannon, Jack C., **148**, 226
Shapiro, Robert, **158**, 344, 365
Sharma, Rajendra K., **139**, 518; **159**, 582
Sharon, Nathan, **138**, 544
Shatsky, Ivan N., **164**, 76
Shatzman, Allan R., **152**, 661
Shay, Jerry W., **151**, 221
Shelanski, Michael, **139**, 824
Shen, C.-C., **159**, 760
Shen, Rong-sen, **142**, 242
Shenolikar, Shirish, **139**, 690
Sheorain, Virender S., **159**, 74
Shepherd, Maxwell G., **160**, 300
Shewale, Jaiprakash G., **160**, 424
Shi, D. J., **167**, 629

Shichiri, Motoaki, **137**, 326
Shiio, Isamu, **143**, 470
Shillito, Raymond D., **153**, 313
Shilo, Moshe, **167**, 616
Shimahara, Kenzo, **161**, 417
Shiman, Ross, **142**, 17
Shimizu, Nobuyoshi, **147**, 382, 387
Shimizu, Shoichi, **160**, 391, 408
Shimizu, Yasuhiko, **137**, 545
Shindo-Okada, Nobuko, **155**, 373
Shing, Yuen, **146**, 42; **147**, 95
Shinmyo, Atsuhiko, **160**, 632
Shinomura, Tamayuki, **144**, 372
Shive, William, **150**, 134
Shively, J. M., **167**, 195, 204
Showell, Henry J., **162**, 653
Showers, Mark O., **159**, 311
Shriner, Carol L., **159**, 339
Shuler, M. L., **135**, 372
Siciliano, A., **149**, 64
Siciliano, Michael J., **151**, 169
Siebers, Annette, **157**, 668
Siegbahn, Nils, **136**, 103
Siegel, Frank L., **139**, 667
Siegmann, Michel, **146**, 362
Siehl, Daniel L., **142**, 422
Siess, Wolfgang, **141**, 176
Sigel, Erwin, **141**, 25
Sigmund, Curt D., **164**, 673
Silverberg, Michael, **163**, 3, 68, 85
Silverman, Philip M., **166**, 435
Silverstein, Karen, **155**, 482
Simmen, Rosalia C. M., **139**, 217, 642
Simmerman, Heather K. B., **157**, 360
Simon, Helmut, **136**, 302
Simonelli, Paul F., **159**, 346
Simons, Guus, **153**, 416
Simpson, Lance L., **165**, 76
Simpson, Ronald, **166**, 114
Singh, Jai Pal, **147**, 13
Singh, Rajeeva, **143**, 129
Sjöholm, Ingvar, **137**, 491
Skeaff, Clark M., **141**, 234
Skiba, William E., **143**, 384
Skinner, Martha, **163**, 523
Skou, Jens Christian, **156**, 1, 43, 278
Skripkin, Evgeny A., **164**, 440
Skriver, Elisabeth, **156**, 80, 430
Skryabin, G. K., **135**, 198
Skudder, Paul J., **136**, 423
Slaughter, Gayle R., **139**, 433
Slavin, Walter, **158**, 117
Sliwkowski, Mark X., **143**, 119
Slonczewski, J. L., **141**, 18
Smith, Arnold J., **167**, 551
Smith, Cassandra L., **151**, 461; **155**, 449
Smith, Charles D., **141**, 92, 261; **162**, 271
Smith, D. Drummond S., **142**, 325
Smith, Gary L., **146**, 270
Smith, Gilbert H., **151**, 530
Smith, Ivan K., **143**, 439
Smith, Jeffrey S., **157**, 480
Smith, John A., **162**, 170
Smith, Lloyd M., **155**, 260
Smith, Margaret, **143**, 144
Smith, Michael, **154**, 329
Smith, Russell L., **167**, 490
Smith, Sandra, **147**, 95
Smith, Thomas Woodward, **156**, 46
Smith, V. L., **139**, 19
Snyder, Fred, **141**, 379
Snyder, I. S., **165**, 137, 399
Snyder, Michael, **154**, 107
Snyderman, Ralph, **141**, 261; **162**, 236, 271, 279
Soberman, Roy J., **163**, 344, 349, 353
Soda, Kenji, **143**, 148, 240, 243, 245, 291, 415, 453, 459, 493, 500
Sodek, J., **145**, 303
Soderquist, Ann Mangelsdorf, **146**, 49
Sodetz, James M., **162**, 551
Sojar, Hakimuddin T., **138**, 341
Sokatch, John R., **166**, 342, 350, 389
Solioz, Marc, **157**, 680
Soll, Jürgen, **148**, 383
Solomon, Beka, **135**, 160
Sommarin, Yngve, **144**, 305, 319
Sonomoto, Kenji, **135**, 230
Sörbo, Bo, **143**, 3, 178
Soskel, Norman T., **144**, 196
Soto-Gil, Rafael W., **161**, 524
Sourkes, Theodore L., **142**, 170
Southwick, Frederick S., **162**, 246
Spanswick, Roger M., **157**, 579
Sparace, S. A., **148**, 338
Spath, Susan, **167**, 667
Speedie, Marilyn K., **142**, 235
Spiegel, Sarah, **138**, 313
Spilberg, Isaias, **162**, 79, 193

Spirin, A. S., **164**, 426, 631
Spitalnik, Steven L., **138**, 195, 492
Spoerel, Nikolaus A., **152**, 588, 598
Spokane, Robert B., **137**, 68
Spragg, Jocelyn, **163**, 96
Sprague, Sallie G., **148**, 319
Sprandel, U., **149**, 301
Spydevold, Øystein, **166**, 476
Squinto, Stephen P., **159**, 225
Stabinsky, Z., **154**, 287
Stabler, Sally P., **166**, 393, 400, 407
Stackebrandt, Erko, **167**, 132
Stadtman, Earl R., **159**, 3
Staehelin, L. Andrew, **148**, 319
Stahl, Ulf, **153**, 366
Stal, Lucas J., **167**, 474
Stam, W. T., **167**, 125
Stanbridge, Eric J., **151**, 313
Stanier (Cohen-Bazire), G., **167**, 157
Stanley, Pamela, **138**, 443
Stark, Michael J. R., **164**, 691
Staros, James V., **150**, 503
Staub, Adrien, **154**, 250
Steen, Rolf, **164**, 691
Steiner, Günter, **164**, 361
Steinmann-Hofmann, B., **136**, 188
Stephany, David A., **150**, 478
Stern, Seth, **164**, 481
Stevens, James L., **143**, 228
Stevens, S. Edward, Jr., **167**, 122
Stewart, Alexander A., **159**, 409
Stewart, Alison C., **167**, 272
Stewart, John C., **160**, 264
Stewart, William D. P., **167**, 410
Stiege, Wolfgang, **164**, 287
Stiles, Charles D., **147**, 64
Stimler-Gerard, Norma P., **162**, 653
Stipanuk, Martha H., **143**, 155
Stirm, Stephan, **138**, 786
Stöffler, Georg, **164**, 64, 503
Stöffler-Meilicke, Marina, **164**, 64, 503
Stone, Dennis K., **139**, 30; **157**, 634
Storm, Daniel R., **139**, 776
Stout, J. Timothy, **151**, 519
Strada, S. J., **159**, 760
Strålfors, Peter, **159**, 390
Straus, Daniel S., **147**, 369
Strömberg, Nicklas, **138**, 220
Stroobant, P., **147**, 40
Struhl, Kevin, **155**, 558, 568
Studebaker, J. F., **137**, 466
Stueber, Dietrich, **155**, 416
Stull, James T., **139**, 188
Stulp, B. K., **167**, 125
Subramanian, Alap R., **164**, 748
Sudan, Ranjan, **137**, 68
Südhof, Thomas C., **139**, 30
Suelter, Clarence H., **142**, 414
Sugden, David, **142**, 590
Sugg, Nancy, **165**, 3
Sugihara, Hidemitsu, **139**, 363
Sullivan, Darryl M., **143**, 7
Sullivan, Richard, **167**, 509
Sullivan, Robert, **146**, 320; **147**, 95
Sun, Anthony M., **137**, 575
Sun, Grace, **164**, 158
Sundaram, P. V., **137**, 288
Sunderman, F. William, Jr., **158**, 382
Sung, Wing L., **153**, 385
Suoranta, Kari, **159**, 777
Suranyi, Gyula, **167**, 622
Surosky, Richard T., **153**, 243
Sussman, Howard H., **147**, 253
Suzuki, Fujio, **146**, 313
Suzuki, Hideo, **136**, 45
Suzuki, Hiroshi, **160**, 200
Suzuki, Koich, **139**, 363
Suzuki, Kunihiko, **138**, 727
Suzuki, Shintaro, **144**, 430, 475
Suzuki, Shuichi, **137**, 668
Suzuki, Takashi, **139**, 518
Svensson, Sigfrid, **136**, 230
Svoboda, Marjorie E., **146**, 207
Swaisgood, Harold E., **135**, 130, 596; **136**, 423; **143**, 119, 504
Swanborg, Robert H., **162**, 413
Swanljung-Collins, Helena, **139**, 105, 137
Swanson, M. E., **139**, 277
Sweadner, Kathleen J., **156**, 65
Sweetser, Douglas, **154**, 107
Swift, Galvin H., **152**, 219
Swillens, Stéphane, **159**, 19
Sykes, P. J., **166**, 350
Szadkowski, Halina, **142**, 386
Szalay, Aladar A., **153**, 336; **154**, 197
Sze, Heven, **148**, 123
Szebenyi, D. M. E., **139**, 585
Sziegoleit, Andreas, **165**, 285
Szoka, Francis C., Jr., **149**, 143

T

Tabata, Masayoshi, **137**, 260
Tabita, F. Robert, **167**, 490
Tada, Michihiko, **157**, 107
Tagawa, Kiyoshi, **160**, 542, 707; **161**, 361
Takada, Yoshikazu, **142**, 273
Takagaki, Yoshio, **163**, 230
Takahashi, Rihei, **160**, 518, 611
Takai, Katsuji, **142**, 195
Takamatsu, Satoru, **136**, 472
Takasaki, Seiichi, **138**, 84
Takata, Isao, **135**, 189; **136**, 455
Takeda, Yoshifumi, **165**, 189
Takiguchi, Yasuyuki, **161**, 417
Takisawa, Haruhiko, **157**, 228
Tam, James P., **146**, 127
Tamkun, John W., **144**, 447
Tamm, V. Y., **160**, 620
Tamura, Shinri, **159**, 437
Tan, Eng M., **150**, 147
Tanahashi, Mitsuhiko, **161**, 101
Tanaka, Atsuo, **135**, 230; **136**, 293
Tanaka, Hidehiko, **143**, 240, 243
Tanaka, Kay, **166**, 155, 360, 374
Tanaka, Masashi, **157**, 321
Tanaka, Michio, **160**, 712
Tanaka, Mitsuo, **160**, 251
Tanaka, Takashi, **136**, 405
Tanaka, Toshio, **139**, 217, 570; **159**, 652
Tanaka, Y., **143**, 164
Tandeau de Marsac, Nicole, **167**, 318, 755, 808
Tang, J.-Y., **154**, 287
Taniguchi, Masayuki, **160**, 251
Tank, A. William, **142**, 71
Tapprich, William E., **164**, 401
Tarentino, Anthony L., **138**, 770
Tash, Joseph S., **139**, 808
Tassanakajohn, Anchalee, **164**, 401
Tatsumi, Masahiro, **159**, 702
Taurog, Joel D., **162**, 339
Tausk, Francisco A., **162**, 626
Taylor, John W., **154**, 473
Taylor, K. A., **157**, 271
Taylor, Loverine P., **153**, 351
Taylor, Mary E. **167**, 272
Taylor, Scott E., **167**, 539
Taylor, Susan, **159**, 325
Teather, Ronald M., **160**, 59
Tegmo-Larsson, I.-M., **149**, 77
Tenne, Meir, **142**, 617
Teramatsu, Takashi, **137**, 545
Terman, David S., **137**, 496
Termine, John D., **145**, 269
Ternynck, Thérèse, **150**, 117
Tetsuo, Koshijima, **161**, 12
Thannhauser, Theodore W., **143**, 115
Thanos, Iordanis, **136**, 302
Theibert, Anne, **159**, 267
Thelestam, Monica, **165**, 278
Thiel, Teresa, **153**, 232
Thoenen, Hans, **147**, 167
Thøgersen, Hans Christian, **153**, 461
Thomas, Daniel, **135**, 454, 554; **137**, 44
Thomas, George, **146**, 362, 369; **164**, 575
Thomas, H. Greg, **146**, 112
Thomas, Kenneth A., **147**, 120
Thomas, Michael J., **163**, 44
Thomas, P. Thomas, **141**, 399
Thomasset, Brigitte, **135**, 454
Thomasset, Monique, **139**, 337
Thompson, Floyd H., **151**, 267
Thompson, James A., **147**, 157
Thompson, LuAnn, **150**, 3
Thompson, W. J., **159**, 760
Thonar, Eugene J.-M. A., **144**, 372
Thorner, Jeremy, **139**, 248
Thotakura, Nageswara R., **138**, 350
Tien, Ming, **161**, 238
Timpl, Rupert, **145**, 363
Ting-Beall, H. P., **157**, 271
Titus, Julie A., **150**, 478
Tobes, Michael C., **142**, 217
Tokuda, Masaaki, **139**, 36, 68
Tollefsen, Sherida, **138**, 536
Toman, Rudolf, **160**, 536
Tomme, Peter, **160**, 187
Tong, Chow Chin, **160**, 300
Tonik, Susan E., **147**, 253
Tonks, Nicholas K., **159**, 427
Torchilin, V. P., **137**, 552, 615
Tosa, Tetsuya, **135**, 189; **136**, 455, 472
Toyama, Seizen, **143**, 500
Toyo'oka, Toshimasa, **143**, 67
Tramper, Johannes, **136**, 254
Tranquillo, Robert T., **162**, 85
Tranum-Jensen, Jørgen, **165**, 285, 357, 374
Traugh, Jolinda A., **141**, 210
Traut, Robert, **164**, 419

Trautwein, Alfred Xaver, **161**, 281
Tremblay, Johanne, **159**, 710
Trent, Jeffrey M., **151**, 267
Triana, Francisco, **159**, 286
Trimble, Robert B., **138**, 763
Tritz, Richard, **166**, 260
Trnka, Maria, **167**, 437
Troen, Bruce R., **151**, 416
Trosko, James E., **141**, 439
Trowbridge, Ian S., **147**, 265
Troy, Frederic A., **138**, 169, 627
Trueheart, Joshua, **154**, 164
Trumbly, Robert J., **138**, 763
Tsang, Hannah Chi-Wan, **149**, 221
Tsipouras, Petros, **145**, 205
Tsong, Tian Yow, **149**, 248; **157**, 240
Tsuiki, Shigeru, **159**, 437
Tuazon, Polygena T., **141**, 210
Tullius, Thomas D., **155**, 537
Tung, H. Y. Lim, **159**, 390
Turano, Carlo, **142**, 179
Turková, Jaroslava, **135**, 341
Turnell, D. C., **143**, 141
Turner, Anthony P. F., **137**, 90
Turunen, Pekka, **137**, 171
Tweten, Rodney K., **165**, 43
Twilfer, Hans, **161**, 281
Tye, Bik-Kwoon, **153**, 243; **155**, 528

U

Ubuka, Toshihiko, **143**, 17
Uchida, Etsuko, **157**, 544
Uchida, Tsuneko, **160**, 712
Uchida, Yasushi, **161**, 501
Ueck, Manfred, **159**, 477
Ueno, Satoki, **159**, 477
Uhler, Michael D., **159**, 299
Uhr, J. W., **150**, 316
Uhteg, Lawrence C., **143**, 350
Ullman, Edwin F., **136**, 93
Ullman, M. D., **138**, 117
Umetsu, Dale T., **150**, 309
Umezawa, Toshiaki, **161**, 200
Underwood, Louis E., **146**, 207, 227
Uno, Isao, **159**, 27
Urabe, Itaru, **136**, 34
Urbanek, Henryk, **160**, 332
Urdea, Mickey S., **146**, 22
Uren, Jack R., **143**, 483
Urist, Marshall R., **146**, 294
Urlaub, Gail, **151**, 449
Urnovitz, Howard B., **151**, 194

V

Vaara, Martti, **167**, 189
Vaara, Timo, **167**, 189
Vaccaro, C., **135**, 577
Vail, W. J., **149**, 64
Vallee, Bert L., **158**, 3, 422
Vanaman, Thomas C., **139**, 50, 417
van Arkel, G. A., **153**, 199
Van Baalen, Chase, **167**, 490
van den Tweel, Will, **135**, 79
Van Dyk, Drew E., **166**, 445
Van Dyk, Tina K., **166**, 97
Van Eldik, Linda J., **139**, 393, 445; **159**, 667
van Gunsteren, W. F., **154**, 430
van Heel, Marin, **164**, 35
van Iersel, Jolanda J. L., **163**, 80
van Knippenberg, Peter H., **164**, 188
Van Montagu, M., **153**, 277
Van Patten, Scott, M., **159**, 253
Van Stolk, Barbara J., **164**, 475
van Tilbeurgh, Herman, **160**, 45
Van Wart, Harold E., **158**, 95
Van Wyk, Judson J., **146**, 207
Varfolomeev, S. D., **137**, 430
Varga, S., **157**, 271
Varshavsky, Alexander, **151**, 551
Vartikar, Jailaxmi V., **164**, 203
Vasiliev, Victor D., **164**, 76
Vaughan, Martha, **159**, 557, 605
Veeger, Cees, **136**, 216
Veillon, Claude, **158**, 56, 334
Verghese, Margrith W., **162**, 279
Vergnolle, Chantal, **148**, 661
Verheijen, J. H., **163**, 302
Verhoeven, Els J. M., **153**, 12
Verschoor, Andriana, **164**, 3
Versieck, Jacques, **158**, 267
Vester, Birte, **164**, 710
Vieira, Jeffrey, **153**, 3
Vigil, Eugene L., **148**, 505
Villafranca, Joseph J., **142**, 50
Villa-Moruzzi, Emma, **159**, 377

Vimr, Eric R., **138**, 169
Vischer, Thomas L., **150**, 109
Vitetta, E. S., **150**, 316
Vitterbo, Alexander, **167**, 561
Vogel, H. J., **135**, 512
Vogeli, G., **152**, 407
Vögeli, Urs, **161**, 479
Volc, J., **161**, 316
Volckaert, Guido, **155**, 231
Volfin, P., **148**, 132
Volpe, Pompeo, **141**, 3
Voltattorni, Carla Borri, **142**, 179
von der Mark, Klaus, **144**, 490
von Hörsten, W., **149**, 301
Voragen, A. G. J., **160**, 243
Vorlop, K. D., **135**, 259
Voss, Helga, **164**, 131
Vršanská, Mária, **160**, 638

W

Wada, Keishiro, **167**, 387
Wade, Margaret H., **141**, 439
Wadström, Torkel, **165**, 200
Wagenknecht, Terence, **164**, 3
Wagh, Premanand V., **138**, 816
Wagner, Fred W., **158**, 21
Wagner, George J., **148**, 55
Wagner, R., **164**, 721
Wahl, Geoffrey M., **152**, 91, 399, 415, 572, 604
Wahl, Sharon M., **163**, 605, 642, 715
Waisman, David Morton, **139**, 36, 68
Wakefield, Lalage M., **146**, 167
Walbot, Virginia, **153**, 351
Walderhaug, Mark O., **157**, 655
Waldren, Charles A., **151**, 121
Walker, David A., **148**, 145
Walker, Ian M., **149**, 51
Walker, Jeremy N. B., **155**, 11
Walker, John E., **167**, 794
Wall, Joseph S., **164**, 49
Wallace, Donald M., **152**, 33, 41
Wallace, R. Bruce, **152**, 432; **154**, 87, 94
Wallace, Robert W., **139**, 846
Walleczek, Jan, **164**, 64
Wallick, Earl T., **156**, 201
Walsby, A. E., **167**, 521, 660
Walseth, Timothy F., **159**, 60
Walsh, Donal A., **159**, 173, 255
Walsh, Michael P., **139**, 88
Walter, Bert, **137**, 394
Walter, Ulrich, **159**, 183
Wandrey, Christian, **136**, 9
Wang, Alice, **154**, 403
Wang, Chen-Yen, **149**, 88
Wang, Daniel I. C., **135**, 318
Wang, Jerry H., **139**, 518; **159**, 582
Wang, Sherry, **157**, 100
Ward, John H., **147**, 247
Ward, Peter A., **162**, 198, 223
Warr, Gregory, W., **150**, 399
Warsh, Jerry J., **142**, 571
Wassef, Nabila M., **141**, 244; **149**, 124
Wasteson, Åke, **147**, 3
Wastyn, Marnik, **167**, 437
Watanabe, Ichiro, **136**, 523
Watanabe, Kiyohiro, **138**, 3
Watanabe, Satoshi, **137**, 545
Watanabe, Takahide, **157**, 233
Waterbury, John B., **167**, 100
Waterfield, M.D., **147**, 40
Waterman, Michael S., **164**, 765
Watterson, D. Martin, **139**, 290; **157**, 328; **159**, 471
Waymack, Parvan P., **166**, 484
Weaver, James C., **137**, 421
Weber, Irene T., **159**, 278
Weber, Wolfgang, **146**, 82
Weckesser, Jürgen, **167**, 173
Wedde, Markus, **164**, 658
Wedding, Randolph T., **143**, 29
Wegener, Adam D., **157**, 360
Weiner, Norman, **142**, 71
Weinstein, Catherine L., **143**, 221, 270, 404
Weinstock, George M., **154**, 156
Weissman, Sherman M., **151**, 461
Weitzmann, Carl J., **164**, 523
Welin, Stefan, **137**, 388
Weller, Peter F., **163**, 31
Wells, Jack N., **159**, 489, 594
Wells, Marilyn S., **143**, 384
Wells, William W., **141**, 92
Welsh, Clement J., **141**, 301
Wennergren, Suzanne, **147**, 3
Weppelman, Roger M., **142**, 608
Werb, Zena, **144**, 288
Wertman, Kenneth F., **152**, 173
Westermark, Bengt, **147**, 3

Westermark, U., **160**, 463
Westley, Aiko, **143**, 235
Westley, John, **143**, 22, 38, 235, 350
Weston, K. M., **155**, 51
Westrick, Mark P., **143**, 281
Wetters, Theodoor van Daalen, **151**, 9
Weyand, Cornelia, **150**, 333
Weyler, Walter, **142**, 627
White, Billie F., **151**, 169
White, Bruce A., **139**, 655
White, Susan A., **164**, 221
Whitesides, George M., **136**, 263; **143**, 129
Whitlock, C. A., **150**, 275
Whorton, A. Richard, **141**, 341
Wickstrom, Eric, **164**, 238
Widell, Susanne, **148**, 558
Wieczorek, Leszek, **157**, 668
Wiegel, Jürgen, **160**, 528
Wienhausen, G., **136**, 82
Wikner, Norman E., **162**, 214
Wilchek, Meir, **135**, 84; **138**, 429; **149**, 119
Wilde, M. W., **141**, 18
Wiley, H. Steven, **146**, 402
Wiley, Ronald G., **165**, 204, 250
Wilkinson, Peter C., **162**, 3, 17, 38, 127, 180
Willey, Joanne M., **167**, 100
Williams, John G. K., **167**, 766
Williams, John P., **148**, 346
Williamson, Patrick, **149**, 281
Willick, Gordon, **160**, 432
Wills, Michael R., **158**, 289
Wilson, David B., **160**, 314
Wilson, Samuel H., **152**, 94
Wingard, Lemuel B., Jr., **137**, 103
Winkler, Martin A., **139**, 505
Winquist, Fredrik, **137**, 232
Wintermeyer, Wolfgang, **164**, 581, 597
Wiseman, Alan, **137**, 675
Witte, O. N., **150**, 275
Witters, Lee A., **141**, 412
Wittmann, H. G., **164**, 95
Wittmann-Liebold, Brigitte, **164**, 542
Woernle, Rolf, **136**, 517
Wold, Finn, **138**, 409
Wolfe, Sidney, **137**, 599
Wolff, J., **139**, 834
Wolf-Heuss, Elisabeth M., **143**, 186
Wolk, C. Peter, **153**, 232; **167**, 93, 747
Wollenzien, Paul L., **164**, 319
Wolnik, Karen A., **158**, 190
Woloski, Barry M. R. N. J., **163**, 594
Wolt, Terril B., **144**, 196
Wong, Chi-Huey, **136**, 263
Wong, Winnie W., **150**, 579
Woo, Savio L. C., **152**, 199, 447
Wood, John L., **143**, 25
Wood, L. L., **136**, 497
Wood, Peter J., **160**, 59
Wood, Phillip G., **149**, 271
Wood, Thomas M., **160**, 3, 19, 87, 187, 216, 221, 398
Wood, William I., **152**, 443
Woodard, Donna, **141**, 379
Woods, Anne, **144**, 394
Wooge, Cynthia Harris, **141**, 429
Wool, Ira G., **164**, 468
Wooley, Paul H., **162**, 361
Woolkalis, Marilyn, **165**, 235, 246
Wootton, Joyce A. M., **145**, 171
Woronick, Charles L., **162**, 223
Wrenn, David S., **144**, 246
Wright, Daniel G., **162**, 538
Wright, Lynda S., **139**, 667
Wright, Woodring E., **151**, 237
Wu, Arnold C. M., **161**, 447
Wu, Gillian E., **150**, 257
Wu, Ray, **152**, 343; **155**, 214
Wu, Tiyun, **152**, 343
Wykle, Robert L., **163**, 44
Wylie, David C., **139**, 50
Wyman, Arlene R., **152**, 173

X

Xie, Xiao-Song, **157**, 634

Y

Yamada, Hideaki, **143**, 474, 486
Yamada, Tomiaki, **136**, 380
Yamagata, Shuzo, **143**, 465, 478
Yamaguchi, Kenji, **143**, 395
Yamaguchi, Motonori, **157**, 233
Yamaizumi, Masaru, **151**, 371
Yamamoto, Kozo, **136**, 455
Yamamoto, Shunsuke, **162**, 140
Yamamoto, Takehiko, **160**, 719

Yamanaka, Shigeru, **136**, 405
Yamane, Kunio, **160**, 200
Yamasaki, Yoshimitsu, **137**, 326
Yamashiro, Donald, **146**, 199
Yamashita, Kamejiro, **139**, 127
Yamashita, Katsuko, **138**, 84
Yamashita, Tetsuro, **157**, 251
Yamazaki, Akio, **159**, 702, 730
Yamazaki, Mitsuo, **135**, 154
Yamazaki, Yoshimitsu, **136**, 21, 45
Yan, Sau-chi Betty, **138**, 413
Yanagawa, Hiroshi, **143**, 516
Yanagawa, Shin-ichi, **147**, 328
Yanagishita, Masaki, **138**, 279
Yang, Funmei, **163**, 452
Yang, Shang Fa, **143**, 426
Yang, Victor C., **137**, 515
Yao, Fei-L., **153**, 385
Yaron, A., **136**, 170
Yasuda, Takeo, **137**, 124
Yasui, Tuneo, **160**, 648, 671, 684, 696
Yasumoto, Kyoden, **143**, 434
Yasuoka-Yabe, Kimiko, **157**, 251
Yatvin, M. B., **149**, 77
Yeaman, Stephen J., **166**, 10, 303
Yee, Boihon C., **167**, 415
Yee, Jiing-Kuan, **149**, 10
Yefenof, E., **150**, 316
Yeoman, Michael M., **135**, 410
Yoda, Kentaro, **137**, 61
Yokota, T., **154**, 3
Yonaha, Kazuo, **143**, 500
Yonath, A., **164**, 95
Yoshida, Ryotaro, **142**, 188
Yoshimoto, Tanihiro, **163**, 353
Yoshitake, Yoshino, **146**, 11
Youdim, Moussa B. H., **142**, 617
Young, Marian F., **145**, 269
Young, Michael, **162**, 101
Young, Richard A., **152**, 359, 458; **154**, 107
Young, William W., Jr., **138**, 125
Yu, Robert K., **138**, 38
Yusupov, M. M., **164**, 426
Yusupova (Tnalina), G. Z., **164**, 631

Z

Zabel, Bernhard U., **151**, 279
Zabriskie, John B., **162**, 373
Zacharova, I. Ya., **160**, 620
Zachrisson, Anders, **135**, 421
Zagursky, Robert, **155**, 139
Zakour, Richard A., **154**, 367
Zalewska-Sobczak, Jadwiga, **160**, 332
Zeikus, J. G., **160**, 351
Zenchoff, Gladys, **164**, 372
Zenk, Meinhart H., **136**, 342
Zeppezauer, Michael, **135**, 546; **158**, 79
Zetter, Bruce R., **147**, 135
Zhang, Bei, **166**, 189
Zhang, Yi-Zheng, **161**, 228
Zhang, Zu-chuan, **141**, 301
Zhong, Li-Chan, **137**, 686
Zhu, Betty C. R., **144**, 420
Ziegler, Daniel M., **143**, 124, 149, 410
Zigmond, Sally H., **141**, 18; **162**, 65, 85
Zikakis, John P., **161**, 490
Zimmer, Warren E., **139**, 290
Zimmermann, Ulrich, **151**, 194
Zimmermann, Wolfgang, **161**, 191
Zoller, Mark J., **154**, 329
Zoon, Robert A., **152**, 25
Zopf, D., **138**, 307
Zorzato, Francesco, **141**, 3
Zullo, John, **147**, 64
Zumstein, Peter, **147**, 64
Zurini, M., **139**, 791
Zysk, John R., **139**, 505
Zyskind, Judith W., **161**, 525